THEORY
OF
MODULES

(An Introduction to the Theory of Module Categories)

ALEXANDRU SOLIAN

THEORY OF MODULES

(An Introduction to the Theory of Module Categories)

Translated from the Romanian by
Mioara Buiculescu

EDITURA ACADEMIEI
București

JOHN WILEY & SONS LIMITED
London • New York • Sydney • Toronto
A Wiley-Interscience Publication

1977

Library of Congress Cataloging in Publication Data:

Solian, Alexandru.
Theory of modules.

Translation of Teoria modulelor.
'A Wiley-Interscience publication'.

Bibliography: p.
Includes index.
1. Modules (Algebra) 2. Categories (Mathematics)
I. Title.
QA247.S6313   512'.522   76-47588

ISBN 0 471 99462 6

This book is the revised English version of the Romanian book
TEORIA MODULELOR (CATEGORII DE MODULE)
published in 1972
by EDITURA ACADEMIEI, Calea Victoriei 125, BUCUREȘTI, 71021

PRINTED IN ROMANIA

Dedicated to the memory of
GRIGORE C. MOISIL

Preface

Modules represent the natural generalization of vector spaces when the domain of operators is an arbitrary ring, not necessarily a skew-field. This generalization is imposed by the very nature of things. Thus, for instance, even when studying the endomorphisms of a vector space over a field K, the consideration of a module structure over the ring of polynomials K[X] leads very easily to the result; further, when a valued field is studied, one is confronted with module structures over the ring of the valuation, which is not a field. Moreover, the notion of a module over a ring contains as a particular case the abelian group. Because of its numerous and essential applications in different fields of Mathematics such as Algebra, Number Theory, Topology, Geometry, Analysis, etc., and through them in Physics, Chemistry, Engineering, etc., the concept of the module imposes itself throughout almost the whole of mathematical literature and the literature of the sciences which use Mathematics. The present volume attempts to be a systematic account of the bases of the general Theory of Modules with a view to enabling the reader to approach other specialized areas. On the other hand, since the Theory of Categories, particularly of abelian categories, draws its fundamental notions and results from the corresponding ones for modules, refining them in the process, this volume gives a parallel account of the bases of the Theory of Categories, paying special attention to abelian categories; each notion for categories is introduced immediately after the development of the corresponding one for modules. Thus, two streams of ideas are interwoven, one of them related simply to modules, the other to the so-called abstract categories. We believe that this manner of exposition is advantageous mainly for the ideas pertaining to modules, which are seen in a new light when presented in the more general setting of categories. Thus, for instance, the notion of the kernel of a homomorphism of modules would be made poorer not only in its essence but also in its applications if it were regarded simply as the set of elements of the domain which are sent into zero and if it were not pointed out that it had the universal property of categorical nature which made it an "instance" of the more general notion of a kernel of a morphism in a preadditive category. As a matter of fact, a systematic treatment of the theory of modules is possible only in the categorical frame. If we referred, for instance, to the notions of projective system and projective limit, we could, of course, have developed them using a preordered set instead of an arbitrary small category of indices. But then, the general notion of projective limit would fail to include that of kernel as a particular case.

Further, it is worth mentioning that almost all the constructions in the theory of modules are in fact constructions of functors mainly adjoint to some forgetful functors (that is, some more complicated structures are neglected) or to immersion functors of certain categories into some more extended ones. The two streams of ideas mentioned above, in constant contact with each other, are unified again in a different manner in the last Section through the exposition of the representation theorems. By increasingly enriching the categories with additional structures what in essence we get are also categories of modules. Accordingly, the present volume contains in fact a treatment of modules from the categorical point of view, and, since everything is thought of as an introduction, the subtitle "An Introduction to the Theory of Module Categories" might be a better title than the principal one "Theory of Modules".

No preliminary knowledge except some elementary acquaintance with integers, rational numbers, etc., and, of course, the usage of mathematical reasoning is required from the reader. Although the book proceeds from the simplest notions and properties, such results as the representability theorems for functors from one category of modules to another one (Section 16) or the representation theorems of abelian categories by categories of modules (Section 20) are obtained. In this connection we would mention that Mitchell's Full Imbedding Theorem and the Group Valued Imbedding Theorem are first proved in weak forms subsequently used in the proof of the plenary forms. The strong forms of Mitchell's Full Imbedding Theorem and of the Group Valued Imbedding Theorem are results of a great deal of knowledge and of a real mathematical beauty; but even the weak forms, as shown at the end of Section 20, may be used for practical purposes if we confine ourselves to small abelian categories already "imbedded" in abelian categories with direct sums and a projective generator.

The book is divided into 21 Sections. Section 0 has an introductory character, its aim being to make the subsequent exposition self-contained. In Sections 1 and 2 the fundamental notions are introduced, i.e. module, linear mapping, computation in a module, addition of linear mappings, as well as the notions of category, preadditive category, functor, additive functor, etc. Sections 3 and 4 are devoted to the study of the notion of submodule, quotient module, and also of subobject and quotient object. They also discuss the notions of kernel, cokernel, image, coimage, canonical factorization of a morphism. Axioms AB1) *and* AB2) *are formulated, and the First and the Third Isomorphism Theorems for modules are given. Section 5 deals with exact sequences and exact functors. The homology modules of a semi-exact sequence are mentioned. A series of famous propositions such as The Five Lemma, the 3×3 Lemma, and the Snake Lemma are proved for modules. Sections 6 and 7 deal with direct products and direct sums for modules and for arbitrary categories, as well as with the connection between these notions. It is shown that in a preadditive category to any direct sum of a finite family of objects a direct product of the family is assigned, and vice versa. The notions of additive category and abelian category are introduced. In Section 8 the intersection and the sum of a family of submodules of a module and of a family of subobjects of an object are considered. It is shown that from the corresponding property for submodules, Grothendieck's celebrated Axiom* AB5) *for abelian categories emerged. Section 9 discusses the decomposition of a module into internal direct sum of submodules and* *the related notions, i.e. direct summand, split exact sequence, etc. Section 10 deals with linearly independent families, bases, and free modules. The notion of adjoint functor is introduced and studied, the construction of the free module generated by a set being one of the important examples of adjoint functor. The problem of the invariance of the cardinality of a basis is also treated. In Section 11 the Jordan-Hölder Theorem and modules of finite length are considered. As a generalization, the notions of Euler-Poincaré mapping and of the Grothendieck group of a Serre class of objects of an abelian category are introduced. In Sections 12, 13, 14 the notions of projective and inductive limit for systems indexed by a small category of indices and the relation between them are studied. The particular cases of kernel and cokernel, direct product and direct sum, intersection and sum of submodules, fibered product and fibered sum are considered. The property of a functor of commuting with projective or inductive limits is studied. Section 15 studies the functor of restriction of scalars when passing from one ring to another by means of a homomorphism. Section 16 treats the case in which on the same abelian group, module structures over two rings are defined. The way in which the supplementary structure (with respect to the second ring) is carried over when applying an additive functor to the category of modules over one of the rings is shown. The Yoneda Lemma is proved, and representable functors are studied, the existence of a bijection between the morphisms* $E \to F$ *and the natural transformations* $h^F \to h^E$, *respectively* $h_E \to h_F$ *being shown. The subrepresentability and representability theorems (Eilenberg-Watts Representability Theorem) for functors between categories of modules are also included. In Section 17 the notions of dual and bidual of a module are studied. In Section 18 projective and injective modules, as well as projective and injective objects in a category are considered. The theorem of imbedding a module in an injective module is given. Section 19 deals with generators, cogenerators, and injective envelopes. Section 20 proceeds by introducing the notion of equivalence between categories. Further, the representation theorems of abelian categories by categories of modules are studied. In addition to Mitchell's Full Imbedding Theorem and the Group Valued Imbedding Theorem (and an example of their application) another type of representation theorem is given, namely the Gabriel-Mitchell Representation Theorem. The Gabriel-Popescu Representation Theorem is presented without proof; a short exposition on quotient categories precedes it.*

The present volume does not include the theory of the tensor product of modules and, in general, Multilinear Algebra. Likewise, the Theory of semi-simple modules and Homological Algebra are not included. They might be the subject of future volumes.

As may be seen, the book was thought of as a reference book in the Theory of Modules. It is intended for a large class of specialists, mathematicians working in different fields of Mathematics, students and teachers of Mathematics, as well as for other specialists who make use of modules.

The Bibliography at the end of the book consists of works (papers and books) which contain material treated in the book, and works which open up other prospects in this field. This Bibliography is by no means intended to be complete. The names of the authors of various results are mentioned only incidentally in the main text.

Thus, even the names of the famous originators of the Theory of Categories, Samuel Eilenberg and Saunders MacLane, are mentioned only in the Bibliography.

In a reference book the possibility of easily finding the definition of a notion or the statement of an important theorem is essential. For this reason, a Subject Index which is as detailed as possible is added at the end of the book.

As a technical detail, attention is drawn to the fact that any matter that concerns the treatment of notions and results related to abstract categories, without particularly concerning modules, is included between asterisks. Furthermore, the end of a definition, remark, proof of a Proposition, etc. is marked by the symbol ▮. Any reference to a definition or result in the same section is made without mentioning the Section, but only the number of the corresponding definition, proposition, etc., for instance, Proposition 1. On the other hand, if the definition or the result to which reference is made is in another section, the latter is also mentioned, for instance, Definition 3, § 12.

Now, that this book has been translated, my thoughts turn to all those who, in one way or another, have supported me in drawing it up and in publishing it, but most of all to the late Professor Gr. C. Moisil, whose advice and support, not only as concerns this book, could never be fully estimated.

I would like to take this opportunity to express my thanks and gratitude to Professor Gh. Galbură and to Professor I. Bucur from the University of Bucharest, whose encouragement and advice have been very valuable, as well as to Professor David B. Scott and to Professor Christian J. Mulvey from Sussex University, who kindly recommended the book to be published in English. Thanks are also due to John Wiley & Sons Ltd. Company for having accepted this book for publication, and to the Publishing House of the Romanian Academy for the contribution to its joint publication.

Last but not least, I would like to thank Miss Mioara Buiculescu, who has undertaken the difficult task of translating the book into English.

Bucharest, April 1975 ALEXANDRU SOLIAN

Contents

§ 0. Preliminary Notions

This Section is not intended to be a systematic account of the treated topics; its aim is to recall some general notions and results in order to make the subsequent exposition as self-contained as possible, and also to give the terminology and the notation employed throughout the book. As a matter of fact, we had no concern with placing ourselves in some axiomatic system of Set Theory.

The notion of a *class* of elements will be taken as a primary notion: it is a collection of elements (in the usual meaning of this word) thought of as a whole; if an element a belongs to the class M, we shall write $a \in M$ or $M \ni a$, and if the element a does not belong to the class M, we shall write $a \notin M$; a class will be determined by its elements, two classes M, N being considered equal or identical, in writing $M = N$, provided $a \in M$ implies $a \in N$ and conversely; otherwise we write $M \neq N$; if for any element $a \in N$ we also have $a \in M$, then we shall say that the class N is *contained* (or *included*) in the class M or that M *contains* N or that N is a *subclass* of M, and we shall write $N \subseteq M$ or $M \supseteq N$.

Sometimes, we may describe a class M by enumerating its elements,

$$M = \{a, b, \ldots\},$$

or by designating a common property $\mathcal{A}$ characteristic of all elements of the class M,

$$M = \{x \mid \mathcal{A}(x)\},$$

relation which is read as: "M is the class of all elements x having the property $\mathcal{A}$"; in particular, the class consisting only of the element a is denoted by $\{a\}$, the class consisting of the elements a and b, $a \neq b$, is denoted by $\{a, b\}$, etc[1].

The reader is assumed to be familiar with the following classes and the elementary notions related to them and their elements:

the class **N** of *natural numbers*,
the class **Z** of (*ordinary*) *integers*,
the class **Q** of *rational numbers*,
the class **R** of *real numbers*,
the class **C** of *complex numbers*.

[1] If the elements a, b of a class are identical, one writes $a = b$; otherwise one writes $a \neq b$.

According to the established notation, if $m, n \in \mathbf{Z}$, $m \leqslant n$, the class of those integers x for which $x \leqslant n$ and $x \geqslant m$ is denoted $\{m, m+1, \ldots, n\}$; this class is also denoted $[m, n]$, and is called the *integer interval* $[m, n]$.

The notion of mapping of classes will also be taken as a primary notion: given two classes M, N, *a mapping* (*of classes*) *defined on* M *and taking values in* N or *a mapping* (*of classes*) *from* M *to* N or *of* M *into* N or *between* M *and* N is a law assigning to each element of M a uniquely determined element of N; as with the term "collection" above, the term "law assigning..." is considered with its usual meaning; denoting this law by f, we shall write $f: M \to N$ or $M \xrightarrow{f} N$, and M, respectively N, will be called the *domain* (or the *source*), respectively the *codomain* (or the *range*) of the mapping f; sometimes we shall use the notation $M \to N$ to designate a mapping (which is not marked) of source M and range N.

In order to indicate the fact that $y \in N$ corresponds to an element $x \in M$ under the mapping $f: M \to N$, we shall write $y = f(x)$ or we shall use the notation $x \mapsto y$; thus, for the mapping f we write $x \mapsto f(x)$, $x \in M$; for instance, the mapping $\mathbf{N} \to \mathbf{N}$ assigning to each natural number n its double is written $n \mapsto 2n$, $n \in \mathbf{N}$.

We shall consider that two mappings of classes f, g are identical, and we shall write $f = g$, when they have the same domain and the same codomain and when for any element x of the common domain we have $f(x) = g(x)$ in the common codomain.

If M, N, P are classes and $f: M \to N$, $g: N \to P$ are mappings, then, by definition, the *composition* (or the *composite mapping*) of the mappings g and f (or of g with f), denoted by $g \circ f$, is the mapping $M \to P$ defined by

$$(g \circ f)(x) = g(f(x)), \quad x \in M;$$

notice that the composite of the mappings g and f is defined if and only if the domain of g coincides with the codomain of f.

For any class M, the mapping $M \to M$ assigning to each element $x \in M$ the element x itself is called the *identity mapping* of the class M, and will be denoted 1_M; obviously, for any class N and any mapping $f: M \to N$ we have

$$f \circ 1_M = f,$$

and for any class P and any mapping $g: P \to M$ we have

$$1_M \circ g = g.$$

If $f: M \to N$ is a mapping of classes and M' is a subclass of M, the *image of the subclass* M' *under* f, denoted $f(M')$, is the subclass of N consisting of the elements $f(x)$, $x \in M'$; if N' is a subclass of N, the *inverse image of* N' *under* f, denoted $f^{-1}(N')$, is the subclass of M consisting of those elements x for which $f(x) \in N'$.

As a particular class we shall consider the *void class* (or the *empty class*), that is, the class containing no elements, and denoted $\emptyset$; for any class M we shall consider that there exists only one mapping $\emptyset \to M$, called the *void mapping* (or the *empty mapping*) into the class M.

If M, N are classes and $f: M \to N$ is a mapping, then f is said to be an *injective mapping* or an *injection* provided $x \neq y$ in M implies $f(x) \neq f(y)$ in N; f is a *surjective mapping* or a *surjection* or a mapping of M *onto* N provided that for any $z \in N$ there exists at least one $x \in M$ such that $f(x) = z$; the mapping f is a *bijective mapping* or a *bijection* if it is both injective and surjective; sometimes, a bijective mapping is referred to as a *one-to-one correspondence*; f is bijective if and only if there exists a mapping $g: N \to M$ such that

$$(1) \qquad g \circ f = 1_M, \quad f \circ g = 1_N$$

(see also the Remark following the Corollary to Proposition 1, § 1); the mapping g with properties (1) is uniquely determined, and is a bijection: it is called the *inverse (bijection) of* f or the bijection *inverse to* f, and it is denoted f^{-1}. For a bijection $f: M \to N$ one also writes $f: M \xrightarrow{\sim} N$.

The composition of two injective, respectively surjective, respectively bijective mappings is an injective, respectively surjective, respectively bijective mapping; given a mapping $f: M \to N$, a mapping $r: N \to M$ such that

$$r \circ f = 1_M$$

is called a *left inverse* (mapping) *of* f or a *retraction associated with* f; a mapping $s: N \to M$ such that

$$f \circ s = 1_N$$

is called a *right inverse of* f or a *section associated with* f; a mapping of classes $f: M \to N$ for which there exists a retraction, respectively a section associated with f is injective, respectively surjective.

If I and M are classes, a *family of elements of* M *indexed by the class* I (*of indices*) is a mapping $I \to M$; in this context, the class I is called *class of indices*, its elements being called *indices*; if the mapping assigns $a_\lambda \in M$ to $\lambda \in I$, then the family is denoted $(a_\lambda)_{\lambda \in I}$ or even $(a_\lambda)_\lambda$; the element a_λ corresponding to the index λ, $\lambda \in I$, is called the *component* of index (or of order) λ of the family; if $I = \emptyset$, then the family of elements of M indexed by I, which is the void mapping into the class M, is called the *void* (or the *empty*) *family*. If $I = \{1, \ldots, n\}$, then a family $(a_\lambda)_{\lambda \in I}$ of elements of the class M indexed by I is also denoted $(a_1, \ldots, a_n)$ or $(a_i)_{1 \leqslant i \leqslant n}$.

If M is a class and N is a subclass of it, the mapping $N \to M$ given by $x \mapsto x$, $x \in N$, is an injective mapping; it is called the *canonical injection* or the *inclusion mapping* of N into M; if, moreover, P is a class and $f: M \to P$ is a mapping, then the mapping $f \circ i$, where i is the canonical injection $N \to M$, is called the *restriction of* f *to* N, and is sometimes denoted $f|_N$; if $(a_\lambda)_{\lambda\in I}$ is a family of elements of the class M indexed by the class I, while J is a subclass of I, then the restriction of the mapping $\lambda \mapsto a_\lambda$, $\lambda \in I$, to J is said to be a *subfamily* of the family $(a_\lambda)_{\lambda\in I}$.

If N is a subclass of the class M, we canonically assign to it a family of elements of M as follows: as class of indices the class N is taken, and as mapping $N \to M$ the canonical injection is taken; hence, the family assigns to any index the index itself considered in M; in this sense, we shall say that "a subclass of M is a particular case of a family of elements of M"; we shall refer to this situation talking of the "family indexed identically by N".

Let $f: M \to N$ be a mapping of classes, and let N' be a subclass of N; if for any $x \in M$ we have $f(x) \in N'$, then there exists a unique mapping $f' : M \to N'$ such that

$$j \circ f' = f,$$

where j is the canonical injection of N' into N; the mapping f' is said to be obtained from f by *restricting the codomain*; next, if a subclass M' of M is also given, if i is the canonical injection $M' \to M$, and if for any $x \in M'$ we have $f(x) \in N'$, then, starting with the restriction $f \circ i$ of f to M', we get a unique mapping $f' : M' \to$ $\to N'$ such that

$$j \circ f' = f \circ i;$$

f' is called the *mapping induced by* f *by restriction to* M' *and* N'.

Let M be a class; the class $\mathcal{P}(M)$ of all subclasses of M may be considered, with the equality given by the equality of classes: we have $N_1 = N_2$, where $N_1, N_2 \in$ $\in \mathcal{P}(M)$, if and only if any element of N_1 belongs to N_2 and conversely; any class M is considered to have as subclasses the *void* (or the *empty*) *subclass* (that is, the subclass containing no element, denoted also $\emptyset$, as the void class) and M itself, which is called the *total subclass* of M. If $N_1, N_2 \in \mathcal{P}(M)$, then one says that the subclass N_1 is *contained* (or *included*) in the subclass N_2 or that the subclass N_2 *contains* the subclass N_1, written $N_1 \subseteq N_2$ or $N_2 \supseteq N_1$, provided that for any element $a \in N_1$ we have also $a \in N_2$; obviously, in this case, if the subclass N_2 is considered by itself, N_1 may be considered as a subclass of N_2.

If N is a subclass of M, the *complement* (or the *complementary subclass*) *of* N in M, denoted $M \setminus N$, is the subclass consisting of those elements $x \in M$ which do not belong to N; thus, the subclass $M \setminus N$ is defined by

$$M \setminus N = \{x \mid x \in M, \quad x \notin N\}.$$

We obviously have $M \setminus (M \setminus N) = N$; next, if $N_1, N_2 \in \mathcal{P}(M)$ and $N_1 \subseteq N_2$, then $M \setminus N_1 \supseteq M \setminus N_2$.[1]

---

[1] Here we could have employed the *symbol of the logical implication*, denoted $\Rightarrow$; if an assertion $\mathcal{A}$ implies an assertion $\mathcal{B}$, one writes $\mathcal{A} \Rightarrow \mathcal{B}$; for instance, the above result may also be written as $N_1 \subseteq N_2 \Rightarrow M \setminus N_1 \supseteq M \setminus N_2$; if the assertion $\mathcal{A}$ implies the assertion $\mathcal{B}$ and conversely, then the two assertions are said to be *equivalent*, and one writes $\mathcal{A} \Leftrightarrow \mathcal{B}$.

Let M be a class, and let $(M_\lambda)_{\lambda\in I}$ be a family of subclasses of it indexed by the class I; the *union* of the family of subclasses $(M_\lambda)_{\lambda\in I}$, denoted $\bigcup_{\lambda\in I} M_\lambda$, is the subclass consisting of all those elements of M which belong to at least one M_λ, $\lambda\in I$; if $I=[1,n]$, then, instead of $\bigcup_{\lambda\in I} M_\lambda$ one also writes $M_1\cup\ldots\cup M_n$; the *intersection* of the family of subclasses $(M_\lambda)_{\lambda\in I}$, denoted $\bigcap_{\lambda\in I} M_\lambda$, is the subclass consisting of all those elements of M which belong to all M_λ, $\lambda\in I$; if $I=[1,n]$, then, instead of $\bigcap_{\lambda\in I} M_\lambda$ one also writes $M_1\cap\ldots\cap M_n$. We have the following properties: $\bigcup_{\lambda\in I} M_\lambda \supseteq M_\lambda$ for any $\lambda\in I$, and if $X\in\mathcal{P}(M)$ has the property $X\supseteq M_\lambda$ for any $\lambda\in I$, then $X\supseteq\bigcup_{\lambda\in I} M_\lambda$; next, $\bigcap_{\lambda\in I} M_\lambda\subseteq M_\lambda$ for any $\lambda\in I$, and if $Y\in\mathcal{P}(M)$ has the property $Y\subseteq M_\lambda$ for any $\lambda\in I$, then $Y\subseteq\bigcap_{\lambda\in I} M_\lambda$.

We have the following relations:

$$N\cap\Big(\bigcup_{\lambda\in I} M_\lambda\Big)=\bigcup_{\lambda\in I}(N\cap M_\lambda),\quad N\cup\Big(\bigcap_{\lambda\in I} M_\lambda\Big)=\bigcap_{\lambda\in I}(N\cup M_\lambda) \tag{2}$$

for any family $(M_\lambda)_{\lambda\in I}$ of subclasses of M and any $N\in\mathcal{P}(M)$; the properties (2) are called the "distributivity properties", the first "of intersection over union", the second "of union over intersection"; next, for any family $(M_\lambda)_{\lambda\in I}$ of elements of $\mathcal{P}(M)$ we have

$$M\setminus\Big(\bigcup_{\lambda\in I} M_\lambda\Big)=\bigcap_{\lambda\in I}(M\setminus M_\lambda),\quad M\setminus\Big(\bigcap_{\lambda\in I} M_\lambda\Big)=\bigcup_{\lambda\in I}(M\setminus M_\lambda); \tag{3}$$

relations (3) are called "de Morgan rules"; furthermore, if $N_1, N_2\in\mathcal{P}(M)$ and $N_1\supseteq N_2$, then

$$N_1\cup N_2=N_1,\quad N_1\cap N_2=N_2; \tag{4}$$

since for any subclass N of M we have $\emptyset\subseteq N$, and since M (as an element of $\mathcal{P}(M)$) contains any subclass N of M, from relations (4) we get in particular

$$N\cup\emptyset=N,\ N\cap\emptyset=\emptyset,\ M\cup N=M,\ M\cap N=N$$

for any $N\in\mathcal{P}(M)$.

Further, if N is a subclass of M, then the subclass $M\setminus N$ is characterized by the relations

$$N\cup(M\setminus N)=M,\quad N\cap(M\setminus N)=\emptyset.$$

Two subclasses N_1, N_2 of M are said to be *disjoint* if $N_1\cap N_2=\emptyset$. A family $(M_\lambda)_{\lambda\in I}$ of subclasses of M is said to be a *partition* of M provided $M=\bigcup_{\lambda\in I} M_\lambda$

and $M_\lambda \cap M_\mu = \emptyset$ for any (λ, μ) with $\lambda \neq \mu$;[1] in particular, for any subclass N of M, the family (N_1, N_2), where $N_1 = N$, $N_2 = M \setminus N$, is a partition of M.

Let M, N be classes, and let $f: M \to N$ be a mapping of classes; if $M_1, M_2 \in \mathcal{P}(M)$, and if $M_1 \subseteq M_2$, then $f(M_1) \subseteq f(M_2)$; likewise, if $N_1, N_2 \in \mathcal{P}(N)$, and if $N_1 \subseteq N_2$, then $f^{-1}(N_1) \subseteq f^{-1}(N_2)$; the first, respectively the second property is called the "monotonicity of taking the image, respectively the inverse image". Moreover, if M' is a subclass of M, then $f^{-1}(f(M')) \supseteq M'$; if $N' \in \mathcal{P}(N)$, then $f(f^{-1}(N')) \subseteq N'$, and if f is a surjective mapping, then $f(f^{-1}(N')) = N'$.

Let $(M_\lambda)_{\lambda \in I}$ be a family of classes indexed by a class I;[2] let a, b be elements (not necessarily distinct) belonging to some of the classes M_λ, $\lambda \in I$, that is, such that there exist some indices $\lambda, \mu \in I$, not necessarily distinct, for which $a \in M_\lambda$, $b \in M_\mu$; by definition, the *ordered pair* (a, b) is the family indexed by the class $J = \{1, 2\}$, of elements of the class $\{a, b\}$ if a and b are distinct elements or of the class $\{a\}$ if $a = b$, for which

$$1 \mapsto a, \quad 2 \mapsto b;$$

it follows that, if we consider also the elements a', b' belonging to some of the classes M_λ, $\lambda \in I$, then the ordered pair (a, b) coincides with the ordered pair (a', b') if and only if a and a' belong to the same class M_λ and in this class $a = a'$ and similarly for b and b'; in particular, $(a, b) = (b, a)$ if and only if $a = b$.

Again let $(M_\lambda)_{\lambda \in I}$ be a family of classes indexed by a class I; the *coproduct* or also the *disjoint union*[3] of the family of classes $(M_\lambda)_{\lambda \in I}$, denoted $\coprod_{\lambda \in I} M_\lambda$, is called the class whose elements are all ordered pairs of the form (x, λ), where $\lambda \in I$ and $x \in M_\lambda$;[4] for any $\lambda \in I$, the mapping $j_\lambda : M_\lambda \to \coprod_{\lambda \in I} M_\lambda$ defined by

$$j_\lambda(x) = (x, \lambda), \quad x \in M_\lambda,$$

is an injection; it is called the *canonical injection* of index λ into the coproduct; we have the following "universal property": if X is a class, and if for any $\lambda \in I$ a mapping $f_\lambda : M_\lambda \to X$ is given, then there exists a unique mapping

$$f: \coprod_{\lambda \in I} M_\lambda \to X$$

with the property that

$$f \circ j_\lambda = f_\lambda$$

---

[1] Nevertheless, the equality of partitions will not be considered as being the equality of families, two partitions $(M_\lambda)_{\lambda \in I}$ and $(N_\nu)_{\nu \in J}$ of M being considered equal provided any subclass M_λ is identical to some N_ν and conversely.

[2] Here we may consider that the given family is a mapping defined on the class I and taking values in the class of the classes M_λ, $\lambda \in I$.

[3] Not to be confounded with the union of a family of subclasses of a class (see the justification given below for the term "disjoint union").

[4] Here the family of classes has been enlarged, adjoining to it with a new index the class I.

for any $\lambda \in I$; the mapping f is defined by

$$f(x, \lambda) = f_\lambda(x),$$

$\lambda \in I$, $x \in M_\lambda$. In the case where $I = [1, n]$, instead of $\coprod_{\lambda \in I} M_\lambda$ one also writes $M_1 \coprod \dots \coprod M_n$.

If we consider the family of subclasses $(j_\lambda(M_\lambda))_{\lambda \in I}$ of $\coprod_{\lambda \in I} M_\lambda$, this is a partition of $\coprod_{\lambda \in I} M_\lambda$; taking into account that for $\lambda \in I$ the class M_λ is sometimes identified by means of j_λ with the subclass $j_\lambda(M_\lambda)$ of $\coprod_{\lambda \in I} M_\lambda$, this justifies the term of "disjoint union" of the family of classes $(M_\lambda)_{\lambda \in I}$, introduced above; the identification mentioned is done in the sequel.

Again let $(M_\lambda)_{\lambda \in I}$ be a family of classes indexed by a class I; we shall admit that it is possible to construct a new class M containing all classes M_λ, $\lambda \in I$, as subclasses, and containing only elements of these classes; here, the elements which may be common to two or several distinct classes are considered only once, and not once for each class M_λ in which they occur, as in the disjoint union; roughly speaking, the class M consists of all elements belonging at least to one of the classes M_λ, $\lambda \in I$, each element being considered only once even if it appears in several classes; a more refined and at the same time more rigorous construction requires us to specify in which sense elements from two different classes are considered equal; in this way we get the notion of *inductive system*, and M could be constructed as the *inductive limit* of this system; we are not going into the details here[1]; we shall refer to the class M as the *union* of the family of classes $(M_\lambda)_{\lambda \in I}$.

Consider again a family $(M_\lambda)_{\lambda \in I}$ of non-void classes indexed by a class I; let M be the union of this family; the *direct product* or the *cartesian product* or simply the *product* of the family of classes $(M_\lambda)_{\lambda \in I}$, denoted $\prod_{\lambda \in I} M_\lambda$, is the class whose elements are those families $(x_\lambda)_{\lambda \in I}$ of elements of M indexed by I for which the element x_λ corresponding to the index $\lambda \in I$, i.e. the component of index λ of the family, is an element of the class M_λ for any $\lambda \in I$; the mapping

$$p_\lambda : \prod_{\lambda \in I} M_\lambda \to M_\lambda$$

assigning to each family $(x_\lambda)_{\lambda \in I}$ its component of index λ is called the *canonical projection of index* λ, $\lambda \in I$, of the direct product; here again we have a universal property, namely: if Y is an arbitrary class, and if for any $\lambda \in I$ a mapping $g_\lambda : Y \to M_\lambda$ is given, then there exists a unique mapping of classes

$$g : Y \to \prod_{\lambda \in I} M_\lambda$$

such that

$$p_\lambda \circ g = g_\lambda$$

[1] For the study of inductive systems of modules and objects in categories, and of their inductive limits, see § 13.

for any $\lambda \in I$; the mapping g is defined by

$$g(\eta) = (g_\lambda(\eta))_{\lambda \in I},$$

$\eta \in Y$. In the case when $I = [1, n]$, instead of $\prod_{\lambda \in I} M_\lambda$ one also writes $M_1 \times \ldots \times M_n$ or $M_1 \prod M_2 \prod \ldots \prod M_n$. In the case of an arbitrary family $(M_\lambda)_{\lambda \in I}$ of classes, if at least for one index $\lambda \in I$ we have $M_\lambda = \emptyset$, then, by definition, $\prod_{\lambda \in I} M_\lambda = \emptyset$.

In particular, if $I = \{1, 2\}$, $M_1 \times M_2$ is the class of all those ordered pairs (a_1, a_2) for which $a_i \in M_i$, $i = 1, 2$.

Another particular case is the one in which the class I is arbitrary, but all the classes M_λ coincide with a fixed class M; in this case the product $\prod_{\lambda \in I} M_\lambda$ is denoted M^I, and is the class of all mappings defined on I with values in M; if, in this case, $I = [1, n]$, then instead of M^I one also writes M^n or $M \times \ldots \times M$ (n times); hence, instead of M^2 one also writes $M \times M$; clearly, M^1 may be identified with M.

Returning to an arbitrary family of classes $(M_\lambda)_{\lambda \in I}$ indexed by a class I, if for any $\lambda \in I$, N_λ is a subclass of M_λ, then there exists a mapping of classes $i: \prod_{\lambda \in I} N_\lambda \to \prod_{\lambda \in I} M_\lambda$ assigning to each family $(x_\lambda)_{\lambda \in I} \in \prod_{\lambda \in I} N_\lambda$ the same family, but in which any element x_λ is considered in the corresponding class M_λ, $\lambda \in I$; obviously, the mapping i is injective, and by means of it $\prod_{\lambda \in I} N_\lambda$ is identified with a subclass of $\prod_{\lambda \in I} M_\lambda$.[1]

Let M be a class; a (*binary*) *relation* on the class M is a subclass R of the cartesian product $M \times M$; if the ordered pair (x, y) is an element of R, one also says that x *is in the relation* R *with* y, and one writes xRy.

A binary relation R on the class M is *reflexive* provided xRx for any $x \in M$; the binary relation R on the class M is by definition *transitive* if xRy, yRz, with $x, y, z \in M$, implies xRz.

A reflexive and transitive binary relation on a class M is called a *preorder relation* or a *preordering*; given a preorder relation R on the class M, this is, if no confusion can arise, generally denoted by $\leqslant$, so that instead of xRy, $x, y \in M$, we write $x \leqslant y$; in this case one says that x *precedes* y or that x *is less than* y (of course, in the sense of the relation $\leqslant$).

If M is a class and R is a binary relation on M, the *inverse* or the *converse* relation of R, denoted by $\overset{-1}{R}$, is the binary relation defined in the following manner: by definition

$$x\overset{-1}{R}y \text{ means } yRx, \ x, y \in M;$$

[1] See also Corollary 1 to Proposition 2, § 6.

if R is a preorder relation, denoted also $\leqslant$, then its inverse $\overset{-1}{R}$ is also a preorder relation, denoted $\geqslant$ in the given context; hence, instead of $x\overset{-1}{R}y$ one writes $x \geqslant y$; in this case one says that x *succeeds* to y or that x *is greater than* y in the sense of the relation $\leqslant$; hence, to say that $x \geqslant y$ is the same to saying that $y \leqslant x$. If $x, y \in M$ and x does not preceed, respectively succeed to y in the sense of the relation $\leqslant$, then one writes $x \nleqslant y$, respectively $x \ngeqslant y$.

A non-void class M endowed with a preorder relation $\leqslant$ is called a *preordered class* (with respect to the relation $\leqslant$) or a class *preordered by* the relation $\leqslant$.

Given the preorder relation $\leqslant$ on the class M, this relation is said to be a *partial order relation* or a *partial ordering* if the relation $\leqslant$ is *antisymmetric*: that is, $x \leqslant y$, $y \leqslant x$ imply $x = y$, $x, y \in M$.

A non-void class M endowed with a partial order relation $\leqslant$ is called a *partially ordered class* (with respect to the relation $\leqslant$) or a class *partially ordered by* the relation $\leqslant$.

For instance, if E is a class, the class $\mathcal{P}(E)$ of all subclasses of E, endowed with the binary relation R given by inclusion (that is, F_1RF_2 means $F_1 \subseteq F_2$, $F_1, F_2 \in \mathcal{P}(E)$) is a partially ordered class [1].

Given on M the partial order relation denoted $\leqslant$, this relation is said to be a *total order relation* or a *linear order relation* or a *linear ordering* if for any $x, y \in M$ we have at least one of the situations

$$x \leqslant y \quad \text{or} \quad y \leqslant x;$$

a non-void class M endowed with a total order relation is called a *totally ordered class* or a *linearly ordered class.*

Let M be a preordered class with respect to the relation R, and let N be a subclass of M; the intersection $R \cap (N \times N) = S$ is a binary relation on the class N, namely a preorder relation; the binary relation S will be called the *preorder relation induced by R on N*, and the class N together with the preordering S will be said to be a *preordered subclass* of M; if $x, y \in N$, then to say that x is in the relation S with y is equivalent to saying that x is in the relation R with y in M; if the preordering R on the class M is denoted by $\leqslant$, then, usually, the preordering S on N is denoted also by $\leqslant$. If the relation R is a partial order relation, then S is also a partial order relation; in this case S is called the *partial order relation induced by R on N*, and the class N together with the partial ordering S is called a *partially ordered subclass* of M; if in this case the partial ordering S is a total order relation, then N together with the partial order relation S is said to be a *totally ordered subclass* of M (even if the binary relation R on M is not a total order relation). As a generalization, let $(a_\lambda)_{\lambda \in I}$ be a family of elements of a class M indexed by the class I, M being partially ordered by the relation $\leqslant$; the family $(a_\lambda)_{\lambda \in I}$ is said to be *totally ordered* (by the relation $\leqslant$) if for any $\lambda, \mu \in I$ we have at least one of the situations

$$a_\lambda \leqslant a_\mu \quad \text{or} \quad a_\mu \leqslant a_\lambda.$$

---

[1] Notice that if $E = \emptyset$, then $\mathcal{P}(E)$ is not the void class, since $\emptyset \in \mathcal{P}(E)$.

Let M be a class endowed with a preorder relation $\leqslant$, and let $(a_\lambda)_{\lambda \in I}$ be a family of elements of M indexed by the class I; an element $a \in M$ is said to be a *lower bound*, respectively an *upper bound* of the family $(a_\lambda)_{\lambda \in I}$ or also a *common lower bound*, respectively a *common upper bound of the elements* a_λ, $\lambda \in I$, if for any $\lambda \in I$ we have

$$a \leqslant a_\lambda, \text{ respectively } a \geqslant a_\lambda;$$

the definition is extended to the case when a subclass N of M and the family indexed identically by N are considered.

If M is a class partially ordered by the relation $\leqslant$, the class M is said to be *inductively ordered* (with respect to the partial order relation $\leqslant$) if for any totally ordered family of elements of M there exists an upper bound $a \in M$; in this case, the partial order relation $\leqslant$ is said to be an *inductive partial order relation.*

If M is a preordered class with respect to the relation $\leqslant$, and $(a_\lambda)_{\lambda \in I}$ is a family of elements of M, an element $a \in M$ is said to be a *greatest lower bound* (or an *infimum*), respectively a *least upper bound* (or a *supremum*) of the family $(a_\lambda)_{\lambda \in I}$ or also a *common greatest lower bound*, respectively a *common least upper bound of the elements* a_λ, $\lambda \in I$, if a is a lower bound, respectively an upper bound of the family $(a_\lambda)_{\lambda \in I}$ and if it is greater, respectively less, in the sense of the relation $\leqslant$, than any lower bound, respectively any upper bound of the family $(a_\lambda)_{\lambda \in I}$; if a and a' are two elements of M with the property of a greatest lower bound of the family $(a_\lambda)_{\lambda \in I}$, then $a \leqslant a'$ and $a' \leqslant a$; the same thing may be said of two elements b, b' with the property of a least upper bound of the family $(a_\lambda)_{\lambda \in I}$: we have $b \leqslant b'$ and $b' \leqslant b$; in particular, if $\leqslant$ is a partial order relation, then a common greatest lower bound, respectively a common least upper bound of the elements a_λ, $\lambda \in I$, whenever exists, is uniquely determined; it is called *the greatest lower bound* (or *the infimum*), respectively *the least upper bound* (or *the supremum*) of the family $(a_\lambda)_{\lambda \in I}$ or also *the greatest common lower bound*, respectively *the least common upper bound of the elements* a_λ, $\lambda \in I$.

In the case of a partially ordered class M, taking instead of the family $(a_\lambda)_{\lambda \in I}$ the total subclass M, it follows that if a lower bound, respectively an upper bound of this class, call it a, respectively b, exists, then it is the greatest common lower bound, respectively the least common upper bound of the elements of this class; the element a, respectively b (uniquely determined by these properties if it exists) is called in this case the *first element*, respectively the *last element* of the partially ordered class M.

For instance, if E is a class, then in the class $\mathcal{P}(E)$ partially ordered by inclusion, the void subclass is the first element and the total subclass is the last element.

Let M be a class preordered by the relation $\leqslant$; an element $a \in M$ is said to be a *minimal element*, respectively a *maximal element* of the class M if for any $x \in M$ for which

$$x \leqslant a, \text{ respectively } x \geqslant a,$$

we have $x \geqslant a$, respectively $x \leqslant a$; in particular, if $\leqslant$ is a partial order relation, to say that a is a minimal, respectively a maximal element of the partiay ordered class M is equivalent to saying that any lower bound, respectively any upper bound of the subclass $\{a\}$ of M coincides with a.

The notion of minimal, respectively maximal element should not be confounded with the notion of first, respectively last element: this is shown by the following

EXAMPLE. 1) Let M be the class $\mathbf{N} \setminus \{0, 1\}$; on the class M we introduce the binary relation denoted $|$ and defined as follows: $a \mid b$, $a, b \in M$, means that a divides b; the relation $|$ is a partial order relation; in the class M partially ordered by the relation $|$, the natural number 2 is a minimal element, but it is not the first element. ∎

Let M be a class preordered by the relation $\leqslant$, and let $(a_\lambda)_{\lambda \in I}$ be a family of elements of M indexed by the class I; the family $(a_\lambda)_{\lambda \in I}$ is said to be *directed to the left*, respectively *to the right* (with respect to $\leqslant$) if for any $\lambda, \mu \in I$ there exists $\nu \in I$ such that

$$a_\nu \leqslant a_\lambda \text{ and } a_\nu \leqslant a_\mu, \text{ respectively } a_\nu \geqslant a_\lambda \text{ and } a_\nu \geqslant a_\mu;$$

of course, the definition is maintained in the particular case when instead of a family we take a subclass N of M; then we speak of a *subclass of M directed to the left*, respectively *to the right*; to say that the subclass N is directed to the left, respectively to the right is equivalent to saying that any family of elements of N indexed by the class $I = \{1, 2\}$ has a lower, respectively an upper bound which is an element of N; if the total subclass M is directed to the left, respectively to the right, M is said to be a *preordered class directed to the left*, respectively *to the right*; in the particular case where $\leqslant$ is a partial order relation, we speak of a *partially ordered class directed to the left*, respectively *to the right*.

In a class M preordered by the relation $\leqslant$, a totally ordered family $(a_\lambda)_{\lambda \in I}$ of elements of M is directed both to the left and to the right.

If E is a class, then in the class $\mathcal{P}(E)$ partially ordered by inclusion, for any family $(F_\lambda)_{\lambda \in I}$ of subclasses indexed by a class I there exist both the greatest lower bound and the least upper bound of the family $(F_\lambda)_{\lambda \in I}$: the first one is $\bigcap_{\lambda \in I} F_\lambda$, and the second one $\bigcup_{\lambda \in I} F_\lambda$. If, in particular, we consider a family of elements of $\mathcal{P}(E)$ indexed by the class $I = \{1, 2\}$, we obtain that the partially ordered class $\mathcal{P}(E)$ is directed both to the left and to the right. To fix the terminology, by *a family of subclasses of E directed to the left*, respectively *to the right*, respectively *totally ordered*, we mean a family of elements of the class $\mathcal{P}(E)$ partially ordered by inclusion, family which is directed to the left, respectively to the right, respectively totally ordered.

Let M be a class preordered by the relation $\leqslant$; given a family $(a_\lambda)_{\lambda \in I}$ of elements of M indexed by the class I, a subfamily $(a_\lambda)_{\lambda \in J}$, where J is a subclass of I, is said to be a *coinitial*, respectively a *cofinal subfamily* of the family $(a_\lambda)_{\lambda \in I}$ if for any $\lambda \in I$ there exists $\mu \in J$ such that

$$a_\mu \leqslant a_\lambda, \text{ respectively } a_\lambda \leqslant a_\mu;$$

the definition is applied also to the particular case when instead of the family $(a_\lambda)_{\lambda \in I}$, a subclass N of M is taken; instead of the subfamily $(a_\lambda)_{\lambda \in J}$, a subclass P of N

is taken, and, provided there is the case, one says that the subclass P is *coinitial*, respectively *cofinal* in the subclass N.

If E is a class, and the class $\mathcal{P}(E)$ partially ordered by inclusion is considered, we may, provided there is the case, talk of a *coinitial*, respectively *cofinal subfamily* of a family of subclasses of E.

Now let M and N be two classes each of them endowed with a preorder relation; both these relations will be denoted by $\leqslant$, since it will be clear from the context which one is meant; a mapping $f: M \to N$ is said to be *monotone* or *preorder preserving*, respectively *antimonotone* or *preorder reversing* provided for any $a_1, a_2 \in M$ such that $a_1 \leqslant a_2$ we have in N

$$f(a_1) \leqslant f(a_2), \text{ respectively } f(a_1) \geqslant f(a_2);$$

it is clear that if P is also a class preordered by a relation denoted also $\leqslant$, and $g: N \to P$ is a monotone mapping, then the composition $g \circ f$ is monotone or antimonotone according as f is monotone or antimonotone; if $g: N \to P$ is an antimonotone mapping, then $g \circ f$ is antimonotone or monotone according as f is monotone or antimonotone. For any preordered class M, the identity mapping 1_M is monotone.

For instance, let E, F be classes, and let $f: E \to F$ be a mapping of classes; the mapping f induces a mapping $\mathcal{P}(E) \to \mathcal{P}(F)$ according to the law

$$E' \mapsto f(E'), \quad E' \in \mathcal{P}(E);$$

if on the classes $\mathcal{P}(E)$ and $\mathcal{P}(F)$ the partial order relations given by inclusion are considered, the above mapping is monotone, as shown by the monotonicity of taking the image; likewise, with the same data, the mapping $\mathcal{P}(F) \to \mathcal{P}(E)$ given by

$$F' \mapsto f^{-1}(F'), \quad F' \in \mathcal{P}(F),$$

is a monotone mapping of the partially ordered class $\mathcal{P}(F)$ into the partially ordered class $\mathcal{P}(E)$.

Let $(M_\lambda)_{\lambda \in I}$ be a family of preordered classes indexed by a class I, the preorder relation on each class being denoted by $\leqslant$, it being clear from the context which one is meant; on the direct product $\prod_{\lambda \in I} M_\lambda$ a binary relation, denoted also by $\leqslant$, is introduced in the following way: if $(x_\lambda)_{\lambda \in I}$, $(y_\lambda)_{\lambda \in I}$ are elements of $\prod_{\lambda \in I} M_\lambda$, then by definition

$$(x_\lambda)_{\lambda \in I} \leqslant (y_\lambda)_{\lambda \in I}$$

means that $x_\lambda \leqslant y_\lambda$ in M_λ for any $\lambda \in I$; the binary relation $\leqslant$ introduced on the class $\prod_{\lambda \in I} M_\lambda$ is also a preorder relation, and the canonical projection of index λ is, for any $\lambda \in I$, a monotone mapping of the preordered class $\prod_{\lambda \in I} M_\lambda$ into the preordered class M_λ; the class $\prod_{\lambda \in I} M_\lambda$ with the preorder relation $\leqslant$ is called the *preordered product class* or the *product* of the family of preordered classes $(M_\lambda)_{\lambda \in I}$; if for any $\lambda \in I$ the relation $\leqslant$ is a partial order relation on the class M_λ, then the relation $\leqslant$ on the class $\prod_{\lambda \in I} M_\lambda$ is also a partial order relation, and in this case one talks of the *partially ordered product class.*

Now let M be a class; a binary relation R on the class M is said to be *symmetric* if xRy, with $x, y \in M$, implies yRx, i.e., if $\overset{-1}{R}$ coincides with R.

A reflexive, symmetric, and transitive binary relation is called an *equivalence relation* or, sometimes, an *equivalence.* Given an equivalence relation R on the class M, if xRy, $x, y \in M$, then x is said to be *equivalent to* y *with respect to* R (or *modulo* R) or, simply, *equivalent to* y if there is no danger of confusion.

On any class M, the binary relation R for which xRy, $x, y \in M$, means that $x = y$ in M is an equivalence; it is called the *identity equivalence relation* or the *identity* on M; the relation $T = M \times M$ is also an equivalence on M; it is called the *total equivalence relation* (on M).

Let the equivalence relation R on the class M be given; if $x \in M$, the subclass of M consisting of those $y \in M$ for which yRx is called the *equivalence class of* x *with respect to* R (or *modulo* R); from xRx it follows that x belongs to its equivalence class modulo R; a subclass N of M is said to be an *equivalence class modulo* R provided it is the equivalence class modulo R of some element $x \in M$ (necessarily belonging to N); if $x_1, x_2 \in M$, then the equivalence classes modulo R of x_1 and x_2 coincide if and only if x_1Rx_2.

If $N_1, N_2 \in \mathcal{P}(M)$ are equivalence classes modulo R, then they coincide or they are disjoint.

Consider the class Q of equivalence classes modulo R (as a subclass of $\mathcal{P}(M)$, that is, a class in which two equivalence classes modulo R are considered equal, respectively distinct provided they are equal, respectively distinct in $\mathcal{P}(M)$); the family of subclasses of M indexed identically by Q is, by the preceding considerations, a partition of M, since any $x \in M$ belongs to the equivalence class of x modulo R; this partition of M is called the *partition associated with* the equivalence R; the class Q of equivalence classes modulo R is called the *quotient class* of M with respect to the equivalence R (or *modulo* R), and is denoted M/R or $\frac{M}{R}$; obviously, if $M \neq \emptyset$, then $M/R \neq \emptyset$ too.

Conversely, any partition $(M_\lambda)_{\lambda \in I}$ of a class M defines on M an equivalence relation R by the law: xRy, $x, y \in M$, means, by definition, that there exists $\lambda \in I$ such that $x \in M_\lambda$, $y \in M_\lambda$; the equivalence relation R thus defined is called the *equivalence relation* on M *associated with* the given partition $(M_\lambda)_{\lambda \in I}$; notice that, for a class M, the two assignments, from an equivalence relation on the class M to the associated partition, and from a partition of the class M to the associated equivalence relation, are inverse to each other.

Let M be a class, and let R be an equivalence relation on M; the mapping $p : M \to M/R$ assigning to an element $x \in M$ the equivalence class of x modulo R is obviously a surjective mapping; it is called the *canonical surjection* of M onto the quotient class of M modulo R.

If M, N are classes, and if on each of these classes an equivalence relation R, respectively S is given, then a mapping of classes $f: M \to N$ is said to be *compatible with the equivalence relations R and S* if $x_1 R x_2$ in M implies $f(x_1) S f(x_2)$ in N; with the same data *minus* the equivalence S, a mapping $f: M \to N$ is said to be *compatible with the equivalence R* if f is compatible with the equivalence R and with the identity equivalence relation on the class N. If M is a class and R is an equivalence on M, then the canonical surjection $p: M \to M/R$ is compatible with R; moreover, we have the following universal property: for any class N and any mapping $f: M \to N$ compatible with the equivalence R there exists a uniquely determined mapping $g: M/R \to N$ such that

$$g \circ p = f;$$

the mapping g is defined as follows: if $C \in M/R$, and if $x \in C$, then the element $f(x) \in N$ does not depend on the chosen element x in the class C, that is, if $y \in C$, then $f(x) = f(y)$; define

$$g(C) = f(x).$$

It follows that if M, N are classes, if on each of these classes the equivalence relations R and S are respectively given, and if $f: M \to N$ is a mapping compatible with the equivalences R and S, then the mapping $q \circ f: M \to N/S$ (where q is the canonical surjection $N \to N/S$) is compatible with the equivalence R, and hence, there exists a uniquely determined mapping $g: M/R \to N/S$ such that

$$g \circ p = q \circ f,$$

where p is the canonical surjection $M \to M/R$; g is called the *mapping induced by f by passing to the quotient* (with respect to R and S).

Let M, N be two non-void classes, and let $f: M \to N$ be a mapping of classes; define on M the following binary relation, denoted R: by definition xRy, $x, y \in M$, means $f(x) = f(y)$; the binary relation R is obviously an equivalence, and the mapping f is compatible with R; the equivalence R is called the *equivalence relation associated with* the mapping f; by the preceding considerations, there exists a unique mapping $g: M/R \to N$ such that $g \circ p = f$, where p is the canonical surjection; in general, g is an injective mapping, and if f is surjective, then g is also surjective, hence bijective.

Now let M be a non-empty class, and let $(M_\lambda)_{\lambda \in I}$ be a partition of it; a subclass N of M is called a *complete system of representatives* of the given partition if the intersection of N with each of the subclasses M_λ, $\lambda \in I$, consists of exactly one element; by extension, given $\lambda \in I$, an element $a \in M_\lambda$ is said to be a *representative* of the class M_λ; starting with an equivalence relation R on M, and considering the partition associated with R, a complete system of representatives of this partition is called a *complete system of representatives of M with respect to the equivalence R* (or *modulo R*); in this way, any $x \in M$ is a representative of the equivalence class of x modulo R.

Let M be a preordered class with respect to the preorder relation $\leqslant$; on M define a binary relation R in the following manner: if $x, y \in M$, then xRy means $x \leqslant y$ and $y \leqslant x$; obviously, R is an equivalence relation; it is called the *equivalence relation associated with* the preorder relation $\leqslant$. If C, $D \in M/R$, let x, y be representatives of C and D, respectively; if $x \leqslant y$ in M, then, whatever the representatives x', y' of C and D, respectively, we have in M, $x' \leqslant y'$; now, define on M/R the binary relation S as follows: we have CSD, C, $D \in M/R$, if and only if for x, y, representatives of C and D, respectively, we have $x \leqslant y$ in M; by the preceding arguments, this definition is independent of the choice of x in C and y in D; the relation S is a partial order relation on the class M/R; it is called the *partial order relation canonically associated with* the preorder relation $\leqslant$ or also the *partial order relation canonically induced by* the preorder relation $\leqslant$. It is clear that, with respect to these preorder relations, the canonical surjection $p : M \to M/R$ is a monotone mapping.

Among classes some special ones, called *small classes* or *sets*, will be distinguished; sets are characterized among classes by the following criteria, established by definition:

1) The empty class $\emptyset$ is a set.

2) If a is an element, the class $\{a\}$ consisting only of the element a is a set.

3) The classes **N**, **Z**, **Q**, **R**, **C** are sets.

4) If M is a set and N is a class for which there exists an injective mappin of classes $i : N \to M$, then N is a set; in particular, if N is a class for which therg exists a bijective mapping of classes $f : N \to M$, where M is a set, then N is a set; also as a particular case, a subclass of a set is a set.

5) If M is a set and P is a class for which there exists a surjection $p : M \to P$, then P is a set; in particular, if M is a set and R is an equivalence relation on the set M, then the quotient class M/R is a set.

6) If $(M_\lambda)_{\lambda \in I}$ is a family of sets indexed by a set I, then the disjoint union $\coprod_{\lambda \in I} M_\lambda$ is a set.

7) If $(M_\lambda)_{\lambda \in I}$ is a family of sets indexed by a set I, then the cartesian product $\prod_{\lambda \in I} M_\lambda$ is a set.

8) If $(M_\lambda)_{\lambda \in I}$ is a family of sets indexed by a set I, then the union of the family of classes $(M_\lambda)_{\lambda \in I}$ is a set.

From these conditions it follows that if M and N are sets, then the class M^N of all mappings of classes defined on the set N and taking values in the set M is a set.

Further, if M is a set, it follows that the class $\mathcal{P}(M)$ of subclasses of M is a set.

The preceding definitions and results given for classes remain valid if we replace the terms "class, subclass, quotient class, mapping of classes", etc., respectively by "*set*, *subset*, *quotient set*, *mapping of sets*", etc.; in particular, a non-void set M endowed with a preorder relation, respectively a partial order relation (denoted $\leqslant$) is called a *preordered set*, respectively a *partially ordered set* (with respect to the relation $\leqslant$); next, for a family of preordered sets (indexed by a set I), the preordered product class is called the *preordered product set*. We mention, however, that the term "equivalence class" with respect to an equivalence relation R is maintained even if the class M on which the equivalence relation R is defined is a set.

We shall admit as valid the following assertion, called

Zermelo Axiom (or **Axiom of Choice**). *Let $(M_\lambda)_{\lambda \in I}$ be a family of non-void sets indexed by the non-void set I; assume that for $\lambda \neq \mu$ in I, no element of M_λ belongs to M_μ; then there exists a set E having in common with each of the sets M_λ, $\lambda \in I$, exactly one element, and besides these, no other elements.*

Using the Axiom of Choice, one may prove that, given a family $(M_\lambda)_{\lambda \in I}$ of non-empty sets indexed by the set $I \neq \emptyset$ without assuming that M_λ and M_μ, $\lambda \neq \mu$, have no common elements, there exists a family $(x_\lambda)_{\lambda \in I}$ of elements of the union of the family $(M_\lambda)_{\lambda \in I}$ such that $x_\lambda \in M_\lambda$ for any $\lambda \in I$; we shall admit in the sequel that this "choice principle" is also valid for arbitrary classes.

If M is a set and R is an equivalence relation on M, then from the Axiom of Choice it follows directly the existence of a complete system N of representatives of M modulo R; hence, we may find a section $s : M/R \to M$ associated with the canonical surjection $p : M \to M/R$; s is the mapping assigning to an equivalence class $C \in M/R$ the element $x \in N \subseteq M$ which is the representative of C in the complete system of representatives N.

If M, M' are sets and $f : M \to M'$ is a surjective mapping, then, by the preceding considerations, there exists a bijective mapping $g : M/R \to M'$ for which $g \circ p = f$, where R is the equivalence relation associated with the mapping f, and p is the canonical surjection $M \to M/R$; if s is a section associated with p, then $s \circ g^{-1}$ is a section associated with the surjective mapping g.

From the Axiom of Choice one may derive the following Proposition, called

Zorn's Lemma. *Any partially ordered (consequently, non-empty) set which is inductively ordered has at least one maximal element.*

REMARK. In fact, we tacitly admitted and even used an assertion analogous to the Zermelo Axiom for arbitrary classes: for instance, when we constructed the product $\prod_{\lambda \in I} M_\lambda$ of a family of classes indexed by a class I, the existence of families $(x_\lambda)_{\lambda \in I}$ for which $x_\lambda \in M_\lambda$, $\lambda \in I$, was taken as valid; if we did not admit the analogue of the Zermelo Axiom for this more general case, it would have been possible that $\prod_{\lambda \in I} M_\lambda$ was the empty class, and this was tacitly excluded; hence, of course, we get the existence of a section associated with a surjective mapping of classes.

The statement of the Axiom of Choice was given mainly for historical reasons and because of its place in general mathematical knowledge, and also because it provides a logical foundation, which should be explicitly stated, for Zorn's Lemma, whose evidence is not so easily admitted. ∎

We shall say that a set M is a *finite set* if there exist a natural number n and a bij ction $M \xrightarrow{\sim} \{1,\ldots, n\}$; a family of elements $(a_\lambda)_{\lambda \in I}$ of a class K will be called a *finite family* if the class I of indices is a finite set; the family denoted before $(a_1 \ldots a_n)$ is thus a finite family; likewise, given a family $(a_\lambda)_{\lambda \in I}$ of elements of a class K indexed by a class I, a *finite subfamily* of this family is the restriction of the mapping $\lambda \mapsto a_\lambda$, $\lambda \in I$, to a finite subset of I.

Given a finite set M, it may be assumed that a determined bijection between M and a certain set $\{1, \ldots, n\}$ for some suitably chosen n has been established; then, denoting by a_i, $i = 1, \ldots, n$, the element of M which is associated under this bijection with $i \in \{1, \ldots, n\}$, the set M may be written

$$M = \{a_1, \ldots, a_n\};$$

this type of writing will be used in the sequel; it turns out that the natural number n is the same no matter how the bijection is chosen; this number n is called the *number of elements* of the finite set M. By definition, the empty set $\emptyset$ is finite, and its number of elements is 0; this definition is justified by the fact that $\emptyset$ may be considered as the set of natural numbers "between 1 and 0".

It follows from the definition that if m and n are integers such that $m \leqslant n$, then the integer interval $[m, n]$ is a finite set whose number of elements is $n - m + 1$.

The following results may be proved: a subset of a finite set is finite; if there exists a bijection between two sets and one of them is finite, then the other one is finite too; if there exists a surjection from a finite set M to a set N, then N is a finite set; the disjoint union and the cartesian product of a finite family of finite sets are finite sets; the union of a finite family of finite sets is a finite set; next, the set of mappings of a finite set into a finite set is a finite set; further, the set of subsets of a finite set is a finite set.

A set which is not finite is said to be an *infinite set*; for instance, the set $\mathbf{N}$ of natural numbers is infinite.

Let M be a set partially ordered by the relation $\leqslant$; if, whatever the pair (i.e. the family indexed by a set consisting of two indices) (a, b) of elements of M, there exists in M a common greatest lower bound, respectively a common least upper bound of the elements a, b, the partially ordered set M is said to be a *meet-semilattice*, respectively a *join-semilattice*; this is equivalent to saying that for any non-void finite family $(a_1, \ldots, a_n)$ of elements of M there exists the greatest lower bound, respectively the least upper bound of the family $(a_1, \ldots, a_n)$; the greatest lower bound, respectively the least upper bound of the family $(a_1, \ldots, a_n)$ is called in this case the *meet*, respectively the *join of the family* (or of the elements $a_1 \ldots a_n$), and is denoted

$$\bigcap_{i=1}^{n} a_i \text{ or } a_1 \cap \ldots \cap a_n, \text{ respectively } \bigcup_{i=1}^{n} a_i \text{ or } a_1 \cup \ldots \cup a_n;$$

in the particular case of the pair (a, b), we have thus the notation $a \cap b$, respectively $a \cup b$. A set M partially ordered by the relation $\leqslant$, which is both a meet and a join-semilattice is called a *lattice*.

A partially ordered set M (with respect to the relation $\leqslant$) having the property that for any non-empty family $(a_\lambda)_{\lambda \in I}$ of elements of M indexed by a set I there exists the greatest lower bound, respectively the least upper bound of the family is called a *complete meet-semilattice*, respectively a *complete join-semilattice*;

the common greatest lower bound, respectively the common least upper bound of the elements a_λ, $\lambda \in I$, which, by the preceding considerations, is uniquely determined, is called in this case the *meet*, respectively the *join* of the family $(a_\lambda)_{\lambda \in I}$ (or of the elements a_λ, $\lambda \in I$), and is denoted

$$\bigcap_{\lambda \in I} a_\lambda, \text{ respectively } \bigcup_{\lambda \in I} a_\lambda.$$

By the preceding considerations, in a complete meet, respectively join-semilattice there exists the first, respectively the last element. A partially ordered set M (with respect to the relation $\leqslant$) which is both a complete meet-semilattice and a complete join-semilattice is called a *complete lattice*; in a complete lattice there exist both the first and the last element; in this way, in a complete lattice we may define the meet, respectively the join of the empty family as being the last, respectively the first element of the complete lattice.

Of course, any complete lattice is a lattice; the same remark is valid for semilattices.

Also by the preceding considerations, it follows that if E is a set, then the set $\mathcal{P}(E)$ partially ordered by inclusion is a complete lattice: the meet, respectively the join, in the sense of the lattice, of a family of subsets is the intersection, respectively the union of the given family as a family of subsets of E.

If M is a lattice, the partial order relation being denoted $\leqslant$, then from $a \leqslant b$, $a, b \in M$, it follows

$$a \cup b = b \quad \text{and} \quad a \cap b = a.$$

A lattice M, with the partial order relation denoted $\leqslant$, which has the property that, whatever $a, b, c \in M$ with $a \leqslant c$, we have

$$(a \cup b) \cap c = a \cup (b \cap c)$$

is called a *modular lattice* or a *Dedekind lattice*.

A complete lattice M is said to be a *completely modular lattice* if, for any pair of families $(a_\lambda)_{\lambda \in I}$, $(b_\lambda)_{\lambda \in I}$ of elements of M indexed by the same set I and such that $a_\lambda \leqslant b_\mu$ whenever $\lambda \neq \mu$, we have

$$(\bigcup_{\lambda \in I} a_\lambda) \cap (\bigcap_{\lambda \in I} b_\lambda) = \bigcup_{\lambda \in I} (a_\lambda \cap b_\lambda).$$

If M is a completely modular complete lattice, then, as a lattice, M is a modular lattice.

It may be shown that if we consider the class having as elements all the sets, or, as often said, the *class of all sets*, this is not a set; in other words, there exist classes which are not sets; such classes will be called *proper classes*; thus the class of all sets is a proper class.

On the class $\mathcal{S}$ of all sets consider the following binary relation, denoted C: if M, N are sets, we have by definition MCN if and only if there exists a bijection of sets $f : M \underset{\sim}{\rightarrow} N$; the relation C is an equivalence on the class $\mathcal{S}$; since $1_M : M \rightarrow \rightarrow M$ is a bijection for any $M \in \mathcal{S}$; next, if $f : M \rightarrow N$ is a bijection, then there exists the inverse bijection $f^{-1} : N \rightarrow M$; finally, if $f : M \rightarrow N$, $g : N \rightarrow P$ are bijections, then $g \circ f : M \rightarrow P$ is a bijection; the equivalence relation C is also called the *cardinal equivalence* or the *Cantor equivalence* of sets, and if $M, N \in \mathcal{S}$ and MCN, then M, N are said to be *cardinally equivalent*. Denote by $\mathcal{N}$ the quotient class $\mathcal{S}/C$; $\mathcal{N}$ is called the *class of cardinal numbers*, and an element of $\mathcal{N}$ is called a *cardinal number* or, simply, a *cardinality*; the equivalence class of a set $M \in \mathcal{S}$ with respect to the equivalence relation C is called the *cardinal number of the set M* or the *cardinality of the set M*; the canonical surjection

$$\mathcal{S} \rightarrow \mathcal{N} = \mathcal{S}/C$$

s denoted Card; hence, the cardinality of a set M is denoted Card (M) or Card M. The class of cardinal numbers is proved to be a proper class. If M is a finite set and the number of elements of M is n, where $n \in \mathbf{N}$, then

$$\operatorname{Card} M = \operatorname{Card}(\{1, \ldots, n\});$$

in this case, one also says that "n is the cardinality of the finite set M"; hence, the cardinality of a finite set is identified with the number of its elements; in particular, 0 is the cardinality of the void set. The cardinality of a finite set is said to be a *finite cardinal number*.

A set which is cardinally equivalent to the set $\mathbf{N}$ of natural numbers is said to be a *countable set*; its cardinality is called the *countable cardinal number*, and is denoted $\aleph_0$ (read "aleph zero").

On the class $\mathcal{N}$ of cardinal numbers define a binary relation, denoted $\leqslant$, as follows: if $\mathfrak{m}, \mathfrak{n} \in \mathcal{N}$, then, by definition, $\mathfrak{m} \leqslant \mathfrak{n}$ means that there exist representatives M and N of $\mathfrak{m}$ and $\mathfrak{n}$, respectively, and an injective mapping $f : M \rightarrow N$; the property does not depend on the chosen representatives M, N; in this case, $\mathfrak{m}$ is said to be *less than or equal to* $\mathfrak{n}$ or $\mathfrak{n}$ *greater than or equal to* $\mathfrak{m}$; the relation $\leqslant$ thus defined is a preorder relation; in fact, it is proved to be an antisymmetric relation, hence a partial order relation; this follows from the following **Cantor-Bernstein Theorem:**

If M, N are sets, and if

$$\operatorname{Card}(M) \leqslant \operatorname{Card}(N) \textit{ and } \operatorname{Card}(N) \leqslant \operatorname{Card}(M),$$

then Card $(M) =$ Card (N).

The relation $\leqslant$ is proved to be in fact a total order relation; hence, if $\mathfrak{m}, \mathfrak{n}$ are cardinal numbers, we have at least one of the situations

$$\mathfrak{m} \leqslant \mathfrak{n} \quad \text{or} \quad \mathfrak{n} \leqslant \mathfrak{m}.$$

On the class $\mathfrak{N}$ another binary relation, denoted $<$, is also introduced as follows: by definition, we have $\mathfrak{m} < \mathfrak{n}$, $\mathfrak{m}, \mathfrak{n} \in \mathfrak{N}$, if $\mathfrak{m} \leqslant \mathfrak{n}$ and $\mathfrak{m} \neq \mathfrak{n}$; if $\mathfrak{m} < \mathfrak{n}$, then $\mathfrak{m}$ is said to be *strictly less than* $\mathfrak{n}$ or $\mathfrak{n}$ *strictly greater than* $\mathfrak{m}$; in this situation one also writes $\mathfrak{n} > \mathfrak{m}$.

We mention also the following results: Let M, N be two sets; if N is a finite set and $\text{Card}(M) \leqslant \text{Card}(N)$, then M is a finite set. Next, if M is an infinite set and N is a finite subset of it, then

$$\text{Card}\,(M \setminus N) = \text{Card}\, M.$$

Now let $(\mathfrak{m}_\lambda)_{\lambda \in I}$ be a family of cardinal numbers indexed by a set I; the *sum* of this family, denoted

$$\sum_{\lambda \in I} \mathfrak{m}_\lambda,$$

is a cardinal number defined in the following manner: for each $\lambda \in I$ a representative M_λ of $\mathfrak{m}_\lambda$ is chosen, and the disjoint union $\coprod_{\lambda \in I} M_\lambda$ of the family $(M_\lambda)_{\lambda \in I}$ of sets is considered; this is also a set by the above condition 6); whatever the family of representatives $(M_\lambda)_{\lambda \in I}$, it turns out that the cardinal number $\text{Card}\,(\coprod_{\lambda \in I} M_\lambda)$ is the same; this is by definition $\sum_{\lambda \in I} \mathfrak{m}_\lambda$. With the same data, we consider the cartesian product $\prod_{\lambda \in I} M_\lambda$, which, by the above condition 7), is also a set; the cardinality of the set $\prod_{\lambda \in I} M_\lambda$ does not depend on the chosen system of representatives $(M_\lambda)_{\lambda \in I}$; this cardinal number, depending only on the given family $(\mathfrak{m}_\lambda)_{\lambda \in I}$ of cardinal numbers, is called the *product* of this family, and is denoted

$$\prod_{\lambda \in I} \mathfrak{m}_\lambda;$$

if $I = [1, r]$, instead of $\sum_{\lambda \in I} \mathfrak{m}_\lambda$, respectively of $\prod_{\lambda \in I} \mathfrak{m}_\lambda$ one writes

$$\mathfrak{m}_1 + \mathfrak{m}_2 + \ldots + \mathfrak{m}_r, \text{ respectively } \mathfrak{m}_1 \cdot \mathfrak{m}_2 \cdot \ldots \cdot \mathfrak{m}_r,$$

and if $r = 2$ and $\mathfrak{m}_1 = \mathfrak{m}$, $\mathfrak{m}_2 = \mathfrak{n}$,

$$\mathfrak{m} + \mathfrak{n}, \text{ respectively } \mathfrak{m} \cdot \mathfrak{n}.$$

Now let $\mathfrak{m}, \mathfrak{n}$ be two cardinal numbers; the cardinal number $\mathfrak{m}^{\mathfrak{n}}$ is defined as follows: a representative M, respectively N, of $\mathfrak{m}$, respectively of $\mathfrak{n}$ is chosen; the class M^N of mappings from N to M is a set, and its cardinality does not depend on the chosen representatives M, N; this cardinality is by definition $\mathfrak{m}^{\mathfrak{n}}$.

It may be shown that if $(\mathfrak{m}_\lambda)_{\lambda\in I}$ is a family of finite or countable cardinal numbers different from 0 indexed by an infinite set I, then

$$\sum_{\lambda\in I}\mathfrak{m}_\lambda = \mathrm{Card}\,(I);$$

further, if $(M_\lambda)_{\lambda\in I}$ is a family (indexed by a set I) of subsets of a set, then

$$\mathrm{Card}\,(\bigcup_{\lambda\in I} M_\lambda) \leqslant \sum_{\lambda\in I}\mathrm{Card}\,(M_\lambda);$$

consequently, if $(M_\lambda)_{\lambda\in I}$ is a family of sets indexed by a set I and such that there might be a convention of identifying some elements from different sets, then the cardinality of their union is less than or equal to $\sum_{\lambda\in I}\mathrm{Card}\,(M_\lambda)$.

In what follows, whenever we talk of a "family indexed by a class I", we shall mean, unless specifically stated otherwise, that the class I of indices is a set.

Let M be a non-void set; a *(binary) internal algebraic operation* or an *internal law of composition* or simply an *operation* on the set M is a mapping

$$\varphi : M\times M \to M.$$

For instance, if M is a meet-semilattice, respectively a join-semilattice, then the mapping $M\times M \to M$ given by

$$(a, b) \mapsto a\cap b, \text{ respectively } (a, b) \mapsto a\cup b,$$

with $a, b \in M$, is an internal law of composition.

Often, if M is a non-empty set endowed with an internal law of composition φ, instead of $\varphi(x, y)$, $x, y\in M$, a special type of writing is used, for instance

$$x+y \quad \text{or} \quad x\cdot y;$$ [1]

in the first case, the given algebraic operation is called *addition*, and $x+y$ is called the *sum* of x and y; one says that the *additive writing* is used; in the second case, the given algebraic operation is called *multiplication*, and $x\cdot y$ is called the *product* of x and y; in this case, one says that the *multiplicative writing* is used; instead of $x\cdot y$ one sometimes writes xy.

Given the internal algebraic operation φ on the set M, this operation is said to be *associative* provided

(5) $$\varphi(x, \varphi(y, z)) = \varphi(\varphi(x, y), z)$$

for any $x, y, z\in M$; in the multiplicative writing, relation (5) becomes

$$x\cdot(yz) = (xy)\cdot z.$$

[1] Similarly, for semilattices we used the writing $x\cap y$ and $x\cup y$.

Let M be a non-empty set endowed with an internal law of composition $\varphi : M\times M \to M$; an element $e \in M$ is called a *neutral element* for the given law of composition if for any $x \in M$ we have

$$\varphi(e, x) = \varphi(x, e) = x;$$

a neutral element, whenever exists, is unique; consequently, we may speak of *the neutral element* provided it exists; if the multiplicative writing is used, the neutral element is also called *unit element*, and is sometimes denoted 1, while if the additive writing is used, the neutral element is also called *zero element*, and is denoted 0.[1]

A non-empty set endowed with an internal law of composition which is associative and for which there exists a neutral element is called a *monoid.*

If M is a non-void set endowed with an internal law of composition φ, this law is said to be *commutative* provided for any $x, y \in M$ we have

$$\varphi(x, y) = \varphi(y, x);$$

a monoid whose law of composition is commutative is called a *commutative monoid.*

If M is a monoid in which the neutral element is denoted e, and if, say, the multiplicative writing is used, then, given $x \in M$, x is said to be an *invertible* element of the monoid provided there exists $x' \in M$ such that

$$x'x = xx' = e; \tag{6}$$

in this case x' is said to be an *inverse* (*element*) *of* x; if x is an invertible element of the monoid M, then $x' \in M$ with the property (6) is uniquely determined; it is called *the inverse element* (or *the inverse*) *of* x, and is denoted x^{-1}; when the additive writing is used, instead of the inverse we speak of the *opposite* of an element $x \in M$, denoted in this case $-x$.

A monoid in which any element is invertible is called a *group*; if the respective monoid is commutative, the group is referred to as a *commutative group* or an *abelian group*; in the case of groups, the additive writing is mainly used for abelian groups, but this does not imply that in any abelian group the additive writing is used.

Let M be a monoid, and let N be a non-void subset of it; as $N\times N$ is a non-void subset of $M\times M$, if for any $x, y \in N$ we have (assuming in M the multiplicative writing was used) $xy \in N$, then the multiplication in M induces by restriction to $N\times N$ and N a mapping $N\times N \to N$; obviously, this internal law of composition on the non-empty set N is also associative; in this situation, N is said to be a *submonoid* of M provided M contains in addition the neutral element of M; in this case, this element is also a neutral element for the law of composition on N; hence, a submonoid of M is a non-void subset which is a monoid under the internal algebraic operation induced by the one on M and for which the neutral element coincides with the neutral element of M; if, in this

[1] Not to be confounded with the natural number 1, respectively 0.

situation, N is a group, N is said to be a *subgroup* of the monoid M; clearly, the definition is maintained if M is a group.

Given two monoids M, N in which we use, say, the multiplicative writing, a mapping $f: M \to N$ is called a *homomorphism of monoids from M to N* or *of M into N* provided[1]

$$f(x \cdot y) = f(x) \cdot f(y) \tag{7}$$

for any $x, y \in M$ and if, moreover, f carries the neutral element of M into the neutral element of N. If M, N are groups, then a homomorphism of monoids $f: M \to N$ is called a *homomorphism of groups of M into N*; in this case, the fact that f carries the neutral element of M into the neutral element of N follows from relation (7): for if e, e' are, respectively, these elements, we have

$$e'f(e) = f(e) = f(e \cdot e) = f(e) \cdot f(e),$$

and multiplying on the right both sides by $(f(e))^{-1}$, we get $f(e) = e'$; next, for any $x \in M$ we have

$$f(x^{-1}) = (f(x))^{-1};$$

if in the groups M and N the additive writing is used, then the fact that $f: M \to N$ is a homomorphism of groups is expressed by the relation

$$f(x + y) = f(x) + f(y)$$

for any $x, y \in M$.

If G and H are groups, a homomorphism of groups $f: G \to H$ is said to be an *isomorphism of groups* provided there exists a homomorphism $g: H \to G$ such that

$$f \circ g = 1_H, \quad g \circ f = 1_G;$$

for the existence of such a g it is necessary and sufficient that the homomorphism f is a bijective mapping (see also the Remark following Proposition 1, § 1); in case g exists, it is uniquely determined and coincides with the inverse bijection of f; in our situation, g is called the *inverse isomorphism* of f, and is denoted also f^{-1}, as in the case of sets; to mark the fact that f is an isomorphism of groups, one writes $f: G \rightleftarrows H$.

A *ring* is a non-empty set A endowed with two internal laws of composition, the additive writing being usually used for the first one and the multiplicative writing for the second, and such that the following conditions are satisfied:

a) A is an abelian group under addition (called the *additive group* of the ring);

b) the multiplication is associative;

c) the multiplication is *distributive* on both sides over addition, that is,

$$x(y + z) = xy + xz, \quad (y + z)x = yx + zx$$

for any $x, y, z \in A$.

---

[1] Of course, the sign between x and y represents the internal algebraic operation on M, while the sign between $f(x)$ and $f(y)$ represents the internal algebraic operation on N.

The neutral element for addition is called the *zero element* of the ring, and is denoted 0.

The ring A is said to be a *commutative ring* provided the multiplication in A is commutative; the ring A is said to be a *ring with unit element* provided there exists a neutral element for the multiplication; in this case, the latter is called the *unit element* of the ring, and is denoted by 1; also in this case, the ring A is a monoid under multiplication, called the *multiplicative monoid of the ring.*

In a set A consisting of the single element a, two algebraic operations may be defined to make a ring of it, setting

$$a + a = a, \quad a \cdot a = a;$$

clearly, in this case the ring A is reduced to its zero element; such a ring is called a *zero ring.*

Given two rings A, B, a mapping $f : A \to B$ is called a *homomorphism of rings of A into B* or *from A to B* or of *source A and range B* if, denoting by "+" both the addition in A and the one in B, and by "·" both the multiplication in A and the one in B, we have

$$f(x + y) = f(x) + f(y) \quad \text{and} \quad f(xy) = f(x) \cdot f(y)$$

whatever $x, y \in A$.

If A and B are rings with unit element whose unit elements are denoted by 1 both for A and B, and if the homomorphism of rings $f : A \to B$ has the additional property

$$f(1) = 1,$$

then f is said to be a *homomorphism of rings with unit element.*

A homomorphism of rings $f : A \to B$ for which there exists a homomorphism of rings $g : B \to A$ such that

$$g \circ f = 1_A, \; f \circ g = 1_B$$

is called an *isomorphism of rings*; for the existence of g with these properties it is necessary and sufficient that the homomorphism of rings f be a bijective mapping, and in this case g is uniquely determined, coincides with the inverse bijection of f, and is denoted f^{-1}; if A and B are rings with unit element, then any isomorphism of rings $f : A \to B$ is necessarily a homomorphism of rings with unit element. To mark that $f : A \to B$ is an isomorphism of rings, one writes $f : A \rightleftarrows B$.

If A and B are zero rings, then there exists a unique isomorphism of rings $f : A \to B$; by means of the corresponding isomorphism all zero rings are identified, and *the* so-called *zero ring*, denoted by 0, is obtained.

A non-zero commutative ring A with unit element and having the property that, whatever the elements $a, b \in A$, from $ab = 0$ it follows $a = 0$ or $b = 0$ is called an *integral domain.*

Let A be a non-zero ring with unit element; an invertible element of the multiplicative monoid of the ring A is called an *invertible element* of the ring; in a ring A with unit element, all invertible elements form a group under the multiplication of the ring, called the *group of invertible elements* of the ring A (with unit element).

A non-zero ring with unit element (hence having at least two elements) in which any non-zero element is invertible is called a *skew-field*; if the ring we started with is in addition commutative, then the ring is called a *field.*

If A is a ring, an element $a \in A$ is said to be an *idempotent* (*element*) of the ring A if $a \cdot a = a$.

Let A be a ring, and let B be a non-void subset of it; since $B \times B$ is a non-void subset of $A \times A$, if for any $x, y \in B$ the elements $x + y$ and xy (the addition and the multiplication being performed in A) are elements of B, then the mappings given by the addition and the multiplication in the ring A respectively induce by restriction to $B \times B$ and B two mappings

$$B \times B \to B;$$

if B is a ring under the operations defined by these two mappings, then B is said to be a *subring* of A. In order that the non-empty subset B of the ring A be a subring of A it is necessary and sufficient that the following conditions be simultaneously satisfied:

d) if $b, b' \in B$, then $b + b'$ (performed in A) is in B;

e) if $b \in B$, then $-b$ (i.e. the opposite of b in the additive group of A) is in B;

f) if $b, b' \in B$, then bb' (performed in A) is in B.

If A is a ring with unit element, and B is a subring containing the unit element of A, then B is said to be a *subring with unit element* of A.

If B is a subring (respectively a subring with unit element) of the ring (respectively of the ring with unit element) A, then the canonical injection of sets $B \to A$ is a homomorphism of rings (respectively of rings with unit element); in the same way as for classes, this homomorphism will be called the *canonical injection* or also the *inclusion homomorphism* of the subring B into the ring A.

Let A be a ring; a non-void subset $\mathfrak{a}$ of A is called a *left*, respectively a *right ideal* of A if $\mathfrak{a}$ is a subgroup of the additive group of the ring A and if, in addition, for any $x \in \mathfrak{a}$ and any $\alpha \in A$ we have $\alpha x \in \mathfrak{a}$, respectively $x\alpha \in \mathfrak{a}$.

If $\mathfrak{a}$ is both a left and a right ideal of the ring A, $\mathfrak{a}$ is called a *two-sided ideal.*

In a commutative ring A the notions of left, right, and two-sided ideal coincide; in this case, we simply talk of an *ideal* of the ring A.

Let A be an arbitrary ring, and let $\mathfrak{a}$ be a two-sided ideal of it; the binary relation R defined on the set A by: xRy, with $x, y \in A$, means that $x - y \in \mathfrak{a}$, is an equivalence relation on A, called the *congruence relation with respect to* $\mathfrak{a}$ (or *modulo* $\mathfrak{a}$); if xRy, x is said to be *congruent to* y *modulo* $\mathfrak{a}$ (or *with respect to* $\mathfrak{a}$); instead of xRy one writes $x \equiv y \pmod{\mathfrak{a}}$ or $x \equiv y(\mathfrak{a})$, $x, y \in A$; the quotient set of A with respect to the congruence relation modulo $\mathfrak{a}$ is denoted $A/\mathfrak{a}$; on the (non-empty) set $A/\mathfrak{a}$ two internal algebraic operations, the first additively, the second multiplicatively written, are defined as follows: if $M, N \in A/\mathfrak{a}$, then representatives x, y of M and N, respectively, are chosen; then, by definition, $M + N$ and $M \cdot N$ are the equivalence classes of $x + y$ and xy modulo R, respectively; these two equivalence classes do not depend on the chosen representatives x, y, hence the two operations on $A/\mathfrak{a}$ are indeed well defined; under the addition and the multiplication thus defined $A/\mathfrak{a}$ is a ring, and the canonical surjection $p : A \to A/\mathfrak{a}$ is a homomorphism of rings; the ring $A/\mathfrak{a}$ is called the *quotient ring* of A with respect to the two-sided ideal $\mathfrak{a}$.

If A is in addition a ring with unit element, denoted 1, then the equivalence class of 1 modulo R is the unit element of the ring $A/\mathfrak{a}$, which in this case is a ring with unit element; the canonical surjection $A \to A/\mathfrak{a}$ is in this case a homomorphism of rings with unit element.

In the sequel we also need the following notion: Let A be a ring; let us consider a new set, denoted A°, whose elements correspond under a given bijection to the elements of A; the element of A corresponding under this bijection to $\alpha \in A$ will be denoted α°, and will be called the *doublet* of α; on A° two internal laws of composition, an addition and a multiplication, are defined as follows: given two elements of A°, they are the doublets α°, β° of the uniquely determined elements α, respectively β of A; we set by definition in A°,

$$\alpha^\circ + \beta^\circ = (\alpha + \beta)^\circ, \quad \alpha^\circ\beta^\circ = (\beta\alpha)^\circ;$$

under these two internal laws of composition, A° is a ring; it is called the *opposite ring of A* or the ring *opposite to A*, and will be denoted also by A°. If A has a unit element, denoted 1, then 1° is the unit element of the ring A°. If we consider the ring $(A^\circ)^\circ$ opposite to A°, then the mapping

$$A \to (A^\circ)^\circ$$

given by

$$\alpha \mapsto (\alpha^\circ)^\circ, \quad \alpha \in A,$$

is obviously an isomorphism of rings (with unit element if A is a ring with unit element); by means of this isomorphism we shall identify A with $(A^\circ)^\circ$; thus we shall consider that $(A^\circ)^\circ = A$. If A is a commutative ring, then the mapping $A \to A^\circ$ given by $\alpha \mapsto \alpha^\circ$, $\alpha \in A$, is an isomorphism of rings; by means of it, A is identified in this case with A°.

By convention, from now on, unless otherwise stated, the term "ring" will designate a non-zero ring with unit element; similarly, by "subring", "homomorphism of rings", etc. we shall mean a subring with unit element, a homomorphism of rings with unit element, etc., respectively.

We recall that if A is a commutative ring (with unit element), then a *polynomial over the ring A in the indeterminates* $X_1, \ldots, X_n$ is an expression of the form

$$f = \sum_{(r_1, \ldots, r_n)} \alpha_{r_1, \ldots, r_n} X_1^{r_1} \ldots X_n^{r_n},$$

where the sum is extended over all $(r_1, \ldots, r_n) \in \mathbf{N}^n$ and the *coefficients* $\alpha_{r_1, \ldots, r_n}$ are elements of A, zero for all except a finite number of the elements $(r_1, \ldots, r_n)$; two such expressions, f above and

$$g = \sum_{(r_1, \ldots, r_n)} \beta_{r_1, \ldots, r_n} X_1^{r_1} \ldots X_n^{r_n},$$

are equal provided for any $(r_1, \ldots, r_n)$ we have

$$\alpha_{r_1, \ldots, r_n} = \beta_{r_1, \ldots, r_n};$$

consider the set of all polynomials over the ring A in the indeterminates $X_1, \ldots, X_n$; define the addition of two polynomials as being their "termwise" addition, and the multiplication by multiplying the terms of the first polynomial with all terms of the second one and then taking the sum of the obtained products, the multiplication of the terms being done by multiplying the coefficients in A and adding the *exponents* r_i of the same *indeterminate* X_i; in this way, a commutative ring, called the *ring of all polynomials over the ring* A *in the indeterminates* $X_1, \ldots, X_n$, and denoted $A[X_1, \ldots, X_n]$, is obtained.

The mapping $A \to A[X_1, \ldots, X_n]$ assigning to $\alpha \in A$ the polynomial f of above for which $\alpha_{0, \ldots, 0} = \alpha$ and all the other coefficients are zero is a homomorphism of rings which is moreover injective; by means of it, A is identified with a subring of $A[X_1, \ldots, X_n]$; in particular, the unit element of the ring A is also the unit element of the ring $A[X_1, \ldots, X_n]$.

Let M and Ω be non-empty sets; an *external law of composition* on the set M and having Ω as *set* (or *domain*) *of operators* is a mapping

$$\psi : \Omega \times M \to M \quad \text{or} \quad \psi : M \times \Omega \to M;$$

the elements of Ω are called the *operators* of the given external law of composition.

Given a non-void set M, an *algebraic structure* (or, in the following, only a *structure*) on the set M consists of one or several internal laws of composition and one or several external laws of composition (each of the latter having a given domain of operators) such that these laws of composition are subject to certain conditions (for instance, the associativity) or are related one to another by other conditions (for instance, the distributivity of one law of composition over another one).

For instance, we may talk of the *structure of a monoid*, or *a group*, or *a ring*, or *a lattice* on a set.

A set M without any law of composition is considered to have the *void* (or the *empty*) *algebraic structure*.

If on a set M an algebraic structure is given, and if only some of the internal or external laws of composition are considered, then from the initial algebraic structure another poorer algebraic structure is obtained, called the corresponding *underlying* algebraic structure of the initial one. For instance, if in a ring A only the addition is considered, then an abelian group is obtained, called the *underlying additive group of the ring*: it is what was called above the "additive group" of the ring A; likewise, the multiplicative monoid of the ring with unit element A is the *underlying multiplicative monoid of the ring* A; likewise, in a lattice L omitting the join (respectively the meet), we get the *underlying meet* (respectively *join*) *semilattice of the lattice L*; in particular, when all the laws of composition of an algebraic structure given on a (non-void) set M are omitted, we get what we shall call the *underlying set of the given structure*; thus, we shall speak of the "underlying set of a group, ring, lattice, etc.", which is exactly the set on which the given structure of group, ring, lattice, etc. is defined.

By a *diagram* of sets, groups, rings, etc. we shall mean, respectively, a family of sets, groups, rings, etc. represented by letters, connected one with another, respectively, by mappings of sets, homomorphism of groups, homomorphisms of rings, etc., mappings represented by arrows which, by their sense, indicate the source and the range of the corresponding mappings; for instance,

(8)
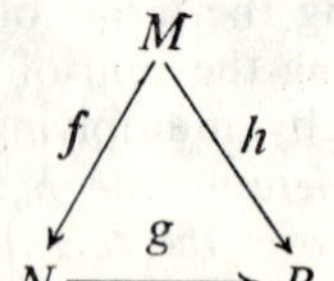

(9)
$$\ldots \to E_n \xrightarrow{u_n} E_{n+1} \xrightarrow{u_{n+1}} E_{n+2} \to \ldots,$$

(10)
$$A \underset{g}{\overset{f}{\rightrightarrows}} B,$$

(11)
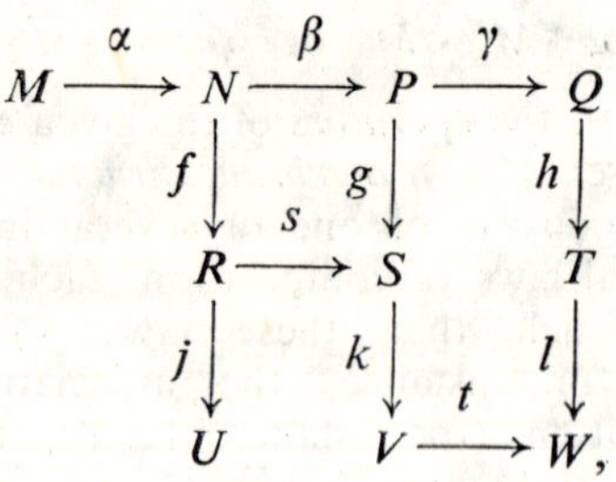

(12)
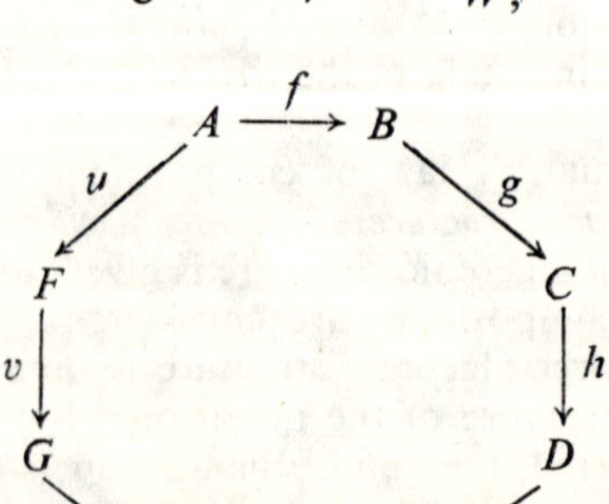

are diagrams, where, if M, N, P in (8) are sets and f, g, h are mappings of sets, then (8) is a diagram of sets, and if in (9) E_n are groups and u_n are homomorphisms of groups, $n \in \mathbf{Z}$, then (9) is a diagram of groups, and so on.

A diagram will be said to be *commutative* if for any pair of letters A, B in the diagram (letters which represent sets, or groups, or rings, etc.), the composition of the mappings given by a sequence of arrows in the diagram whose successive composition makes sense and for which the first has A as source and the last has B as range does not depend on the sequence of mappings chosen with the stated properties: for instance, diagram (8) (say, of sets) is commutative if $g \circ f = h$; diagram (10) is commutative if $f = g$, and diagram (12) is commutative if $j \circ h \circ g \circ f = w \circ v \circ u$.

A diagram is commutative if and only if all its simple "polygons" are commutative diagrams: for instance, diagram (11) is commutative if and only if

$$s \circ f = g \circ \beta \quad \text{and} \quad t \circ k \circ g = l \circ h \circ \gamma.$$

Notice that it makes no sense to talk of the commutativity of diagram (9).

The horizontal and vertical lines in a diagram will be called *rows* and *columns*, respectively: thus, in diagram (8) we have the row

$$N \xrightarrow{g} P,$$

but we have no columns; diagram (9) is itself a row; the rows in diagram (11) are the diagrams

$$M \xrightarrow{\alpha} N \xrightarrow{\beta} P \xrightarrow{\gamma} Q,$$

$$R \xrightarrow{s} S,$$

$$V \xrightarrow{t} W,$$

and the columns in diagram (11) are the diagrams

$$\begin{array}{ccccc} N & & P & & Q \\ f\downarrow & & g\downarrow & & h\downarrow \\ R & & S & & T \\ j\downarrow & & k\downarrow & & l\downarrow \\ U, & & V, & & W. \end{array}$$

§ 1. Modules and Linear Mappings

Definition 1. Given a ring A, a *left module over A* or briefly a *left A-module* is a set E endowed with two laws of composition: an internal law of composition, i.e. a mapping

$$s : E \times E \to E,$$

and an external law of composition having A as domain of operators, i.e. a mapping

$$p : A \times E \to E,$$

these two laws being subject to the following axioms:

a) Setting $s(x, y) = x + y$, $x, y \in E$, E is an *abelian group* with respect to this law; the neutral element of this group will be called the *zero element* of the module and will be denoted by 0, while the opposite of x will be denoted by $-x$.

Denoting $p(\alpha, x) = \alpha x$ (or $\alpha \cdot x$), we have

b) $\alpha(x + y) = \alpha x + \alpha y$,

c) $(\alpha + \beta)x = \alpha x + \beta x$,

d) $(\alpha\beta)x = \alpha(\beta x)$,

e) $1x = x$

for any $\alpha, \beta \in A$, $x, y \in E$; 1 is the unit element of A.

A *right A-module* is a set E endowed with an internal law of composition satisfying axiom *a*) above, and an external law of composition

$$p : E \times A \to E$$

such that, denoting $p(x, \alpha) = x\alpha$, we have

b') $(x + y)\alpha = x\alpha + y\alpha$,

c') $x(\alpha + \beta) = x\alpha + x\beta$,

d') $x(\alpha\beta) = (x\alpha)\beta$,

e') $x \cdot 1 = x$,

$\alpha, \beta \in A$, $x, y \in E$.

When the (left or right) A-module E is given, the elements of A are called *scalars*; the external law of composition is called *scalar multiplication* (by elements of A). The internal law of composition will be also called *addition*.

For a left (respectively right) A-module E we also say that A or the operators from A *act* in E on the left (respectively on the right).

If the considered ring is a skew-field K, a (left, respectively right) K-module is called (left, respectively right) *vector space* over K or *K-vector space.* The elements of a vector space are called, in opposition with the scalars, *vectors.* ▮

The abelian group obtained from an A-module E by omitting the scalar multiplication in the A-module E will be called the *underlying abelian group* (or the *additive group*) of the A-module E.

REMARK. The distinction between left and right A-modules is not merely a difference of writing: so, if we rewrite axiom d') of Definition 1 with the scalars on the left, we get

$$(\alpha\beta)x = \beta(\alpha x),$$

which is not axiom d) for left modules.

If, however, we consider the ring A° opposite to A, then, defining on the left A-module E the external law of composition

$$E \times A^\circ \to E$$

by

$$(x, \alpha^\circ) \mapsto x\alpha^\circ = \alpha x, \quad \alpha \in A, \ x \in E,$$

we get

$$x(\beta^\circ\alpha^\circ) = x(\alpha\beta)^\circ = (\alpha\beta)x = \alpha(\beta x) = (x\beta^\circ)\alpha^\circ,$$

which is axiom d') for right modules; the other axioms are trivially satisfied; consequently, by means of this transformation, any left A-module may be considered as a right A°-module, and conversely. In the particular case of a *commutative* ring A, it is known (see § 0) that A may be identified with A° by identifying α with α°, $\alpha \in A$; then any left A-module is also a right A-module according to the law

$$x\alpha = \alpha x,$$

and conversely. In this case we shall speak no more of "left" or "right" A-modules, but simply of *A-modules.* ▮

Henceforth, unless explicitly mentioned otherwise, by an "A-module" we mean a left A-module, having in mind that the corresponding notions and properties are to be automatically transposed to right modules. Of course, the same convention is assumed to be valid for vector spaces over a skew-field K.

EXAMPLES. 1) If A is a ring, then A may be thought of as a left A-module, taking the addition in A as internal law of the module, and as external law

$$A \times A \to A$$

the multiplication in A: in other words, $\alpha \cdot \beta$, where α is thought of as a scalar and β as an element of the module A, is the product $\alpha\beta$ performed in the ring A; this left A-module will be denoted A_s.

Similarly, one may define on A a structure of right A-module, which will be denoted A_d.

2) Let A be a commutative ring, and let $A[X]$ be the set of all polynomials over A in the indeterminate X; taking the usual addition of polynomials as internal law, and as external law the multiplication of polynomials by elements of A obtained by multiplying all coefficients by such an element, we get a structure of A-module on $A[X]$.

3) If A is a ring and M is a non-empty set, then on the set A^M of all mappings f defined on M and taking values in A, a structure of A-module (say on the left) may be defined taking as sum $f + g$ of two such mappings the mapping given by

$$x \mapsto f(x) + g(x), \quad x \in M,$$

and as result αf of the scalar multiplication, $\alpha \in A$, $f \in A^M$, the mapping given by

$$x \mapsto \alpha \cdot f(x), \quad x \in M.$$

In particular, if $M = \{1, \ldots, n\}$, then we write A^n instead of A^M, and the mappings $M \to A$ are represented as the families $(\alpha_1, \ldots, \alpha_n)$ of n elements of A; according to the above definition, the addition and the scalar multiplication in A^n are to be done "componentwise". ∎

From the axioms of A-modules one concludes that if E is an A-module, then

$$0 \cdot x = \alpha \cdot 0 = 0,$$

$$\text{(1)} \qquad \alpha(-x) = (-\alpha)x = -(\alpha x),$$

$$(\alpha - \beta)x = \alpha x - \beta x, \quad \alpha(x - y) = \alpha x - \alpha y;$$

here $\alpha, \beta \in A$, $x, y \in E$, and the first 0 is the zero element of A, while the second and the third ones represent the zero element of E.

For instance, the first relation (1) is proved as follows:

$$0 \cdot x = (0 + 0)x = 0x + 0x,$$

and adding to both sides the opposite of $0 \cdot x$, we get $0 = 0x$. The proof of the other stated relations is left to the reader.

Let G be an abelian group whose law of composition is additively written; if $(a_1, \ldots, a_n)$ is a finite family of its elements, the *sum of the family* $(a_1, \ldots, a_n)$, denoted

$$a_1 + \ldots + a_n = \sum_{i=1}^{n} a_i,$$

is defined by induction on n as follows: if $n = 0$, i.e., if the family is empty, then by definition the sum is 0; if $n > 0$ and the sum $a_1 + \ldots + a_{n-1}$ has been defined, then by definition

$$a_1 + \ldots + a_n = (a_1 + \ldots + a_{n-1}) + a_n.$$

Of course, if the family is indexed by an arbitrary finite set J instead of $\{1, \ldots, n\}$, the sum is still defined, since if $J = \{\lambda_1, \ldots, \lambda_n\}$, then, putting $a_i = a_{\lambda_i}$, $i = 1, \ldots, n$, we fall back upon the previous case. By the associativity of the addition in G, one can show that if $0 \leqslant k \leqslant n$, then

$$a_1 + \ldots + a_n = (a_1 + \ldots + a_k) + (a_{k+1} + \ldots + a_n).$$

In view of the commutativity of the addition, it follows that if σ is an arbitrary permutation of the set $\{1, \ldots, n\}$, then

$$\sum_{i=1}^{n} a_{\sigma(i)} = \sum_{i=1}^{n} a_i,$$

which may be obtained by induction on the number n of terms.

In the same way, if $(a_{ij})_{i,j}$ is a family of elements of G, where i ranges over $\{1, \ldots, m\}$ and j over $\{1, \ldots, n\}$, then

$$\sum_{i=1}^{m}\left(\sum_{j=1}^{n} a_{ij}\right) = \sum_{j=1}^{n}\left(\sum_{i=1}^{m} a_{ij}\right);$$

in other words, "the order of summation may be inverted": this may be proved by induction on both m and n; we leave it to the reader to formulate and prove that the sum is invariant by changing the order of summation for finite families indexed by k-size indices, $k \geqslant 1$.[1]

In particular, if all the elements of the family $(a_i)_i$ are equal to the same element a, then we set by definition

$$\sum_{i=1}^{n} a_i = n \cdot a;$$

it follows that

$$0 \cdot a = 0;$$

if n is a negative integer, then $-n > 0$, and we define

$$n \cdot a = (-n) \cdot (-a) = -((-n) \cdot a).$$

[1] The above considerations are also valid if instead of the abelian group G, a commutative monoid is considered.

These considerations lead us to the following

EXAMPLE. 4) If G is an additively written abelian group, then, taking as internal law the addition in G, and as external law the mapping $\mathbf{Z} \times G \to G$ given by

$$(n, a) \mapsto n \cdot a, \quad n \in \mathbf{Z},\ a \in G,$$

where $n \cdot a$ has the above meaning, we get a $\mathbf{Z}$-module structure on G. ▮

Clearly, if E is an A-module, all the preceding considerations are still valid for the underlying abelian group; moreover, the "distributivity" laws *b*) and *c*) of Definition 1 are extended by induction as follows:

$$\begin{aligned} &\alpha(x_1 + \ldots + x_n) = \alpha x_1 + \ldots + \alpha x_n, \\ (2) \qquad & \\ &(\alpha_1 + \ldots + \alpha_n)\, x = \alpha_1 x + \ldots + \alpha_n x, \end{aligned}$$

where $\alpha, \alpha_1, \ldots, \alpha_n \in A$, $x, x_1, \ldots, x_n \in E$, and $n \geqslant 0$. In particular, if $\alpha \in A$, $n \in \mathbf{Z}$, and $x \in E$, then $\alpha(n \cdot x) = (n \cdot \alpha)\, x = n(\alpha x)$.

Now, let I be an arbitrary index set and $(x_\lambda)_{\lambda \in I}$ a family indexed by I of elements from the union of a collection of rings, or abelian groups, or modules, etc.; by the *support* of this family we shall mean the subset $I_0 \subseteq I$ consisting of those $\lambda \in I$ for which $x_\lambda \neq 0$ in the corresponding ring, or group, etc. If the support of the family is finite, then we shall refer to the latter as a *family with finite support* (or a *family having finite support* or, briefly, an *fs-family*). If G is an abelian group, in particular the underlying group of a module, and $(x_\lambda)_{\lambda \in I}$ is an fs-family of elements of G indexed by I and whose support is I', then the *sum* of the family $(x_\lambda)_{\lambda \in I}$, denoted $\sum_{\lambda \in I} x_\lambda$, may be defined by

$$\sum_{\lambda \in I} x_\lambda = \sum_{\lambda \in I'} x_\lambda,$$

the right hand side sum making sense according to what precedes. If $(x_\lambda)_{\lambda \in I}$, $(y_\lambda)_{\lambda \in I}$ are two fs-families of elements of G indexed by I, then the *sum family* $(x_\lambda + y_\lambda)_{\lambda \in I}$ is also an fs-family: in fact, if λ belongs neither to the support of $(x_\lambda)_\lambda$ nor to that of $(y_\lambda)_\lambda$, then we have $x_\lambda = y_\lambda = 0$, hence $x_\lambda + y_\lambda = 0$, and λ does not belong to the support of the sum family; it follows that the support of the family $(x_\lambda + y_\lambda)_\lambda$ is contained in the union of the supports of the families $(x_\lambda)_\lambda$ and $(y_\lambda)_\lambda$. In this case we have

$$\sum_{\lambda \in I} x_\lambda + \sum_{\lambda \in I} y_\lambda = \sum_{\lambda \in I} (x_\lambda + y_\lambda),$$

which is easily seen from the definition.

This last relation may be generalized as follows: if $(x_{\lambda\mu})_{(\lambda,\mu) \in I \times J}$ is a family of elements of the abelian group G indexed by the cartesian product $I \times J$, with J

finite, and if for each $\mu \in J$ the family $(x_{\lambda\mu})_{\lambda \in I}$ has finite support, then the family $(\sum_{\mu \in J} x_{\lambda\mu})_{\lambda \in I}$ has finite support, and we have

$$\sum_{\lambda \in I} (\sum_{\mu \in J} x_{\lambda\mu}) = \sum_{\mu \in J} (\sum_{\lambda \in I} x_{\lambda\mu}); \tag{3}$$

in other words, in this case the order of summation may be still inverted: since if I_μ, $\mu \in J$, is the support of the family $(x_{\lambda\mu})_{\lambda \in I}$, then the support of the family $(\sum_{\mu \in J} x_{\lambda\mu})_{\lambda \in I}$ is contained in the finite set $I' = \bigcup_{\mu \in J} I_\mu$; but then, for the proof of (3) it suffices to prove the "finite" relation

$$\sum_{\lambda \in I'} (\sum_{\mu \in J} x_{\lambda\mu}) = \sum_{\mu \in J} (\sum_{\lambda \in I'} x_{\lambda\mu}),$$

which is true in view of the above arguments.

If $I = \bigcup_{\mu \in J} I_\mu$, with $I_\mu \cap I_{\mu'} = \emptyset$ for $\mu \neq \mu'$, is a partition of the index set I, then for an fs-family $(x_\lambda)_{\lambda \in I}$ of elements of G, the family $(x_\lambda)_{\lambda \in I_\mu}$ has finite support for each $\mu \in J$, and the support of the family $(\sum_{\lambda \in I_\mu} x_\lambda)_{\mu \in J}$ is also finite; moreover, we have

$$\sum_{\lambda \in I} x_\lambda = \sum_{\mu \in J} (\sum_{\lambda \in I_\mu} x_\lambda).$$

If $(x_\lambda)_{\lambda \in I}$ is an fs-family of elements of the A-module E and $\alpha \in A$, then the family $(\alpha x_\lambda)_{\lambda \in I}$ has finite support too; likewise, if $(\alpha_\lambda)_{\lambda \in I}$ is an fs-family of elements of A and $x \in E$, then the family $(\alpha_\lambda x)_{\lambda \in I}$ has finite support; in these last two cases, relations (2) may be generalized, respectively, to

$$\begin{aligned} \alpha(\sum_{\lambda \in I} x_\lambda) &= \sum_{\lambda \in I} \alpha x_\lambda, \\ (\sum_{\lambda \in I} \alpha_\lambda)\, x &= \sum_{\lambda \in I} \alpha_\lambda x, \end{aligned} \tag{4}$$

which can be seen reducing the sums to finite ones.

If, moreover, in the A-module E two families, the first $(\alpha_\lambda)_{\lambda \in I}$ of scalars, the other $(x_\lambda)_{\lambda \in I}$ of elements of E are given, and if at least one of these two families has finite support, then this is true also for the family $(\alpha_\lambda x_\lambda)_{\lambda \in I}$; this follows from the first relation (1).

If $(x_\lambda)_{\lambda \in I}$ is a family of elements of the A-module E, an element $x \in E$ is said to be a *linear combination* of the family $(x_\lambda)_{\lambda \in I}$ (with coefficients in A) if there exists an fs-family $(\alpha_\lambda)_{\lambda \in I}$ of elements of A such that

$$x = \sum_{\lambda \in I} \alpha_\lambda x_\lambda.$$

According to the preceding convention, any linear combination of the empty family is 0. If M is a subset of E, then, by definition, a *linear combination of* (*the elements of*) *the subset* M (with coefficients in A) is a linear combination of the family indexed identically by M. In particular, if $M = \{a\}$, then any linear combination of the subset M is of the form αa, with $\alpha \in A$.

If $(x_\lambda)_{\lambda \in I}$, $(y_\mu)_{\mu \in J}$ are families of elements of the A-module E, and *if each* x_λ, $\lambda \in I$, *is a linear combination of the family* $(y_\mu)_{\mu \in J}$, *then any linear combination of the family* $(x_\lambda)_{\lambda \in I}$ *is also a linear combination of the family* $(y_\mu)_{\mu \in J}$: since let $x = \sum_{\lambda \in I} \alpha_\lambda x_\lambda$, with $(\alpha_\lambda)_\lambda$ a family with finite support; by hypothesis we have for any $\lambda \in I$,

$$x_\lambda = \sum_{\mu \in J} \beta_{\mu\lambda} y_\mu,$$

with $(\beta_{\mu\lambda})_{\mu \in J}$ a family with finite support; if J' is the union of the supports of the families $(\beta_{\mu\lambda})_{\mu \in J}$, the union being extended to all λ in the support I' of $(\alpha_\lambda)_\lambda$, then J' is a finite set, and we have for every $\lambda \in I'$,

$$x_\lambda = \sum_{\mu \in J'} \beta_{\mu\lambda} y_\mu;$$

as $x = \sum_{\lambda \in I'} \alpha_\lambda x_\lambda$, we have

$$x = \sum_{\lambda \in I'} \sum_{\mu \in J'} \alpha_\lambda \beta_{\mu\lambda} y_\mu = \sum_{\mu \in J'} \Big(\sum_{\lambda \in I'} \alpha_\lambda \beta_{\mu\lambda}\Big) y_\mu = \sum_{\mu \in J} \gamma_\mu y_\mu,$$

where we put $\gamma_\mu = \sum_{\lambda \in I'} \alpha_\lambda \beta_{\mu\lambda}$ if $\mu \in J'$ and $\gamma_\mu = 0$ if $\mu \in J \setminus J'$.

In particular, *if* M *and* N *are subsets of* E *such that* $M \subseteq N$, since each element of M is a linear combination of N, *it follows that if* $x \in E$ *is a linear combination of* M, *then* x *is also a linear combination of* N.

Definition 2. Given two left A-modules E and F, an *A-linear mapping* or a *homomorphism of left A-modules* or an *A-homomorphism* or a *linear mapping* (if confusion is not possible) *of E into F* or *defined on E and taking values in F* or *between E and F* or *from E to F* is a mapping $u : E \to F$ with the properties

a) $u(x + y) = u(x) + u(y)$,

b) $u(\alpha x) = \alpha \cdot u(x)$

for any $x, y \in E$, $\alpha \in A$.

E will be called the *domain* or *source*, and F the *codomain* or *range* of the mapping u.

If E and F are right A-modules, a *homomorphism of right A-modules* of E into F is a mapping $u : E \to F$ satisfying *a*) and

b') $u(x\alpha) = u(x) \cdot \alpha$

for any $x \in E$, $\alpha \in A$.

In the particular case where $F = E$, a linear mapping $E \to E$ is called an *endomorphism* of the (left, respectively right) A-module E. ▌

The set of all A-linear mappings of the module E into the module F, in which two mappings u, u' are considered identical if for any $x \in E$ we have $u(x) = u'(x)$, is denoted $\mathrm{Hom}_A(E, F)$ or also $\mathrm{Hom}(E, F)$ if there is no possibility of confusion. The set of all endomorphisms of E, i.e. $\mathrm{Hom}(E, E)$, is sometimes also denoted $\mathrm{End}_A(E)$ or $\mathrm{End}(E)$.

On the contrary, if two linear mappings have not the same domain or the same codomain, they will be considered distinct.

From the definition it follows readily that if u is an A-linear mapping $E \to F$, then

(5) $$u(0) = 0, \; u(-x) = -u(x), \; u(x-y) = u(x) - u(y),$$

where the first 0 is the zero element of E, the second 0 is the zero element of F, and $x, y \in E$; for instance, the second relation (5) is proved as follows:

$$u(x) + u(-x) = u(x + (-x)) = u(0) = 0,$$

that is, $u(-x)$ is the opposite of $u(x)$. Also, if $(x_\lambda)_{\lambda \in I}$ is a family of elements of E and $(\alpha_\lambda)_{\lambda \in I}$ is a family of elements of A such that the family $(\alpha_\lambda x_\lambda)_{\lambda \in I}$ has finite support, then the family $(\alpha_\lambda u(x_\lambda))_{\lambda \in I}$ is an fs-family of elements of F, and we have

$$u(\sum_{\lambda \in I} \alpha_\lambda x_\lambda) = \sum_{\lambda \in I} \alpha_\lambda u(x_\lambda);$$

this may be seen as reducing the sums to finite ones, which in fact they are; for such sums, an induction on the number of terms is to be applied starting from the very definition.

EXAMPLES. 5) If A is a ring and E is an A-module, the *identity mapping* of E, denoted 1_E, i.e. the mapping $E \to E$ defined by

$$1_E(x) = x, \quad x \in E,$$

is an A-linear mapping; in the same way, if E and F are two A-modules, the *zero mapping* $E \to F$, denoted 0 (there is no possibility of confusion, since it is clearly understood from the context which 0 is meant), defined by

$$0(x) = 0, \quad x \in E$$

(the second 0 is the zero element of F), is an A-linear mapping; it will be also referred to as the *zero homomorphism* $E \to F$.

6) Let A be a ring, E an A-module, and let a be a fixed element of E; the mapping $\theta_a : A_s \to E$ defined by

$$\theta_a(\alpha) = \alpha a, \quad \alpha \in A,$$

is a homomorphism of left A-modules.

7) If $I = (\alpha, \beta)$ is an open interval of the real line $\mathbf{R}$, if E is the vector space over $\mathbf{R}$ consisting of all functions $I \to \mathbf{R}$ derivable on I, and if $F = \mathbf{R}^I$ is the $\mathbf{R}$-vector space of all functions $I \to \mathbf{R}$ (see Example 3)), then the mapping $E \to F$ assigning to each derivable function on I its derivative is an $\mathbf{R}$-linear mapping.

8) If we consider a commutative ring A and the A-module $A[X]$ of all polynomials in the indeterminate X (see Example 2)), the mapping from $A[X]$ to itself assigning to each polynomial its derivative is an endomorphism of $A[X]$.

9) Let A be a commutative ring; let E be an A-module and α a fixed element of A; the mapping $h_\alpha : E \to E$ defined by

$$h_\alpha(x) = \alpha x, \quad x \in E,$$

is an endomorphism of E; it is called the *homothety of ratio* α in E.

10) Let G and H be abelian groups; if $f : G \to H$ satisfies the condition

$$f(x + x') = f(x) + f(x') \tag{6}$$

for any $x, x' \in G$, we say that f is a *homomorphism of abelian groups*; we have however seen (Example 4)) that every abelian group is canonically endowed with a $\mathbf{Z}$-module structure; taking this structure into account, any homomorphism of abelian groups $f : G \to H$ is a $\mathbf{Z}$-homomorphism: since, for $n > 0$, the relation

$$f(n \cdot x) = n \cdot f(x), \qquad x \in G, \tag{7}$$

is verified by induction on n starting from (6); for $n = 0$, (7) is true because, by (6), $f(0) = 0$; as for negative n, the validity of (7) is a consequence of its validity for positive n and of the fact that $f(-x) = -f(x)$, which follows from (6). ▮

REMARK. The homothety h_α of ratio α, $\alpha \in A$, may be defined as in Example 9) in a module E over an arbitrary ring A, not necessarily commutative; but in this case it is no more an endomorphism of the A-module E but only an endomorphism of the $\mathbf{Z}$-module corresponding to the underlying abelian group. ▮

Let A be a ring and E, F, G three A-modules; let

$$u : E \to F \text{ and } v : F \to G$$

be two A-linear mappings.

DEFINITION 3. The *composition* of (the linear mappings) v and u or the *composition* of v with u or the *composite mapping* of v and u or the *composite* of v with u, denoted $v \circ u$, is the mapping

$$v \circ u : E \to G$$

defined by

$$(v \circ u)(x) = v(u(x)), \quad x \in E. \text{ ▮}$$

Note that the composite mapping is the mapping obtained by the composition of v and u considered as mappings of sets. But here it can be shown that $v \circ u$ is a linear mapping $E \to G$: since, for instance, if $\alpha \in A$, $x \in E$, we have

$$(v \circ u)(\alpha x) = v(u(\alpha x)) = v(\alpha \cdot u(x)) = \alpha \cdot v(u(x)) = \alpha \cdot (v \circ u)(x).$$

We leave to the reader to check condition *a*) of Definition 2 for $v \circ u$.

If, moreover, H is a fourth A-module, and if $w : G \to H$ is an A-homomorphism, then we have in $\operatorname{Hom}_A(E, H)$,

$$(w \circ v) \circ u = w \circ (v \circ u), \tag{8}$$

which is readily verified.

The property (8) is called the "associativity of the composition of homomorphisms".

If $E_1, \ldots, E_{n+1}$, $n \geqslant 1$, is a family of A-modules, and if

$$u_i : E_i \to E_{i+1}, \quad i = 1, \ldots, n,$$

is a homomorphism, the *composition* (or the *composite mapping*)

$$u_n \circ u_{n-1} \circ \ldots \circ u_1 \tag{9}$$

is defined recursively as follows: if $n = 1$, then this composition is u_1; if $n > 1$, then, by definition, the composition (9) is

$$u_n \circ (u_{n-1} \circ \ldots \circ u_1).$$

From the associativity property (8) we get that for each k such that $1 \leqslant k \leqslant n-1$ we have

$$u_n \circ \ldots \circ u_1 = (u_n \circ \ldots \circ u_{k+1}) \circ (u_k \circ \ldots \circ u_1).$$

If $u : E \to F$ is a linear mapping, then in $\operatorname{Hom}(E, F)$ we have the equality

$$u \circ 1_E = 1_F \circ u = u, \tag{10}$$

which is easy to verify.

*Modules over a ring A and A-linear mappings are subordinated to a more general concept, whose formal definition is the following:

DEFINITION 4. A *category* $\mathscr{C}$ consists of the following data:

1) A non-empty class (not necessarily a set) of elements, denoted $\operatorname{Ob} \mathscr{C}$, whose elements will be called the *objects of the category*.

2) For each pair $E, F \in \text{Ob}\,\mathcal{C}$, a set of elements, denoted $\text{Hom}_{\mathcal{C}}(E, F)$ or $\text{Hom}(E, F)$ if no confusion may arise, whose elements are called *morphisms* or *arrows* having E as *domain* or *source* and F as *codomain* or *range* or, alternatively, *morphisms from E to F* or *of E into F* or *between E and F*.

3) For any three objects E, F, G, a mapping

$$\varphi(E, F, G) : \text{Hom}_{\mathcal{C}}(E, F) \times \text{Hom}_{\mathcal{C}}(F, G) \rightarrow \text{Hom}_{\mathcal{C}}(E, G).$$

We shall write with equivalent meaning

$$u \in \text{Hom}_{\mathcal{C}}(E, F) \text{ or } u : E \rightarrow F \text{ or } E \xrightarrow{u} F;$$

in the same way, if we have

$$E \xrightarrow{u} F, \quad F \xrightarrow{v} G,$$

which may also be written as

$$E \xrightarrow{u} F \xrightarrow{v} G,$$

then we shall denote

$$\varphi(E, F, G)\,(u, v) = v \circ u,$$

and $v \circ u$ will be called the *composition* or the *composite morphism* of v and u (or of v with u).

We assume that the data 1), 2), 3) are subject to the following axioms:

a) If $E, F, G, H \in \text{Ob}\,\mathcal{C}$, and if we have the situation

$$E \xrightarrow{u} F \xrightarrow{v} G \xrightarrow{w} H,$$

then

$$w \circ (v \circ u) = (w \circ v) \circ u.$$

b) For any $E \in \text{Ob}\,\mathcal{C}$ there exists in $\text{Hom}_{\mathcal{C}}(E, E)$ an element, denoted 1_E, called the *identity morphism* of E, having the property that for any pair F, G of objects of $\mathcal{C}$, any $u \in \text{Hom}_{\mathcal{C}}(E, F)$, and any $v \in \text{Hom}_{\mathcal{C}}(G, E)$ we have

$$u \circ 1_E = u \quad \text{and} \quad 1_E \circ v = v.$$

c) If the pairs (E, F) and (E', F') of objects of $\mathcal{C}$ are distinct, that is to say, if either of $E \neq E'$ or $F \neq F'$ or both are valid, then

$$\text{Hom}_{\mathcal{C}}(E, F) \cap \text{Hom}_{\mathcal{C}}(E', F') = \emptyset.$$

A category $\mathcal{C}$ whose object class Ob $\mathcal{C}$ is a set is referred to as a *small category*.

A small category $\mathcal{C}$ for which the set Ob $\mathcal{C}$ is finite and $\mathrm{Hom}_{\mathcal{C}}(E, F)$ is a finite set for any pair E, F of objects of $\mathcal{C}$ is called a *finite category*. ▮

We notice that a small category $\mathcal{C}$ is finite if and only if the set $\coprod_{E,F} \mathrm{Hom}_{\mathcal{C}}(E,F)$ is finite.

The notion of *diagram*, respectively *commutative diagram*, given in § 0 is now extended to a category, the vertices being objects and the arrows morphisms.*

Now, let A be a ring; let us consider as class of objects the class of all left A-modules, in which we suppose that we know when two A-modules are identical; if E and F are two A-modules, let the set $\mathrm{Hom}_A(E, F)$ of linear mappings from E to F stand for the set of morphisms of source E and range F; if

$$u : E \to F, \quad v : F \to G$$

are given, where E, F, G are A-modules, we take for $v \circ u$ the linear mapping which is the composition of v and u; for the identity morphism of the A-module E we take the identity mapping 1_E; from what precedes it follows that in this way we have obtained a category; in the sequel, we shall refer to it as to the *category of left A-modules*, and will denote it $\mathbf{Mod}_s(A)$ or, if any possibility of confusion is excluded, $\mathbf{Mod}(A)$. (Thus, we have $\mathrm{Hom}_{\mathbf{Mod}(A)}(E, F) = \mathrm{Hom}_A(E, F)$.) Similarly, if the right A-modules are considered, we obtain the *category of right A-modules*, which will be denoted $\mathbf{Mod}_d(A)$. If A is a commutative ring, the categories $\mathbf{Mod}_s(A)$ and $\mathbf{Mod}_d(A)$ are to be identified in an obvious manner.

EXAMPLES. 11) If the class of abelian groups stands for class of objects, and for any pair G, H of abelian groups the set of morphisms between G and H is the set of homomorphisms of abelian groups $G \to H$, the composition of two morphisms being the one defined for modules, we get a category, called the *category of abelian groups* and denoted in what follows **Ab**. Since any abelian group is endowed with a **Z**-module structure (see Example 4)), since with this structure any homomorphism of abelian groups is a **Z**-homomorphism (Example 10)), and since, conversely, to any **Z**-module the underlying abelian group is assigned, the two procedures involved being inverse to each other, one may identify **Ab** with **Mod(Z)**, identification which is supposed to be done in the sequel.

12) Considering as class of objects the class of all sets, taking for any pair of sets M, N as set of arrows $M \to N$ the set of all mappings from M to N, the composition being as above, by successive application, we get a new category, called the *category of sets* and denoted **Ens**.

13) Let M be a preordered set with a preorder relation denoted by $\leqslant$; we consider M as class of objects; if $a, b \in M$, we take for Hom(a, b) the empty set or a set consisting of a single element, denoted $\overset{a}{\underset{b}{\to}}$, according as $a \nleqslant b$ or $a \leqslant b$; if $a \leqslant b \leqslant c$, then $a \leqslant c$, and, by definition, $\overset{b}{\underset{c}{\to}} \circ \overset{a}{\underset{b}{\to}} = \overset{a}{\underset{c}{\to}}$; in this way, a

category, called the *category associated with the preordered set M* and also denoted M, is obtained; by an abuse of notation, instead of $a \xrightarrow[b]{a} b$ we shall write $a \leqslant b$.

14) Taking the class of all rings (with unit element) as class of objects, and for any pair of rings A, B as set of morphisms from A to B the set of ring homomorphisms $A \to B$ (carrying the unit element of A into the unit element of B), the composition being done as in **Ens**, we get a category, called the *category of rings* (with unit element) and denoted **Ann**. ▮

*Definition 5. Given a category $\mathcal{C}$, a morphism $u : E \to F$ is said to be an *isomorphism* (of $\mathcal{C}$ or in $\mathcal{C}$) *from E to F* or *between E and F* if there exists a morphism $v : F \to E$ such that

$$v \circ u = 1_E, \quad u \circ v = 1_F.$$

If there exists an isomorphism from E to F, E is said to be *isomorphic* to F. ▮

Note that if $w : F \to E$ is another morphism satisfying the conditions imposed on v in Definition 5, then

$$w \circ u \circ v = 1_E \circ v = v,$$
$$= w \circ 1_F = w,$$

i.e., $w = v$, and v from Definition 5 is uniquely determined by u; according to its definition, v is an isomorphism too; it is called the *isomorphism inverse to u* or the *inverse of u*, and is denoted by u^{-1}; we have $(u^{-1})^{-1} = v^{-1} = u$; sometimes, one also says that u and v are *isomorphisms inverse to each other.*

From the preceding considerations it follows also that *if E is isomorphic to F, then F is isomorphic to E.*

Since for any object E of the category we have $1_E \circ 1_E = 1_E$, it follows that 1_E is itself an isomorphism $E \to E$, with $(1_E)^{-1} = 1_E$, and thus that *E is isomorphic to itself.*

In the same manner, if $u : E \to F$ and $u' : F \to G$ are isomorphisms in the category $\mathcal{C}$, then $u' \circ u : E \to G$ is also an isomorphism, and $(u' \circ u)^{-1} = u^{-1} \circ u'^{-1}$: since,

$$u' \circ u \circ u^{-1} \circ u'^{-1} = u' \circ 1_F \circ u'^{-1} = u' \circ u'^{-1} = 1_G,$$

and, similarly, $u^{-1} \circ u'^{-1} \circ u' \circ u = 1_E$; hence, *if E is isomorphic to F, and F is isomorphic to G, then E is isomorphic to G.*

Consequently, the relation "isomorphic to" is an equivalence relation on the class Ob $\mathcal{C}$; owing to this, we can say, if E is isomorphic to F, that E and F are *isomorphic objects* of the category $\mathcal{C}$.

To point out that $u : E \to F$ is an isomorphism in the category $\mathcal{C}$, we sometimes write $u : E \xrightarrow{\sim} F$, and the fact that E and F are isomorphic objects is indicated by $E \approx F$.*

If A is a ring, an isomorphism $u : E \to F$ in the category $\mathbf{Mod}_s(A)$ (or $\mathbf{Mod}_d(A)$) is called an *isomorphism of A-modules* or an *A-isomorphism*. If E and F are isomorphic A-modules, we also say that they are *A-isomorphic modules*. An isomorphism $u : E \rightrightarrows E$, where E is an A-module, is called an *automorphism* of E.

The isomorphisms of modules are characterized in the following manner:

PROPOSITION 1. *An A-linear mapping $u : E \to F$ is an isomorphism if and only if it is bijective.*

Proof. The condition is necessary: Denote $u^{-1} = v$; if for $x, x' \in E$ we have $u(x) = u(x')$, then, since $v \circ u = 1_E$, we obtain

$$x = 1_E(x) = v(u(x)) = v(u(x')) = 1_E(x') = x',$$

hence u is injective; since $u \circ v = 1_F$, we have $y = 1_F(y) = u(v(y))$ for any $y \in F$, and u is surjective.

The condition is sufficient: If $y \in F$, there is a uniquely determined element $x_y \in E$ such that $u(x_y) = y$; defining

$$v(y) = x_y, \quad y \in F,$$

we obtain a mapping $v : F \to E$ with the property that $v \circ u = 1_E$, $u \circ v = 1_F$; these relations imply the linearity of v since, for instance,

$$u(v(x) + v(x')) = u(v(x)) + u(v(x')) = x + x' = u(v(x + x')),$$

and as u is injective, $v(x) + v(x') = v(x + x')$. ∎

DEFINITION 6. If for the A-linear mapping $u : E \to F$ there exists a linear mapping $v : F \to E$ such that $v \circ u = 1_E$ (respectively $u \circ v = 1_F$), then u is said to be *left* (respectively *right*) *invertible*, and v is called a *linear mapping left* (respectively *right*) *inverse to u* or a *left* (respectively *right*) *inverse of u* or also a *linear retraction* (respectively a *linear section*) *associated with u*.

*The definition may be given in any category for a morphism $u : E \to F$; but then one refers to v as a *left* (respectively *right*) *inverse morphism of u*.* ∎

From the proof of necessity in Proposition 1 we get the following

COROLLARY. *If u is a linear mapping for which a left inverse exists, then u is injective; if u admits a right inverse, then u is surjective.* ∎

REMARK. Proposition 1 and its Corollary hold also for mappings between sets, as one can see adapting the proof of Proposition 1; from the proof of Proposition 1 it also follows that for a linear mapping $u : E \to F$ to be an isomorphism it is sufficient to find a mapping $v : F \to E$ which is not *a priori* linear but for which we have $v \circ u = 1_E$, $u \circ v = 1_F$; v turns to be linear, and we have $v = u^{-1}$. ∎

§ 2. Groups of Homomorphisms

Let A be a ring, and let E, F be two A-modules; on the set $\mathrm{Hom}(E, F)$ (see § 1) we shall define an internal law of composition additively written, in the following way: if $u, v \in \mathrm{Hom}_A(E, F)$, then $u + v$ is the mapping $E \to F$ given by

$$x \mapsto u(x) + v(x), \qquad x \in E;$$

i.e.,

$$(u + v)(x) = u(x) + v(x).$$

It is not difficult to see that, $u + v$ is an A-homomorphism; let us check, for instance, the condition *a*) in Definition 2, § 1:[1]

$$(u + v)(x + y) = u(x + y) + v(x + y) = ux + uy + vx + vy =$$
$$= ux + vx + uy + vy = (u + v)(x) + (u + v)(y).$$

DEFINITION 1. The linear mapping $u + v$ defined above is called the *sum* of the given linear mappings u and v. ∎

It is easily verified that, with respect to this addition, $\mathrm{Hom}_A(E, F)$ is an abelian group; for instance, the zero element of this group is the zero mapping $E \to F$ (see Example 5), § 1); in the same way, if $u \in \mathrm{Hom}(E, F)$, its *opposite* $-u$ is the mapping $E \to F$ given by

$$x \mapsto -u(x), \qquad x \in E,$$

which is immediately verified to be A-linear.

REMARK. Generally, $\mathrm{Hom}_A(E, F)$ is not endowed in the manner we would immediately think of with a (left or right) A-module structure: for if we tried to endow it, as would seem natural, with the A-module structure given by

$$(\alpha u)(x) = \alpha \cdot u(x), \quad \alpha \in A, \ u \in \mathrm{Hom}_A(E, F), \quad x \in E,$$

the axiom *b*) in Definition 2, § 1, would not be verified; indeed, we would obtain

$$(\alpha u)(\beta x) = \alpha \cdot u(\beta x) = \alpha(\beta \cdot u(x)) = (\alpha\beta) \cdot u(x),$$

$$\beta \cdot ((\alpha u)(x)) = \beta \cdot (\alpha \cdot u(x)) = (\beta\alpha) \cdot u(x),$$

---

[1] Sometimes, instead of $u(x)$ we shall write ux.

where $\alpha, \beta \in A$, $x \in E$, $u \in \mathrm{Hom}(E, F)$; but generally the two results do not coincide; if, however, A is a *commutative* ring, then $\alpha\beta = \beta\alpha$, and the condition is verified (see also § 16). ▮

Now let E, F, G be three A-modules and $u, u_1, u_2 \in \mathrm{Hom}_A(E, F)$, $v, v_1, v_2 \in \mathrm{Hom}_A(F, G)$; we have

$$(1) \qquad \begin{aligned} v\circ(u_1 + u_2) &= v\circ u_1 + v\circ u_2, \\ (v_1 + v_2)\circ u &= v_1\circ u + v_2\circ u. \end{aligned}$$

For instance, the first relation (1) is verified as follows: we have

$$(v\circ(u_1 + u_2))(x) = v((u_1 + u_2)(x)) = v(u_1x + u_2x) = v(u_1x) + v(u_2x) =$$

$$= (v\circ u_1)(x) + (v\circ u_2)(x) = (v\circ u_1 + v\circ u_2)(x)$$

for any $x \in E$.

Relations (1) represent the "distributivity of the composition of homomorphisms over their addition".

From relations (1) we now get

$$(2) \qquad \begin{aligned} v\circ 0 &= 0\circ u = 0, \\ v\circ(-u) &= (-v)\circ u = -(v\circ u), \end{aligned}$$

where, in the first relation (2), the first 0 is the zero mapping $E \to F$, the second, the zero mapping $F \to G$, while the third, the zero mapping $E \to G$, $u \in \mathrm{Hom}_A(E, F)$, $v \in \mathrm{Hom}_A(F, G)$.

Particularly, if $G = F = E$, the associativity of the composition of linear mappings (see § 1) and the distributivity of the composition over addition imply that, with respect to the two internal laws defined on $\mathrm{Hom}_A(E, E)$, namely the addition defined above and the composition (now acting between two elements of $\mathrm{Hom}_A(E, E)$ and giving another from the same set), the set $\mathrm{Hom}_A(E, E)$ of all endomorphisms of E is a ring; it is called the *ring of endomorphisms* of the A-module E, and is also denoted $\mathrm{End}_A(E)$ or $\mathrm{End}(E)$ (see § 1). This ring has a unit element: it is the identity mapping 1_E of E (see Example 5), § 1), as follows from relations (10), § 1.

The automorphisms of E are, according to their definition, the invertible elements of $\mathrm{End}(E)$; with respect to composition they form a group, denoted $\mathrm{Aut}(E)$ or also $\mathbf{GL}(E)$, and called the *general linear group* relative to E.

If E and F are two A-modules, then, defining on the abelian group $\mathrm{Hom}_A(E, F)$ the external law of composition

$$\mathrm{End}_A(F)\times \mathrm{Hom}_A(E, F) \to \mathrm{Hom}_A(E, F)$$

given by

$$(v, u) \mapsto v \circ u,$$

with $v \in \text{End}(F)$ and $u \in \text{Hom}_A(E, F)$, we get on $\text{Hom}_A(E, F)$ a left $\text{End}_A(F)$-module structure, as it is seen from the associativity and the distributivity.

In the same way, the external law

$$\text{Hom}_A(E, F) \times \text{End}_A(E) \to \text{Hom}_A(E, F)$$

given by

$$(u, v) \mapsto u \circ v,$$

with $u \in \text{Hom}_A(E, F)$ and $v \in \text{End}_A(E)$, defines on $\text{Hom}_A(E, F)$ together with the addition a structure of right $\text{End}_A(E)$-module.

Now let E, E', F, F' be four A-modules and

$$u: E' \to E, \quad v: F \to F',$$

A-homomorphisms; if $f \in \text{Hom}_A(E, F)$, then the composite mapping[1]

$$E' \xrightarrow{u} E \xrightarrow{f} F \xrightarrow{v} F'$$

is a linear mapping $E' \to F'$; let us denote by $\text{Hom}_A(u, v)$ or $\text{Hom}(u, v)$ the mapping

$$\text{Hom}_A(E, F) \to \text{Hom}_A(E', F')$$

given by

$$f \mapsto v \circ f \circ u, \qquad f \in \text{Hom}_A(E, F);$$

in view of relations (1), $\text{Hom}(u, v)$ turns out to be, for the structures of abelian group above defined on $\text{Hom}_A(E, F)$ and $\text{Hom}_A(E', F')$, a *homomorphism of abelian groups*; since if $f, g \in \text{Hom}_A(E, F)$, then

$$\text{Hom}(u, v)(f + g) = v \circ (f + g) \circ u = v \circ f \circ u + v \circ g \circ u =$$

$$= \text{Hom}(u, v)(f) + \text{Hom}(u, v)(g).$$

Suppose that E'', F'' are some other two A-modules, and consider also the linear mappings

$$u': E'' \to E', \qquad v': F' \to F'';$$

---

[1] This is an alternative form for "the composite mapping $v \circ f \circ u$"; we shall use this form in the sequel too.

then we have for any $f \in \mathrm{Hom}(E, F)$,

$$\mathrm{Hom}\,(u \circ u',\ v' \circ v)\,(f) = v' \circ v \circ f \circ u \circ u' = v' \circ (\mathrm{Hom}\,(u, v)\,(f)) \circ u' =$$

$$= (\mathrm{Hom}\,(u', v') \circ \mathrm{Hom}\,(u, v))\,(f).$$

Consequently, we have

$$\mathrm{Hom}(u \circ u',\ v' \circ v) = \mathrm{Hom}(u', v') \circ \mathrm{Hom}(u,\ v). \tag{3}$$

This may be displayed on a diagram of abelian groups:

$$\begin{array}{ccc} \mathrm{Hom}_A(E, F) & \xrightarrow{\mathrm{Hom}(u,\, v)} & \mathrm{Hom}_A(E', F') \\ & {\scriptstyle \mathrm{Hom}(u \circ u',\ v' \circ v)} \searrow & \downarrow {\scriptstyle \mathrm{Hom}(u',\, v')} \\ & & \mathrm{Hom}_A(E'', F''). \end{array} \tag{4}$$

Relation (3) expresses the commutativity of diagram (4).

In particular, if $E'' = E' = E$, and if $u = u' = 1_E$, then from (3) we get

$$\mathrm{Hom}(1_E,\ v' \circ v) = \mathrm{Hom}\,(1_E,\ v') \circ \mathrm{Hom}(1_E,\ v); \tag{5}$$

if, however, $F = F' = F''$ and $v = v' = 1_F$, then

$$\mathrm{Hom}(u \circ u',\ 1_F) = \mathrm{Hom}(u',\ 1_F) \circ \mathrm{Hom}(u,\ 1_F). \tag{6}$$

If $E = E'$, $F = F'$, $u = 1_E$, $v = 1_F$, then

$$\mathrm{Hom}(1_E,\ 1_F) = 1_{\mathrm{Hom}(E,\, F)}, \tag{7}$$

which is a consequence of relations (10), § 1.

In the same way, if $u, u_1, u_2 \in \mathrm{Hom}_A(E', E)$, $v, v_1, v_2 \in \mathrm{Hom}_A(F, F')$, then

$$\begin{aligned} &\mathrm{Hom}(u,\ v_1 + v_2) = \mathrm{Hom}(u,\ v_1) + \mathrm{Hom}(u,\ v_2), \\ &\mathrm{Hom}(u_1 + u_2,\ v) = \mathrm{Hom}(u_1,\ v) + \mathrm{Hom}(u_2,\ v). \end{aligned} \tag{8}$$

The equalities and the sums on the right hand side of (8) are to be considered in the abelian group $\mathrm{Hom}_{\mathbf{Z}}(\mathrm{Hom}_A(E, F),\ \mathrm{Hom}_A(E', F'))$: indeed, $\mathrm{Hom}_A(E, F)$ and $\mathrm{Hom}_A(E', F')$ are abelian groups, i.e. **Z**-modules, and one may take the group of **Z**-homomorphisms of the first one into the second one.

Relations (8) are checked taking into account the distributivity properties (1): really, for the first, if $f \in \mathrm{Hom}_A(E, F)$, then

$$\mathrm{Hom}(u, v_1 + v_2)(f) = (v_1 + v_2)\circ f\circ u = v_1\circ f\circ u + v_2\circ f\circ u =$$

$$= \mathrm{Hom}(u, v_1)(f) + \mathrm{Hom}(u, v_2)(f) = (\mathrm{Hom}(u, v_1) + \mathrm{Hom}(u, v_2))(f).$$

In the usual way, we get now from (8)

$$\mathrm{Hom}(u, 0) = \mathrm{Hom}(0, v) = 0,$$

(9)

$$\mathrm{Hom}(-u, v) = \mathrm{Hom}(u, -v) = -\mathrm{Hom}(u, v),$$

the meaning of zeros and other symbols being clear.

* DEFINITION 2. Let $\mathcal{C}$ and $\mathcal{C}'$ be two categories; a *functor* T *from the category* $\mathcal{C}$ *to the category* $\mathcal{C}'$ or *defined on the category* $\mathcal{C}$ *and taking values in the category* $\mathcal{C}'$ or *of* $\mathcal{C}$ *into* $\mathcal{C}'$ or *between* $\mathcal{C}$ *and* $\mathcal{C}'$, denoted $T\colon \mathcal{C} \to \mathcal{C}'$, consists of the following data:

1) To each object E of $\mathcal{C}$ there is assigned an object $T(E)$ of $\mathcal{C}'$;

2) To each pair of objects E, F of $\mathcal{C}$, a mapping of sets

$$T(E, F)\colon \mathrm{Hom}_{\mathcal{C}}(E, F) \to \mathrm{Hom}_{\mathcal{C}'}(T(E), T(F))$$

is assigned; if we have $u\colon E \to F$ in $\mathcal{C}$, then instead of $T(E, F)(u)$ we shall write, without any risk of confusion, $T(u)$.

These data are assumed to satisfy the following conditions:

a) If in $\mathcal{C}$ we have the situation described by the diagram

$$E \xrightarrow{u} F \xrightarrow{v} G,$$

then in $\mathcal{C}'$,

$$T(v\circ u) = T(v)\circ T(u);$$

b) For any object E of $\mathcal{C}$, $T(1_E) = 1_{T(E)}$. ∎

If $T\colon \mathcal{C} \to \mathcal{C}'$ is a functor and $u\colon E \xrightarrow{\sim} F$ is an isomorphism in the category $\mathcal{C}$, then $T(u) : T(E) \to T(F)$ is an isomorphism in the category $\mathcal{C}'$: since if u^{-1} is the isomorphism inverse to u, we have

$$u\circ u^{-1} = 1_F \text{ and } u^{-1}\circ u = 1_E;$$

then, according to Definition 2,

$$T(u)\circ T(u^{-1}) = 1_{T(F)} \text{ and } T(u^{-1})\circ T(u) = 1_{T(E)};$$

this implies also that

$$T(u^{-1}) = T(u)^{-1}.$$

If, in addition, the category $\mathcal{C}''$ is also given and $U\colon \mathcal{C}' \to \mathcal{C}''$ is a functor, then the mappings assigning to each object E of $\mathcal{C}$ the object $U(T(E))$ of $\mathcal{C}''$ and to each morphism $u\colon E \to F$ of $\mathcal{C}$ the morphism $U(T(u))\colon U(T(E)) \to U(T(F))$ of $\mathcal{C}''$ define a functor $V\colon \mathcal{C} \to \mathcal{C}''$.

Also, for any category $\mathcal{C}$, the mappings assigning to each object E of $\mathcal{C}$ the object E and to each morphism $u\colon E \to F$ of $\mathcal{C}$ the morphism u define a functor $W\colon \mathcal{C} \to \mathcal{C}$.

DEFINITION 3. The functor $V\colon \mathcal{C} \to \mathcal{C}''$ defined above is called the *composition* of the functor U with the functor T or the *composition* of the functors U and T or the *composite functor* of U and T (or of U with T), and is denoted by $U \circ T$.

The functor $W\colon \mathcal{C} \to \mathcal{C}$ defined above is called the *identity functor* of the category $\mathcal{C}$, and is denoted $1_{\mathcal{C}}$. ∎

It is easy to see that the composition of functors is associative. Analogously, whenever there is the case, the definition of the composite functor of a finite family of functors may be given. It is also easily seen that if $T\colon \mathcal{C} \to \mathcal{C}'$ is a functor, then the composition $1_{\mathcal{C}'} \circ T$ and $T \circ 1_{\mathcal{C}}$ are defined, and we have $1_{\mathcal{C}'} \circ T = T$ and $T \circ 1_{\mathcal{C}} = T$, the equality of functors having the meaning that they are defined on the same category, that they take their values in the same category, and that they assign to each object, respectively morphism from the first category the same object, respectively morphism from the second one.*

If a ring A is given, then to each A-module E one may assign the underlying set, and to each A-homomorphism $u\colon E \to F$ the mapping of sets u considered as being defined on the underlying set of E and taking values in the underlying set of F; these assignments define a functor $\mathbf{Mod}(A) \to \mathbf{Ens}$, which will be called the *forgetful functor*; in particular, we have the forgetful functor $\mathbf{Ab} = \mathbf{Mod}(\mathbf{Z}) \to \mathbf{Ens}$; also, for any ring A we have the forgetful functor $\mathbf{Mod}_d(A) \to \mathbf{Ens}$; analogously, there exists the forgetful functor $\mathbf{Mod}(A) \to \mathbf{Ab}$, etc.

If A is a ring and E a fixed A-module, then, assigning to each A-module F the abelian group $\mathrm{Hom}_A(E, F)$, and to each A-homomorphism $v : F \to F'$ the homomorphism of abelian groups

$$\mathrm{Hom}(1_E, v) : \mathrm{Hom}_A(E, F) \to \mathrm{Hom}_A(E, F'),$$

we get, as relations (5) and (7) show, a functor $\mathbf{Mod}(A) \to \mathbf{Ab}$, which we shall denote h^E in the sequel; thus we have

$$h^E(F) = \mathrm{Hom}_A(E, F), \quad h^E(v) = \mathrm{Hom}(1_E, v).$$

* Let $\mathcal{C}$ be a category; starting with $\mathcal{C}$, we define a new category $\mathcal{C}^\circ$ as follows:

The objects of $\mathcal{C}^\circ$ are the *doublets* of the objects of the category $\mathcal{C}$, i.e. elements which are in a fixed one-to-one correspondence with the objects of $\mathcal{C}$ but distinct from these; the doublet of the object E of $\mathcal{C}$ will be denoted E°.

If E°, F° are such doublets, for $\mathrm{Hom}_{\mathcal{C}^\circ}(E^\circ, F^\circ)$ a set of doublets of the elements of $\mathrm{Hom}_{\mathcal{C}}(F, E)$ is taken; the doublet of $u : F \to E$ will be denoted $u^\circ : E^\circ \to F^\circ$.

If we have the situation described by the diagram

$$E^\circ \xrightarrow{u^\circ} F^\circ \xrightarrow{v^\circ} G^\circ, \tag{10}$$

this means that in $\mathcal{C}$ we have the situation

$$G \xrightarrow{v} F \xrightarrow{u} E; \tag{11}$$

we define in $\mathcal{C}^\circ$

$$v^\circ \circ u^\circ = (u \circ v)^\circ.$$

It turns out that these data define a category; if $E \in \mathrm{Ob}\,\mathcal{C}$, then $(1_E)^\circ = 1_{E^\circ}$.

DEFINITION 4. The category $\mathcal{C}^\circ$ defined above starting from the category $\mathcal{C}$ is called the *dual (category) of the category* $\mathcal{C}$ or the category *dual to* $\mathcal{C}$. ∎

If $(\mathcal{C}^\circ)^\circ$ is the dual of the category $\mathcal{C}^\circ$, then, denoting, as set up, by $(E^\circ)^\circ$ the doublet in $(\mathcal{C}^\circ)^\circ$ of the object E° of $\mathcal{C}^\circ$ and by $(u^\circ)^\circ$ the doublet of the morphism u° of $\mathcal{C}^\circ$, suppose the situation described by diagram (11) is given; then in $\mathcal{C}^\circ$ we have the situation described by diagram (10), while in $(\mathcal{C}^\circ)^\circ$ we have the situation

$$(G^\circ)^\circ \xrightarrow{(v^\circ)^\circ} (F^\circ)^\circ \xrightarrow{(u^\circ)^\circ} (E^\circ)^\circ,$$

and

$$(u^\circ)^\circ \circ (v^\circ)^\circ = (v^\circ \circ u^\circ)^\circ = ((u \circ v)^\circ)^\circ. \tag{12}$$

In the same way, for any object E of $\mathcal{C}$ we have

$$((1_E)^\circ)^\circ = (1_{E^\circ})^\circ = 1_{(E^\circ)^\circ}. \tag{13}$$

Relations (12) and (13) show that by identifying the object $(E^\circ)^\circ$ of $(\mathcal{C}^\circ)^\circ$ with the object E of $\mathcal{C}$ and the morphism $(u^\circ)^\circ$ of $(\mathcal{C}^\circ)^\circ$ with the morphism u of $\mathcal{C}$, these identifications do not destroy the identity of objects and morphisms, and that they are compatible with the composition and the identity morphisms, the meaning of the last assertion being obvious. These identifications are supposed to be done in the sequel, and on their basis *we shall identify the category* $(\mathcal{C}^\circ)^\circ$ *with the category* $\mathcal{C}$.

Now let $\mathcal{C}$, $\mathcal{C}'$ be categories and $T : \mathcal{C} \to \mathcal{C}'$ a functor; define a functor $T^\circ : \mathcal{C}^\circ \to \mathcal{C}'^\circ$ as follows: if $E^\circ \in \mathrm{Ob}\,\mathcal{C}^\circ$, then, by definition, $T^\circ(E^\circ) = (T(E))^\circ$; if $u^\circ \in$

$\in \mathrm{Hom}_{\mathcal{C}^\circ}(E^\circ, F^\circ)$, then $u \in \mathrm{Hom}_{\mathcal{C}}(F, E)$ and $T(u) \in \mathrm{Hom}_{\mathcal{C}'}(T(F), T(E))$; we set by definition $T^\circ(u^\circ) = (T(u))^\circ \in \mathrm{Hom}_{\mathcal{C}'^\circ}(T^\circ(E^\circ), T^\circ(F^\circ))$; conditions *a*) and *b*) of Definition 2 obviously hold; we shall, by an abuse of language, call the functor T° the *dual (functor) of T* or the functor *dual to T*; if we identify as above $(\mathcal{C}^\circ)^\circ$, respectively $(\mathcal{C}'^\circ)^\circ$, with $\mathcal{C}$, respectively $\mathcal{C}'$, then for any functor $T: \mathcal{C} \to \mathcal{C}'$ we have with the above notation used for the dual of a functor, $(T^\circ)^\circ = T$.*

Let A be a ring and F a fixed A-module; to any object E° of the category $(\mathbf{Mod}(A))^\circ$ we assign the abelian group $\mathrm{Hom}_A(E, F)$; if $u^\circ : E^\circ \to E'^\circ$ is a morphism of the category $(\mathbf{Mod}(A))^\circ$, then this is the doublet of the A-homomorphism $u: E' \to E$; assign to u° the homomorphism of abelian groups $\mathrm{Hom}(u, 1_F)$; these data define a functor

$$(\mathbf{Mod}(A))^\circ \to \mathbf{Ab},$$

which we shall denote by h_F; thus we have

$$h_F(E^\circ) = \mathrm{Hom}_A(E, F), \qquad h_F(u^\circ) = \mathrm{Hom}(u, 1_F).$$

To see that h_F is indeed a functor, consider for instance condition *a*) of Definition 2: suppose we have the situation

$$E^\circ \xrightarrow{u^\circ} E'^\circ \xrightarrow{u'^\circ} E''^\circ$$

in $(\mathbf{Mod}(A))^\circ$; then in $\mathbf{Mod}(A)$ we have the situation

$$E'' \xrightarrow{u'} E' \xrightarrow{u} E;$$

according to relation (6) and to the definition of the composition in the dual category,

$$h_F(u'^\circ \circ u^\circ) = h_F((u \circ u')^\circ) = \mathrm{Hom}(u \circ u', 1_F) =$$

$$= \mathrm{Hom}(u', 1_F) \circ \mathrm{Hom}(u, 1_F) = h_F(u'^\circ) \circ h_F(u^\circ).$$

* Let $(\mathcal{C}_\lambda)_{\lambda \in I}$ be a family of categories indexed by a set $I \neq \emptyset$; from this family we shall construct a new category $\mathcal{D}$ as follows:

The objects of $\mathcal{D}$ are families of objects $(E_\lambda)_{\lambda \in I}$ indexed by I such that $E_\lambda \in \mathrm{Ob}\,\mathcal{C}_\lambda$ for any $\lambda \in I$.

If $E = (E_\lambda)_\lambda$ and $F = (F_\lambda)_\lambda$ are two objects of $\mathcal{D}$, the set $\mathrm{Hom}_{\mathcal{D}}(E, F)$ consists of the families $(u_\lambda)_{\lambda \in I}$, where $u_\lambda \in \mathrm{Hom}_{\mathcal{C}_\lambda}(E_\lambda, F_\lambda)$ for any $\lambda \in I$; in other words, we have $\mathrm{Hom}_{\mathcal{D}}(E, F) = \prod_\lambda \mathrm{Hom}_{\mathcal{C}_\lambda}(E_\lambda, F_\lambda)$.

If another object $G = (G_\lambda)_\lambda$ of $\mathcal{D}$ is also given, if $u = (u_\lambda)_\lambda \in \mathrm{Hom}_{\mathcal{D}}(E, F)$, and $v = (v_\lambda)_\lambda \in \mathrm{Hom}_{\mathcal{D}}(F, G)$, then the composition $v \circ u$ is defined as the family $(v_\lambda \circ u_\lambda)_\lambda \in \mathrm{Hom}_{\mathcal{D}}(E, G)$.

Clearly, these data define a category provided for any object $E=(E_\lambda)_\lambda$ of $\mathcal{D}$, the family $(1_{E_\lambda})_{\lambda\in I}$ stands for 1_E.

DEFINITION 5. The category $\mathcal{D}$ defined above is called the *product category* or the *product* of the given family of categories $(\mathcal{C}_\lambda)_{\lambda\in I}$, and is denoted $\prod_{\lambda\in I}\mathcal{C}_\lambda$. If I is finite consisting of the natural numbers $1,\ldots,n$, then instead of $\prod_{\lambda\in I}\mathcal{C}_\lambda$ we sometimes also write $\mathcal{C}_1\times\mathcal{C}_2\times\ldots\times\mathcal{C}_n$; there is of course a corresponding notation in the case of another finite index set. For arbitrary I, if for all $\lambda\in I$, $\mathcal{C}_\lambda$ coincides with the same category $\mathcal{C}$, then instead of $\prod_{\lambda\in I}\mathcal{C}_\lambda$ we also write $\mathcal{C}^I$. ∎

If $I=\{1\}$, then $\prod_{\lambda\in I}\mathcal{C}_\lambda$ will be identified with $\mathcal{C}_1$.*

Now let A be a ring; to each object (E°, F) of the product category $(\mathbf{Mod}(A))^\circ\times\mathbf{Mod}(A)$ assign the abelian group $\mathrm{Hom}_A(E, F)$; if in the product category the morphism (u°, v): $(E^\circ, F)\to(E'^\circ, F')$ is given, then we have the homomorphisms of modules

$$u\colon E'\to E,\quad v\colon F\to F';$$

assign to the morphism (u°, v) the homomorphism of abelian groups $\mathrm{Hom}(u, v)$. Certainly, these data define, according to relations (3) and (7), a functor

$$(\mathbf{Mod}(A))^\circ\times\mathbf{Mod}(A)\to\mathbf{Ab},$$

which will be referred to in what follows as the functor "Hom".

* DEFINITION 6. A *preadditive category* is a category $\mathcal{C}$ satisfying the following conditions:

a) For any pair of objects E, F of $\mathcal{C}$, on $\mathrm{Hom}_{\mathcal{C}}(E, F)$ a fixed structure of abelian group additively written is given; the zero element of this group is called the *zero morphism* from E to F, and is denoted 0, the ambiguity being eliminated by the context.

b) If E, F, G are objects of $\mathcal{C}$, and if $u, u_1, u_2\in\mathrm{Hom}_{\mathcal{C}}(E, F)$ and $v, v_1, v_2\in\mathrm{Hom}_{\mathcal{C}}(F, G)$, then

$$v\circ(u_1+u_2)=v\circ u_1+v\circ u_2,\quad (v_1+v_2)\circ u=v_1\circ u+v_2\circ u,$$

where "+" denotes the addition in the corresponding groups.

c) In $\mathcal{C}$ there exists at least one object X such that $\mathrm{Hom}_{\mathcal{C}}(X, X)$ is reduced to its zero element. ∎

One may show that, in the presence of *a*) and *b*), condition *c*) in Definition 6 is equivalent to each of the following conditions:

c') There exists at least one object X for which 1_X is the zero element of the group $\mathrm{Hom}_{\mathcal{C}}(X, X)$;

c'') There exists at least one object X for which $\mathrm{Hom}_{\mathcal{C}}(E, X)$ is reduced to its zero element for any $E\in\mathrm{Ob}\,\mathcal{C}$;

$c''')$ There exists at least one object X for which $\mathrm{Hom}_{\mathcal{C}}(X, F)$ is reduced to its zero element for any $F \in \mathrm{Ob}\,\mathcal{C}$.

Moreover, the same X satisfies all these conditions provided it satisfies one of them.

The equivalence of the statements $c) - c''')$ is proved using the fact that the composition of two morphisms one of which is the zero morphism is the corresponding zero morphism; thus, the implication $c') \Rightarrow c)$ is obtained as follows: if $u \in \mathrm{Hom}_{\mathcal{C}}(X, X)$, then

$$u = u \circ 1_X = u \circ 0 = 0;$$

the implications $c') \Rightarrow c'')$ and $c') \Rightarrow c''')$ are obtained in a similar way, while $c) \Rightarrow c')$, $c'' \Rightarrow c)$, and $c''') \Rightarrow c)$ are obvious.

An object X of a preadditive category $\mathcal{C}$, enjoying the properties given by the equivalent conditions $c) - c''')$ is called a *zero object* of the category; two zero objects X, Y of $\mathcal{C}$ are isomorphic, since taking the composition of the unique morphisms $X \to Y$ and $Y \to X$ in both directions, we get the identity morphisms. By means of the unique isomorphism between two zero objects, all zero objects are sometimes identified, the object obtained in this way being called *the zero object* of the preadditive category $\mathcal{C}$; all zero objects are generally denoted 0, this leading to no undesirable consequence.

Notice that if E and F are two objects of the preadditive category $\mathcal{C}$, the zero morphism $E \to F$ is the composition of the unique morphisms

$$E \to 0, \quad 0 \to F.$$

If $\mathcal{C}$ is a preadditive category, then for any $E \in \mathrm{Ob}\,\mathcal{C}$, the set $\mathrm{Hom}_{\mathcal{C}}(E, E)$ is endowed, as in the case of categories of modules, with a ring structure, the addition being given according to Definition 6, *a*) while the multiplication is the composition; the unit element of this ring is the identity morphism 1_E; this ring is denoted $\mathrm{End}_{\mathcal{C}}(E)$ or $\mathrm{End}(E)$. In the same way as for modules, for any pair of objects E, F of $\mathcal{C}$, the abelian group $\mathrm{Hom}_{\mathcal{C}}(E, F)$ is endowed with an additional structure of left $\mathrm{End}(F)$-module and with one of right $\mathrm{End}(E)$-module, the external law of composition being given in both cases by the composition of morphisms.

If $\mathcal{C}$ is an arbitrary category and E, E', F, F' are four objects, while $u: E' \to E$, $v: F \to F'$ are morphisms, a mapping of sets

$$\mathrm{Hom}_{\mathcal{C}}(u, v): \mathrm{Hom}_{\mathcal{C}}(E, F) \to \mathrm{Hom}_{\mathcal{C}}(E', F')$$

is defined by

$$\mathrm{Hom}_{\mathcal{C}}(u, v)(f) = v \circ f \circ u$$

for any $f \in \mathrm{Hom}_{\mathcal{C}}(E, F)$. Sometimes, provided any confusion is excluded, we write $\mathrm{Hom}(u, v)$ instead of $\mathrm{Hom}_{\mathcal{C}}(u, v)$. Moreover, if $\mathcal{C}$ is a preadditive category, then

$\mathrm{Hom}(u, v)$ becomes a homomorphism of abelian groups, as seen from conditions *b*) of Definition 6.

In the same way as for categories of modules one may show that if E is any fixed object of $\mathcal{C}$, the mappings

$$F \mapsto \mathrm{Hom}_{\mathcal{C}}(E, F), \qquad v \mapsto \mathrm{Hom}(1_E, v),$$

with $F \in \mathrm{Ob}\,\mathcal{C}$, $v \in \mathrm{Hom}_{\mathcal{C}}(F, F')$, define a functor $\mathcal{C} \to \mathbf{Ens}$, denoted h^E. Also, given an object F of the category $\mathcal{C}$, the mappings

$$E^\circ \mapsto \mathrm{Hom}_{\mathcal{C}}(E, F), \quad u^\circ \mapsto \mathrm{Hom}(u, 1_F),$$

with $E^\circ \in \mathrm{Ob}\,\mathcal{C}^\circ$, $u^\circ \in \mathrm{Hom}_{\mathcal{C}^\circ}(E^\circ, E'^\circ)$, define a functor $\mathcal{C}^\circ \to \mathbf{Ens}$, denoted h_F. In the same way, the mappings

$$(E^\circ, F) \mapsto \mathrm{Hom}_{\mathcal{C}}(E, F), \; (u^\circ, v) \mapsto \mathrm{Hom}(u, v),$$

where $E, F \in \mathrm{Ob}\,\mathcal{C}$, $u \in \mathrm{Hom}_{\mathcal{C}}(E', E)$, $v \in \mathrm{Hom}_{\mathcal{C}}(F, F')$, define a functor $\mathcal{C}^\circ \times \mathcal{C} \to$ $\to \mathbf{Ens}$, referred to as the functor "Hom".

If $\mathcal{C}$ is a preadditive category, then h^E, h_F, and Hom take their values in the category $\mathbf{Ab}$, i.e., they are the composition of the forgetful functor $\mathbf{Ab} \to \mathbf{Ens}$ with some uniquely defined functors $\mathcal{C} \to \mathbf{Ab}$, $\mathcal{C}^\circ \to \mathbf{Ab}$, $\mathcal{C}^\circ \times \mathcal{C} \to \mathbf{Ab}$, respectively, which are determined by the preadditive category structure of $\mathcal{C}$; by an abuse of notation, in this case, these latter functors will be denoted also by h^E, h_F, Hom, respectively.

If $\mathcal{C}$ is a preadditive category and $E, F \in \mathrm{Ob}\,\mathcal{C}$, then, defining on $\mathrm{Hom}_{\mathcal{C}^\circ}(E^\circ, F^\circ)$ a law of addition by

$$u_1^\circ + u_2^\circ = (u_1 + u_2)^\circ,$$

where $u_1, u_2, u_1 + u_2 \in \mathrm{Hom}_{\mathcal{C}}(F, E)$, we get a structure of abelian group on $\mathrm{Hom}_{\mathcal{C}^\circ}(E^\circ, F^\circ)$; also, taking the doublet of a zero object of $\mathcal{C}$, we get a zero object of $\mathcal{C}^\circ$; with these definitions, $\mathcal{C}^\circ$ turns out to be a preadditive category; in this sense we say that *the dual of a preadditive category is a preadditive category.*

Passing to $(\mathcal{C}^\circ)^\circ$, we have

$$(u_1^\circ)^\circ + (u_2^\circ)^\circ = (u_1^\circ + u_2^\circ)^\circ = ((u_1 + u_2)^\circ)^\circ,$$

hence the identification of $(\mathcal{C}^\circ)^\circ$ with $\mathcal{C}$ is also compatible with the addition of morphisms with respect to the defined canonical structure of preadditive category of $(\mathcal{C}^\circ)^\circ$. Consequently, in the case of a preadditive category $\mathcal{C}$, the identification $(\mathcal{C}^\circ)^\circ = \mathcal{C}$ is an identification of preadditive categories.

In what follows, on some categories $\mathcal{C}$ we shall also define some other additional structures which will be, whenever it is the case, carried over to the dual categories $\mathcal{C}^\circ$; we leave it to the reader, without further discussion, to show that the identification $(\mathcal{C}^\circ)^\circ = \mathcal{C}$ preserves, in all these cases, the additional structures of $\mathcal{C}$ and $(\mathcal{C}^\circ)^\circ$.

If $(\mathcal{C}_\lambda)_{\lambda\in I}$ is a family of categories, then for any object $E = (E_\lambda)_{\lambda\in I}$ and any morphism $u = (u_\lambda)_{\lambda\in I}$ of $\prod_{\lambda\in I} \mathcal{C}_\lambda$, the object E° and the morphism u° of $(\prod_{\lambda\in I} \mathcal{C}_\lambda)^\circ$ are, respectively, identified with the object $(E_\lambda^\circ)_{\lambda\in I}$ and the morphism $(u_\lambda^\circ)_{\lambda\in I}$ of $\prod_{\lambda\in I}(\mathcal{C}_\lambda^\circ)$; it is not difficult to see that these identifications do not destroy the identity of objects and morphisms, and that they are compatible with the source and the range as well as with the composition and the identity morphism; also, as can be verified at the appropriate time, they are compatible with some additional structures eventually defined on $\mathcal{C}_\lambda$, $\lambda \in I$, and which are carried over to $\mathcal{C}_\lambda^\circ$, $\lambda \in I$, $\prod_{\lambda\in I}\mathcal{C}_\lambda$, etc. By means of these identifications *one identifies* $(\prod_{\lambda\in I}\mathcal{C}_\lambda)^\circ$ *with the category* $\prod_{\lambda\in I}(\mathcal{C}_\lambda^\circ)$.*

If A is a ring, the category **Mod**(A) has a zero object: it is an A-module X containing a single element 0, and for which

$$0 + 0 = 0, \qquad \alpha\cdot 0 = 0, \qquad \alpha \in A;$$

indeed, $\mathrm{Hom}_A(X, X)$ is in this case reduced to its only element 0; such a module is called a *zero A-module*, and will also be denoted by 0, though the notation is ambiguous; identifying the zero A-modules, we get *the zero A-module*, denoted also by 0. The plurality of meanings of this notation does not lead to any undesirable consequence.

In view of the previous arguments, for any ring A *the categories* **Mod**$_s(A)$ *and* **Mod**$_d(A)$ *are preadditive categories.*

* DEFINITION 7. Let $\mathcal{C}$, $\mathcal{C}'$ be two preadditive categories; a functor $T\colon \mathcal{C} \to \mathcal{C}'$ is said to be an *additive functor* if for any pair of objects E, F, the mapping $T(E, F)$ in Definition 2 is a homomorphism of abelian groups, i.e., if for any $u, v \in \mathrm{Hom}_{\mathcal{C}}(E, F)$ we have

$$T(E, F)(u + v) = T(E, F)\,u + T(E, F)\,v$$

or, briefly,

$$T(u + v) = T(u) + T(v). \blacksquare$$

An additive functor T carries the zero morphism $E \to F$ into the zero morphism $T(E) \to T(F)$; similarly, the opposite morphism into the opposite morphism. Consequently, if X is a zero object of the category $\mathcal{C}$, then from $1_X = 0$ it follows that $1_{T(X)} = T(1_X) = T(0) = 0$, and $T(X)$ is a zero object of the category $\mathcal{C}'$.

Obviously, the composition of two (or several) additive functors is an additve functor.

REMARK. The definition of the additive functor may be given for categories $\mathcal{C}, \mathcal{C}'$ not necessarily preadditive but satisfying only conditions *a*) and *b*) of Definition 6. ▮ *

For any ring A and any A-module E, *the functor* h^E *is*, by relations (8), *an additive functor* from the preadditive category **Mod**(A) to the preadditive category **Ab** = **Mod**(**Z**); likewise, for any A-module F, *the functor* h_F *is an additive functor* $(\mathbf{Mod}(A))^\circ \to \mathbf{Ab}$.

* If $\mathcal{C}$ is a preadditive category, then for any object E of $\mathcal{C}$, the functor $h^E : \mathcal{C} \to \mathbf{Ab}$ is an additive functor, as well as the functor $h_F : \mathcal{C}^\circ \to \mathbf{Ab}$ for any $F \in \mathrm{Ob}\,\mathcal{C}$.*

§ 3. Submodules and Quotient Modules

Let A be a ring, E an A-module, and M a non-void subset of it; if the images of $M \times M$ and $A \times M$ under the internal and external laws of composition, respectively, are contained in M, then these two laws induce by restriction two mappings

$$(1) \qquad M \times M \to M, \quad \text{respectively} \quad A \times M \to M;$$

they define on M an A-module structure: indeed, the first mapping is nothing else but the addition in E restricted to M, and it induces an internal law on M; this is associative and commutative, being so in E; the second mapping is the scalar multiplication in E by elements of A, restricted to M; since there exists $x \in M$, it follows that $0 \cdot x$ (i.e. 0 of E) belongs to M, and it is obviously a neutral element for the internal law on M; also, if $x \in M$, then $(-1) \cdot x = -x$ (taken in E) is in M, and it is an opposite of x with respect to the internal law on M; the second law (1) defines on M a scalar multiplication by elements of A; axioms *b*), *c*), *d*), *e*) of Definition 1, § 1, are satisfied in M since they are satisfied in E.

Definition 1. Given a (left or right) A-module E, a *submodule* of E or an *A-submodule* of E is a subset M of E which is a module with respect to the restrictions to M of the module operations on E. When A is a skew-field K, and hence E is a vector space over K, a submodule is called a *vector subspace* of E or, briefly, a *subspace* of E. ▮

According to the above considerations, in order that the non-void subset M of the A-module E be a submodule it is sufficient (as a matter of fact, also necessary) that if $x, y \in M$, then the sum $x + y$ taken in E belong to M, and if $\alpha \in A$, $x \in M$, then αx taken in E belong to M.

Notice that if M, M' are submodules of E and $M' \subseteq M$, then M' is a submodule of M.

Examples. 1) If E is an arbitrary A-module, the subset consisting only of 0 is a submodule of E; it is called the *zero submodule* of E, and is denoted by 0; as an independent module it is a zero module. Also, the subset of all elements of E is trivially a submodule of E, called the *total submodule* of E. The total and the zero submodules will be called the *improper submodules* of E; the other submodules of E will be called *proper submodules.* If A is a skew-field, we speak of the *zero subspace*, the *total subspace*, and so on.

2) If A is a ring, the submodules of A_s (see Example 1), § 1) are the left ideals of A; the submodules of A_d are the right ideals of A.

3) Let G be an abelian group additively written and H a non-void subset of G with the property that the sum in G of two elements of H is in H and that the opposite in G of an element of H is in H; H is then an abelian group with respect to the restriction to H of the addition in G; H is called a *subgroup* of the abelian group G. We know (Example 4), § 1) that any abelian group G is canonically endowed with a **Z**-module structure; it is easily seen that any subgroup H of the abelian group G is a **Z**-submodule.

4) If I is an open interval of the real line, and if the vector space $\mathbf{R}^I$ over **R** (see Example 3), § 1) of all functions $I \to \mathbf{R}$ is considered, then the subset C of $\mathbf{R}^I$ consisting of all **R**-valued continuous functions defined on I is a subspace of $\mathbf{R}^I$; the subset D of all derivable functions from I to **R** is an **R**-subspace of C; this subset is also an **R**-subspace of $\mathbf{R}^I$. ▮

Now let A be a ring, E an A-module, and M a submodule of E; denote by i the mapping $M \to E$ which assigns to any $x \in M$ the element x considered in E; the mapping i turns out to be an injective linear mapping; it is called the *canonical injection* or the *inclusion mapping* of the submodule M into the module E; obviously, $i(M) \subseteq M$. We have the following "universal property": if F is an A-module and u: $F \to E$ a linear mapping for which $u(F) \subseteq M$, then there exists a uniquely determined linear mapping v: $F \to M$ such that the diagram

(2)

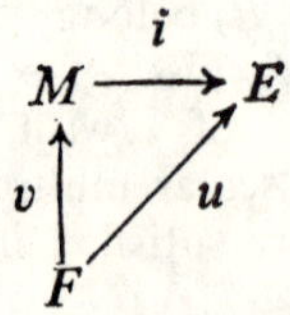

commutes, i.e. such that $i\circ v = u$ (or, more synthetically, u "factors" uniquely through i): indeed, define

$$v(y) = u(y)$$

for each $y \in F$; v is thus the mapping obtained from u by restricting the range, but now v turns out to be linear, as can be readily verified; if v' is a linear mapping $F \to M$ for which $i\circ v' = a$, then for any $y \in F$,

$$v'(y) = i(v'(y)) = u(y) = v(y),$$

and hence $v' = v$; therefore v is unique with the property of rendering diagram (2) commutative [1].

As a consequence, let u: $E \to F$ be an A-homomorphism, and let M, respectively N, be a submodule of E, respectively F, such that $u(M) \subseteq N$; denoting

[1] Another argument for the uniqueness of v will be given later on, when speaking of *monomorphisms*.

by i, respectively j, the canonical injections $M \to E$, respectively $N \to F$, we have the composite mapping of

$$M \xrightarrow{i} E \xrightarrow{u} F;$$

since $(u \circ i)(M) = u(M) \subseteq N$, it follows by the universal property of the submodule that there exists a unique homomorphism $u'\colon M \to N$ such that $j \circ u' = u \circ i$; if $x \in M$, then

$$u'(x) = j(u'(x)) = u(i(x)) = u(x);$$

hence u' acts in the same way as u; u' is said to be the *homomorphism induced by u by restriction (to M and N).*

Let A be a ring, E an A-module, and F an A-submodule of E; define on E a binary relation, called the *congruence relation with respect to F* or the *congruence relation modulo F*, as follows: for $x, y \in E$ we say that *x is congruent to y modulo F* and write

$$x \equiv y \,(\mathrm{mod}\ F) \text{ or } x \equiv y\,(F)$$

if $x - y \in F$; this relation is an equivalence relation: for $x - x = 0 \in F$, hence $x \equiv x(F)$; if $x \equiv y(F)$, then $x - y \in F$, hence $-(x - y) = -x + y \in F$, and $y \equiv x(F)$; if $x \equiv y(F)$, $y \equiv z(F)$, then $x - y \in F$, $y - z \in F$, hence $x - z = (x - y) + (y - z) \in F$, and hence $x \equiv z(F)$. According to the general theory of equivalence relations (see § 0), the set E is divided into equivalence classes with respect to this relation, classes which are now called *cosets modulo F* or *cosets with respect to the submodule F*: a coset consists of all elements which are congruent to a given element x; x may be arbitrarily picked up in the corresponding class, and if we denote by $\hat{x}$ the coset containing x, we have $\hat{x} = \hat{y}$ if and only if $x \equiv y(F)$.

The congruence relation is *compatible* with the module operations: i.e., if

(3) $$x \equiv y\,(F) \quad \text{and} \quad x' \equiv y'\,(F),$$

then

$$x + x' \equiv y + y'\,(F),$$

and if the first relation (3) is valid, then

(4) $$\alpha x \equiv \alpha y\,(F)$$

for any $\alpha \in A$; let us verify, for instance, the last implication: from the first relation (3) we have $x - y \in F$, whence $\alpha x - \alpha y = \alpha(x - y) \in F$, hence (4) follows.

Let us now consider the set, denoted by E/F or $\dfrac{E}{F}$, of all cosets of E modulo F, in which the equality is the identity of subsets: it is the quotient set of E modulo the congruence relation with respect to F. In E/F an A-module structure (in our case on the left, but if we considered right A-modules, we would get on E/F a right

A-module structure) is defined as follows: consider two cosets, and in each of them choose a representative, say x and x', respectively; thus the two considered cosets are, in the above notation, $\hat{x}$ and $\hat{x}'$; the sum of the two cosets is defined as the coset containing $x + x'$, i.e., as $\widehat{x + x'}$; this coset does not depend on the choice of the representatives, since picking up in the same cosets the elements y, respectively y', we have (3), thus, by the preceding considerations, $\widehat{x + x'} = \widehat{y + y'}$; likewise, considering a coset and one of its representatives, say x, the result of the scalar multiplication of $\alpha \in A$ with this coset is by definition $\widehat{\alpha x}$; as above, this result does not depend on the chosen representative x. It is easily verified, and we leave it to the reader as an exercise, that E/F becomes in this way an A-module; we mention only that the zero element is the coset $\hat{0}$ (which, as a subset of E, coincides with F), while the opposite of the coset $\hat{x}$ is the coset $\widehat{-x}$.

DEFINITION 2. If E is a (left or right) A-module and F is a submodule of it, the A-module E/F $\left(\text{or } \dfrac{E}{F}\right)$ set up as above is called the *quotient module* of the module E by (with respect to) the submodule F; if A is a skew-field, and therefore E, F are vector spaces, E/F is called the *quotient vector space* or the *quotient space* of the space E by (with respect to) the subspace F. ▮

If A is a ring, E is an A-module, and F is a submodule of E, denote by p the mapping $E \to E/F$ sending an element $x \in E$ into the coset modulo F containing it (briefly, "its coset modulo F"); in other words, $p(x) = \hat{x}$; p is linear by the definition of the operations in E/F; moreover, p is surjective, since any element of E/F is a coset containing at least one element x, and thus of the form $p(x)$; p is called the *canonical surjection* of E onto E/F; from what precedes we deduce that if $x \equiv y \pmod F$, then $p(x) = p(y)$. We have the following "universal property": if G is an A-module and $u\colon E \to G$ is a linear mapping enjoying the property that $x \equiv y \pmod F$ implies $u(x) = u(y)$ (i.e. a linear mapping *compatible* with the congruence relation modulo F), then there exists a uniquely determined linear mapping $v : E/F \to G$ such that the following diagram

(5)
$$\begin{array}{ccc} E & \xrightarrow{\;p\;} & E/F \\ & {\scriptstyle u}\searrow & \downarrow{\scriptstyle v} \\ & & G \end{array}$$

commutes, i.e. such that $v \circ p = u$: indeed, let $\hat{x} \in E/F$ (this implies a representative has been chosen); define

$$v(\hat{x}) = u(x);$$

from the compatibility of u with the congruence relation modulo F it follows that v depends only on the coset and not on the choice of a representative in it; it is easily verified that v is linear and that $v \circ p = u$: for instance,

$$v(\hat{x} + \hat{x}') = v(\widehat{x + x'}) = u(x + x') = ux + ux' = v(\hat{x}) + v(\hat{x}');$$

if there were another linear mapping v': $E/F \to G$ with the same properties as v, we would have

$$v'(\hat{x}) = v'(p(x)) = u(x) = v(\hat{x}),$$

and as $\hat{x}$ is arbitrary in E/F, $v' = v$.[1]

As an application, let u: $E \to F$ be an A-homomorphism, let M, respectively N, be a submodule of E, respectively F, such that $u(M) \subseteq N$; denoting by p, respectively q, the canonical surjection $E \to E/M$, respectively $F \to F/N$, we have the composite homomorphism

$$E \xrightarrow{u} F \xrightarrow{q} F/N;$$

if $x \equiv y \pmod{M}$, then $x - y \in M$, hence $u(x - y) \in N$, and $(q \circ u)(x - y) = q(u(x - y)) = 0$, i.e., $(q \circ u)(x) = (q \circ u)(y)$; according to the universal property of the quotient module, there exists a uniquely determined linear mapping v: $E/M \to F/N$ with the property that $v \circ p = q \circ u$; if $\hat{x}$, respectively $\hat{y}$, are the cosets of $x \in E$, respectively $y \in F$, with respect to M, respectively N, then

$$v(\hat{x}) = v(p(x)) = q(u(x)) = \widehat{u(x)};$$

v is said to be the *homomorphism induced by u by passing to the quotient module* (or *to the quotient*).

In particular, if M and N are two submodules of the A-module E and $M \subseteq N$, then, taking in the preceding considerations $F = E$, $u = 1_E$, we get a linear mapping

$$v\colon\ E/M \to E/N,$$

which is the only one with the property that the diagram

$$\begin{array}{ccc} E & & \\ {\scriptstyle p}\downarrow & \searrow{\scriptstyle q} & \\ E/M & \xrightarrow{v} & E/N \end{array}$$

---

[1] We shall provide another argument for the uniqueness of v later on, when speaking of *epimorphisms*.

commutes, p and q being the canonical surjections; we have $v(\hat{x}) = \check{x}$, where $\hat{x}$, respectively $\check{x}$, represent the cosets of x modulo M, respectively N; v is surjective, and it is called the *canonical surjection* $E/M \to E/N$.

Combining the restriction with the passing to the quotient, we get the following result: Let $u: E \to F$ be an A-homomorphisms, let M, M' (respectively N, N') be submodules of E (respectively F) such that $M' \subseteq M$ (respectively $N' \subseteq N$); assume $u(M) \subseteq N$ and $u(M') \subseteq N'$; then by restriction to M and N we get a homomorphism $u': M \to N$ induced by u; we have, however, $u'(M') = u(M') \subseteq N'$, hence by passing to the quotient we get a homomorphism

$$\bar{u}: M/M' \to N/N';$$

if $\hat{x}$, respectively $\hat{y}$, are the cosets of $x \in M$, respectively $y \in N$, with respect to M', respectively N', then

$$\bar{u}(\hat{x}) = \widehat{u'(x)} = \widehat{u(x)}.$$

In this situation we shall say that $\bar{u}$ is the *homomorphism* $M/M' \to N/N'$ *induced by u.*

Notice that if $F = E$, then any two elements $x, y \in E$ are congruent modulo F; in other words, the congruence relation modulo E is the total equivalence relation on E; hence we have only one coset of E modulo E, so that E/E reduces to a single element, and, consequently, *it is a zero A-module*; the converse is also true, since if E/F is reduced to a single element, we have for any $x \in E$, $x = x - 0 \in F$, hence $E = F$; in the other extreme case, when F is the zero submodule of E, from $x \equiv y \pmod F$ it follows $x = y$, hence the congruence relation with respect to the zero submodule is the identity equivalence relation on E; thus each coset contains only one element; consequently, the canonical surjection $E \to E/0$ is also injective, i.e. an isomorphism (Proposition 1, § 1); conversely, if the canonical surjection $E \to E/F$ is injective, then $x \in F$ implies $x = 0$; hence, in this case, $F = 0$.

Next, *if E is an A-module and M, N are submodules of E such that* $N \subseteq M$, *the quotient module M/N may be thought of as a submodule of E/N*: for if $\hat{x}$ is any element of M/N, with $x \in M$, then any element $y \in E$ congruent to x modulo N is of the form $y = x + z$, with $z \in N$, hence $y \in M$; consequently, denoting by $\check{x}$ the coset from E/N containing x, we have $\check{x} = \hat{x}$, the equality being the equality of subsets of E; thus the elements of M/N, considered as subsets of E, are also elements of E/N, and so M/N may be considered as a subset of E/N; as for the fact that the addition and the scalar multiplication in M/N are induced by the corresponding operations in E/N, this is a consequence of the very definition of the quotient module. In this identification the elements of M/N are characterized among the elements of E/N as being those cosets from E/N which contain elements of M.

Let A be a ring, E, F two A-modules, and $u: E \to F$ a linear mapping; for a subset M of E, let us consider the subset $u(M)$ of F, i.e. the *image of M under u*; if M is a submodule of E, $u(M)$ is a submodule of F, as is readily verified. In particular, *u(E) is a submodule of F.* Likewise, if N is a subset of F, one may consider the subset $u^{-1}(N)$, i.e. the *inverse image of N under u*; if N is a submodule of F, $u^{-1}(N)$ is a submodule of E; let us verify this last assertion: as N contains $0 \in F$, it follows that $0 \in E$ belongs to $u^{-1}(N)$, since $u(0) = 0$ (relation (5), § 1); hence $u^{-1}(N) \neq \emptyset$; if $x, x' \in u^{-1}(N)$, then $u(x), u(x') \in N$, hence $u(x + x') = u(x) + u(x') \in N$, and $x + x' \in u^{-1}(N)$; similarly, $\alpha \in A$ and $x \in u^{-1}(N)$ imply $\alpha x \in u^{-1}(N)$. In particular, *$u^{-1}(0)$ is a submodule of E.*

DEFINITION 3. Given an A-linear mapping $u: E \to F$, the submodule $u^{-1}(0)$ of E is called the *kernel* of u, and is denoted $\mathrm{Ker}(u)$ or $\mathrm{Ker}\, u$; the submodule $u(E)$ of F is called the *image* of u, and is denoted $\mathrm{Im}(u)$ or $\mathrm{Im}\, u$; the quotient module $E/\mathrm{Ker}\, u$ is called the *coimage* of u, and is denoted $\mathrm{Coim}(u)$ or $\mathrm{Coim}\, u$; the quotient module $F/\mathrm{Im}\, u$ is called the *cokernel* of u, and is denoted $\mathrm{Coker}(u)$ or $\mathrm{Coker}\, u$. ▮

Notice that if $i: M \to E$ is the canonical injection of the submodule M into the A-module E, then $\mathrm{Coker}(i)$ is $E/i(M) = E/M$; in particular, if $u: E \to F$ is a linear mapping, then $E/\mathrm{Ker}\,(u) = \mathrm{Coim}(u)$ is the cokernel of the canonical injection $i: \mathrm{Ker}(u) \to E$ of the kernel of E; one says, by an abuse of language, that *the coimage is the cokernel of the kernel.* Also, if N is a submodule of the A-module F and the canonical surjection $p: F \to F/N$ is considered, the kernel of p is N: for, the zero element of F/N is the coset of 0, which as a set coincides with N; in particular, if $u: E \to F$ is an A-linear mapping, the kernel of the canonical surjection $p: F \to F/\mathrm{Im}(u) = \mathrm{Coker}(u)$ of F onto the cokernel is $\mathrm{Im}(u)$; one also says, by an abuse of language, that *the image is the kernel of the cokernel.*

PROPOSITION 1. *Let A be a ring, and let $u: E \to F$ be an A-homomorphism; if i is the canonical injection of* $\mathrm{Ker}\, u$ *into E, then $u \circ i = 0$, and we have the universal property: for any A-module X and any $f \in \mathrm{Hom}_A(X, E)$ for which $u \circ f = 0$ there exists a homomorphism $g: X \to$* Ker *u unique with the property of making the following diagram commute:*

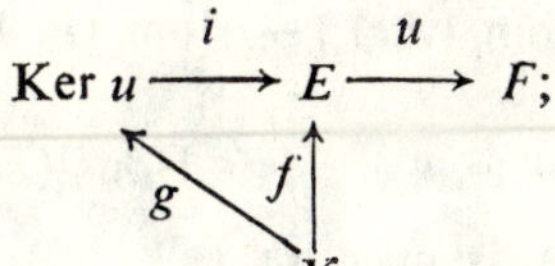

if p is the canonical surjection $F \to$ Coker *u, then $p \circ u = 0$, and the pair* $(\mathrm{Coker}(u), p)$ *is universal with this property, i.e., for any A-module Y and any linear mapping $h: F \to Y$ such that $h \circ u = 0$ there exists a linear mapping j:* Coker *$u \to Y$ uniquely determined by the property of rendering commutative the diagram*

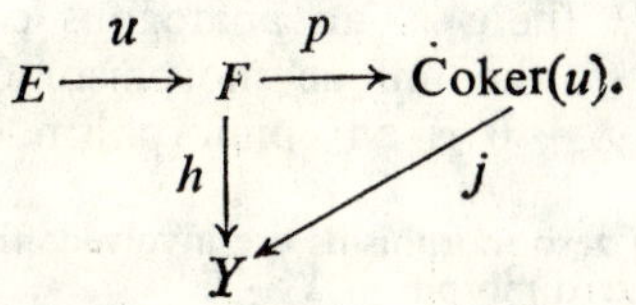

Proof. To prove the first part, let $x \in \mathrm{Ker}(u)$; then $(u \circ i)\, x = u(x) = 0$, hence $u \circ i = 0$; if X and f are as above, then $f(X) \subseteq \mathrm{Ker}(u)$, since for any $t \in X$ we have $u(f(t)) = 0$; according to the universal property of the submodule, we have the required result.

For the second part, if $x \in E$, then $u(x) \in \mathrm{Im}(u) = \mathrm{Ker}(p)$; hence $p(u(x)) = 0$, and $p \circ u = 0$; given Y and h as above, if $y_1 \equiv y_2 \pmod{\mathrm{Im}(u)}$, with $y_1, y_2 \in F$, we have $y_1 - y_2 = u(x)$ for some $x \in E$, hence $h(y_1 - y_2) = (h \circ u)(x) = 0$, and $hy_1 = hy_2$; according to the universal property of the quotient module, we get the desired result. ▮

* DEFINITION 4. Let $\mathcal{C}$ be a category; a morphism $u \in \mathrm{Hom}_{\mathcal{C}}(E, F)$ is said to be a *monomorphism* of (or in) the category $\mathcal{C}$ if for any object X of $\mathcal{C}$ and any pair of morphisms of the category, $X \underset{v_2}{\overset{v_1}{\rightrightarrows}} E$, such that $u \circ v_1 = u \circ v_2$ we have $v_1 = v_2$; briefly, if u may be cancelled on the left. A morphism $u \in \mathrm{Hom}_{\mathcal{C}}(E, F)$ is said to be an *epimorphism* of (or in) the category $\mathcal{C}$ if for any $Y \in \mathrm{Ob}\,\mathcal{C}$ and any $w_1, w_2 \in \mathrm{Hom}_{\mathcal{C}}(F, Y)$ such that $w_1 \circ u = w_2 \circ u$ we have $w_1 = w_2$; briefly, if u may be cancelled on the right. ▮

It is readily seen that $u \in \mathrm{Hom}_{\mathcal{C}}(E, F)$ is a monomorphism if and only if for any $X \in \mathrm{Ob}\,\mathcal{C}$, the mapping

$$\mathrm{Hom}_{\mathcal{C}}(X, E) \to \mathrm{Hom}_{\mathcal{C}}(X, F)$$

given by

$$v \mapsto u \circ v, \qquad v \in \mathrm{Hom}_{\mathcal{C}}(X, E),$$

i.e. the mapping $\mathrm{Hom}(1_X, u)$, is injective; also, $u \in \mathrm{Hom}_{\mathcal{C}}(E, F)$ is an epimorphism if and only if for any $Y \in \mathrm{Ob}\,\mathcal{C}$, the mapping

$$\mathrm{Hom}_{\mathcal{C}}(F, Y) \to \mathrm{Hom}_{\mathcal{C}}(E, Y)$$

given by

$$w \mapsto w \circ u, \qquad w \in \mathrm{Hom}_{\mathcal{C}}(F, Y),$$

i.e. the mapping $\mathrm{Hom}(u, 1_Y)$, is injective.

Any isomorphism u of the category $\mathcal{C}$ is both a monomorphism and an epimorphism: for if $u \circ v_1 = u \circ v_2$, then the composition of both sides with u^{-1} gives $v_1 = v_2$; also, the composition of two monomorphisms, respectively epimorphisms, is a monomorphism, respectively an epimorphism; finally, if $u \circ v$ is a monomorphism, then v is a monomorphism, and if $u \circ v$ is an epimorphism, then u is an epimorphism.

If $\mathcal{C}$ is a preadditive category, the fact that $u\colon E \to F$ is a monomorphism amounts to the fact that for any $X \in \mathrm{Ob}\,\mathcal{C}$ and any $v \in \mathrm{Hom}_{\mathcal{C}}(X, E)$ such that $u \circ v = 0$ we have $v = 0$;[1] there is an analogous condition for epimorphisms. Note that for any $E \in \mathrm{Ob}\,\mathcal{C}$, the unique morphism $0 \to E$ is a monomorphism, and the unique morphism $E \to 0$ is an epimorphism.

[1] Of course, two different zero morphisms are involved: the first one is the zero morphism $X \to F$, while the second is the zero morphism $X \to E$.

REMARK. If u is a monomorphism of the category $\mathcal{C}$, then the doublet u° of u in the dual category $\mathcal{C}^\circ$ (see Definition 4, § 2) is an epimorphism of $\mathcal{C}^\circ$, and conversely: if u is an epimorphism of $\mathcal{C}$, u° is a monomorphism of $\mathcal{C}^\circ$; one says that the concept of epimorphism is *dual* to that of monomorphism, and conversely. Generally, the concept, respectively property, *dual* to another concept, respectively property, is the corresponding concept, respectively property, in the dual category, carried back to the given category by returning to the initial objects; roughly speaking, the dual notion is obtained by "reversing the arrows", since a morphism $u^\circ \colon E^\circ \to F^\circ$ of $\mathcal{C}^\circ$ is obtained as a doublet of a morphism $u \colon F \to E$ of $\mathcal{C}$. ▮ *

The monomorphisms and epimorphisms of the category **Mod**(A) are characterized as follows:

PROPOSITION 2. *Let A be a ring and $u \in \mathrm{Hom}_A(E, F)$, with E and F, A-modules; the following assertions are equivalent*:

i) u is an injective homomorphism;

ii) u is a monomorphism of the category **Mod**(A);

iii) $\mathrm{Ker}(u) = 0$.

The following statements are also equivalent:

iv) u is a surjective homomorphism;

v) u is an epimorphism of the category **Mod**(A);

vi) $\mathrm{Coker}\,(u) = 0$.

Proof. i) $\Rightarrow$ *ii)*: Let $X \underset{v_2}{\overset{v_1}{\rightrightarrows}} E$ be two homomorphisms such that $u \circ v_1 = u \circ v_2$; if $t \in X$, then

$$u(v_1(t)) = u(v_2(t)),$$

and since u is injective, $v_1(t) = v_2(t)$; hence $v_1 = v_2$.

ii) $\Rightarrow$ *iii)*: Let us consider the canonical injection $i \colon \mathrm{Ker}(u) \to E$; according to Proposition 1 we have

$$u \circ i = 0 = u \circ 0,$$

hence, as u is a monomorphism, $i = 0$; but then we have for any $x \in \mathrm{Ker}\, u$,

$$x = i(x) = 0(x) = 0,$$

and $\mathrm{Ker}\,(u)$ consists only of its zero element.

iii) $\Rightarrow$ *i)*: If $u(x) = u(y)$ for $x, y \in E$, then $u(x - y) = 0$, and $x - y \in \mathrm{Ker}\,(u) = 0$; hence $x = y$, and u is injective.

The equivalences *iv)* $\Leftrightarrow$ *v)* $\Leftrightarrow$ *vi)* are analogously proved and are left to the reader as an exercise. ▮

It is now clear why v is unique with the property of making diagram (2) commute: i being injective, it is a monomorphism, and since $i \circ v' = u$ implies

$$i \circ v = i \circ v',$$

we have $v = v'$; a similar argument holds for v rendering diagram (5) commutative.

We mention also the fact that, given the A-homomorphisms

$$E \xrightarrow{u} F \xrightarrow{v} G,$$

where v is a monomorphism, then

$$\operatorname{Ker}(v \circ u) = \operatorname{Ker}(u);$$

for, the inclusion $\operatorname{Ker}(u) \subseteq \operatorname{Ker}(v \circ u)$ is obviously generally true, while the inverse one is a consequence of the fact that if $(v \circ u)\,x = 0$, with $x \in E$, then, v being injective, we have $u(x) = 0$.

Similarly, in the same situation, if u is an epimorphism, then

$$\operatorname{Im}(v \circ u) = \operatorname{Im}(v);$$

the inclusion $\operatorname{Im}(v \circ u) \subseteq \operatorname{Im}(v)$ is generally valid, while the inverse one follows from the fact that if $z \in G$ takes on the form $z = v(y)$, with $y \in F$, then, u being surjective, y is of the form $y = u(x)$ for some $x \in E$; whence $z = v(u(x))$. ∎

* The notions of submodule and quotient module are recovered in the Theory of Categories as follows: from the characteristic properties of the notion of submodule only that of being a module endowed with the canonical injection is considered; the notion of injective mapping making no sense for an arbitrary category, it is recovered by the notion of monomorphism according to Proposition 2 above; similarly for the quotient module; more precisely, one proceeds as follows:

Let $\mathscr{C}$ be a category; let $E \in \operatorname{Ob} \mathscr{C}$; given two pairs (M, i), (M', i'), where M, M' are objects of $\mathscr{C}$ and $i\colon M \to E$, $i'\colon M' \to E$ are monomorphisms of range E, these two pairs are said to be *equivalent* provided there exists some morphisms $v\colon M \to M'$, $v'\colon M' \to M$ rendering the diagrams

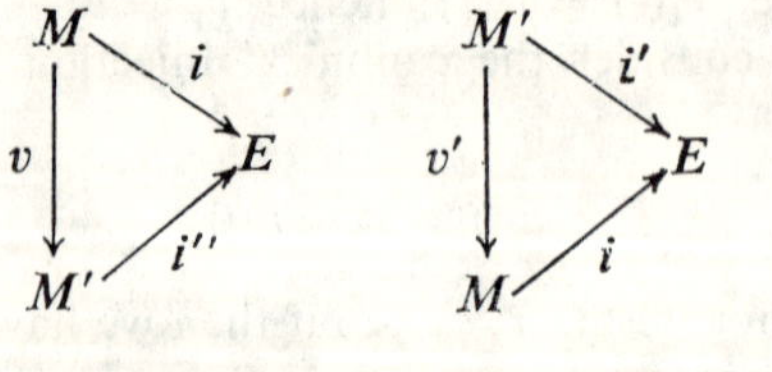

commutative. The morphisms v, v' turn out to be isomorphisms inverse to each other and the defined relation to be indeed an equivalence relation on the class of these pairs. Proceeding in the same way for the pairs (P, p), where P is an object and p an epimorphism $E \to P$, we define an equivalence relation dual to the first one: two pairs (P, p) and (P', p') as above which are in this relation with each other are said to be, as in the dual case, *equivalent*.

Definition 5. A *subobject* of the object E of the category $\mathscr{C}$ is an equivalence class of pairs (M, i) as above; sometimes, instead of the whole class, by a subobject of E only a representative (M, i) of such a class is meant. A *quotient object* of the object E of the category $\mathscr{C}$ is an equivalence class of pairs (P, p) as above; sometimes, instead of the whole class, by a quotient object of E only a representative (P, p) of this class is meant. ∎*

§ 4. Kernels, Cokernels, the Canonical Factorization of a Morphism

* The concepts of kernel, cokernel, etc. are recovered in a preadditive category as follows:

DEFINITION 1. Let $\mathscr{C}$ be a preadditive category, let $E, F \in \mathrm{Ob}\,\mathscr{C}$ and $u \in \mathrm{Hom}_{\mathscr{C}}(E, F)$.

A *kernel* of the morphism u (in the category $\mathscr{C}$) is a pair (M, i) consisting of an object M of $\mathscr{C}$ and a morphism $i : M \to E$ such that the following conditions are satisfied:

a) $u \circ i = 0$;

b) for any object X of $\mathscr{C}$ and any $f : X \to E$ such that $u \circ f = 0$ there exists a unique morphism $g : X \to M$ such that $i \circ g = f$.

A *cokernel* of the morphism u (in the category $\mathscr{C}$) is a pair (P, p), where P is an object of $\mathscr{C}$ and p is a morphism $F \to P$, this pair having the properties:

c) $p \circ u = 0$;

d) for any object Y of $\mathscr{C}$ and any morphism $h : F \to Y$ such that $h \circ u = 0$ there exists a unique morphism $j : P \to Y$ such that $j \circ p = h$.

An *image* of the morphism u (in the category $\mathscr{C}$) is a kernel (in the category $\mathscr{C}$) of the morphism p, where (P, p) is a cokernel of u. A *coimage* of the morphism u (in the category $\mathscr{C}$) is a cokernel of the morphism i, where (M, i) is a kernel of u. ▮

From the definition it follows that if the pair (M, i) is a kernel of the morphism u, then i is a monomorphism, hence (M, i) defines a subobject of E; it is easily seen that two pairs (M, i), (M', i') which are kernels of u are equivalent in the sense of Definition 5, § 3; thus, the subobject determined by the kernel is unique; the term "kernel" is sometimes used to designate such a class of equivalent pairs (M, i); even when considering only representatives, on the basis of the mentioned equivalence we say, by an abuse of language, that *there exists a unique kernel* (provided it exists), and one refers accordingly to "*the kernel*" of the morphism u and not to "a kernel"; either the class or a pair (M, i) chosen as representative (or even, by an abuse of language, only the object M of) it is denoted $\mathrm{Ker}(u)$ or $\mathrm{Ker}\, u$. Similarly, it may be shown that if the pair (P, p) is a cokernel of the morphism u, then p is an epimorphism, and all pairs with this property are equivalent in the sense of Definition 5, § 3, thus defining a unique quotient object of F; this equivalence class or a representative (P, p) of it or, by an abuse of language, just the object P of such a representative is called *the cokernel* of u, and is denoted $\mathrm{Coker}(u)$ or $\mathrm{Coker}\, u$. Moreover, if two pairs (P, p), (P', p') are cokernels of u, then any pair (I, j) which is a kernel of p is equivalent to any pair (I', j') which is a kernel of p';

therefore we have only one class of such pairs, which is a subobject of F, and this class or a representative (I, j) of it or even just the object I of such a representative is called *the image* of u, and is denoted $\mathrm{Im}(u)$ or $\mathrm{Im}\, u$. This is also valid for coimages, and the unique class (which is a quotient object of E) or a representative of it or just the object of such a representative is called *the coimage* of u, and is denoted $\mathrm{Coim}\,(u)$ or $\mathrm{Coim}\, u$.

REMARK. The kernel and the cokernel are notions *dual* to each other (see the Remark following Definition 4, § 3); the image and the coimage are also concepts dual to each other. ▮

It follows directly from the definition that if the morphism $u : E \to F$ of the preadditive category $\mathcal{C}$ is a monomorphism, then the pair $(0, 0)$ (the first 0 is the zero object of the category, while the second is the unique monomorphism $0 \to E$) is a kernel of u; conversely, if a morphism u has $(0, 0)$ as kernel, then u is a monomorphism. By duality, $u : E \to F$ is an epimorphism if and only if $(0, 0)$ (the second 0 is the unique morphism $F \to 0$) is a cokernel of u. In this sense, we may talk of the validity of the equivalence of assertions *ii*) and *iii*) of Proposition 2, § 3; similarly, of the equivalence of assertions *v*), *vi*) of the same Proposition.

We also notice that the morphism $0 : E \to F$ has $(E, 1_E)$ as kernel and $(F, 1_F)$ as cokernel; in particular, the unique morphism $0 \to F$ has $(0, 0)$ as kernel and $(F, 1_F)$ as cokernel, while the unique morphism $E \to 0$ has $(E, 1_E)$ as kernel and $(0, 0)$ as cokernel.*

From Proposition 1, § 3, and the considerations preceding it, the kernel, the cokernel, the image, and the coimage defined for modules (Definition 3, § 3) with the corresponding canonical homomorphisms (injections, respectively surjections) turn out to be instances of the homonymous notions in the category **Mod**(A), in the sense of Definition 1.

* Let $\mathcal{C}$ be a preadditive category satisfying the Axiom:

AB1) *For any morphism u of the category*[1] $\mathcal{C}$ *there exists in the category* $\mathcal{C}$ *a kernel and a cokernel, in the sense of Definition* 1.

Then any morphism has an image and a coimage.

Let $u \in \mathrm{Hom}_{\mathcal{C}}(E, F)$; let (M, i) be a kernel of u, (P, p) a cokernel of u, (C, q) a coimage of u, and (I, j) an image of u (see the diagram below); we have $u \circ i = 0$, hence, as (C, q) is a cokernel of i, there exists $v : C \to F$ uniquely determined by the property that

$$v \circ q = u;$$

on the other hand, we have

$$0 = p \circ u = p \circ v \circ q;$$

but since q is an epimorphism, it follows that $p \circ v = 0$, hence, as (I, j) is a kernel of p, we get a uniquely determined morphism $\bar{u} : C \to I$ such that

$$j \circ \bar{u} = v;$$

---

[1] We formulate it like this instead of using the more precise statement: "For any pair $E, F \in \mathrm{Ob}\, \mathcal{C}$ and any $u \in \mathrm{Hom}_{\mathcal{C}}(E, F)$ ...".

the situation is completely described by saying that the following diagram commutes:

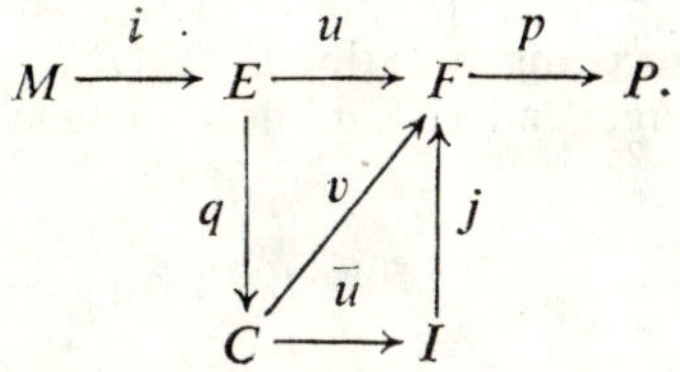

Consequently, we have

(1) $$u = j \circ \bar{u} \circ q,$$

and $\bar{u}$ is uniquely determined by this relation, taking into account that j is a monomorphism and q is an epimorphism.

By an abuse of language [1], $\bar{u}$ is called the *canonical morphism* from the coimage of u to the image of u; the factorization (1) is called a *canonical factorization* or also, if we fix the considered pairs, *the canonical factorization* of the morphism u.

If (C', q'), respectively (I', j') is another coimage, respectively image of u, and if for this choice we denote by $\bar{u}'$ the canonical morphism $C' \to I'$, then there exist isomorphisms $l : C \xrightarrow{\sim} C'$, $m_1 : I' \xrightarrow{\sim} I$ such that

$$l \circ q = q', \qquad j \circ m_1 = j';$$

but then the following diagram is commutative:

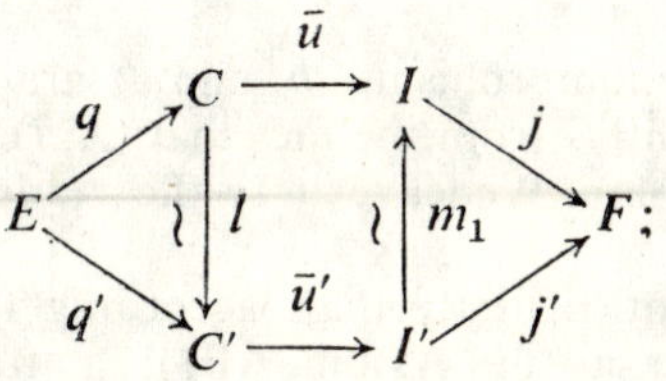

indeed, we have

$$j \circ m_1 \circ \bar{u}' \circ l \circ q = j' \circ \bar{u}' \circ q' = u = j \circ \bar{u} \circ q,$$

hence

$$m_1 \circ \bar{u}' \circ l = \bar{u};$$

hence $\bar{u}$ is obtained from $\bar{u}'$ by composition with isomorphisms.

---

[1] The abuse is committed twice: firstly, we considered a particular image and a particular coimage of u; it would have been more adequate to say that u is *the canonical morphism associated with the given coimage and image*; secondly, we identified the image and the coimage with the component objects of the considered pairs, respectively.

We notice that, since the kernel and the cokernel are notions dual to each other, if a preadditive category $\mathcal{C}$ satisfies Axiom AB1), then its dual $\mathcal{C}^\circ$ also satisfies AB1); if $u^\circ : F^\circ \to E^\circ$ is a morphism in $\mathcal{C}^\circ$, namely the doublet of $u : E \to F$ in $\mathcal{C}$, then, in the preceding notation for u, (P°, p°) is a kernel, (M°, i°) is a cokernel, (I°, j°) is a coimage, and (C°, q°) is an image of u° in the category $\mathcal{C}^\circ$; therefore, as

$$u^\circ = q^\circ \circ \bar{u}^\circ \circ j^\circ,$$

it follows that $\bar{u}^\circ$ is the canonical morphism from the coimage of u° to the image of u°.

If in the preadditive category $\mathcal{C}$ satisfying Axiom AB1), *for each monomorphism $u : E \to F$, $E, F \in \mathrm{Ob}\,\mathcal{C}$, there exists some morphism $f : F \to M$, $M \in \mathrm{Ob}\,\mathcal{C}$, such that (E, u) is a kernel of f, then for each monomorphism $u : E \to F$, the fact that (G, v) is a cokernel of u implies that (E, u) is a kernel of v* : since let $f : F \to M$ be such that (E, u) is a kernel of f; let (K, i) be a kernel of v; since $f \circ u = 0$ and (G, v) is a cokernel of u, it follows by the definition of the cokernel that there exists $g : G \to M$ such that $g \circ v = f$; we have $f \circ i = g \circ v \circ i = 0$, hence, by the definition of the kernel, there exists $k : K \to E$ such that $u \circ k = i$; since $v \circ u = 0$, it follows, again by the definition of the kernel, that there exists $h : E \to K$ such that $u = i \circ h$; the situation is displayed on the following diagram:

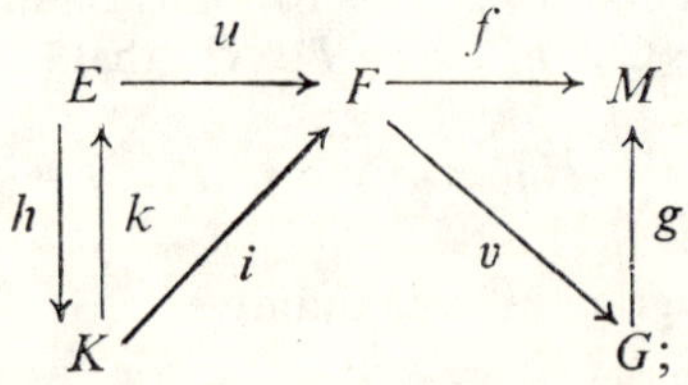

now, since u and i are monomorphisms, h and k are isomorphisms inverse to each other, and, consequently, from the fact that (K, i) is a kernel of v we obtain that (E, u) also is a kernel of v. We leave it to the reader to formulate and prove the dual result.*

If A is a ring, we notice firstly that, according to the arguments exhibited in § 3, the category **Mod**(A) satisfies Axiom AB1); in attempting to apply the above general considerations to the category **Mod**(A), let E and F be A-modules and u a homomorphism $E \to F$; taking $u^{-1}(0)$ for M, $u(E)$ for I, E/M for C, $F/u(E)$ for P, and for i, q, j, p the canonical injections, respectively surjections, as in § 3, we see that $\bar{u}$ (called in this case the *canonical homomorphism* from the coimage of u to the image of u) acts as follows: if $\hat{x} \in E/M$, then

$$\bar{u}(\hat{x}) = j(\bar{u}(\hat{x})) = j(\bar{u}(q(x))) = u(x);$$

if $y \in u(E)$, then y is of the form $u(x) = \bar{u}(x)$ with some $x \in E$; hence $\bar{u}$ is surjective; if $\bar{u}(\hat{x}) = 0$, then $u(x) = 0$, and so $x \in u^{-1}(0) = M$, thus $\hat{x} = 0$, and $\bar{u}$ is injective (Proposition 2, § 3); hence $\bar{u}$ is an A-isomorphism (Proposition 1, § 1).

Thus, *for modules the canonical homomorphism* Coim$(u) \to$ Im(u) *is an isomorphism.*

* Let $\mathcal{C}$ be a preadditive category satisfying Axiom AB1); assume that the following Axiom is also satisfied:

AB2) *For any morphism u of the category $\mathcal{C}$, the canonical morphism $\bar{u}$ from the coimage of u to the image of u is an isomorphism*[1].

Then *any morphism of the category $\mathcal{C}$ which is both a monomorphism and an epimorphism*[2] *is an isomorphism*: for let

$$E \xrightarrow{q} C \xrightarrow{\bar{u}} I \xrightarrow{j} F$$

be a canonical factorization of u; since u is a monomorphism, Ker(u) is the pair $(0, 0)$; hence we can take $(E, 1_E)$ as an image of u (cokernel of 0); q being obtained from 1_E by composition with an isomorphism, it is itself an isomorphism; by dual considerations it follows that j is an isomorphism too; therefore u is also an isomorphism.

Let $\mathcal{C}$ be again a preadditive category satisfying Axioms AB1) and AB2); let $E, F \in \text{Ob}\, \mathcal{C}$, let $u \in \text{Hom}_{\mathcal{C}}(E, F)$, and let (M, i), (P, p), (C, q), (I, j), and $\bar{u}$ have their previous meanings; suppose u admits also another factorization

$$u = j' \circ \bar{u}' \circ q',$$

where j', $\bar{u}'$, q' are morphisms in the situation displayed on the diagram[3]

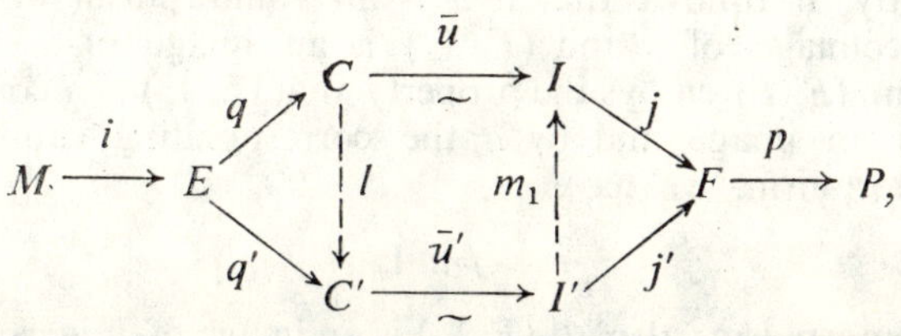

and where we assume that q' is an epimorphism, $\bar{u}'$ an isomorphism, and j' a monomorphism.

We shall show that *in these conditions, (C', q'), respectively (I', j') is a coimage, respectively an image of u*: indeed, we have

$$j' \circ \bar{u}' \circ q' \circ i = u \circ i = 0,$$

---

[1] It would have been more adequate to put it like this: "For any morphism u there exists a coimage (C, q) and an image (I, j) such that the canonical morphism corresponding to them is an isomorphism"; nevertheless, in view of the above considerations, for any other coimage and any other image, the corresponding canonical morphism is also an isomorphism.

[2] Such a morphism is called a *bimorphism* of the given category.

[3] This diagram is filled in at two different moments: the first one, corresponding to the moment where it is placed, when the solid arrows are filled in; the second one, later on, when the morphisms represented by the dashed arrows are filled in; that is, at this very moment only the vertices and solid arrows are to be considered; this convention will be constantly put into effect in the sequel.

and hence, as $j' \circ \bar{u}'$ is a monomorphism, $q' \circ i = 0$; (C, q) being a cokernel of i, the existence of a morphism $l : C \to C'$ uniquely determined by the property $l \circ q = q'$ follows; q' being an epimorphism, l is an epimorphism too; by dual considerations we get the existence of a unique morphism $m_1 : I' \to I$ with the property that $j \circ m_1 = j'$, and this is a monomorphism; as above,

$$m_1 \circ \bar{u}' \circ l = \bar{u};$$

since $\bar{u}$ is a monomorphism and an epimorphism, l turns out to be a monomorphism and m_1 an epimorphism; hence l and m_1 are isomorphisms; but then, since $q' \circ i = 0$, and since for $\xi : E \to X$, $X \in \text{Ob}\,\mathcal{C}$, with $\xi \circ i = 0$ we can find out a unique morphism $\eta : C \to X$ with the property that $\eta \circ q = \xi$ and consequently the (unique) morphism $\eta \circ l^{-1} : C' \to X$ such that

$$\eta \circ l^{-1} \circ q' = \eta \circ q = \xi,$$

it follows that (C', q') is a cokernel of i, hence a coimage of u; likewise, (I', j') is an image of u.

Hence, it follows that if $u \in \text{Hom}_{\mathcal{C}}(E, F)$ is a monomorphism of the category $\mathcal{C}$, then (E, u) is an image of u and $(E, 1_E)$ is a coimage of u: for, u admits the trivial factorization

$$u = u \circ 1_E \circ 1_E,$$

where 1_E is successively an isomorphism and an epimorphism while u is a monomorphism; similarly, it follows that if u is an epimorphism of the category $\mathcal{C}$, then (F, u) is a coimage of u and $(F, 1_F)$ is an image of u. Conversely, if a morphism $u \in \text{Hom}_{\mathcal{C}}(E, F)$ enjoys the property that $(E, 1_E)$ is a coimage of it, then, denoting by (I, j) an image and by $\bar{u}$ the corresponding canonical morphism, which is an isomorphism, we have

$$u = j \circ \bar{u} \circ 1_E,$$

hence u is a monomorphism; also, if $(F, 1_F)$ is an image of u, u is an epimorphism; the fact that (E, u) is an image of u trivially implies that u is a monomorphism, since in a pair giving the image, the morphism is always a monomorphism; similarly, the fact that (F, u) is a coimage of u implies the fact that u is an epimorphism.

We also obtain that *if* $u : E \to F$ *is a monomorphism, and if* (I, p) *is a cokernel of it, then* (E, u) *is a kernel of* p, since it is an image of u; also, *if* $u : E \to F$ *is an epimorphism and* (M, i) *is a kernel of it, then* (F, u) *is a cokernel of* i.

Notice also that, since the doublet in the dual category of an isomorphism of $\mathcal{C}$ is an isomorphism of $\mathcal{C}^\circ$, and since the doublet of the canonical morphism $\text{Coim}(u) \to \text{Im}(u)$ is the canonical morphism $\text{Coim}(u^\circ) \to \text{Im}(u^\circ)$ (for a given choice of the coimage and of the image), it follows that *the dual of a preadditive category* (in the sense of § 2) *satisfying* AB1) *and* AB2) *is a preadditive category satisfying* AB1) *and* AB2).*

From the preceding considerations it follows that if A is a ring, the category **Mod**(A), which is a preadditive category satisfying AB1), also satisfies AB2); the above results put together give:

PROPOSITION 1. *If A is a ring and $u \in \mathrm{Hom}_A(E, F)$, in the canonical factorization* (I) *of u (with the above meaning of notations), the homomorphism $\bar{u}: E/u^{-1}(0) \to$ $\to u(E)$ is an isomorphism; if a factorization of u is given,*

$$\begin{array}{ccc} E & \xrightarrow{\quad u \quad} & F \\ {\scriptstyle q'}\downarrow & & \uparrow{\scriptstyle j'} \\ C' & \xrightarrow{\quad \bar{u}' \quad} & I' \end{array}$$

(the diagram being commutative), with j' an injective homomorphism, $\bar{u}'$ an isomorphism, q' a surjective homomorphism, then there exist two isomorphisms

$$l : C = E/u^{-1}(0) \xrightarrow{\sim} C', \; m : I = u(E) \xrightarrow{\sim} I'$$

which as homomorphisms are unique with the properties

$$l \circ q = q', \qquad j' \circ m = j;$$

moreover, the pair (C', q') is a coimage of u (in the sense of Definition 1) in **Mod**(A), *and the pair (I', j') an image.* ▮

The fact that for any homomorphism of A-modules $u : E \to F$ there exists an isomorphism $\bar{u} : C \xrightarrow{\sim} I$ such that $j \circ \bar{u} \circ q = u$ (where C, I, q, j have the above precise meaning, i.e., $C = E/u^{-1}(0)$, etc.) represents the **First Isomorphism Theorem for modules.**

COROLLARY 1. *Let A be a ring and $u : E \to F$ a homomorphism of A-modules; there exists a unique A-homomorphism $\hat{u} : E/\mathrm{Ker}\,(u) \to F$ such that $\hat{u} \circ q = u$, q being the canonical surjection $E \to E/\mathrm{Ker}(u)$; in the general case, $\hat{u}$ is injective; if u is surjective, then $\hat{u}$ is an isomorphism.*

Proof. In the preceding notation, we have only to put

$$\hat{u} = j \circ \bar{u};$$

the last assertion follows from the fact that, provided u is surjective, we have $u(E) = F$, so that $j = 1_F$ and $\hat{u} = \bar{u}$. ▮

COROLLARY 2. *Let A be a ring, and let $u: E \to F$ be a homomorphism of A-modules; there exists a unique A-homomorphism $u' : E \to u(E)$ such that $j \circ u' = u$, j being the canonical injection $u(E) \to F$; in the general case, u' is surjective; if u is injective, then u' is an isomorphism.*

Proof. We have only to set, in the preceding notation, $u' = \bar{u} \circ q$; the last statement is a consequence of the fact that $\mathrm{Ker}(u) = 0$ provided u is injective, and hence $q : E \to E/\mathrm{Ker}(u)$ is injective according to the arguments of § 3. ▮

Sometimes, the surjective homomorphism $u' = \bar{u} \circ q$ will be called the *canonical surjection* $E \rightarrow \mathrm{Im}(u)$.

Remark. The existence of the homomorphisms $\hat{u}$ and u' given by Corollaries 1 and 2 is a particular case of the operations of passing to the quotient and restriction given in § 3; indeed, in the first case, we have $u(\mathrm{Ker}(u)) = 0$, and hence we get the homomorphism $E/\mathrm{Ker}(u) \rightarrow F/0$; but $F/0$ is canonically isomorphic to F, and we have only to take the composition of the two mappings to get $\hat{u}$; in the second case, we have trivially $u(E) \subseteq u(E)$, and u' is obtained by restriction. ▮

If $u : E \rightarrow F$ is an A-homomorphism, the image under u of a submodule of E and the inverse image under u of a submodule of F may be characterized in terms of the kernel and of the image of a homomorphism in the following way[1]:

Let M be a submodule of E; then, to say that $y \in F$ is an element of $u(M)$ is equivalent to the statement that y is of the form

$$y = u(i(x)) = (u \circ i)(x),$$

with $x \in M$, where i is the canonical injection of M into E; hence $u(M) = \mathrm{Im}(u \circ i)$.

In particular, if M is $\mathrm{Im} f$, where $f \in \mathrm{Hom}_A(E', E)$, then $u(\mathrm{Im} f) = \mathrm{Im}(u \circ i) = \mathrm{Im}(u \circ f)$, since f admits, according to Corollary 2, the factorization $f = i \circ f'$, with f' surjective (see § 3).

Let N be a submodule of F; if $x \in E$, to say that x is an element of $u^{-1}(N)$ amounts to $u(x) \in N$ or, denoting by q the canonical surjection of F onto F/N, it is equivalent to the fact that $0 = q(u(x)) = (q \circ u)(x)$; hence $u^{-1}(N) = \mathrm{Ker}(q \circ u)$.

In particular, if N is of the form $N = \mathrm{Ker}\, g$, with $g \in \mathrm{Hom}_A(F, F')$, then $u^{-1}(N) = \mathrm{Ker}(q \circ u) = \mathrm{Ker}(g \circ u)$, since according to Corollary 1, g admits the factorization $g = \hat{g} \circ q$, with $\hat{g}$ injective.

* Let $\mathcal{C}$ be a preadditive category satisfying Axioms AB1) and AB2); let $E, F, E', F' \in \mathrm{Ob}\, \mathcal{C}$, and let the commutative diagram in $\mathcal{C}$ be given

$$\begin{array}{ccc} E & \xrightarrow{u} & F \\ {\scriptstyle f}\downarrow & & \downarrow{\scriptstyle g} \\ E' & \xrightarrow{u'} & F'. \end{array} \tag{2}$$

Let (M, i), respectively (M', i'), be a kernel of u, respectively u'; let (C, q), respectively (C', q'), be a coimage of u, respectively u'; let (I, j), respectively (I', j'), be an image of u, respectively u'; let (P, p), respectively (P', p'), be a cokernel

---

[1] See also the end of § 8, where the corresponding notions for (abelian) categories are defined.

of u, respectively u'; we have then the following diagram in $\mathscr{C}$ (here $\bar{u}$, respectively $\bar{u}'$ represent the corresponding canonical morphisms):

$$
\begin{array}{ccccccccccc}
M & \xrightarrow{i} & E & \xrightarrow{q} & C & \xrightarrow{\bar{u}} & I & \xrightarrow{j} & F & \xrightarrow{p} & P \\
\downarrow{\scriptstyle f^*} & & \downarrow{\scriptstyle f} & & \downarrow{\scriptstyle \hat{f}} & & \downarrow{\scriptstyle g^*} & & \downarrow{\scriptstyle g} & & \downarrow{\scriptstyle \hat{g}} \\
M' & \xrightarrow{i'} & E' & \xrightarrow{q'} & C' & \xrightarrow{\bar{u}'} & I' & \xrightarrow{j'} & F' & \xrightarrow{p'} & P'.
\end{array}
\tag{3}
$$

As $j'\circ\bar{u}'\circ q'\circ f\circ i = u'\circ f\circ i = g\circ u\circ i = 0$, it follows $q'\circ f\circ i = 0$, since $j'\circ\bar{u}'$ is a monomorphism; (C, q) being a cokernel of i, there exists a unique morphism $\hat{f}: C \to C'$ such that

$$\hat{f}\circ q = q'\circ f; \tag{4}$$

also, as $u'\circ f\circ i = 0$, there exists $f^* : M \to M'$ unique with the property

$$i'\circ f^* = f\circ i; \tag{5}$$

by dual considerations, there exist some morphisms $g^* : I \to I'$ and $\hat{g}: P \to P'$ uniquely determined by the properties

$$j'\circ g^* = g\circ j, \text{ respectively } p'\circ g = \hat{g}\circ p. \tag{6}$$

But then we also have

$$j'\circ\bar{u}'\circ\hat{f}\circ q = j'\circ\bar{u}'\circ q'\circ f = u'\circ f = g\circ u = g\circ j\circ\bar{u}\circ q = j'\circ g^*\circ\bar{u}\circ q;$$

j' being a monomorphism and q an epimorphism, it follows

$$\bar{u}'\circ\hat{f} = g^*\circ\bar{u}; \tag{7}$$

relations (4), (5), (6), (7) show that diagram (3) is commutative.

If we also consider another commutative diagram of the same type as (2),

$$
\begin{array}{ccc}
E' & \xrightarrow{u'} & F' \\
\downarrow{\scriptstyle h} & & \downarrow{\scriptstyle l} \\
E'' & \xrightarrow{u''} & F'',
\end{array}
$$

and if we do for u'' what was done for u and u', denoting by $h^*, h, l^*, \hat{l}, (h\circ f)^*, \widehat{(h\circ f)}, (l\circ g)^*, \widehat{(l\circ g)}$ the morphisms which correspond in an obvious way to $h, l, h\circ f, l\circ g,$

respectively, then one can show that

(8) $$(h\circ f)^* = h^*\circ f^*,\ \widehat{(h\circ f)} = \hat{h}\circ\hat{f},\ (l\circ g)^* = l^*\circ g^*,\ \widehat{(l\circ g)} = \hat{l}\circ\hat{g};$$

moreover, if $E' = E, F' = F, f = 1_E, g = 1_F$, and if we take $M' = M, i' = i, C' = C, q' = q$, etc., then

(9) $$f^* = 1_M,\ \hat{f} = 1_C,\ g^* = 1_I,\ \hat{g} = 1_P.$$

Relations (8) and (9) express a certain functor character of the assignments $u \mapsto \operatorname{Ker}(u)$, $u \mapsto \operatorname{Coim}(u)$, etc.; this character may be formalized by adequately defining the categories the respective functors are defined on and those they take on their values in; this is done as follows:

Let $\mathscr{C}$ be an arbitrary category; a new category, denoted **Morph**($\mathscr{C}$), is defined taking as objects the morphisms of $\mathscr{C}$, and, if $u : E \to F$, $u' : E' \to F'$ are two such objects, taking as morphisms $u \to u'$ in **Morph** ($\mathscr{C}$) those pairs (f, g) of morphisms of $\mathscr{C}$, $f : E \to E'$, $g : F \to F'$, which render diagram (2) commutative; the composite morphism of $(h, l) : u' \to u''$ as above, with (f, g) is $(h\circ f, l\circ g)$ (note that in order to maintain the validity of the axiom *c*) of Definition 4, § 1, if diagram (2) is rendered commutative by some morphisms u_1, u_1' instead of u, u', respectively, with $(u, u') \neq (u_1, u_1')$, then the morphism $(f, g) : u_1 \to u_1'$ will be considered distinct from the morphism $(f, g) : u \to u')$.

Now, if $\mathscr{C}$ is a preadditive category satisfying AB1) and AB2), then, assigning to the morphism $u \in \operatorname{Hom}_{\mathscr{C}}(E, F)$ the object M, and to the pair (f, g) as above the morphism f^*, where the meaning of the notation is maintained, then we get a functor Ker : **Morph**($\mathscr{C}$) $\to \mathscr{C}$; in the same way the functors Coim, Im, and Coker are obtained, all of them being defined on **Morph**($\mathscr{C}$) and taking values in $\mathscr{C}$.*

For the category **Mod**(A), taking for (M, i), (C, q), (I, j), (P, p), the "determinations" for the kernel, coimage, image, cokernel given in Definition 3, § 3, and the same for (M', i'), etc., we have

$$f(M) \subseteq M' \quad \text{and} \quad g(I) \subseteq I';$$

indeed, if $x \in M = u^{-1}(0)$, we have

$$u'(f(x)) = g(u(x)) = 0,$$

hence $f(x) \in u'^{-1}(0) = M'$; likewise, if $y \in I$, then $y = u(x)$, $x \in E$, and hence

$$g(y) = g(u(x)) = u'(f(x)),$$

and $g(y) \in u'(E') = I'$. Then, from the commutativity of diagram (3) it follows that f^*, respectively g^* are the homomorphisms induced by f, respectively g by restriction, and that $\hat{f}$, respectively $\hat{g}$ are the homomorphisms induced by f, respectively g by passing to the quotient.

PROPOSITION 2 (**Third Isomorphism Theorem for modules**). *Let A be a ring and $u : E \to F$ a surjective A-homomorphism; if $M = u^{-1}(0)$, and if we denote* *by $\mathcal{S}_M(E)$, respectively by $\mathcal{S}(F)$, the set of submodules of E containing M, respectively the set of all submodules of F, then the mapping*

(10)
$$\mathcal{S}_M(E) \to \mathcal{S}(F)$$
given by

$$E' \mapsto u(E'),\ E' \in \mathcal{S}_M(E),$$

is a bijection whose inverse is given by

$$F' \mapsto u^{-1}(F'), \quad F' \in \mathcal{S}(F).$$

The mapping (10) *and its inverse preserve the inclusion relations, and if in $\mathcal{S}_M(E)$ we have $E' \subseteq E''$, then u induces an isomorphism*

$$E''/E' \xrightarrow{\sim} u(E'')/u(E').$$

Proof. If $E' \in \mathcal{S}_M(E)$, then $u(E')$ is a submodule of F; if $F' \in \mathcal{S}(F)$, then $u^{-1}(F')$ is a submodule of E, and we have

$$u^{-1}(F') \supseteq u^{-1}(0) = M$$

according to the monotonicity of taking the inverse images (see § 0); according to the same property and to the monotonicity of taking the images, the mapping (10) and the mapping stated as its inverse, which were shown to be well defined, preserve inclusions. We have to show that for any $E' \in \mathcal{S}_M(E)$ and any $F' \in \mathcal{S}(F)$ we have

(11)
$$u^{-1}(u(E')) = E' \quad \text{and} \quad u(u^{-1}(F')) = F';$$

the inclusion $u^{-1}(u(E')) \supseteq E'$ follows by definition; conversely, if $x \in u^{-1}(u(E'))$, then $ux = ux'$ for some $x' \in E'$, hence $u(x - x') = 0$, and

$$x - x' \in u^{-1}(0) = M \subseteq E';$$

if $x - x' = z$, then $x = x' + z \in E'$; the inclusion $u(u^{-1}(F')) \subseteq F'$ follows also by definition; conversely, if $y \in F'$, then, by the surjectivity of u, there exists $x \in E$ such that $u(x) = y \in F'$; hence $x \in u^{-1}(F')$, and $y = u(x) \in u(u^{-1}(F'))$.

Finally, if we have in $\mathcal{S}_M(E)$ the inclusion $E' \subseteq E''$, then as we have trivially $u(E'') \subseteq u(E'')$, u induces by restriction a homomorphism

$$u^* : E'' \to u(E'')$$

which is surjective; by composition with the canonical surjection $q : u(E'') \to$ $\to u(E'')/u(E')$ we get a surjective homomorphism

$$q \circ u^* : E'' \to u(E'')/u(E');$$

if $x \in \operatorname{Ker}(q \circ u^*)$, then $0 = q(u^*(x)) = q(u(x))$, and so $u(x) \in u(E')$; hence $x \in u^{-1}(u(E')) = E'$; conversely, if $x \in E'$, then $(q \circ u^*)x = q(u(x)) = 0$, since $u(x) \in u(E')$; it follows that

$$E' = \operatorname{Ker}(q \circ u^*);$$

by Corollary 1 to Proposition 1 we get an isomorphism

$$\hat{u} : E''/E' \rightleftarrows u(E'')/u(E')$$

for which

$$\hat{u} \circ p = q \circ u^*, \tag{12}$$

where p is the canonical surjection $E'' \to E''/E'$; but then $\hat{u}$ is induced by u, in the sense defined in § 3. ▮

COROLLARY. *Let E be an A-module and M, N two submodules such that $M \subseteq N$; then there exists a canonical isomorphism*

$$E/N \rightleftarrows (E/M)/(N/M).$$

Proof. We have only to take in Proposition 2 $F = E/M$, $E' = N$, $E'' = E$, and for u the canonical surjection $E \to E/M$; then $u(E) = E/M$, $u(N) = N/M$. ▮

REMARK. In attempting to extend to categories the notions of kernel, cokernel, etc. defined for modules by Definition 3, § 3, we see that instead of the corresponding modules we take classes of equivalent objects (endowed with morphisms), which, when returning to the categories of modules, provide classes of equivalent modules for which the ones previously introduced represent only particular "instances"; for this reason, in a category of modules **Mod**(A), notions as kernel, cokernel, etc. might have an ambiguous meaning. We therefore establish that, unless otherwise explicitly mentioned, by kernel, cokernel, etc. in a category of modules, the notions given by Definition 3, § 3, should be meant. ▮

§ 5. Exact Sequences

Definition 1. Given a ring A and the diagram of A-modules and A-homomorphisms

(1) $$E \xrightarrow{u} F \xrightarrow{v} G,$$

we say that this diagram is an *exact sequence* of A-modules and A-homomorphisms provided

$$v^{-1}(0) = u(E),$$

i.e., provided the kernel of v coincides with the image of u; in this case we say also that *the pair* (u, v) *is an exact sequence.*

Given the diagram of A-modules and A-homomorphisms

(2) $$\dots \to E_{n-1} \xrightarrow{u_{n-1}} E_n \xrightarrow{u_n} E_{n+1} \to \dots,$$

where an infinity of terms may occur to the left and to the right, or only to the left, or only to the right, or neither to the left nor to the right (but where we have at least three modules and two homomorphisms), we say that diagram (2) is an *exact sequence at* E_n if the pair (u_{n-1}, u_n) is an exact sequence. One simply says that diagram (2) is an *exact sequence* of A-modules and A-homomorphisms if it is an exact sequence at any of its terms except of the extremities provided they exist. ▮

If diagram (1) is an exact sequence, then

(3) $$v \circ u = 0:$$

since, for any $x \in E$ we have $u(x) \in \operatorname{Im}(u) = \operatorname{Ker}(v)$; note that here we used only the inclusion

(4) $$\operatorname{Im}(u) \subseteq \operatorname{Ker}(v).$$

Conversely, the inclusion (4) follows readily from (3); thus, in order that diagram (1) should satisfy (3) it is necessary and sufficient to have (4); a diagram (1) satisfying (3) is called a *semi-exact sequence* or a *0-sequence* or a *complex* (of A-modules and A-homomorphisms); the notion may obviously be generalized to diagrams of the type (2), requiring that the composition of any two successive homomorphisms be 0. *Any exact sequence is a 0-sequence.*

Proposition 1. *If E and F are A-modules and u is an A-homomorphism $E \to F$, the following statements are true (here the mappings with domain or codomain 0 are the uniquely possible linear mappings):*

a) u is a monomorphism (i.e. an injective homomorphism) if and only if the diagram

$$0 \to E \xrightarrow{u} F$$

is an exact sequence;

b) u is an epimorphism (i.e. a surjective homomorphism) if and only if the diagram

$$E \xrightarrow{u} F \to 0$$

is an exact sequence;

c) u is an isomorphism (i.e. a bijective homomorphism) if and only if the diagram

$$0 \to E \xrightarrow{u} F \to 0$$

is an exact sequence;

d) if F is a submodule of E, and i, p represent, respectively, the canonical injection and surjection, then the diagram

$$0 \to F \xrightarrow{i} E \xrightarrow{p} E/F \to 0$$

is an exact sequence;

e) the diagram

$$0 \to u^{-1}(0) \xrightarrow{i} E \xrightarrow{u} F \xrightarrow{p} F/u(E) \to 0,$$

where i and p are, respectively, the canonical injection and surjection, is an exact sequence.

Proof. a) According to Proposition 2, § 3, the fact that u is injective is equivalent to the fact that $u^{-1}(0) = 0$; but the image of the homomorphism $0 \to E$ is 0.

b) The fact that u is surjective means that $u(E) = F$; but the kernel of $F \to 0$ is F.

c) It follows combining *a)* and *b)*.

d) The diagram is an exact sequence at E and E/F according to *a)* and *b)*; but $p^{-1}(0) = F = i(F)$ (see § 3).

e) The diagram is an exact sequence at $u^{-1}(0)$ and $F/u(E)$ according to *a)* and *b)*; at E it is an exact sequence since $i(u^{-1}(0)) = u^{-1}(0)$; at F it is an exact sequence since $p^{-1}(0) = u(E)$. ∎

* Let $\mathscr{C}$ be a preadditive category satisfying Axioms AB1) and AB2); if diagram (1) is given as a diagram of objects and morphisms of $\mathscr{C}$, then this

diagram is said to be an *exact sequence* (of, or in the category $\mathcal{C}$) or also *the pair* (u, v) *is said to be an exact sequence* provided, in the sense of Definition 1, § 4,

$$\operatorname{Ker}(v) = \operatorname{Im}(u)$$

as subobjects of F; in the same manner, the notion of *exact sequence* and that of *exact sequence at the term* E_n are defined for a more general diagram of the type (2) if the A-modules and A-homomorphisms are replaced by objects and morphisms of $\mathcal{C}$. From the results of § 4 it can be seen that the previous results, including those of Proposition 1, concerning exact sequences in categories of modules remain valid in an appropriate form in a preadditive category satisfying AB1) and AB2); for instance, the fact that $v \circ u = 0$ in diagram (1) is shown like this: a kernel (M, j) of v is considered; this being an image of u, by the canonical factorization of u there exists a morphism $q : E \to M$ such that $u = j \circ q$; since $v \circ j = 0$, it follows that $v \circ u = v \circ j \circ q = 0$. Proposition 1, *a*) is a consequence of the fact that the image of the morphism $0 \to E$ is $(0, 0)$ and this pair is a kernel of u if and only if u is a monomorphism; Proposition 1, *e*) may be adapted by saying that if (M, i) is a kernel and (P, p) a cokernel of the morphism $u \in \operatorname{Hom}_{\mathcal{C}}(E, F)$, then the diagram

$$0 \to M \xrightarrow{i} E \xrightarrow{u} F \xrightarrow{p} P \to 0$$

is an exact sequence of the category $\mathcal{C}$; it is true that for the moment the inclusion $\operatorname{Im}(u) \subseteq \operatorname{Ker}(v)$ makes no sense, but will receive a precise meaning in § 8, when the partial order relation of the subobjects of an object is discussed; at that moment it will be possible to show that for diagram (1), the relation $\operatorname{Im}(u) \leqslant \operatorname{Ker}(v)$ is equivalent to $v \circ u = 0$.

It is possible to show also that *if diagram* (1) *is an exact sequence of the category* $\mathcal{C}$, *then the diagram*

$$G^\circ \xrightarrow{v^\circ} F^\circ \xrightarrow{u^\circ} E^\circ$$

is an exact sequence of the category $\mathcal{C}^\circ$.

We also have

PROPOSITION 2. *Suppose we have a preadditive category* $\mathcal{C}$ *satisfying Axioms* AB1) *and* AB2); *then*:

a) *the diagram*

(5) $$0 \to F \xrightarrow{u} E \xrightarrow{v} G$$

is an exact sequence of the category $\mathcal{C}$ *if and only if* (E, u) *is a kernel of* v, *in the sense of Definition* 1, § 4;

b) *the diagram*

(6) $$E \xrightarrow{u} F \xrightarrow{v} G \to 0$$

is an exact sequence of the category $\mathcal{C}$ *if and only if* (G, v) *is a cokernel of* u, *in the sense of Definition* 1, § 4.

Proof. *a*) If (5) is an exact sequence, then, according to Proposition 1, u is a monomorphism; by the considerations of § 4, (E, u) is an image of u, hence, owing to the exactness, it is also a kernel of v; conversely, if (E, u) is a kernel of v, then u is a monomorphism, thus diagram (5) is an exact sequence at E; u being a monomorphism, (E, u) is an image of u; as it is also a kernel of v, diagram (5) is an exact sequence also at F.

b) If diagram (6) is an exact sequence, then, according to the preceding arguments, we have $v \circ u = 0$; if $Y \in \mathrm{Ob}\,\mathcal{C}$, and if $h \in \mathrm{Hom}_{\mathcal{C}}(F, Y)$ is such that $h \circ u = 0$, then let (I, j) be an image of u; by the canonical factorization of u, there exists an epimorphism $q \colon E \to I$ such that $u = j \circ q$; q being an epimorphism, we have $h \circ j = 0$; diagram (6) being an exact sequence, it follows that (I, j) is a kernel of v and that v is an epimorphism; as shown in § 4, (G, v) is a cokernel of j, and hence, there exists $l \colon G \to Y$, unique with the property $l \circ v = h$; conversely, if (G, v) is a cokernel of u, then v is an epimorphism, hence diagram (6) is an exact sequence at G; if (I, j) is an image of u, it follows easily that (G, v) is a cokernel of j, hence that (I, j) is a kernel of v, and diagram (6) is an exact sequence at F.[1] ▮ *

If the diagram (1) of A-modules and A-homomorphisms is a complex, then, as $\mathrm{Im}(u) \subseteq \mathrm{Ker}(v)$, we may consider the quotient module $\mathrm{Ker}(v)/\mathrm{Im}(u)$; this is called the *homology module* of the complex (1); it gives a "measure" of the deviation from the exactness of the semi-exact sequence (1), because it is the zero module if any only if (1) is an exact sequence.

If diagram (2) is a complex, the modules $\mathrm{Ker}(u_n)/\mathrm{Im}(u_{n-1})$, where n runs through the given set of indices (eventually excluding the left or/and right extremities provided such extremities exist) are called the *homology modules* of the complex (2).

If in the category **Mod**(A) the commutative diagram

$$\begin{array}{ccccc} E & \xrightarrow{u} & F & \xrightarrow{v} & G \\ {\scriptstyle f}\downarrow & & \downarrow{\scriptstyle g} & & \downarrow{\scriptstyle h} \\ E' & \xrightarrow{u'} & F' & \xrightarrow{v'} & G' \end{array}$$

is given in which the rows are complexes, then, as shown in § 4 when discussing diagram (2), we have

$$g(\mathrm{Ker}\ v) \subseteq \mathrm{Ker}\ v', \qquad g(\mathrm{Im}\ u) \subseteq \mathrm{Im}\ u';$$

then, according to § 3, g induces a homomorphism

$$g^{\natural} \colon \mathrm{Ker}\ v/\mathrm{Im}\ u \to \mathrm{Ker}\ v'/\mathrm{Im}\ u';$$

---

[1] We could have given for *b*) a proof based on *a*) by passing to the dual category; we preferred however a direct proof, without making use of the above result concerning the exactness of the "dual" of an exact sequence.

similar considerations can be made for a larger commutative diagram, of the form

$$\begin{array}{ccccccccc}
\cdots & \longrightarrow & E_{n-1} & \xrightarrow{u_{n-1}} & E_n & \xrightarrow{u_n} & E_{n+1} & \longrightarrow & \cdots \\
 & & \Big\downarrow{\scriptstyle f_{n-1}} & & \Big\downarrow{\scriptstyle f_n} & & \Big\downarrow{\scriptstyle f_{n+1}} & & \\
\cdots & \longrightarrow & E'_{n-1} & \xrightarrow{u'_{n-1}} & E'_n & \xrightarrow{u'_n} & E'_{n+1} & \longrightarrow & \cdots,
\end{array}$$

in which the rows are complexes.

* Definition 2. Let $\mathcal{C}, \mathcal{C}'$ be two preadditive categories satisfying Axioms AB1) and AB2); an additive functor $T : \mathcal{C} \to \mathcal{C}'$ is said to be an *exact functor* if, whenever the diagram

$$0 \to E \xrightarrow{u} F \xrightarrow{v} G \to 0 \tag{7}$$

is an exact sequence in $\mathcal{C}$, the diagram

$$0 \to T(E) \xrightarrow{T(u)} T(F) \xrightarrow{T(v)} T(G) \to 0 \tag{8}$$

is an exact sequence in $\mathcal{C}'$. The additive functor T is said to be a *left exact*, respectively *right exact functor* if, whenever the diagram (7) is an exact sequence in $\mathcal{C}$, the diagram (8) is an exact sequence in the category $\mathcal{C}'$ at the terms $T(E)$ and $T(F)$, respectively $T(F)$ and $T(G)$. ∎

Clearly, an additive functor $T : \mathcal{C} \to \mathcal{C}'$, where $\mathcal{C}, \mathcal{C}'$ are preadditive categories satisfying AB1) and AB2), is an exact functor if and only if it is both left and right exact, or if and only if it is right exact and carries the monomorphisms of $\mathcal{C}$ into monomorphisms of $\mathcal{C}'$, etc.*

For the categories of modules we give the following useful forms of the preceding definition:

Proposition 3. *Let A and B be two rings, and let $T :$ **Mod**$(A) \to$ **Mod**(B) be an additive functor*[1]. *The following assertions are equivalent:*

i) T is an exact functor;

ii) if the diagram

$$E \xrightarrow{u} F \xrightarrow{v} G \tag{9}$$

is an exact sequence of the category **Mod**(A), *then the diagram*

$$T(E) \xrightarrow{T(u)} T(F) \xrightarrow{T(v)} T(G) \tag{10}$$

[1] According to the established convention (see § 1) we deal here with the categories of *left* A-modules and *left* B-modules; we could have considered any other situation, for instance the category of *right* A-modules and the category of *left* B-modules, and, so on.

is an exact sequence of the category **Mod**(B);

iii) if the diagram

(11) $$\ldots \to E_{n-1} \xrightarrow{u_{n-1}} E_n \xrightarrow{u_n} E_{n+1} \to \ldots$$

is an exact sequence of the category **Mod**(A) *(where, as before, the situation concerning the finiteness of the number of terms is arbitrary), then the diagram*

$$\ldots \to T(E_{n-1}) \xrightarrow{T(u_{n-1})} T(E_n) \xrightarrow{T(u_n)} T(E_{n+1}) \to \ldots$$

is an exact sequence of **Mod**(B).

The following assertions are also equivalent:

iv) T is a left exact functor;

v) if the diagram

(12) $$0 \to E \xrightarrow{u} F \xrightarrow{v} G$$

is an exact sequence in **Mod**(A), *then the diagram*

(13) $$0 \to T(E) \xrightarrow{T(u)} T(F) \xrightarrow{T(v)} T(G)$$

is an exact sequence in **Mod**(B).

We have also the equivalence of the following assertions:

iv') T is a right exact functor;

v') if the diagram

$$E \xrightarrow{u} F \xrightarrow{v} G \to 0$$

is an exact sequence in **Mod**(A), *then the diagram*

$$T(E) \xrightarrow{T(u)} T(F) \xrightarrow{T(v)} T(G) \to 0$$

is an exact sequence in **Mod**(B).

Proof. i) $\Rightarrow$ *ii)*: Let the exact sequence (9) be given; we have the diagrams

$$0 \to \operatorname{Ker} u^* \xrightarrow{i} E \xrightarrow{u^*} u(E) \to 0,$$

$$0 \to u(E) \xrightarrow{j} F \xrightarrow{v^*} v(F) \to 0,$$

$$0 \to v(F) \xrightarrow{k} G \xrightarrow{p} \operatorname{Coker}(k) \to 0,$$

where i, j, k are the canonical injections, p is the canonical surjection, while u^* and v^* are the linear mappings obtained from u and v, respectively, according to Corollary 2 to Proposition 1, § 4; these diagrams are exact sequences: this follows immediately except perhaps for the exactness at F of the second sequence; but this is a consequence of the relations

$$\operatorname{Ker}(v^*) = \operatorname{Ker}(v) = u(E) = \operatorname{Im}(j).$$

By hypothesis i), the following diagrams are exact sequences:

$$0 \to T(\operatorname{Ker} u^*) \xrightarrow{T(i)} T(E) \xrightarrow{T(u^*)} T(u(E)) \to 0,$$

(14) $$0 \to T(u(E)) \xrightarrow{T(j)} T(F) \xrightarrow{T(v^*)} T(v(F)) \to 0,$$

$$0 \to T(v(F)) \xrightarrow{T(k)} T(G) \xrightarrow{T(p)} T(\operatorname{coker}(k)) \to 0.$$

Since

$$j \circ u^* = u \quad \text{and} \quad k \circ v^* = v,$$

we have

$$T(j) \circ T(u^*) = T(u) \quad \text{and} \quad T(k) \circ T(v^*) = T(v);$$

by Proposition 1, the exactness of the first sequence (14) implies that $T(u^*)$ is a surjection, thus, as shown in § 3,

$$\operatorname{Im}(T(u)) = \operatorname{Im}(T(j));$$

in the same way, from the exactness of the third sequence (14), $T(k)$ turns to be a monomorphism, hence

$$\operatorname{Ker}(T(v)) = \operatorname{Ker}(T(v^*));$$

finally, from the exactness of the second sequence (14) it follows

$$\operatorname{Ker}(T(v^*)) = \operatorname{Im}(T(j)),$$

and the sequence (10) is proved to be exact.

ii) ⇒ *i*): Obvious, taking into account that $T(0) = 0$.

ii) ⇒ *iii*): Obvious.

iii) ⇒ *ii*): Obvious if we take into account that diagram (9) is a particular case of the diagram (11).

iv) ⇒ *v*): Let the exact sequence (12) of **Mod**(A) be given; if v^* represents the linear mapping $F \to v(F)$ obtained from v according to Corollary 2 to Proposition 1, § 4, then we have the diagrams

$$0 \to E \xrightarrow{u} F \xrightarrow{v^*} v(F) \to 0,$$

$$0 \to v(F) \xrightarrow{k} G \xrightarrow{p} \operatorname{Coker}(k) \to 0,$$

where k, respectively p, is the canonical injection, respectively surjection; these diagrams are exact sequences since $k \circ v^* = v$ and, as k is injective, it follows that $\mathrm{Ker}(v^*) = \mathrm{Ker}(v) = \mathrm{Im}(u)$. Applying hypothesis *iv*) we get the exact sequences

$$(15) \qquad \begin{gathered} 0 \to T(E) \xrightarrow{T(u)} T(F) \xrightarrow{T(v^*)} T(v(F)), \\ 0 \to T(v(F)) \xrightarrow{T(k)} T(G) \xrightarrow{T(p)} T(\mathrm{Coker}(k)); \end{gathered}$$

since from the exactness of the second sequence (15) $T(k)$ turns to be injective, and since $T(k) \circ T(v^*) = T(v)$, we have from the exactness of the first sequence (15),

$$\mathrm{Ker}(T(v)) = \mathrm{Ker}(T(v^*)) = \mathrm{Im}(T(u));$$

it follows that diagram (13) is an exact sequence.

v) ⇒ *iv*): Obvious.

The equivalence *iv'*) ⇔ *v'*) is proved similarly; the details are left to the reader as an exercise. ∎

REMARK. Given the exact sequence of A-modules and A-homomorphisms

$$0 \to E \xrightarrow{u} F \xrightarrow{v} G \to 0,$$

it is said to be a *short exact sequence*; the triple (F, u, v) is also said to be an *extension of the module G by the module E.* ∎

PROPOSITION 4. *Let A be a ring, and let E be an A-module; if the diagram of A-modules and A-homomorphisms*

$$(16) \qquad 0 \to F' \xrightarrow{u} F \xrightarrow{v} F''$$

is an exact sequence, then the diagram of abelian groups and homomorphisms of abelian groups

$$(17) \qquad 0 \to \mathrm{Hom}_A(E, F') \xrightarrow{\mathrm{Hom}(1_E, u)} \mathrm{Hom}_A(E, F) \xrightarrow{\mathrm{Hom}(1_E, v)} \mathrm{Hom}_A(E, F'')$$

is an exact sequence; conversely, given the diagram (16) *in* **Mod**(A), *if for any A-module E diagram* (17) *is an exact sequence of the category* **Ab**, *then diagram* (16) *is an exact sequence.*

Similarly, for any A-module F and any exact sequence of A-modules and A-homomorphisms

$$(18) \qquad E' \xrightarrow{w} E \xrightarrow{t} E'' \to 0,$$

the diagram in **Ab**

$$(19)\quad 0 \to \mathrm{Hom}_A(E'', F) \xrightarrow{\mathrm{Hom}(t, 1_F)} \mathrm{Hom}_A(E, F) \xrightarrow{\mathrm{Hom}(w, 1_F)} \mathrm{Hom}_A(E', F)$$

is an exact sequence; conversely, given the diagram (18) *in* **Mod**(A), *if diagram* (19) *is an exact sequence for any A-module F, then* (18) *is an exact sequence.*

Proof. Let the exact sequence (16) be given; if for $f_1, f_2 \in \mathrm{Hom}_A(E, F')$ we have

$$\mathrm{Hom}(1_E, u)(f_1) = \mathrm{Hom}(1_E, u)(f_2),$$

then $u \circ f_1 = u \circ f_2$, and as u is a monomorphism (Proposition 1, *a*)), $f_1 = f_2$; hence $\mathrm{Hom}(1_E, u)$ is injective, and diagram (17) is an exact sequence at $\mathrm{Hom}_A(E, F')$.

As $v \circ u = 0$, we have

$$\mathrm{Hom}(1_E, v) \circ \mathrm{Hom}(1_E, u) = \mathrm{Hom}(1_E, v \circ u) = 0$$

according to relations (9), § 2; it follows

$$(20)\qquad \mathrm{Im}(\mathrm{Hom}(1_E, u)) \subseteq \mathrm{Ker}(\mathrm{Hom}(1_E, v));$$

if $g \in \mathrm{Ker}(\mathrm{Hom}(1_E, v))$, then $v \circ g = 0$; since, according to Proposition 2, *a*), the pair (F', u) is a kernel of v in the sense of Definition 1, § 4, there exists an A-homomorphism $f : E \to F'$ such that

$$g = u \circ f = \mathrm{Hom}(1_E, u)(f),$$

hence $g \in \mathrm{Im}(\mathrm{Hom}(1_E, u))$. Thus we have the inclusion inverse to (20), and (17) is an exact sequence at $\mathrm{Hom}_A(E, F)$.

Now let diagram (16) be given, and suppose that (17) is an exact sequence for any A-module E; the fact that u is a monomorphism, hence that the sequence (16) is exact at F' (Proposition 1, *a*)) follows by the argument converse to the one used at the beginning. Next, for any A-module E we have

$$0 = \mathrm{Hom}(1_E, v) \circ \mathrm{Hom}(1_E, u) = \mathrm{Hom}(1_E, v \circ u);$$

taking F' for E, we obtain

$$0 = \mathrm{Hom}(1_{F'}, v \circ u)(1_{F'}) = v \circ u,$$

and $\mathrm{Im}(u) \subseteq \mathrm{Ker}(v)$; let $E = \mathrm{Ker}(v)$, and let i be the canonical injection $\mathrm{Ker}(v) \to F$; we have

$$0 = v \circ i = \mathrm{Hom}(1_E, v)(i);$$

the exactness of the sequence (17) for the chosen E implies that there exists $f : \mathrm{Ker}\, v \to F'$ such that $i = \mathrm{Hom}(1_E, u)(f) = u \circ f$; hence, if $y \in \mathrm{Ker}(v)$, we have

$$y = i(y) = u(f(y)) \in \mathrm{Im}(u),$$

and $\mathrm{Ker}(v) \subseteq \mathrm{Im}(u)$, which means that (16) is an exact sequence also at F.

The second part of the Proposition is proved by dual arguments; we leave the details to the reader, giving as a hint that, in the proof of the exactness of (18) provided (19) is an exact sequence, in order to show that $\mathrm{Ker}(t) \subseteq \mathrm{Im}(w)$, one should take $F = \mathrm{Coker}(w)$ and consider the canonical surjection $p : E \to \mathrm{Coker}(w)$. ∎

COROLLARY. *Let A be a ring; for any A-module E, the functor*

$$h^E : \mathbf{Mod}(A) \to \mathbf{Ab}$$

is left exact; also, for any A-module F, the functor

$$h_F : (\mathbf{Mod}(A))^\circ \to \mathbf{Ab}$$

is left exact.

Proof. Only the second part might need an explanation: if the diagram

$$0 \to (E'')^\circ \xrightarrow{t^\circ} E^\circ \xrightarrow{w^\circ} (E')^\circ$$

is an exact sequence in $(\mathbf{Mod}(A))^\circ$, then, in view of the preceding arguments, the diagram

$$((E')^\circ)^\circ \xrightarrow{(w^\circ)^\circ} (E^\circ)^\circ \xrightarrow{(t^\circ)^\circ} ((E'')^\circ)^\circ \to 0$$

is an exact sequence in the category $((\mathbf{Mod}(A))^\circ)^\circ$; by the canonical identification $(\mathcal{C}^\circ)^\circ = \mathcal{C}$, this diagram becomes diagram (18), which is an exact sequence in the category $\mathbf{Mod}(A)$; we need only interpret now diagram (19) in terms of the functor h_F. ∎

* Propositions 3 and 4 and the Corollary to Proposition 4 remain valid when replacing the categories $\mathbf{Mod}(A)$ and $\mathbf{Mod}(B)$, respectively, by preadditive categories $\mathcal{C}$, $\mathcal{C}'$ satisfying Axioms AB1) and AB2); also, the term "*short exact sequence*" is kept in such categories.

Hence, it follows in particular that the composition of two (or of any finite number of) exact, respectively left exact, respectively right exact functors is also an exact, respectively left exact, respectively right exact functor.

From the same arguments and from Proposition 2 it follows that an additive functor $T : \mathcal{C} \to \mathcal{C}'$, where $\mathcal{C}, \mathcal{C}'$ are preadditive categories satisfying AB1) and AB2), is left exact (respectively right exact) if and only if for any morphism u of the category $\mathcal{C}$, whenever (M, i) is a kernel of u (respectively (P, p) is a cokernel of u) in the category $\mathcal{C}$ in the sense of Definition 1, § 4, $(T(M), T(i))$ is a kernel (respectively $(T(P), T(p))$ is a cokernel) of $T(u)$ in the category $\mathcal{C}'$, i.e., it "preserves kernels" (respectively "preserves cokernels"); consequently, the additive functor $T : \mathcal{C} \to \mathcal{C}'$ is exact if and only if it has both properties, i.e., it "preserves kernels and cokernels"; hence, if the additive functor $T : \mathcal{C} \to \mathcal{C}'$ is exact, and if $u \in \mathrm{Hom}_{\mathcal{C}}(E, F)$, with $E, F \in \mathrm{Ob}\,\mathcal{C}$, then, denoting by (M, i), respectively $(\mathcal{C}, q)$, a kernel, respectively a coimage of u in the category $\mathcal{C}$, it follows that $(T(M), T(i))$

is a kernel of $T(u)$ and $(T(C), T(q))$ is a cokernel of $T(i)$ hence a coimage of $T(u)$ in the category $\mathscr{C}'$, and T "preserves coimages"; likewise, an exact functor "preserves images".*

To end with, some famous propositions concerning exact sequences and which are useful in applications will be given.

PROPOSITION 5 **(The Five Lemma).** *Let the following diagram of A-modules and A-homomorphisms be given:*

$$\begin{array}{ccccccccc} E_1 & \xrightarrow{u_1} & E_2 & \xrightarrow{u_2} & E_3 & \xrightarrow{u_3} & E_4 & \xrightarrow{u_4} & E_5 \\ \downarrow{\scriptstyle f_1} & & \downarrow{\scriptstyle f_2} & & \downarrow{\scriptstyle f_3} & & \downarrow{\scriptstyle f_4} & & \downarrow{\scriptstyle f_5} \\ F_1 & \xrightarrow{v_1} & F_2 & \xrightarrow{v_2} & F_3 & \xrightarrow{v_3} & F_4 & \xrightarrow{v_4} & F_5; \end{array}$$

assume that the diagram is commutative and that the rows in it are exact sequences; then:

a) if f_1 is an epimorphism, and f_2, f_4 are monomorphisms, then f_3 is a monomorphism;

b) if f_5 is a monomorphism, and f_2, f_4 are epimorphisms, then f_3 is an epimorphism;

c) if f_1 is an epimorphism, f_2 and f_4 are isomorphisms, and f_5 is a monomorphism, then f_3 is an isomorphism.

Proof. a) Suppose $x_3 \in E_3$ has the property $f_3(x_3) = 0$; it follows

$$0 = (v_3 \circ f_3)(x_3) = f_4(u_3(x_3));$$

since f_4 is injective, we have $u_3(x_3) = 0$; the top row in the diagram being an exact sequence, there exists $x_2 \in E_2$ such that $x_3 = u_2(x_2)$; we have

$$0 = f_3(x_3) = (f_3 \circ u_2)(x_2) = v_2(f_2(x_2));$$

hence there exists $y_1 \in F_1$ such that $f_2(x_2) = v_1(y_1)$; as f_1 is surjective, there exists $x_1 \in E_1$ such that $f_1(x_1) = y_1$; consequently,

$$f_2(u_1(x_1)) = v_1(f_1(x_1)) = f_2(x_2),$$

and as f_2 is injective, $u_1(x_1) = x_2$; since $u_2 \circ u_1 = 0$, we get $x_3 = u_2(x_2) = (u_2 \circ u_1)(x_1) = 0$.

b) is proved by arguments similar to those used in the proof of *a)*.

c) follows from *a)* and *b)*. ∎

PROPOSITION 6 (**3 × 3 Lemma**). *Let the commutative diagram of A-modules and A-homomorphisms be given:*

$$\begin{array}{ccccccccc}
 & & 0 & & 0 & & 0 & & \\
 & & \downarrow & & \downarrow & & \downarrow & & \\
0 & \longrightarrow & E' & \xrightarrow{u'} & E & \xrightarrow{u} & E'' & \longrightarrow & 0 \\
 & & \downarrow{\scriptstyle f'} & & \downarrow{\scriptstyle f} & & \downarrow{\scriptstyle f''} & & \\
0 & \longrightarrow & F' & \xrightarrow{v'} & F & \xrightarrow{v} & F'' & \longrightarrow & 0 \\
 & & \downarrow{\scriptstyle g'} & & \downarrow{\scriptstyle g} & & \downarrow{\scriptstyle g''} & & \\
0 & \longrightarrow & G' & \xrightarrow{w'} & G & \xrightarrow{w} & G'' & \longrightarrow & 0; \\
 & & \downarrow & & \downarrow & & \downarrow & & \\
 & & 0 & & 0 & & 0 & &
\end{array}$$

if the columns and the two bottom rows in this diagram are exact sequences, then the top row is also an exact sequence.

Proof. According to Proposition 1, it is sufficient to prove that u' is injective, u is surjective, and $\text{Ker}(u) = \text{Im}(u')$.

As $f \circ u' = v' \circ f'$, and as f', v' are monomorphisms, it follows, according to § 4, that u' is also a monomorphism. Next, we have $f'' \circ u \circ u' = v \circ v' \circ f' = 0$, and since f'' is a monomorphism, it follows $u \circ u' = 0$, hence $\text{Im}(u') \subseteq \text{Ker}(u)$; conversely, if $x \in \text{Ker}\, u$, then

$$v(f(x)) = f''(u(x)) = 0,$$

and there exists $y' \in F'$ such that $f(x) = v'(y')$; we have

$$w'(g'(y')) = g(v'(y')) = g(f(x)) = 0,$$

and as w' is injective, $g'(y') = 0$; consequently, there exists $x' \in E'$ such that $y' = f'(x')$; but then

$$f(x) = v'(y') = v'(f'(x')) = f(u'(x')),$$

and as f is injective, it follows $x = u'(x')$; hence we also have $\text{Ker}(u) \subseteq \text{Im}(u')$. Finally, let $x'' \in E''$; v being surjective, there exists $y \in F$ such that $f''(x'') = v(y)$; we have

$$w(g(y)) = g''(v(y)) = g''(f''(x'')) = 0,$$

hence there exists $z' \in G'$ such that $g(y) = w'(z')$; as g' is surjective, there exists $y' \in F'$ such that $z' = g'(y')$; we have

$$(g(v'(y')) = w'(g'(y')) = g(y),$$

hence, since $y - v'(y') \in \operatorname{Ker}(g)$, there exists $x \in E$ such that

$$y - v'(y') = f(x);$$

but then

$$f''(u(x)) = v(f(x)) = v(y) - v(v'(y')) = v(y) = f''(x''),$$

and since f'' is injective, we have $x'' = u(x)$; hence u is surjective. ▮

PROPOSITION 7 (**Snake Lemma**). *Let the commutative diagram of A-modules and A-homomorphisms in which the rows are exact sequences be given*:

$$\text{(21)}\qquad \begin{array}{ccccccc} & & E' & \xrightarrow{u'} & E & \xrightarrow{u} & E'' & \longrightarrow 0; \\ & & \downarrow{\scriptstyle f'} & & \downarrow{\scriptstyle f} & & \downarrow{\scriptstyle f''} & \\ 0 & \longrightarrow & F' & \xrightarrow{v'} & F & \xrightarrow{v} & F'' & \end{array}$$

then, denoting by u'^* and u^* the homomorphism $\operatorname{Ker} f' \to \operatorname{Ker} f$, *respectively* $\operatorname{Ker} f \to \operatorname{Ker} f''$, *obtained from u', respectively u, by applying the functor* Ker *of* § 4,[1] *and, likewise, denoting by $\hat{v}'$ and $\hat{v}$ the homomorphism* $\operatorname{Coker} f' \to \operatorname{Coker} f$, *respectively* $\operatorname{Coker} f \to \operatorname{Coker} f''$, *obtained from v', respectively v, by applying the functor* Coker *of* § 4, *one may canonically construct a homomorphism of A-modules*

$$\delta : \operatorname{Ker} f'' \to \operatorname{Coker} f'$$

such that the following diagram is an exact sequence:

$$\text{(22)}\qquad \operatorname{Ker} f' \xrightarrow{u'^*} \operatorname{Ker} f \xrightarrow{u^*} \operatorname{Ker} f'' \xrightarrow{\delta} \operatorname{Coker} f' \xrightarrow{\hat{v}'} \operatorname{Coker} f \xrightarrow{\hat{v}} \operatorname{Coker} f''.$$

Proof. The construction of δ is as follows: let $x'' \in \operatorname{Ker} f''$; u being surjective, there exists $x \in E$ such that $x'' = u(x)$; we have

$$v(f(x)) = f''(u(x)) = 0,$$

hence, from the exactness of the bottom row in diagram (21) we get an $y' \in F'$ such that $v'(y') = f(x)$. We claim that the image of y' under the canonical surjection $F' \to F'/\operatorname{Im} f' = \operatorname{Coker} f'$ does not depend on the chosen x: indeed, let $x_1 \in E$ be such that $u(x_1) = x''$; from $u(x_1) = u(x)$ it follows that $x_1 - x \in \operatorname{Ker} u$, hence there exists $x' \in E'$ such that

$$x_1 - x = u'(x');$$

if $y_1' \in F'$ has the property $v'(y_1') = f(x_1)$, then

$$v'(y_1' - y') = f(x_1 - x) = v'(f'(x')),$$

[1] See diagram (2), § 4, and the considerations following it.

and as v' is a monomorphism, we have $y_1' - y' = f'(x') \in \operatorname{Im} f'$; hence the mapping δ defined by

$$\delta(x'') = \hat{y}',$$

where $\hat{y}'$ is the coset of y' with respect to $\operatorname{Im} f'$, is well defined. The fact that δ is an A-homomorphism follows by its construction: for instance, if $\alpha \in A$, in the previous notation, we have $u(\alpha x) = \alpha x''$ and $v'(\alpha y') = f(\alpha x)$; consequently,

$$\delta(\alpha x'') = \widehat{\alpha y'} = \alpha \hat{y}' = \alpha \delta(x'').$$

To show that (22) is an exact sequence, we take into account that, by virtue of the considerations made in § 4, u'^* and u^* are obtained from u', respectively u, by restriction, while $\hat{v}'$ and $\hat{v}$ are obtained from v', respectively v, by passing to the quotient; therefore $u^* \circ u'^* = 0$ and $\hat{v} \circ \hat{v}' = 0$. Now let $x \in \operatorname{Ker} f$ be such that $u^*(x) = 0$; we get $u(x) = 0$, and the exactness of the top row in (21) implies the existence of some $x' \in E'$ such that $u'(x') = x$; but

$$v'(f'(x')) = f(u'(x')) = 0,$$

and as v' is injective, $f'(x') = 0$, hence $x' \in \operatorname{Ker} f'$; now $u'^*(x') = u'(x') = x$, and the sequence (22) is exact at $\operatorname{Ker} f$. If $x'' \in \operatorname{Ker} f''$ is of the form $x'' = u^*(x)$, with $x \in \operatorname{Ker} f$, then by the definition of δ we have

$$\delta(x'') = \hat{y}',$$

where $v'(y') = f(x) = 0$; as v' is a monomorphism, it follows $y' = 0$; hence $\delta \circ u^* = 0$; conversely, if $x'' \in \operatorname{Ker} \delta$, using the above notation, we get $y' \in \operatorname{Im} f'$; consequently, there exists $x' \in E'$ such that $y' = f'(x')$; but

$$f(x) = v'(y') = v'(f'(x')) = f(u'(x')),$$

and $x - u'(x') \in \operatorname{Ker} f$; we have

$$u^*(x - u'(x')) = u(x - u'(x')) = u(x) = x'',$$

and hence $x'' \in \operatorname{Im} u^*$; this proves the exactness of the sequence (22) at $\operatorname{Ker} f''$. Consider now $x'' \in \operatorname{Ker} f''$; in the above notation, we have

$$\hat{v}'(\delta(x'')) = \hat{v}'(\hat{y}') = \widehat{v'(y')} = \widehat{f(x)} = 0,$$

where the notation "^" generally designates the corresponding coset with respect to $\operatorname{Im} f'$, $\operatorname{Im} f$, etc.; it follows $\hat{v}' \circ \delta = 0$; let now $\hat{y}' \in \operatorname{Ker}(\hat{v}')$, with $y' \in F'$; it follows $v'(y') \in \operatorname{Im} f$; hence there exists $x \in E$ such that $f(x) = v'(y')$; we have

$$f''(u(x)) = v(f(x)) = v(v'(y')) = 0,$$

hence $x'' = u(x) \in \operatorname{Ker} f''$, and, by the definition of δ, $\delta(x'') = \hat{y}'$; thus the sequence (22) turns to be exact at $\operatorname{Coker} f'$ also. Finally, let $\hat{y} \in \operatorname{Ker} \hat{v}$, with $y \in F$; we have $v(y) \in \operatorname{Im} f''$, hence there exists $x'' \in E''$ such that $f''(x'') = v(y)$; u being an epimorphism, there exists $x \in E$ such that $u(x) = x''$; but then

$$v(y - f(x)) = f''(x'') - v(f(x)) = f''(x'' - u(x)) = 0;$$

from the exactness of the bottom row in diagram (21) it follows the existence of $y' \in F'$ such that $v'(y') = y - f(x)$; hence $\hat{v}'(\hat{y}') = \hat{y}$, and the sequence (22) is exact at $\operatorname{Coker} f$ also. ▮

§ 6. Direct Products

Let A be a ring and $(E_\lambda)_{\lambda \in I}$ a family of A-modules. Let E be the cartesian product of the sets E_λ, $\lambda \in I$; the elements of E are thus families $(x_\lambda)_{\lambda \in I}$ indexed by I such that $x_\lambda \in E_\lambda$, $\lambda \in I$; we define on E an A-module structure in the following manner: if $x = (x_\lambda)_\lambda$, $y = (y_\lambda)_\lambda$ are elements of E and $\alpha \in A$, then, by definition,

$$x + y = (x_\lambda + y_\lambda)_{\lambda \in I}, \tag{1}$$

$$\alpha x = (\alpha x_\lambda)_{\lambda \in I}, \tag{2}$$

the addition and the scalar multiplication being performed for each component as in the corresponding E_λ; if instead of left A-modules, right A-modules are considered, then the external law of composition on E is defined by

$$x\alpha = (x_\lambda \alpha)_{\lambda \in I}. \tag{2'}$$

In any case, the addition and the scalar multiplication are performed "componentwise". One easily verifies that with respect to the laws of composition (1) and (2), respectively (1) and (2′), E is an A-module, respectively a right A-module: the zero element of E is the family whose components are the zero elements of the modules E_λ, $\lambda \in I$, while the opposite of $x = (x_\lambda)_{\lambda \in I}$ is the family $(-x_\lambda)_{\lambda \in I}$; let us also verify, for instance, that

$$\alpha(x + y) = \alpha x + \alpha y, \quad \alpha \in A, \quad x, y \in E:$$

if x and y are as above, then

$$\alpha(x + y) = \alpha \cdot (x_\lambda + y_\lambda)_\lambda = (\alpha(x_\lambda + y_\lambda))_\lambda = (\alpha x_\lambda + \alpha y_\lambda)_\lambda =$$

$$= (\alpha x_\lambda)_\lambda + (\alpha y_\lambda)_\lambda = \alpha x + \alpha y.$$

The mapping

$$p_\lambda : E \to E_\lambda, \qquad \lambda \in I,$$

assigning to each family $(x_\lambda)_\lambda$ its component of index λ, i.e. x_λ, is obviously linear.

It is also surjective, since if $x_\lambda \in E_\lambda$ for a given λ, we may consider in E the family x whose component of index μ, $\mu \neq \lambda$, is $0 \in E_\mu$ and the one of index λ is x_λ; we obviously have $p_\lambda(x) = x_\lambda$.

DEFINITION 1. The A-module E constructed above is called the *direct product* or the *product* or also the *product module* of the given family of A-modules $(E_\lambda)_{\lambda \in I}$, and it is denoted $\prod_{\lambda \in I} E_\lambda$; if $I = \{1, 2, \ldots, n\}$, then, instead of $\prod_{\lambda \in I} E_\lambda$ we also write $E_1 \times \ldots \times E_n$ or also $\prod_{i=1}^{n} E_i$; if $I = \emptyset$, then $\prod_{\lambda \in I} E_\lambda$ is the zero A-module 0.

Considering the general case again, the linear mapping

$$p_\lambda : \prod_{\lambda \in I} E_\lambda \to E_\lambda$$

defined above[1] is called the *canonical projection of index* (or *of order*) λ, $\lambda \in I$. ▮

If $I = \{1\}$, then $\prod_{\lambda \in I} E_\lambda$ is identified with E_1.

We have the following universal property of the direct product:

PROPOSITION 1. *Let* $E = \prod_{\lambda \in I} E_\lambda$ *be the product of the family of A-modules* $(E_\lambda)_{\lambda \in I}$, *with the canonical projections* p_λ, $\lambda \in I$; *then, for any A-module X and any family of linear mappings* $f_\lambda : X \to E_\lambda$, $\lambda \in I$, *there exists a unique A-homomorphism* $f : X \to E$ *with the property of rendering commutative the diagram*

(3)

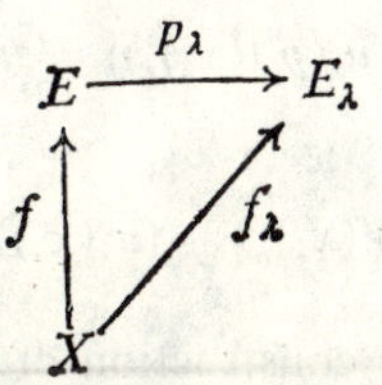

for any $\lambda \in I$, *i.e., such that* $p_\lambda \circ f = f_\lambda$, $\lambda \in I$.

Proof. If $x \in X$, define $f(x)$ as being the family $(f_\lambda(x))_{\lambda \in I}$; the mapping f defined in this way is linear: for instance,

$$f(x + y) = (f_\lambda(x + y))_\lambda = (f_\lambda x + f_\lambda y)_\lambda = (f_\lambda x)_\lambda + (f_\lambda y)_\lambda = f(x) + f(y).$$

We have

$$p_\lambda(f(x)) = p_\lambda((f_\lambda(x))_\lambda) = f_\lambda(x), \qquad \lambda \in I,$$

and hence f renders diagram (3) commutative for any $\lambda \in I$. If there were another homomorphism $g : X \to E$ which put instead of f would make diagram (3) com-

---

[1] Note that we denoted by λ both a *current* index of I and a *given* index of I. This is an abuse of notation already used, and we shall do the same in the sequel.

mute, then, as $p_\lambda(g(x))$, $x \in X$, is the component of index λ of $g(x)$, $\lambda \in I$, we would have that the component of index λ of $g(x)$ coincides with $f_\lambda(x)$, which is the component of index λ of $f(x)$, $\lambda \in I$; hence $g(x) = f(x)$, $x \in X$, and $g = f$. ▮

COROLLARY 1. *Let $(E_\lambda)_{\lambda \in I}$, $(F_\lambda)_{\lambda \in I}$ be two families of A-modules indexed by the same set I; assume that for any $\lambda \in I$ an A-homomorphism $u_\lambda : E_\lambda \to F_\lambda$ is given; under these conditions, there exists a unique homomorphism $u : \prod_{\lambda \in I} E_\lambda \to \prod_{\lambda \in I} F_\lambda$ with the property that for any $\lambda \in I$, the diagram*

$$\begin{array}{ccc} \prod_{\lambda \in I} E_\lambda & \xrightarrow{\ u\ } & \prod_{\lambda \in I} F_\lambda \\ {\scriptstyle p_\lambda}\downarrow & & \downarrow{\scriptstyle q_\lambda} \\ E_\lambda & \xrightarrow{\ u_\lambda\ } & F_\lambda \end{array}$$

commutes, p_λ and q_λ being the canonical projections of index λ.

Proof. We have only to apply Proposition 1, with $(F_\lambda)_\lambda$, $\prod_{\lambda \in I} E_\lambda$, $(u_\lambda \circ p_\lambda)_\lambda$ standing, respectively, for $(E_\lambda)_\lambda$, X, $(f_\lambda)_\lambda$ from the aforementioned Proposition. ▮

REMARK. Under the conditions of Corollary 1, if $(x_\lambda)_\lambda \in \prod_{\lambda \in I} E_\lambda$, then

$$q_\lambda(u((x_\lambda)_\lambda)) = u_\lambda(p_\lambda((x_\lambda)_\lambda)) = u_\lambda(x_\lambda), \quad \lambda \in I,$$

hence

$$u((x_\lambda)_\lambda) = (u_\lambda(x_\lambda))_\lambda,$$

which may be seen also from the construction of f in Proposition 1. ▮

DEFINITION 2. Under the conditions of Corollary 1, the linear mapping $u : \prod_{\lambda \in I} E_\lambda \to \prod_{\lambda \in I} F_\lambda$ is called the *direct product* or also the *product* of the given family of linear mappings $(u_\lambda)_{\lambda \in I}$, and is denoted $\prod_{\lambda \in I} u_\lambda$; if $I = \{1, \ldots, n\}$, we shall write instead of $\prod_{\lambda \in I} u_\lambda$ also $u_1 \times \ldots \times u_n$. ▮

COROLLARY 2. *Let three families of A-modules $(E_\lambda)_{\lambda \in I}$, $(F_\lambda)_{\lambda \in I}$, $(G_\lambda)_{\lambda \in I}$ indexed by the same set I, and for any $\lambda \in I$ two A-homomorphisms $u_\lambda : E_\lambda \to F_\lambda$ and $v_\lambda : F_\lambda \to G_\lambda$ be given; then*

$$(\prod_{\lambda \in I} v_\lambda) \circ (\prod_{\lambda \in I} u_\lambda) = \prod_{\lambda \in I} (v_\lambda \circ u_\lambda); \tag{4}$$

moreover,

$$\prod_{\lambda \in I} 1_{E_\lambda} = 1_{\prod_{\lambda \in I} E_\lambda}; \tag{5}$$

if for any $\lambda \in I$ the A-homomorphism $u'_\lambda\colon E_\lambda \to F_\lambda$ is also given, then

$$(\prod_{\lambda\in I} u_\lambda) + (\prod_{\lambda\in I} u'_\lambda) = \prod_{\lambda\in I} (u_\lambda + u'_\lambda). \tag{6}$$

Proof. Let us consider, for each $\lambda \in I$, the diagram

$$\begin{array}{ccccc}
\prod_{\lambda\in I} E_\lambda & \xrightarrow{\prod_{\lambda\in I} u_\lambda} & \prod_{\lambda\in I} F_\lambda & \xrightarrow{\prod_{\lambda\in I} v_\lambda} & \prod_{\lambda\in I} G_\lambda \\
\downarrow{\scriptstyle p_\lambda} & & \downarrow{\scriptstyle q_\lambda} & & \downarrow{\scriptstyle r_\lambda} \\
E_\lambda & \xrightarrow{u_\lambda} & F_\lambda & \xrightarrow{v_\lambda} & G_\lambda,
\end{array}$$

$p_\lambda, q_\lambda, r_\lambda$ being the canonical projections; by the definition of $\prod_\lambda u_\lambda$ and $\prod_\lambda v_\lambda$, these diagrams are commutative; hence

$$r_\lambda \circ (\prod_\lambda v_\lambda) \circ (\prod_\lambda u_\lambda) = v_\lambda \circ u_\lambda \circ p_\lambda, \qquad \lambda \in I;$$

but we have also

$$r_\lambda \circ (\prod_\lambda (v_\lambda \circ u_\lambda)) = v_\lambda \circ u_\lambda \circ p_\lambda, \qquad \lambda \in I,$$

and by the uniqueness of $\prod_{\lambda\in I} (v_\lambda \circ u_\lambda)$ with this property, we obtain (4). Relations (5) and (6) are proved analogously. ▮

COROLLARY 3. *Let $(E_\lambda)_{\lambda\in I}$ be a family of A-modules; let*

$$I = \bigcup_{\mu\in M} J_\mu$$

be a partition of the index set I; then there exists a canonical A-isomorphism

$$\prod_{\lambda\in I} E_\lambda \rightleftarrows \prod_{\mu\in M} (\prod_{\lambda\in J_\mu} E_\lambda).$$

Proof. Let $\mu \in M$; for any $\lambda \in J_\mu$, let us consider the canonical projection $p_\lambda : \prod_{\lambda\in I} E_\lambda \to E_\lambda$; by Proposition 1, there exists an A-homomorphism

$$f_\mu : \prod_{\lambda\in I} E_\lambda \to \prod_{\lambda\in J_\mu} E_\lambda$$

uniquely determined by the property $q_\lambda \circ f_\mu = p_\lambda$, $\lambda \in J_\mu$, where q_λ is the canonical projection $\prod_{\lambda\in J_\mu} E_\lambda \to E_\lambda$; applying Proposition 1 again to the family of linear mappings $(f_\mu)_{\mu \in M}$, we get an A-homomorphism

$$f: \prod_{\lambda \in I} E_\lambda \to \prod_{\mu \in M} \Big(\prod_{\lambda \in J_\mu} E_\lambda \Big)$$

uniquely determined by the property $r_\mu \circ f = f_\mu$, $\mu \in M$, r_μ being the canonical projection $\prod_{\mu \in M} \big(\prod_{\lambda \in J_\mu} E_\lambda \big) \to \prod_{\lambda \in J_\mu} E_\lambda$.

We are going to show that f is an isomorphism: to do it, let $\lambda \in I$; there exists $\mu \in M$ uniquely determined by the property $\lambda \in J_\mu$; we have the composite homomorphism $q_\lambda \circ r_\mu$; according to Proposition 1, there exists an A-homomorphism

$$g: \prod_{\mu \in M} \Big(\prod_{\lambda \in J_\mu} E_\lambda \Big) \to \prod_{\lambda \in I} E_\lambda$$

uniquely determined by the property $p_\lambda \circ g = q_\lambda \circ r_\mu$, where $\lambda \in J_\mu$.

If $\lambda \in J_\mu$, the situation is described by the diagram

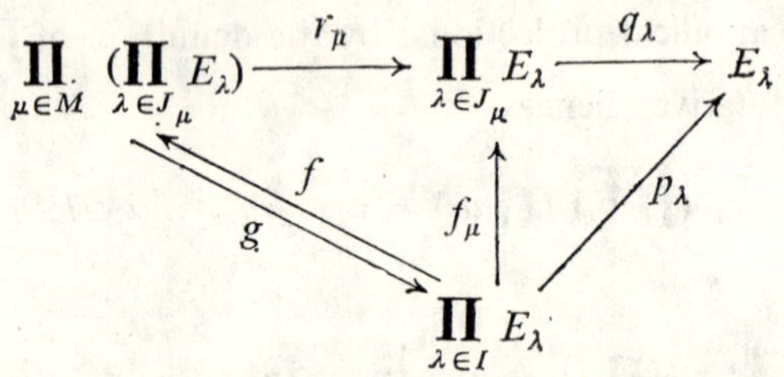

For any $\mu \in M$ we have

$$r_\mu \circ f \circ g = f_\mu \circ g; \tag{7}$$

but if $\lambda \in J_\mu$, then $q_\lambda \circ f_\mu \circ g = p_\lambda \circ g = q_\lambda \circ r_\mu$, whence, by virtue of the uniqueness, we have

$$f_\mu \circ g = r_\mu = r_\mu \circ 1_{\prod_\mu (\prod_\lambda E_\lambda)}, \qquad \mu \in M; \tag{8}$$

from (7) and (8), again by virtue of the uniqueness, it follows

$$f \circ g = 1_{\prod_\mu (\prod_\lambda E_\lambda)}.$$

On the other hand, for any $\lambda \in I$, if $\lambda \in J_\mu$, we have $p_\lambda \circ g \circ f = q_\lambda \circ r_\mu \circ f = q_\lambda \circ f_\mu = p_\lambda$, and, again by the uniqueness of the linear mapping obtained according to Proposition 1, we have

$$g \circ f = 1_{\prod_\lambda E_\lambda}.$$

Hence f is an isomorphism, and g is its inverse. ∎

The property that for any partition of the index set there exists the canonical isomorphism given by Corollary 3 is called the "associativity of the direct product".

REMARK. If, using the notations of Corollary 3, $(x_\lambda)_{\lambda\in I}$ is an arbitrary element of $\prod\limits_{\lambda\in I} E_\lambda$, then

$$f((x_\lambda)_{\lambda\in I}) = ((x_\lambda)_{\lambda\in J_\mu})_{\mu\in M},$$

as it is easily seen from the definition of f. ∎

If in the family $(E_\lambda)_{\lambda\in I}$ all modules E_λ, $\lambda\in I$, are equal to the same A-module E, then $\prod\limits_{\lambda\in I} E_\lambda$ is denoted E^I or E^n if $I=\{1,\ldots,n\}$; note that E^1 is identified with E; the elements of E^I are then the mappings $I\to E$, while the addition and the scalar multiplication are defined "pointwise". Also in this case, the mapping

$$E\to E^I$$

obtained from the universal property of the direct product taking in Proposition 1 $f_\lambda = 1_E$, $\lambda\in I$, carries an element $x\in E$ into the family whose components are all x; it is called the *diagonal mapping* $E\to E^I$.

* DEFINITION 3. Let $\mathscr{C}$ be a category; let $(E_\lambda)_{\lambda\in I}$ be a family of objects of $\mathscr{C}$ indexed by the set I; a *direct product* or a *product* in the category $\mathscr{C}$ of the given family of objects is a system $(E,(p_\lambda)_{\lambda\in I})$ consisting of an object E of $\mathscr{C}$ and of a family of morphisms $p_\lambda : E\to E_\lambda$, $\lambda\in I$, such that, for any object X of $\mathscr{C}$ and for any family of morphisms $f_\lambda : X\to E_\lambda$, $\lambda\in I$, there exists a unique morphism $f: X\to E$ with the property that

$$p_\lambda\circ f = f_\lambda \qquad \text{for any } \lambda\in I. \ ∎$$

Using the definition, it follows that if $(E',(p'_\lambda)_{\lambda\in I})$ is another direct product of the same family of objects $(E_\lambda)_{\lambda\in I}$, then there exists a unique morphism

$$u : E\to E'$$

with the property of rendering commutative the diagram

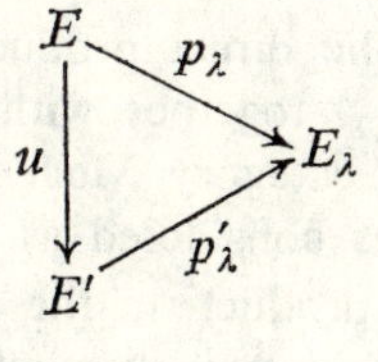

for any $\lambda\in I$; moreover, u is an isomorphism of the category $\mathscr{C}$.

We next give a proof of this fact; this is actually a standard proof in the Theory of Categories, and can be used *mutatis mutandis* also in other circumstances (kernel, direct sum, projective limit, etc.); we are not going to give it again for the other constructions, all the details involved there being left to the reader.

Taking E' for X of Definition 3, and p'_λ for f_λ, $\lambda \in I$, we get a unique morphism $v : E' \to E$ with the property that

$$p_\lambda \circ v = p'_\lambda \qquad \text{for any } \lambda \in I;$$

interchanging the roles of $(E, (p_\lambda)_{\lambda \in I})$ and $(E', (p'_\lambda)_{\lambda \in I})$, we get a unique morphism $u : E \to E'$ with the property that

$$p'_\lambda \circ u = p_\lambda \qquad \text{for any } \lambda \in I;$$

but then, by substitution, we get

$$p_\lambda \circ v \circ u = p_\lambda = p_\lambda \circ 1_E \qquad \text{for any } \lambda \in I;$$

taking now in Definition 3 $X = E$ and $f_\lambda = p_\lambda$, $\lambda \in I$, it follows that for the family of mappings $(p_\lambda)_{\lambda \in I}$ there exist two morphisms, namely $v \circ u$ and 1_E, satisfying the condition imposed on f in Definition 3; by virtue of the uniqueness, it follows that

$$v \circ u = 1_E;$$

in the same way, we get

$$u \circ v = 1_{E'},$$

and hence u and v are isomorphisms inverse to each other.

Having in mind this property, if for the family $(E_\lambda)_{\lambda \in I}$ of objects of $\mathscr{C}$ there exists at least one direct product, then, from all the direct products of this family, one is chosen, say $(E, (p_\lambda)_{\lambda \in I})$, which is called *the direct product* of the family $(E_\lambda)_{\lambda \in I}$; the object E is denoted in this case $\prod_{\lambda \in I} E_\lambda$, and it is this object which is sometimes called, by an abuse of language, the *direct product* of the given family of objects; the morphism p_λ, $\lambda \in I$, corresponding to this choice is called the *canonical projection* of index λ, $\prod_{\lambda \in I} E_\lambda \to E_\lambda$.*

Proposition 1 shows that the direct product, in the sense of Definition 1, of a family of A-modules $(E_\lambda)_{\lambda \in I}$, together with the family of canonical projections, is a direct product in the category **Mod**(A), in the sense of Definition 3, of the given family of A-modules considered as objects of the category **Mod**(A).

Another example of direct product, in the sense of Definition 3, is the one in the category **Ann** of all rings (with unit element): given a family of rings $(A_\lambda)_{\lambda \in I}$, the cartesian product A of the underlying sets A_λ, $\lambda \in I$, is considered; a structure of ring is defined on A putting

$$(\alpha_\lambda)_{\lambda \in I} + (\beta_\lambda)_{\lambda \in I} = (\alpha_\lambda + \beta_\lambda)_{\lambda \in I},$$

$$(\alpha_\lambda)_{\lambda \in I} \cdot (\beta_\lambda)_{\lambda \in I} = (\alpha_\lambda \beta_\lambda)_{\lambda \in I},$$

where α_λ, $\beta_\lambda \in A_\lambda$, $\lambda \in I$; it follows immediately that, under these operations, A turns out to be a ring, the unit element of it being the family whose component of index λ is the unit element of A_λ, $\lambda \in I$; moreover, the mapping $p_\lambda : A \to A_\lambda$, $\lambda \in I$, which carries a family $(\alpha_\lambda)_{\lambda \in I}$ into its component of index λ is a homomorphism of rings with unit element; the system $(A, (p_\lambda)_{\lambda \in I})$ turns out to be a direct product, in the sense of Definition 3, in the category **Ann** of the family of rings $(A_\lambda)_{\lambda \in I}$; the ring A is denoted $\prod_{\lambda \in I} A_\lambda$, while the mapping p_λ is called, according to the general terminology, the *canonical projection of index* λ or the canonical projection $\prod_{\lambda \in I} A_\lambda \to A_\lambda$, $\lambda \in I$.

* Returning to a category $\mathcal{C}$, let us assume that the following Axiom is satisfied:

AB3°) *For any family $(E_\lambda)_{\lambda \in I}$ of objects of $\mathcal{C}$, I being an arbitrary set, there exists in the category $\mathcal{C}$ at least one direct product, in the sense of Definition* 3.

When this is the case, $\mathcal{C}$ is said to be a *category with direct products* or a *category with products*; if only finite sets I are considered, $\mathcal{C}$ is said to be a *category with finite (direct) products.*

Then, keeping the notations employed in the case of modules, the assertions given by Corollaries 1 and 3 to Proposition 1, as well as those given by formulae (4) and (5) of Corollary 2 for A-modules remain valid when replacing A-modules and A-homomorphisms by objects, respectively morphisms of the category $\mathcal{C}$, and where $\prod_{\lambda \in I} E_\lambda$ and the canonical projections have the *determined* meaning mentioned above.

It is to be understood that at the proper moment we have formulated Definition 2 for the category $\mathcal{C}$, the notation $\prod_{\lambda \in I} u_\lambda$ preserving its meaning.

From the statements analogous to those given by formulae (4) and (5) of Corollary 2 we get that, the index set I, $I \neq \emptyset$, being fixed, the mappings

$$(E_\lambda)_{\lambda \in I} \mapsto \prod_{\lambda \in I} E_\lambda, \quad (u_\lambda)_{\lambda \in I} \mapsto \prod_{\lambda \in I} u_\lambda$$

define a functor from the category $\mathcal{C}^I$ (see Definition 5, § 2) to the category $\mathcal{C}$, called the *direct product functor.* *

PROPOSITION 2. *Let A be a ring, and let $(E_\lambda)_{\lambda \in I}$, $(F_\lambda)_{\lambda \in I}$, $(G_\lambda)_{\lambda \in I}$ be three families of A-modules indexed by the same set I; we assume that for any $\lambda \in I$ the A-homomorphisms*

$$E_\lambda \xrightarrow{u_\lambda} F_\lambda \xrightarrow{v_\lambda} G_\lambda \tag{9}$$

are given; then the diagram

$$\prod_{\lambda\in I} E_\lambda \xrightarrow{\prod_{\lambda\in I} u_\lambda} \prod_{\lambda\in I} F_\lambda \xrightarrow{\prod_{\lambda\in I} v_\lambda} \prod_{\lambda\in I} G_\lambda \tag{10}$$

is an exact sequence if and only if diagram (9) *is an exact sequence for any* $\lambda \in I$.

Proof. Let $y = (y_\lambda)_{\lambda\in I}$ be an element of $\prod_{\lambda\in I} F_\lambda$; the statement that y belongs to $\mathrm{Ker}(\prod_{\lambda\in I} v_\lambda)$ amounts to $v_\lambda y_\lambda = 0$ for any $\lambda \in I$; hence, if (9) is an exact sequence for any $\lambda \in I$, this is equivalent to saying that $y_\lambda = u_\lambda x_\lambda$ with some $x_\lambda \in E_\lambda$ for any $\lambda \in I$, hence that

$$y = (y_\lambda)_{\lambda\in I} = (\prod_{\lambda\in I} u_\lambda)((x_\lambda)_{\lambda\in I}) \in \mathrm{Im}(\prod_{\lambda\in I} u_\lambda);$$

thus, if (9) is an exact sequence, (10) is also an exact sequence. Conversely, if (10) is an exact sequence, let $\lambda \in I$; then to say that y_λ belongs to $\mathrm{Ker}(v_\lambda)$ is equivalent to saying that the family $y \in \prod_{\lambda\in I} F_\lambda$ whose component of index λ is y_λ while all the other components are 0 is in the kernel of $\prod_{\lambda\in I} v_\lambda$, which is equivalent to saying that y belongs to $\mathrm{Im}(\prod_{\lambda\in I} u_\lambda)$, hence equivalent to the fact that y_λ belongs to $\mathrm{Im}(u_\lambda)$. ∎

COROLLARY 1. *If* $(E_\lambda)_{\lambda\in I}$, $(F_\lambda)_{\lambda\in I}$, $(u_\lambda)_{\lambda\in I}$ *have the same meaning as in Proposition* 2, *and if for any* $\lambda \in I$, u_λ *is an injective, respectively surjective, respectively bijective mapping, then* $\prod_{\lambda\in I} u_\lambda$ *is an injective, respectively surjective, respectively bijective mapping.*

Proof. We have only to apply Proposition 1, § 5, and Proposition 2 taking into account that the product of a family of zero modules is the zero module. ∎

In particular, if $(E_\lambda)_{\lambda\in I}$ is a family of A-modules, and if for any $\lambda \in I$, M_λ is a submodule of E_λ, with the canonical injection i_λ, then $\prod_{\lambda\in I} i_\lambda$ is an injective homomorphism $\prod_{\lambda\in I} M_\lambda \to \prod_{\lambda\in I} E_\lambda$; by this homomorphism $\prod_\lambda M_\lambda$ is identified with the submodule $(\prod_\lambda i_\lambda)(\prod_\lambda M_\lambda)$ of $\prod_\lambda E_\lambda$; under this identification $\prod_\lambda i_\lambda$ becomes the canonical injection.

Relations (11) of Corollary 2 below as well as Corollary 3 are now given having in mind the above identification.

COROLLARY 2. *If* $(E_\lambda)_{\lambda\in I}$, $(F_\lambda)_{\lambda\in I}$, $(u_\lambda)_{\lambda\in I}$ *have the same meaning as in Proposition* 2, *then*

$$\mathrm{Ker}(\prod_{\lambda\in I} u_\lambda) = \prod_{\lambda\in I} \mathrm{Ker}(u_\lambda),\quad \mathrm{Im}(\prod_{\lambda\in I} u_\lambda) = \prod_{\lambda\in I} \mathrm{Im}(u_\lambda), \tag{11}$$

and there exist canonical isomorphisms

$$\mathrm{Coker}\,(\prod_{\lambda\in I} u_\lambda) \rightleftarrows \prod_{\lambda\in I} \mathrm{Coker}\,(u_\lambda),\quad \mathrm{Coim}\,(\prod_{\lambda\in I} u_\lambda) \rightleftarrows \prod_{\lambda\in I} \mathrm{Coim}(u_\lambda). \tag{12}$$

Proof. Let $M_\lambda = \text{Ker}(u_\lambda)$ and $i_\lambda : M_\lambda \to E$ be the canonical injection; we have for any $\lambda \in I$ the exact sequence

$$0 \to M_\lambda \xrightarrow{i_\lambda} E_\lambda \xrightarrow{u_\lambda} F_\lambda;$$

according to Proposition 2 we get the exact sequence

$$0 \to \prod_{\lambda\in I} M_\lambda \xrightarrow{\prod_\lambda i_\lambda} \prod_{\lambda\in I} E_\lambda \xrightarrow{\prod_\lambda u_\lambda} \prod_{\lambda\in I} F_\lambda;$$

hence $\text{Ker}(\prod_\lambda u_\lambda) = (\prod_\lambda i_\lambda)(\prod_\lambda M_\lambda) = \prod_\lambda M_\lambda$.

For the second equality (11) we argue in the same way, but considering this time the exact sequences

$$0 \to I_\lambda \xrightarrow{j_\lambda} F_\lambda \xrightarrow{p_\lambda} P_\lambda, \quad \lambda \in I,$$

where $I_\lambda = \text{Im}(u_\lambda)$, $P_\lambda = \text{Coker}(u_\lambda)$, while j_λ and p_λ represent, respectively, the canonical injection and surjection.

Next, for any $\lambda \in I$ we have the exact sequence

$$E_\lambda \xrightarrow{u_\lambda} F_\lambda \xrightarrow{p_\lambda} P_\lambda \to 0,$$

whence we derive the exact sequence

$$\prod_{\lambda\in I} E_\lambda \xrightarrow{\prod_\lambda u_\lambda} \prod_{\lambda\in I} F_\lambda \xrightarrow{\prod_\lambda p_\lambda} \prod_{\lambda\in I} P_\lambda \to 0;$$

by Proposition 2, *b*), § 5, there exists an isomorphism

$$v : (\prod_\lambda F_\lambda)/((\prod_\lambda u_\lambda)(\prod_\lambda E_\lambda)) = \text{Coker}\,(\prod_\lambda u_\lambda) \rightleftarrows \prod_\lambda P_\lambda = \prod_\lambda \text{Coker}\,(u_\lambda)$$

with the property that $v \circ p = \prod_\lambda p_\lambda$, p being the canonical surjection $\prod_\lambda F_\lambda \to$ $\to \text{Coker}\,(\prod_\lambda u_\lambda)$; this isomorphism is in fact the one referred to in the statement of the Corollary.

For the second isomorphism (12) we proceed as before, considering the exact sequences

$$M_\lambda \xrightarrow{i_\lambda} E_\lambda \xrightarrow{q_\lambda} C_\lambda \to 0, \quad \lambda \in I,$$

where $C_\lambda = \text{Coim}(u_\lambda)$ and q_λ is the canonical surjection. ∎

COROLLARY 3. *If $(E_\lambda)_{\lambda\in I}$ is a family of A-modules, and if for any $\lambda \in I$, F_λ is a submodule of E_λ, then we have a canonical isomorphism*

$$\prod_{\lambda\in I} (E_\lambda/F_\lambda) \rightleftarrows (\prod_{\lambda\in I} E_\lambda)/(\prod_{\lambda\in I} F_\lambda).$$

Proof. We need only apply Corollary 2 to the case when u_λ is the canonical injection $F_\lambda \to E_\lambda$; the stated isomorphism is the inverse of the first isomorphism (12) for this case. ▮

COROLLARY 4. *If $(E_\lambda)_{\lambda\in I}$, $(F_\lambda)_{\lambda\in I}$, $(u_\lambda)_{\lambda\in I}$ are as in Proposition 2, then $\prod_{\lambda\in I} u_\lambda$ is, respectively, an injective, surjective, bijective, zero mapping if and only if for any $\lambda \in I$, u_λ is, respectively, an injective, surjective, bijective, zero mapping.*

Proof. The first three assertions are consequences of Corollaries 1 and 2 above; as for the zero mapping, the assertion follows directly from the way in which $\prod_\lambda u_\lambda$ acts. ▮

* Let $(\mathcal{C}_\lambda)_{\lambda\in I}$, $I \neq \emptyset$, be a family of preadditive categories, and let $\mathcal{C} = \prod_{\lambda\in I} \mathcal{C}_\lambda$ (see Definition 5, § 2). If $E = (E_\lambda)_{\lambda\in I}$ and $F = (F_\lambda)_{\lambda\in I}$ are two objects of $\mathcal{C}$, we may define on $\mathrm{Hom}_{\mathcal{C}}(E, F)$ an abelian group structure as follows: according to the definition of the product category, $\mathrm{Hom}_{\mathcal{C}}(E, F)$ is, when considered as a set, exactly the cartesian product of the sets $\mathrm{Hom}_{\mathcal{C}_\lambda}(E_\lambda, F_\lambda)$, $\lambda \in I$; each of these sets is endowed with a structure of abelian group, i.e. of **Z**-module (see Example 4), § 1); in turn, $\mathrm{Hom}_{\mathcal{C}}(E, F)$ may be endowed with the **Z**-module structure obtained as the product of the modules $\mathrm{Hom}_{\mathcal{C}_\lambda}(E_\lambda, F_\lambda)$, $\lambda \in I$, and it is this abelian group structure of $\mathrm{Hom}_{\mathcal{C}}(E, F)$ which is considered; it is not difficult to see that, with this definition, for any other object $G = (G_\lambda)_{\lambda\in I}$ of $\mathcal{C}$ and any family of morphisms $v, v', v'' \in \mathrm{Hom}_{\mathcal{C}}(F, G)$, $u, u', u'' \in \mathrm{Hom}_{\mathcal{C}}(E, F)$ we have the relations

$$v\circ(u' + u'') = v\circ u' + v\circ u'', \quad (v' + v'')\circ u = v'\circ u + v''\circ u.$$

It is also easily seen that if 0_λ is a zero object of the category $\mathcal{C}_\lambda$, then the object $0 = (0_\lambda)_{\lambda\in I}$ is a zero object of $\mathcal{C}$. In this sense, we say that *the product of a family of preadditive categories is a preadditive category.*

If, moreover, each of the categories $\mathcal{C}_\lambda$, $\lambda \in I$, satisfies Axiom AB1) (respectively AB1) and AB2)), then $\mathcal{C} = \prod_{\lambda\in I} \mathcal{C}_\lambda$ also satisfies AB1) (respectively AB1) and AB2)): for instance, if, in the above notations, the pair (M_λ, i_λ) is a kernel of u_λ in $\mathcal{C}_\lambda$, $\lambda \in I$, then the pair $((M_\lambda)_{\lambda\in I}, (i_\lambda)_{\lambda\in I})$ is a kernel of $u = (u_\lambda)_{\lambda\in I}$ in $\mathcal{C}$. The fact that the diagram

$$(E_\lambda)_{\lambda\in I} \xrightarrow{(u_\lambda)_{\lambda\in I}} (F_\lambda)_{\lambda\in I} \xrightarrow{(v_\lambda)_{\lambda\in I}} (G_\lambda)_{\lambda\in I}$$

is an exact sequence in $\mathcal{C}$ amounts to the fact that the diagram

$$E_\lambda \xrightarrow{u_\lambda} F_\lambda \xrightarrow{v_\lambda} G_\lambda$$

is an exact sequence in $\mathcal{C}_\lambda$ for any $\lambda \in I$.

Now, let $\mathcal{C}$ be a preadditive category satisfying Axiom AB3°), and let I be a non-empty set; then, under the conditions of Corollary 2 to Proposition 1, the assertion given by relation (6) is still true when replacing the A-modules and A-homomorphisms, respectively, by objects and morphisms of the category $\mathcal{C}$, and where $\prod_{\lambda} E_{\lambda}$ and the canonical projections have the *determined* meaning, in the already mentioned sense. Consequently, under these conditions, the direct product functor $\mathcal{C}^I \to \mathcal{C}$ is an additive functor.*

These being specified, Proposition 2 also has the following Corollary:

COROLLARY 5. *The direct product functor*

$$(\mathbf{Mod}(A))^I \to \mathbf{Mod}(A),\ I \neq \emptyset,$$

is an exact functor. ∎

§ 7. Direct Sums

Let A be a ring, and let $(E_\lambda)_{\lambda \in I}$ be a family of A-modules; consider in the module $\prod_{\lambda \in I} E_\lambda$ the subset E of all fs-families $(x_\lambda)_{\lambda \in I}$ (see § 1); if $x = (x_\lambda)_\lambda$, $y = (y_\lambda)_\lambda$ are two such families, then their sum in $\prod_{\lambda \in I} E_\lambda$ is also such a family, since the support of $x + y$ is contained in the union of the two supports; likewise, the support of αx is, for any $\alpha \in A$, contained in the support of x; also, the zero element of $\prod_{\lambda \in I} E_\lambda$ is in E, this being the family whose support is empty; whence E is a submodule of $\prod_{\lambda \in I} E_\lambda$.

DEFINITION 1. The module E considered by itself is called the *external direct sum* or also the *direct sum* of the given family of A-modules $(E_\lambda)_{\lambda \in I}$, and is denoted $\bigoplus_{\lambda \in I} E_\lambda$; if $I = \{1, 2, \ldots, n\}$, then instead of $\bigoplus_{\lambda \in I} E_\lambda$ we write also $E_1 \oplus E_2 \oplus \ldots \oplus E_n$ or $\bigoplus_{i=1}^{n} E_i$. ▮

Since $\bigoplus_{\lambda \in I} E_\lambda$ is a submodule of $\prod_{\lambda \in I} E_\lambda$, it follows that for $I = \emptyset$ we have $\bigoplus_{\lambda \in I} E_\lambda = 0$; from the definition we obtain also that if I is a finite set, then $\bigoplus_{\lambda \in I} E_\lambda = \prod_{\lambda \in I} E_\lambda$: indeed, in this case any element of $\prod_{\lambda \in I} E_\lambda$ is an fs-family.

Returning to the general case and using the above notation, let us consider for any $\lambda \in I$ the mapping

$$j_\lambda \colon E_\lambda \to E$$

which carries an element $x_\lambda \in E_\lambda$ into the family (element of $\prod_{\lambda \in I} E_\lambda$) whose component of index λ is x_λ and whose component of index μ, $\mu \neq \lambda$, is 0; this family is an element of E; obviously, j_λ is an injective linear mapping; it is called the *canonical injection of index* (or of *order*) λ.[1]

If E, F are two A-modules, then, considering their direct sum $E \oplus F$, which is a direct product as well, and denoting by j_E and j_F the canonical injections and by p_E and p_F the canonical projections, the diagram

$$0 \to E \xrightarrow{j_E} E \oplus F \xrightarrow{p_F} F \to 0$$

---

[1] This should not be confounded with the canonical injection of a submodule into a module (see § 3).

is an exact sequence, since for $(x, y) \in E \oplus F$ we have $p_F(x, y) = 0$ if and only if $(x, y) = j_E(x)$.

LEMMA 1. *Any element* $x \in \bigoplus_{\lambda \in I} E_\lambda = E$ *may be uniquely written in the form*

$$x = \sum_{\lambda \in I} j_\lambda(x_\lambda), \tag{1}$$

where $x_\lambda \in E_\lambda$, $\lambda \in I$, *and the family* $(x_\lambda)_{\lambda \in I}$ *is an fs-family.*

Proof. We first notice that the sum of the right side of (1) makes sense, since $(x_\lambda)_{\lambda \in I}$ being an fs-family, it follows that $(j_\lambda(x_\lambda))_{\lambda \in I}$ is an fs-family of elements of E.

Now, if J is the support of the family $x = (x_\lambda)_{\lambda \in I} \in E$, then J is also the support of the family $(j_\lambda(x_\lambda))_{\lambda \in I}$, and we have

$$\sum_{\lambda \in I} j_\lambda(x_\lambda) = \sum_{\lambda \in J} j_\lambda(x_\lambda); \tag{2}$$

let $\mu \in J$; the component of index μ of $j_\mu(x_\mu)$ is x_μ, while for $\lambda \in J$, $\lambda \neq \mu$, the component of index μ of $j_\lambda(x_\lambda)$ is 0; hence the component of index μ of the element of the right hand side of (2) is x_μ; if $\mu \in I \setminus J$, the component of index μ of $j_\lambda(x_\lambda)$ is 0 for any $\lambda \in J$; in this way, we proved (1), in which the family $(x_\lambda)_\lambda$ is x itself.

If besides (1) we have also $x = \sum_{\lambda \in I} j_\lambda(x'_\lambda)$, where $(x'_\lambda)_{\lambda \in I}$ is an fs-family, then, by an argument similar to the one employed above, it follows $x = (x'_\lambda)_{\lambda \in I}$; hence $x_\lambda = x'_\lambda$ for any $\lambda \in I$, and this completes the proof. ▮

We have the following universal property of the direct sum:

PROPOSITION 1. *Let* $E = \bigoplus_{\lambda \in I} E_\lambda$ *be the direct sum of the family* $(E_\lambda)_{\lambda \in I}$ *of A-modules, with the canonical injections* j_λ, $\lambda \in I$; *then, for any A-module* Y *and any family of linear mappings* $g_\lambda : E_\lambda \to Y$, $\lambda \in I$, *there exists a unique A-homomorphism* $g : E \to Y$ *rendering commutative the diagram*

(3)

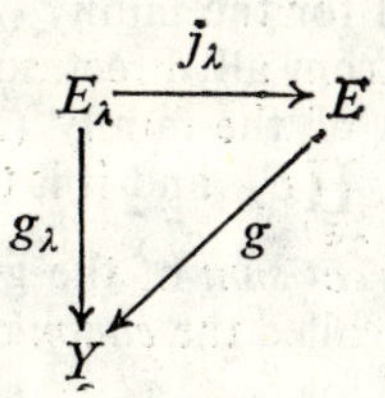

for any $\lambda \in I$; *hence for which* $g \circ j_\lambda = g_\lambda$, $\lambda \in I$.

Proof. Let $x = (x_\lambda)_{\lambda \in I} \in E$; we define

$$g((x_\lambda)_\lambda) = \sum_{\lambda \in I} g_\lambda(x_\lambda),$$

the right hand side sum making sense since $(x_\lambda)_{\lambda\in I}$ is an fs-family, and, consequently, $(g_\lambda(x_\lambda))_{\lambda\in I}$ is also an fs-family. The mapping g thus defined is linear: let us check, for instance, that

$$g(\alpha(x_\lambda)_{\lambda\in I}) = \alpha\cdot g((x_\lambda)_{\lambda\in I});$$

we have

$$g(\alpha(x_\lambda)_\lambda) = g((\alpha x_\lambda)_\lambda) = \sum_{\lambda\in I} g_\lambda(\alpha x_\lambda) = \sum_{\lambda\in I} \alpha g_\lambda(x_\lambda) = \alpha\cdot g((x_\lambda)_\lambda).$$

Next, if $\lambda\in I$ and $x_\lambda\in E_\lambda$, the support of $j_\lambda(x_\lambda)$ consists at most of λ, hence $g(j_\lambda(x_\lambda)) = g_\lambda(x_\lambda)$, and the commutativity of diagram (3) follows for any $\lambda\in I$.

Let $h: E\to Y$ be another linear mapping rendering diagram (3) commutative when replacing g, and let $x\in E$; according to Lemma 1, x may be uniquely written as $x = \sum_{\lambda\in I} j_\lambda(x_\lambda)$, where $x_\lambda\in E_\lambda$ and $(x_\lambda)_{\lambda\in I}$ is an fs-family; then we have

$$h(x) = \sum_{\lambda\in I} h(j_\lambda(x_\lambda)) = \sum_{\lambda\in I} g_\lambda(x_\lambda) = g(x),$$

and $h = g$.[1] ▮

* DEFINITION 2. Let $\mathcal{C}$ be a category; let $(E_\lambda)_{\lambda\in I}$ be a family of objects of $\mathcal{C}$ indexed by the set I; a *direct sum* or a *coproduct* in the category $\mathcal{C}$ of the family $(E_\lambda)_{\lambda\in I}$ is a system $(E, (j_\lambda)_{\lambda\in I})$ consisting of an object E of $\mathcal{C}$ and a family of morphisms $j_\lambda: E_\lambda\to E$, $\lambda\in I$, such that for any object Y of $\mathcal{C}$ and any family of morphisms $g_\lambda: E_\lambda\to Y$, $\lambda\in I$, there exists a unique morphism $g: E\to Y$ with the property that

$$g\circ j_\lambda = g_\lambda \quad \text{for any } \lambda\in I. ▮$$

As seen from the definition, the notions of direct product and direct sum are dual to each other (see § 3).

In the same way as for direct products, if $(E', (j'_\lambda)_{\lambda\in I})$ is another direct sum of the family of objects $(E_\lambda)_{\lambda\in I}$, then there exists a unique morphism $v: E\to E'$ such that $v\circ j_\lambda = j'_\lambda$ for any $\lambda\in I$; moreover, v is an isomorphism of the category $\mathcal{C}$. Taking this fact into account, if for the family $(E_\lambda)_{\lambda\in I}$ of objects of $\mathcal{C}$ there exists at least one direct sum, then from all direct sums one is chosen, say $(E, (j_\lambda)_{\lambda\in I})$, which is called *the direct sum* of the family $(E_\lambda)_{\lambda\in I}$; the object E is denoted in this case $\bigoplus_{\lambda\in I} E_\lambda$ or, alternatively, $\coprod_{\lambda\in I} E_\lambda$, and it is this object which is sometimes called, by an abuse of language, the *direct sum* of the given family; the morphism j_λ, $\lambda\in I$, corresponding to this choice is called the *canonical injection* of index λ, $E_\lambda\to\bigoplus_{\lambda\in I} E_\lambda$.*

---

[1] Let us show, at least in this case, the way we applied the homomorphism h to the "infinite" sum $\sum_{\lambda\in I} j_\lambda(x_\lambda)$: let J be the support of the family $(x_\lambda)_\lambda$; then the support of $(j_\lambda(x_\lambda))_\lambda$ is also J, and we have $\sum_{\lambda\in I} j_\lambda(x_\lambda) = \sum_{\lambda\in J} j_\lambda(x_\lambda)$; but J being finite, we get $h(\sum_{\lambda\in I} j_\lambda(x_\lambda)) = h(\sum_{\lambda\in J} j_\lambda(x_\lambda)) = \sum_{\lambda\in J} h(j_\lambda(x_\lambda))$; now, the support of the family $(h(j_\lambda(x_\lambda)))_{\lambda\in I}$ is contained in J, hence $\sum_{\lambda\in I} h(j_\lambda(x_\lambda)) = \sum_{\lambda\in J} h(j_\lambda(x_\lambda))$.

Proposition 1 shows that the direct sum, in the sense of Definition 1, of a family $(E_\lambda)_{\lambda \in I}$ of A-modules, together with the family of canonical injections defined after Definition 1, is a direct sum in the category **Mod**(A), in the sense of Definition 2, of the given family of A-modules considered as objects of the category **Mod**(A).

Corollaries 1, 2, 3 to Proposition 1, given below, can be proved by arguments dual, respectively, to Corollaries 1, 2, 3 to Proposition 1, § 6.

COROLLARY 1. *Let $(E_\lambda)_{\lambda \in I}$, $(F_\lambda)_{\lambda \in I}$ be two families of A-modules indexed by the same set I; suppose for any $\lambda \in I$ an A-homomorphism $u_\lambda : E_\lambda \to F_\lambda$ is given; under these conditions, there exists a homomorphism $u : \bigoplus_{\lambda \in I} E_\lambda \to \bigoplus_{\lambda \in I} F_\lambda$ uniquely determined by the property that for any $\lambda \in I$, the following diagram commutes:*

$$\begin{array}{ccc} E_\lambda & \xrightarrow{u_\lambda} & F_\lambda \\ {\scriptstyle j_\lambda}\downarrow & & \downarrow{\scriptstyle k_\lambda} \\ \bigoplus_{\lambda \in I} E_\lambda & \xrightarrow{u} & \bigoplus_{\lambda \in I} F_\lambda , \end{array}$$

j_λ and k_λ being the canonical injections of index λ. ▮

REMARK. Under the conditions of Corollary 1, let $x = (x_\lambda)_{\lambda \in I} \in \bigoplus_{\lambda \in I} E_\lambda$; then, according to Lemma 1, we have

$$x = (x_\lambda)_{\lambda \in I} = \sum_{\lambda \in I} j_\lambda(x_\lambda);$$

hence

$$u(x) = \sum_{\lambda \in I} u(j_\lambda(x_\lambda)) = \sum_{\lambda \in I} k_\lambda(u_\lambda(x_\lambda)) = (u_\lambda(x_\lambda))_{\lambda \in I}.$$

This shows that u is the homomorphism induced by $\prod_{\lambda \in I} u_\lambda$ by restriction to $\bigoplus_{\lambda \in I} E_\lambda$ and $\bigoplus_{\lambda \in I} F_\lambda$ (see § 3): indeed, if $(x_\lambda)_\lambda$ is an element of $\bigoplus_{\lambda \in I} E_\lambda$, i.e. an fs-family from $\prod_{\lambda \in I} E_\lambda$, then $(\prod_\lambda u_\lambda)((x_\lambda)_\lambda) = (u_\lambda(x_\lambda))_\lambda$ is an fs-family from $\prod_{\lambda \in I} F_\lambda$, i.e. an element of $\bigoplus_{\lambda \in I} F_\lambda$. ▮

DEFINITION 3. Under the conditions of Corollary 1, the linear mapping

$$u : \bigoplus_{\lambda \in I} E_\lambda \to \bigoplus_{\lambda \in I} F_\lambda$$

is called the *direct sum* of the given family of linear mappings $(u_\lambda)_{\lambda \in I}$, and is denoted $\bigoplus_{\lambda \in I} u_\lambda$; if $I = \{1, \ldots, n\}$, then, instead of $\bigoplus_{\lambda \in I} u_\lambda$ one writes also $u_1 \oplus \ldots \oplus u_n$ or $\bigoplus_{i=1}^{n} u_i$, the notation being adequately modified in the case of another *finite* set of indices. ∎

Corollary 2. *Given three families of A-modules $(E_\lambda)_{\lambda \in I}$, $(F_\lambda)_{\lambda \in I}$, $(G_\lambda)_{\lambda \in I}$, and for any $\lambda \in I$ two A-homomorphisms $u_\lambda \colon E_\lambda \to F_\lambda$ and $v_\lambda \colon F_\lambda \to G_\lambda$, then*

$$(4) \qquad (\bigoplus_{\lambda \in I} v_\lambda) \circ (\bigoplus_{\lambda \in I} u_\lambda) = \bigoplus_{\lambda \in I} (v_\lambda \circ u_\lambda);$$

moreover,

$$(5) \qquad \bigoplus_{\lambda \in I} 1_{E_\lambda} = 1_{\underset{\lambda}{\oplus} E_\lambda};$$

if, in addition, for any $\lambda \in I$ the A-homomorphism $u'_\lambda \colon E_\lambda \to F_\lambda$ is given, then

$$(6) \qquad (\bigoplus_{\lambda \in I} u_\lambda) + (\bigoplus_{\lambda \in I} u'_\lambda) = \bigoplus_{\lambda \in I} (u_\lambda + u'_\lambda). \blacksquare$$

Corollary 3. *Let $(E_\lambda)_{\lambda \in I}$ be a family of A-modules, and let*

$$I = \bigcup_{\mu \in M} J_\mu$$

be a partition of the index set I; then there exists a canonical A-isomorphism

$$\bigoplus_{\mu \in M} (\bigoplus_{\lambda \in J_\mu} E_\lambda) \underset{\sim}{\rightarrow} \bigoplus_{\lambda \in I} E_\lambda. \blacksquare$$

Remark. In Corollary 3 the isomorphism is given by

$$((x_\lambda)_{\lambda \in J_\mu})_{\mu \in M} \mapsto (x_\lambda)_{\lambda \in I}. \blacksquare$$

If, in addition, $(F_\lambda)_{\lambda \in I}$ is another family of A-modules indexed by the same set I, and if, for any $\lambda \in I$, the homomorphism $u_\lambda : E_\lambda \to F_\lambda$ is given, then, under the conditions of Corollary 3, the following diagram commutes:

$$\begin{array}{ccc}
\bigoplus_{\mu \in M} (\bigoplus_{\lambda \in J_\mu} E_\lambda) & \xrightarrow[\sim]{\hspace{2em}} & \bigoplus_{\lambda \in I} E_\lambda \\
{\scriptstyle \bigoplus_{\mu \in M} (\bigoplus_{\lambda \in J_\mu} u_\lambda)} \downarrow & & \downarrow {\scriptstyle \bigoplus_{\lambda \in I} u_\lambda} \\
\bigoplus_{\mu \in M} (\bigoplus_{\lambda \in J_\mu} F_\lambda) & \xrightarrow[\sim]{\hspace{2em}} & \bigoplus_{\lambda \in I} F_\lambda,
\end{array}$$

where the horizontal arrows represent the canonical isomorphisms [1].

---

[1] Naturally, there is a similar commutative diagram for direct products.

The property that for any partition of the index set there exists a canonical isomorphism obtained according to Corollary 3 and making commutative the diagram above is called the "associativity of the direct sum".

If in the family $(E_\lambda)_{\lambda \in I}$ all modules E_λ, $\lambda \in I$, are equal to the same A-module E, then $\bigoplus_{\lambda \in I} E_\lambda$ is denoted $E^{(I)}$ (see further, § 10): the elements of $E^{(I)}$ are thus the mappings $I \to E$ having finite support. The mapping $E^{(I)} \to E$ obtained by the universal property of the direct sum taking in Proposition 1 $Y = E$ and $g_\lambda = 1_E$, $\lambda \in I$, carries an element $(x_\lambda)_{\lambda \in I} \in E^{(I)}$ into the element $\sum_{\lambda \in I} x_\lambda$ (the sum making sense since $(x_\lambda)_\lambda$ is an fs-family); this mapping is called the *codiagonal mapping* $E^{(I)} \to E$.

* Let $\mathcal{C}$ be a category satifsying the Axiom:

AB3) *For any family* $(E_\lambda)_{\lambda \in I}$ *of objects of* $\mathcal{C}$, *there exists in* $\mathcal{C}$ *at least one direct sum of the family, in the sense of Definition* 2.

When this is the case, $\mathcal{C}$ is said to be a *category with direct sums* or a *category with coproducts*; if only finite sets I are considered, $\mathcal{C}$ is said to be a *category with finite direct sums.* Obviously, the dual of a category with direct products is a category with direct sums, and conversely.

In a category with direct sums, all assertions of Corollaries 1 and 3 to Proposition 1, as well as those given by relations (4) and (5) of Corollary 2 for A-modules remain valid when replacing A-modules and A-homomorphisms by objects, respectively morphisms of $\mathcal{C}$, and where $\bigoplus_{\lambda \in I} E_\lambda$ and the canonical injections have the *determined* meaning mentioned above. It is to be understood that, when necessary, we inserted Definition 3, preserving the meaning of the notation $\bigoplus_{\lambda \in I} u_\lambda$.

From the assertions analogous to those given by formulae (4) and (5) of Corollary 2 it can be seen that if the index set I, $I \neq \emptyset$, is fixed, the mappings

$$(E_\lambda)_{\lambda \in I} \mapsto \bigoplus_{\lambda \in I} E_\lambda, \quad (u_\lambda)_{\lambda \in I} \mapsto \bigoplus_{\lambda \in I} u_\lambda$$

define a functor from the category $\mathcal{C}^I$ (see Definition 5, § 2) to the category $\mathcal{C}$, called the *direct sum functor*.

Moreover, if $\mathcal{C}$ is a preadditive category, then the assertion given by relation (6) of Corollary 2 to Proposition 1 remains true provided the appropriate replacements have been made and the determined meaning for $\bigoplus_{\lambda \in I} E_\lambda$ and for the canonical injections has been maintained. Consequently, in this case, the direct sum functor $\mathcal{C}^I \to \mathcal{C}$ is an additive functor ($\mathcal{C}^I$ is a preadditive category, as shown in § 6).*

PROPOSITION 2. *Let A be a ring, and let $(E_\lambda)_{\lambda \in I}$, $(F_\lambda)_{\lambda \in I}$, $(G_\lambda)_{\lambda \in I}$ be three families of A-modules indexed by the same index set I; suppose for any $\lambda \in I$ the diagram of A-modules and A-homomorphisms*

$$E_\lambda \xrightarrow{u_\lambda} F_\lambda \xrightarrow{v_\lambda} G_\lambda \tag{7}$$

is given; then the diagram

$$(8)\qquad \bigoplus_{\lambda\in I} E_\lambda \xrightarrow{\bigoplus_\lambda u_\lambda} \bigoplus_{\lambda\in I} F_\lambda \xrightarrow{\bigoplus_\lambda v_\lambda} \bigoplus_{\lambda\in I} G_\lambda$$

is an exact sequence if and only if diagram (7) *is an exact sequence for any* $\lambda\in I$.

Proof. Let $y=(y_\lambda)_{\lambda\in I}\in\bigoplus_{\lambda\in I} F_\lambda$; to say that y belongs to $\operatorname{Ker}(\bigoplus_\lambda v_\lambda)$ is equivalent to saying that $v_\lambda y_\lambda=0$ for any $\lambda\in I$; if (7) is an exact sequence for any $\lambda\in I$, the last assertion amounts to the fact that $y_\lambda=u_\lambda(x_\lambda)$ for any $\lambda\in I$, with some $(x_\lambda)_\lambda\in\bigoplus_{\lambda\in I} E_\lambda$, since if λ does not belong to the support of $(y_\lambda)_\lambda$, we may take $x_\lambda=0$. Hence, the exactness of the sequence (7) for any $\lambda\in I$ implies the exactness of the sequence (8).

The converse assertion is proved similarly to the corresponding part of the proof of Proposition 2, § 6, taking into account that a family $y=(y_\lambda)_\lambda\in\prod_{\lambda\in I} F_\lambda$ having all components zero except perhaps one of them is an element of $\bigoplus_{\lambda\in I} F_\lambda$. ∎

Corollaries 1, 2, 3, 4 which follow can be proved similarly to Corollaries 1, 2, 3, 4 to Proposition 2, § 6, respectively, the canonical isomorphisms involved being constructed as in those Corollaries:

COROLLARY 1. *If* $(E_\lambda)_{\lambda\in I}$, $(F_\lambda)_{\lambda\in I}$, $(u_\lambda)_{\lambda\in I}$ *have the same meaning as in Proposition* 2, *and if for any* $\lambda\in I$, u_λ *is, respectively, an injective, surjective, bijective mapping, then* $\bigoplus_{\lambda\in I} u_\lambda$ *is an injective, surjective, bijective mapping, respectively.* ∎

In the same way as for direct products (see § 6), if for any $\lambda\in I$, M_λ is a submodule of the A-module E_λ, with the canonical injection i_λ, then by means of the injective homomorphism $\bigoplus_{\lambda\in I} i_\lambda$, the module $\bigoplus_{\lambda\in I} M_\lambda$ is identified with a submodule of $\bigoplus_{\lambda\in I} E_\lambda$; this identification is used in the formulation of Corollaries 2 and 3 below.

COROLLARY 2. *If* $(E_\lambda)_{\lambda\in I}$, $(F_\lambda)_{\lambda\in I}$, $(u_\lambda)_{\lambda\in I}$ *are as in Proposition* 2, *then*

$$\operatorname{Ker}(\bigoplus_{\lambda\in I} u_\lambda)=\bigoplus_{\lambda\in I}\operatorname{Ker}(u_\lambda),\ \operatorname{Im}(\bigoplus_{\lambda\in I} u_\lambda)=\bigoplus_{\lambda\in I}\operatorname{Im}(u_\lambda),$$

and there exist canonical isomorphisms

$$\operatorname{Coker}(\bigoplus_{\lambda\in I} u_\lambda)\rightleftarrows\bigoplus_{\lambda\in I}\operatorname{Coker}(u_\lambda),\ \operatorname{Coim}(\bigoplus_{\lambda\in I} u_\lambda)\rightleftarrows\bigoplus_{\lambda\in I}\operatorname{Coim}(u_\lambda).\ ∎$$

COROLLARY 3. *If* $(E_\lambda)_{\lambda\in I}$ *is a family of A-modules, and if for any* $\lambda\in I$, F_λ *is a submodule of* E_λ, *then there exists a canonical isomorphism*

$$\bigoplus_{\lambda\in I}(E_\lambda/F_\lambda)\rightleftarrows(\bigoplus_{\lambda\in I} E_\lambda)/(\bigoplus_{\lambda\in I} F_\lambda).\ ∎$$

COROLLARY 4. *If* $(E_\lambda)_{\lambda\in I}$, $(F_\lambda)_{\lambda\in I}$, $(u_\lambda)_{\lambda\in I}$ *are as in Proposition* 2, *then* $\bigoplus_{\lambda\in I} u_\lambda$ *is, respectively, an injective, surjective, bijective, zero mapping if and only if for any* $\lambda\in I$, u_λ *is, respectively, an injective, surjective, bijective, zero mapping.* ∎

Taking into account the fact that the direct sum functor $(\mathbf{Mod}(A))^I \to \mathbf{Mod}(A)$ is an additive functor, from Proposition 2 we derive also the following Corollary:

COROLLARY 5. *The direct sum functor*

$$(\mathbf{Mod}(A))^I \to \mathbf{Mod}(A), \qquad I\neq\emptyset,$$

is an exact functor. ∎

In the remainder of this Section we shall combine the properties of direct products with those of direct sums.

PROPOSITION 3. *Let* $(E_\lambda)_{\lambda\in I}$, $(F_\mu)_{\mu\in J}$ *be two families of A-modules; there exists a canonical isomorphisms of abelian groups*

$$\mathrm{Hom}_A(\bigoplus_{\lambda\in I} E_\lambda, \prod_{\mu\in J} F_\mu) \rightleftarrows \prod_{(\lambda,\mu)\in I\times J} \mathrm{Hom}_A(E_\lambda, F_\mu).$$

Proof. Let u be a linear mapping $\bigoplus_{\lambda\in I} E_\lambda \to \prod_{\mu\in J} F_\mu$; the stated isomorphism is given by

$$u \mapsto (p_\mu\circ u\circ j_\lambda)_{(\lambda,\mu)\in I\times J}, \tag{9}$$

where j_λ is the canonical injection $E_\lambda \to \bigoplus_{\lambda\in I} E_\lambda$ and p_μ the canonical projection $\prod_{\mu\in J} F_\mu \to F_\mu$. The mapping defined by (9) is a homomorphism of abelian groups, since

$$u+v \mapsto (p_\mu\circ(u+v)\circ j_\lambda)_{(\lambda,\mu)} = (p_\mu\circ u\circ j_\lambda)_{(\lambda,\mu)} + (p_\mu\circ v\circ j_\lambda)_{(\lambda,\mu)}.$$

It is also bijective, since let $(u_{\lambda\mu})_{(\lambda,\mu)\in I\times J}$ be an arbitrary element of $\prod_{(\lambda,\mu)\in I\times J} \mathrm{Hom}(E_\lambda, F_\mu)$; μ being fixed in J, the familly of mappings $(u_{\lambda\mu})_{\lambda\in I}$ defines by the universal property of the direct sum, a linear mapping

$$v_\mu\colon \bigoplus_{\lambda\in I} E_\lambda \to F_\mu$$

which is unique with the property that $v_\mu\circ j_\lambda = u_{\lambda\mu}$ for any $\lambda\in I$ and for the fixed μ; if this is done for all $\mu\in J$, we get, by virtue of the universal property of the direct product (Proposition 1, § 6), a linear mapping

$$u\colon \bigoplus_{\lambda\in I} E_\lambda \to \prod_{\mu\in J} F_\mu$$

which is unique with the property $p_\mu \circ u = v_\mu$ for any $\mu \in J$; we have for any $\lambda \in I$ and $\mu \in J$,

$$p_\mu \circ u \circ j_\lambda = v_\mu \circ j_\lambda = u_{\lambda\mu},$$

hence the mapping (9) is surjective. The injectivity follows from the uniqueness of the construction of u. ∎

* DEFINITION 4. Let $\mathcal{C}$, $\mathcal{C}'$ be two categories, and let $T: \mathcal{C} \to \mathcal{C}'$ be a functor; we say that *the functor T commutes with direct products* or that *T preserves direct products* if, whenever $(E, (p_\lambda)_{\lambda \in I})$ is a direct product in $\mathcal{C}$, in the sense of Definition 3, § 6, of the family of objects $(E_\lambda)_{\lambda \in I}$ of $\mathcal{C}$, $(T(E), (T(p_\lambda))_{\lambda \in I})$ is a direct product in $\mathcal{C}'$ of the family of objects $(T(E_\lambda))_{\lambda \in I}$ of $\mathcal{C}'$; analogously, we say that *the functor T commutes with direct sums* or that *T preserves direct sums* if, whenever $(F, (j_\lambda)_{\lambda \in I})$ is a direct sum in $\mathcal{C}$, in the sense of Definition 2, of the family of objects $(E_\lambda)_{\lambda \in I}$ of $\mathcal{C}$, $(T(F), (T(j_\lambda))_{\lambda \in I})$ is a direct sum in $\mathcal{C}'$ of the family of objects $(T(E_\lambda))_{\lambda \in I}$ of $\mathcal{C}'$.

If only finite index sets I are considered, then we say that *T commutes with* (or *preserves*) *finite direct products*, respectively *finite direct sums*. ∎*

It is a consequence of Proposition 3 the fact that for any A-module E, *the functor* $h^E: \mathbf{Mod}(A) \to \mathbf{Ab}$ *commutes with direct products*: since let $(F_\mu)_{\mu \in J}$ be a family of A-modules, and let $(F, (p_\mu)_{\mu \in J})$ be a direct product of this family in the category $\mathbf{Mod}(A)$; according to the arguments of § 6, F is isomorphic to $\prod_{\mu \in J} F_\mu$ considered in the sense of Definition 1, § 6; the isomorphism involved has the property of carrying by composition the mapping p_μ into the canonical projection of index μ, $\mu \in J$; for this reason we shall confine ourselves to the case $F = \prod_{\mu \in J} F_\mu$, p_μ being the canonical projection, the reduction of the general case to this one being left to the reader as an easy exercise. Let E then be any A-module; let us show that $(h^E(\prod_{\mu \in J} F_\mu), (h^E(p_\mu))_{\mu \in J})$ is a direct product of the family of abelian groups $(h^E(F_\mu))_{\mu \in J}$: let X be an abelian group, and let $f_\mu: X \to \mathrm{Hom}_A(E, F_\mu) = h^E(F_\mu)$, $\mu \in J$, be a family of $\mathbf{Z}$-homomorphisms; for any $\mu \in J$ we have the diagram

(10)

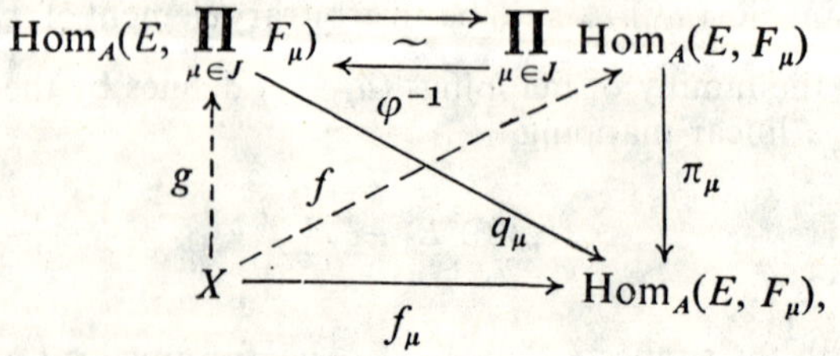

where φ is the isomorphism constructed in Proposition 3 for $I = \{1\}$ and $E_1 = E$, π_μ is the canonical projection of index μ, while $q_\mu = h^E(p_\mu)$; diagram (10) is commutative since, according to Proposition 3, if $u \in \mathrm{Hom}_A(E, \prod_{\mu \in J} F_\mu)$, then

$$\pi_\mu(\varphi(u)) = p_\mu \circ u = \mathrm{Hom}(1_E, p_\mu)(u);$$

by the universal property of the direct product, there exists a **Z**-homomorphism

$$f: X \to \prod_{\mu \in J} \mathrm{Hom}_A(E, F_\mu)$$

which is unique with the property that $\pi_\mu \circ f = f_\mu$ for any $\mu \in J$; the homomorphism $g = \varphi^{-1} \circ f$ has then the property $q_\mu \circ g = \pi_\mu \circ \varphi \circ \varphi^{-1} \circ f = \pi_\mu \circ f = f_\mu$ for any $\mu \in J$; if $g': X \to \mathrm{Hom}_A(E, \prod_\mu F_\mu)$ has the same property, then $\pi_\mu \circ \varphi \circ g' = q_\mu \circ g' = f_\mu$ for any $\mu \in J$, hence, from the uniqueness of f it follows $\varphi \circ g' = f$, whence $g' = \varphi^{-1} \circ f = g$.

Analogously, taking into account the fact that the direct sum is the notion dual to that of direct product, hence that the doublet in the dual category of a direct sum is the direct product of the doublets, and using the isomorphism given by Proposition 3 for $J = \{1\}$ and $F_1 = F$, it follows that *the functor* $h_F: (\mathbf{Mod}(A))^\circ \to \mathbf{Ab}$ *commutes with direct products* for any A-module F.

Remark.* Given two categories $\mathscr{C}, \mathscr{C}'$, if $(E_\lambda)_{\lambda \in I}$, respectively $(E'_\mu)_{\mu \in J}$, is a family of objects from $\mathscr{C}$, respectively $\mathscr{C}'$, then if $(E_\lambda)_{\lambda \in I}$, respectively $(E'_\mu)_{\mu \in J}$, has a product $(\prod_\lambda E_\lambda, (p_\lambda)_{\lambda \in I})$, respectively $(\prod_\mu E'_\mu), (p'_\mu)_{\mu \in J})$, in $\mathscr{C}$, respectively $\mathscr{C}'$, it follows that $((\prod_\lambda E_\lambda, \prod_\mu E'_\mu), (p_\lambda, p'_\mu)_{(\lambda, \mu) \in I \times J})$ is a product of the family $(E_\lambda, E'_\mu)_{(\lambda, \mu) \in I \times J}$ in the product category $\mathscr{C} \times \mathscr{C}'$. *

Consequently, if $(E_\lambda)_{\lambda \in I}$ and $(F_\mu)_{\mu \in J}$ are two families of A-modules, then, using the notations of Proposition 3, $(((\bigoplus_{\lambda \in I} E_\lambda)^\circ, \prod_{\mu \in J} F_\mu), (j^\circ_\lambda, p_\mu)_{(\lambda, \mu) \in I \times J})$ is a direct product of the family $(E^\circ_\lambda, F_\mu)_{(\lambda, \mu) \in I \times J}$ in the category $(\mathbf{Mod}(A))^\circ \times \mathbf{Mod}(A)$. Then, following the scheme of the proof above and applying Proposition 3 in its general form, we get that the system $(\mathrm{Hom}_A(\bigoplus_\lambda E_\lambda, \prod_\mu F_\mu), (\mathrm{Hom}(j_\lambda, p_\mu))_{(\lambda, \mu) \in I \times J})$ is a direct product in the category **Ab** of the family of objects $(\mathrm{Hom}_A(E_\lambda, F_\mu))_{(\lambda, \mu) \in I \times J}$; however, it does not follow thence that the functor $\mathrm{Hom}: (\mathbf{Mod}(A))^\circ \times \mathbf{Mod}(A) \to \mathbf{Ab}$ commutes with direct products, since the index set $I \times J$ is not arbitrary. ∎

* Proposition 3 with the appropriate modifications, the properties of the functors h^E and h_F of commuting with direct products, as well as the preceding Remark are valid for any preadditive category $\mathscr{C}$ instead of $\mathbf{Mod}(A)$ and for arbitrary objects E and F, as may be seen repeating almost word for word the previous proofs. *

We leave it to the reader to show that if $(E'_\lambda)_{\lambda \in I}$, $(F'_\mu)_{\mu \in J}$ are two other families of A-modules indexed respectively by the same sets I and J, and if

$$u_\lambda: E'_\lambda \to E_\lambda, \qquad v_\mu: F_\mu \to F'_\mu,$$

$\lambda \in I$, $\mu \in J$, are linear mappings, then the following diagram is commutative:

$$
\begin{array}{ccc}
\mathrm{Hom}_A\left(\bigoplus_{\lambda \in I} E_\lambda, \prod_{\mu \in J} F_\mu\right) & \xrightarrow[\sim]{} & \prod_{(\lambda,\mu) \in I \times J} \mathrm{Hom}_A(E_\lambda, F_\mu) \\
\Big\downarrow {\scriptstyle \mathrm{Hom}(\bigoplus_\lambda u_\lambda, \prod_\mu v_\mu)} & & \Big\downarrow {\scriptstyle \prod_{(\lambda,\mu)} (\mathrm{Hom}(u_\lambda, v_\mu))} \\
\mathrm{Hom}_A\left(\bigoplus_{\lambda \in I} E'_\lambda, \prod_{\mu \in J} F'_\mu\right) & \xrightarrow[\sim]{} & \prod_{(\lambda,\mu) \in I \times J} \mathrm{Hom}_A(E'_\lambda, F'_\mu),
\end{array}
\tag{11}
$$

the horizontal arrows representing the canonical isomorphisms given by Proposition 3.

We shall return later to this fact (see § 10) in order to interpret it from the categorical point of view.

* We have seen previously that in the category **Mod**(A), for a finite set of indices, the sum and the product coincide; let us see how things turn out for an arbitrary preadditive category:

Let $\mathscr{C}$ be a preadditive category, and let $(E_i)_{i\in I}$ be a *finite* family of objects of $\mathscr{C}$; suppose there exists a direct sum $(S, (j_i)_{i\in I})$ of this family, in the sense of Definition 2; let $i \in I$; we have the morphisms

$$1_{E_i}: E_i \to E_i, \qquad 0: E_{i'} \to E_i, \quad i' \neq i;$$

according to the definition of the direct sum, there exists a morphism

$$p_i: S \to E_i$$

uniquely determined by the properties

$$p_i \circ j_i = 1_{E_i}, \qquad p_i \circ j_{i'} = 0, \qquad i \neq i'; \tag{12}$$

doing this for any $i \in I$, we get the system $(E, (p_i)_{i\in I})$; let us show that this system is a direct product of the family $(E_i)_{i\in I}$, in the sense of Definition 3, § 6: to begin with, we show that

$$\sum_{i\in I} j_i \circ p_i = 1_S \tag{13}$$

(the sum being performed in the group $\mathrm{Hom}_{\mathscr{C}}(S, S)$); to prove (13) it is however sufficient to show that for any $i' \in I$ we have

$$\left(\sum_{i\in I} j_i \circ p_i\right) \circ j_{i'} = j_{i'}, \tag{14}$$

since then, the family of mappings $j_{i'}: E_{i'} \to S$, $i' \in I$, may be factored in two ways through the mappings $j_{i'}$, $i' \in I$, and these two ways must coincide; but relations (14) are an immediate consequence of relations (12).

Let now X be an object of $\mathcal{C}$, and let

$$f_i: X \to E_i, \qquad i \in I,$$

be a family of morphisms; let us consider the morphism $f: X \to S$ given by

$$f = \sum_{i \in I} j_i \circ f_i;$$

we have for any $i' \in I$,

$$p_{i'} \circ f = \sum_{i \in I} p_{i'} \circ j_i \circ f_i = f_{i'};$$

if $g: X \to S$ is another morphism such that $p_i \circ g = f_i$, $i \in I$, then, according to (13),

$$g = (\sum_{i \in I} j_i \circ p_i) \circ g = \sum_{i \in I} j_i \circ f_i = f;$$

these show that $(S, (p_i)_{i \in I})$ is a direct product of the family $(E_i)_{i \in I}$.

Suppose now that for the finite family of objects $(E_i)_{i \in I}$, a system $(S, (j_i)_{i \in I})$ is given, where $j_i: E_i \to S$, $i \in I$, and such that there exists a family of morphisms

$$p_i: S \to E_i, \qquad i \in I,$$

satisfying (12) and (13); then $(S, (j_i)_{i \in I})$ is a direct sum of the family $(E_i)_{i \in I}$: for let $Y \in \operatorname{Ob} \mathcal{C}$, and let

$$g_i: E_i \to Y, \qquad i \in I,$$

be a family of morphisms of the category $\mathcal{C}$; define a morphism $g: S \to Y$ by

$$g = \sum_{i \in I} g_i \circ p_i;$$

as above, we have $g \circ j_i = g_i$ for any $i \in I$, and g is uniquely determined by these conditions.

Thus we have proved:

PROPOSITION 4. *In the category $\mathcal{C}$ let a finite family of objects $(E_i)_{i \in I}$ and a system $(S, (j_i)_{i \in I})$ be given, where $S \in \operatorname{Ob} \mathcal{C}$, while $j_i: E_i \to S$, $i \in I$, are morphisms of the category; then $(S, (j_i)_{i \in I})$ is a direct sum of the family of objects $(E_i)_{i \in I}$*

in the category $\mathcal{C}$ *if and only if there exists a family of morphisms* $p_i\colon S \to E_i$, $i \in I$, *such that*

$$p_i \circ j_i = 1_{E_i} \text{ for } i \in I,\ p_i \circ j_{i'} = 0 \text{ for } i \neq i', \tag{15}$$

$$\sum_{\lambda \in I} j_i \circ p_i = 1_S;$$

under these conditions, the family of morphisms $(p_i)_{i \in I}$ *is uniquely determined by the first two conditions in* (15), *and the system* $(S, (p_i)_{i \in I})$ *is a direct product of the family* $(E_i)_{i \in I}$.

Indeed, the uniqueness of the family $(p_i)_{i \in I}$ satisfying the first two conditions (15) may be obtained as follows: let $i \in I$ be fixed; then p_i is the unique morphism $S \to E_i$ such that $p_i \circ j_i = 1_E$, $p_i \circ j_{i'} = 0$, $i' \neq i$; this procedure is applied to all $i \in I$. ∎

By dual arguments one may prove a proposition differing from Proposition 4 by the fact that the arrows are reversed and that the sums and the products interchange their roles; we notice however that relations (15) remain valid, hence that they are "selfdual".

Given a finite family of objects $(E_i)_{i \in I}$ of the preadditive category $\mathcal{C}$ and a direct sum $(S, (j_i)_{i \in I})$ of it, the direct product $(S, (p_i)_{i \in I})$ of this family of objects constructed in Proposition 4 starting from $(S, (j_i)_{i \in I})$ will be called the *direct product associated with* the given direct sum; in the dual situation we shall talk of the *direct sum associated with* the given direct product. Of course, passing from the direct product $(S, (p_i)_{i \in I})$ associated with the direct sum $(S, (j_i)_{i \in I})$ to the direct sum associated with $(S, (p_i)_{i \in I})$, we again get $(S, (j_i)_{i \in I})$, since the new morphisms, say $\bar{j}_i$, $i \in I$, must satisfy the same conditions (15) as j_i, $i \in I$, hence by virtue of the uniqueness, they coincide; we argue in a similar way for the dual situation; whenever these conditions are satisfied, we may refer to *the direct product and the direct sum of the finite family* $(E_i)_{i \in I}$ *associated with each other*.

From Proposition 4 and its dual we obtain the following

COROLLARY. *If* $\mathcal{C}, \mathcal{C}'$ *are preadditive categories, and if* $T : \mathcal{C} \to \mathcal{C}'$ *is an additive functor, then* T *commutes with finite direct sums and with finite direct products.* ∎

Hence, by the preceding arguments, *if a family of two objects has a direct sum* $(S, (j_i)_{i=1,2})$ *in the preadditive category* $\mathcal{C}$, *then it has also a direct product in* $\mathcal{C}$, and the object S in the sum is also object in the product, and conversely, interchanging the words "sum" and "product".

DEFINITION 5. A preadditive category $\mathcal{C}$ for which any pair of objects (a family of two objects) of $\mathcal{C}$ has a direct sum, in the sense of Definition 2, or, equivalently, any pair of objects of $\mathcal{C}$ has a direct product, in the sense of Definition 3, § 6, is called an *additive category*.

An additive category satisfying Axioms AB1) and AB2) of § 4 is called an *abelian category*. ∎

Obviously, taking into account the associativity of direct products and of direct sums, in an additive category any finite family of objects has a direct sum and a direct product. Likewise, a preadditive category $\mathcal{C}$ satisfying Axiom AB3°) of § 6 (or Axiom AB3)) is an additive category, because in this case any family of objects of $\mathcal{C}$ has a direct product (respectively a direct sum). However, in the general case, the Axioms AB3°) and AB3) are not equivalent even for preadditive categories, i.e., there exist preadditive categories satisfying one axiom but not the other.

If, however, $\mathcal{C}$ is a preadditive category satisfying both AB3°) and AB3), then if $(E_\lambda)_{\lambda\in I}$ is a family of objects of $\mathcal{C}$, we consider for any $\lambda \in I$ the family of morphisms $(f_{\mu\lambda})_{\mu\in I}$,

$$f_{\mu\lambda}: E_\mu \to E_\lambda,$$

such that

$$f_{\mu\lambda} = \begin{cases} 1_{E_\lambda} & \text{if } \mu = \lambda \\ 0 & \text{if } \mu \neq \lambda. \end{cases}$$

This family induces, according to the definition of the direct sum, a unique morphism

$$g_\lambda: \bigoplus_{\mu\in I} E_\mu \to E_\lambda$$

with the property that

$$g_\lambda \circ j_\mu = f_{\mu\lambda}, \ \mu \in I,$$

where j_μ is the canonical injection $E_\mu \to \bigoplus_{\mu\in I} E_\mu$. If this is done for all $\lambda \in I$, then we get, according to the definition of the direct product, a morphism

$$u: \bigoplus_{\lambda\in I} E_\lambda \to \prod_{\lambda\in I} E_\lambda$$

which is unique with the property

$$p_\lambda \circ u = g_\lambda, \qquad \lambda \in I,$$

p_λ being the canonical projection of index λ. The morphism u is called, again by an abuse of language, the *canonical morphism* from the direct sum of the family $(E_\lambda)_{\lambda\in I}$ to the direct product of this family.

Notice that *the dual of an additive category $\mathcal{C}$ is an additive category*, since the existence of the direct sum of any two objects of $\mathcal{C}$ amounts to the existence of the direct product of any two objects of $\mathcal{C}°$; from this result and from that obtained in § 4 concerning the dual of a preadditive category satisfying Axioms AB1) and AB2) it follows that *the dual of an abelian category is an abelian category.**

Since for any ring A the preadditive categories with AB1) and AB2) **Mod**$_s(A)$ and **Mod**$_d(A)$ satisfy both AB3°) and AB3), it follows that *these categories are abelian categories*. Then, if $(E_\lambda)_{\lambda\in I}$ is a family of A-modules, with the meaning of $\bigoplus_{\lambda\in I} E_\lambda$ given by Definition 1, and of $\prod_{\lambda\in I} E_\lambda$ given by Definition 1, § 6, the canonical morphism from the direct sum to the direct product is exactly the inclusion mapping, as one can easily see.

§ 8. Intersections and Sums of Submodules

Let A be a ring, let E be an A-module, and let $(M_\lambda)_{\lambda\in I}$ be a family of submodules of E; consider the intersection of the family of subsets $(M_\lambda)_{\lambda\in I}$; this is non-void, since $0 \in M_\lambda$ for any $\lambda \in I$ (see § 3); further, if $x, y \in \bigcap_{\lambda\in I} M_\lambda$, then $x, y \in M_\lambda$ for any $\lambda \in I$, hence $x+y \in M_\lambda$ for any $\lambda \in I$, whence $x+y \in \bigcap_{\lambda\in I} M_\lambda$; likewise, if $x \in \bigcap_{\lambda\in I} M_\lambda$ and $\alpha \in A$, then $\alpha x \in \bigcap_{\lambda\in I} M_\lambda$; it follows that $\bigcap_{\lambda\in I} M_\lambda$ *is a submodule of* E.

Definition 1. The submodule thus obtained is called the *intersection submodule* or the *intersection* of the family of submodules $(M_\lambda)_{\lambda\in I}$, and is denoted, the same as the underlying set is, by $\bigcap_{\lambda\in I} M_\lambda$; if $I = \{1, \ldots, n\}$, then instead of $\bigcap_{\lambda\in I} M_\lambda$ we write also $M_1 \cap \ldots \cap M_n$ or also $\bigcap_{i=1}^{n} M_i$; if $I = \emptyset$, then, by definition, $\bigcap_{\lambda\in I} M_\lambda = E$.

Given a set $\mathcal{M}$ of submodules, the *intersection* of this set of submodules is the intersection of the family indexed identically by the set $\mathcal{M}$. ∎

Obviously, $\bigcap_{\lambda\in I} M_\lambda$ is the greatest lower bound of the submodules M_λ, $\lambda \in I$, in the sense of the partial order relation given by inclusion, i.e. the greatest submodule contained in all M_λ, $\lambda \in I$; consequently, *the partially ordered set of all submodules of a module E is a complete meet-semilattice*, the meet operation in the semillatice being the intersection of submodules. If $(N_\lambda)_{\lambda\in I}$ is also a family of submodules of E such that $N_\lambda \subseteq M_\lambda$ for any $\lambda \in I$, then $\bigcap_{\lambda\in I} N_\lambda \subseteq \bigcap_{\lambda\in I} M_\lambda$; this property will be called the "monotonicity of the intersection with respect to inclusion".

Let now f_λ be the canonical surjection $E \to E/M_\lambda$, $\lambda \in I$; by the universal property of the direct product (Proposition 1, § 6) we obtain an A-homomorphism

$$f : E \to \prod_{\lambda\in I} (E/M_\lambda)$$

uniquely determined by the property that

$$p_\lambda \circ f = f_\lambda, \qquad \lambda \in I,$$

p_λ being the canonical projection of index λ.

PROPOSITION 1. *We have the exact sequence of A-modules and A-homomorphisms*

$$(1) \qquad 0 \to \bigcap_{\lambda\in I} M_\lambda \xrightarrow{i} E \xrightarrow{f} \prod_{\lambda\in I} (E/M_\lambda),$$

i being the canonical injection; in other words, $\mathrm{Ker}(f) = \bigcap_{\lambda\in I} M_\lambda$.

Proof. The sequence (1) is exact at $\bigcap_{\lambda\in I} M_\lambda$ since i is injective; if $x \in E$, then from the proof of Proposition 1, § 6, we have

$$f(x) = (f_\lambda(x))_{\lambda\in I};$$

hence, to say that x belongs to $\mathrm{Ker}\, f$ is equivalent to saying that $f_\lambda(x) = 0$ for any λ, or also that x belongs to M_λ for any λ. ▮

Now, let E be an A-module, and let M be a subset of E.

DEFINITION 2. By the *submodule generated* or *spanned* by M in E we mean the intersection F of all submodules of E containing M; in this situation, we also say that M is a *set of generators* or a *system of generators* of F or that M *generates F* or *spans F* (over A or with respect to A). If F has a set of generators M which is a finite set, we also say that F is a *finitely generated module* (respectively *submodule*) or also a *module* (respectively *submodule*) *of finite type*; if F admits a set of generators consisting of a single element a, then F is said to be a *cyclic module* (respectively *submodule*), and a is a *generator* of F.

By the *submodule generated by a family* $(a_\lambda)_{\lambda\in I}$ of elements of E is meant the submodule generated by the set of all "values" of the mapping defining the family. The notion of *system of generators* is also extended to this case.

If the ring A is a skew-field, and so A-modules become A-vector spaces, then instead of "submodule generated by a subset", "finitely generated module", etc. we speak, respectively, of the *subspace generated* by a subset, *finitely generated vector space*, etc. ▮

Obviously, the submodule generated by a set M contains M; the submodule generated by $\emptyset$ is the zero submodule; moreover, if M is a submodule, then the submodule generated by M is M. From the last assertion it follows that each module E has at least a set of generators, namely the set E itself. Also, if M and N are subsets of E and $M \subseteq N$, then the submodule generated by M in E is contained in the submodule generated by N in E. If the finite subset $M = \{a_1, \ldots, a_n\}$ of the module E generates the submodule F, one says, by an abuse of language, that the elements $a_1, \ldots, a_n$ *generate* F or that F is the *submodule generated by* $a_1, \ldots, a_n$.

PROPOSITION 2. *Let E be an A-module, and let M be a subset of E; the submodule generated by M in E consists of all linear combinations of elements of M.*

Proof. Indeed, the set of all linear combinations $\sum_{x\in M} \alpha_x x$, where $(\alpha_x)_{x\in M}$ is an fs-family of elements of A, is a submodule of E containing M: we need only consider the above expression in which $\alpha_x = 1$ for a certain $x \in M$ and $\alpha_y = 0$ for $y \in M$, $y \neq x$, and we obtain x. Herefrom it follows that the set of all linear combinations of elements of M contains the submodule generated by M.

Conversely, any submodule containing M contains all linear combinations of elements of M, since these are in fact finite sums of elements obtained from elements of M by scalar multiplication; but a submodule is closed under finite sums (i.e., contains the finite sums of elements of it) and under scalar multiplication. It follows that the submodule generated by M also contains all these linear combinations. ▮

REMARK. Likewise, the submodule generated in E by a family $(a_\lambda)_{\lambda \in I}$ of elements of E consists of all linear combinations of the family $(a_\lambda)_{\lambda \in I}$. ▮

In particular, if a is an element of the A-module E, then the submodule generated by $\{a\}$ in E consists of all elements of the form αa, with $\alpha \in A$; we shall denote this submodule by Aa. So, if E is a cyclic A-module, and $a \in E$ is a generator of E, then $E = Aa$.

COROLLARY 1. *Let $u : E \to F$ be an A-linear mapping; if M is a subset of E, and the submodule generated by M in E is E', then the submodule generated by $u(M)$ in F is $u(E')$.*

Proof. The elements of $u(E')$ are the linear combinations with coefficients in A of the elements of $u(M)$. ▮

COROLLARY 2. *Let $u: E \to F$ be an A-linear mapping; if E' is a finitely generated submodule of E, then $u(E')$ is finitely generated, and if E' admits a system of generators consisting of n elements, then $u(E')$ admits a system of generators consisting of at most n elements.*

Proof. We need only take in Corollary 1 the set M finite, respectively finite and consisting of n elements; then $u(M)$ will be finite, respectively finite and consisting of at most n elements. ▮

In particular, if E is a cyclic A-module and u is surjective, then F is also cyclic; this may be applied, for instance, when F is a quotient module E/M of the cyclic module E by a submodule M of it and u is the canonical surjection; hence *any quotient module of a cyclic module is a cyclic module.*

However, *not any submodule of a cyclic module is cyclic*: this is shown by the following

EXAMPLE. 1) Let $A = K[X, Y]$ be the ring of polynomials in the indeterminates X, Y over the field K, and let $\mathfrak{a}$ be the ideal generated in A by the subset $\{X, Y\}$; this is the ideal consisting of all polynomials having 0 as the coefficient of $X^0 Y^0$; the A-module A is cyclic, being generated by 1; however, the submodule $\mathfrak{a}$ is not cyclic: for if $f \in \mathfrak{a}$ were a generator of it, then there would exist $g, h \in A$ such that

$$X = f \cdot g, \qquad Y = f \cdot h;$$

these relations show that neither X nor Y occur effectively in f; as $f \in \mathfrak{a}$, we have $f = 0$, which is a contradiction, since $\mathfrak{a} \neq 0$. ▮

COROLLARY 3. *Let $u : E \to F$ be an A-linear mapping; if M is a subset of E, and E' is the submodule generated by M in E, then the fact that u vanishes at any element of M implies that u vanishes at any element of E'.*

Proof. We have $u(M) = 0$, hence $u(E')$ is the submodule generated by 0, hence the zero submodule. ∎

COROLLARY 4. *Let the short exact sequence of A-modules and A-homomorphisms*

$$0 \to E \xrightarrow{u} F \xrightarrow{v} G \to 0 \tag{2}$$

be given; if M is a set of generators of E, if N is a set of generators of G, and if N' is a subset of F with the property $v(N') = N$, then $u(M) \cup N'$ is a set of generators of F; in particular, if in the sequence (2) E and G are finitely generated, then so is F.

Proof. To prove the first assertion, let y be an element of F; there exists an fs-family $(\alpha_z)_{z \in N}$ of elements of A such that

$$v(y) = \sum_{z \in N} \alpha_z z;$$

there exists a subset N_1' of N' so that v induces a bijection $N_1' \rightleftarrows N$; it follows that

$$v(y) = \sum_{y' \in N_1'} \alpha_{v(y')}\, v(y') = \sum_{y' \in N'} \beta_{y'}\, v(y'),$$

where we have taken $\beta_{y'} = \alpha_{v(y')}$ if $y' \in N_1'$ and $\beta_{y'} = 0$ if $y' \in N' \setminus N_1'$; it follows that $y - \sum_{y' \in N'} \beta_{y'} y' \in \operatorname{Ker}(v) = \operatorname{Im}(u)$, hence that there exists an fs-family $(\gamma_x)_{x \in M}$ such that

$$y = \sum_{y' \in N'} \beta_{y'} y' + \sum_{x \in M} \gamma_x u(x),$$

and so $u(M) \cup N'$ is a set of generators of F.

For the second assertion, if E and G are of finite type, we can, in the above notation, take M, N finite and N' also finite; it follows that $u(M) \cup N'$ is finite. ∎

Let now E be an A-module and $(M_\lambda)_{\lambda \in I}$ a family of submodules of E.

DEFINITION 3. The *sum* in E of the family of submodules $(M_\lambda)_{\lambda \in I}$, denoted $\sum_{\lambda \in I} M_\lambda$, is the submodule generated by the subset $\bigcup_{\lambda \in I} M_\lambda$ of E; if $I = \{1, \ldots, n\}$, then instead of $\sum_{\lambda \in I} M_\lambda$ we write also $M_1 + \ldots + M_n$ or $\sum_{i=1}^{n} M_i$; there is a corresponding notation in the case of another finite index set.

If a set $\mathcal{M}$ of submodules is given, then the *sum* of this set of submodules, denoted $\sum_{M \in \mathcal{M}} M$, is the sum of the family indexed identically by $\mathcal{M}$. ∎

Obviously, $\sum_{\lambda \in I} M_\lambda$ is the least upper bound of the submodules M_λ, $\lambda \in I$, in the sense of the partial order relation given by the inclusion, i.e. the least submodule containing all M_λ, $\lambda \in I$; consequently, *the partially ordered set of all submodules of a module is a complete join-semilattice*, the join operation in the semilattice being the sum of submodules. It follows also that if $(M_\lambda)_{\lambda \in J}$, with $J \subseteq I$, is a cofinal subfamily with respect to inclusion of the family $(M_\lambda)_{\lambda \in I}$, then $\sum_{\lambda \in J} M_\lambda = \sum_{\lambda \in I} M_\lambda$; in particular, if $M_1 \supseteq M_2$, then $M_1 + M_2 = M_1$, and hence for any submodule M of E we have $M + 0 = M$, $M + E = E$, $M + M = M$. Hencefrom we get also that if $I = \bigcup_{\mu \in L} I_\mu$, then $\sum_{\lambda \in I} M_\lambda = \sum_{\mu \in L} (\sum_{\lambda \in I_\mu} M_\lambda)$.

We have the "monotonicity of the sum with respect to inclusion" analogous to that for intersection.

From Proposition 2 also follows

COROLLARY 5. *Let E be an A-module, and let $(M_\lambda)_{\lambda \in I}$ be a family of submodules of E; $\sum_{\lambda \in I} M_\lambda$ consists of all sums $\sum_{\lambda \in I} x_\lambda$, where $(x_\lambda)_{\lambda \in I}$ is an fs-family of elements of E such that $x_\lambda \in M_\lambda$, $\lambda \in I$; moreover, we have the exact sequence*

$$\bigoplus_{\lambda \in I} M_\lambda \xrightarrow{f} E \xrightarrow{p} E/(\sum_{\lambda \in I} M_\lambda) \to 0,$$

where f is obtained by the universal property of the direct sum starting with the canonical injections $i_\lambda : M_\lambda \to E$, $\lambda \in I$, while p is the canonical surjection.

Proof. According to Proposition 2 and to the definition of $\sum_{\lambda \in I} M_\lambda$, the elements of $\sum_{\lambda \in I} M_\lambda$ are the linear combinations of elements of $M = \bigcup_{\lambda \in I} M_\lambda$; but in such a linear combination $\sum_{x \in M} \alpha_x x$ putting together all the terms $\alpha_x x$ for which x belongs to the same M_λ (here the choice of λ such that $x \in M_\lambda$ is not uniquely determined), we get a sum of linear combinations of elements, in each linear combination all the elements belonging to the same M_λ; M_λ being a submodule, contains all linear combinations of its elements; but only a finite number of such linear combinations are non-zero, since in the linear combination we started with only a finite number of terms were non-zero.

For the second assertion, if $(x_\lambda)_{\lambda \in I} \in \bigoplus_{\lambda \in I} M_\lambda$, then the stated f acts as follows (see the proof of Proposition 1, § 7):

$$f((x_\lambda)_\lambda) = \sum_{\lambda \in I} x_\lambda;$$

from the first assertion we obtain that $\mathrm{Im}(f) = \sum_{\lambda \in I} M_\lambda = \mathrm{Ker}(p)$. ∎

REMARK. According to Corollary 5 above, if M and N are two submodules of the A-module E, then $M + N$ consists of the sums $x + y$, with $x \in M$, $y \in N$ (of course, the representation as $x + y$ is not unique, i.e., we might have $x + y = x' + y'$, with $x, x' \in M$, $y, y' \in N$, without necessarily having $x = x'$, $y = y'$); extending the notation, if M and N are two arbitrary subsets, not necessarily submodules, of the A-module E, then $M + N$ denotes the set of all sums $x + y$, with

$x \in M$, $y \in N$; generally, this is not a submodule; certainly, as for submodules, we have $M + N = N + M$; in the particular case where M consists of a single element a, instead of $\{a\} + N$ one writes also $a + N$; likewise, instead of $M + \{a\}$ one writes $M + a$. As an application, let E be an A-module, and let F be a submodule of it; let $\hat{x}$ be the coset of x modulo F, $x \in E$; then

$$\hat{x} = x + F:$$

for if $y \in \hat{x}$, then $y - x = z \in F$, hence $y = x + z \in x + F$; conversely, if $y \in x + F$, then $y = x + z$, $z \in F$, and $y \equiv x \pmod F$, hence $y \in \hat{x}$; in particular, if $x \in F$, then, as $\hat{x} = \hat{0} = F$, we have $x + F = F$. ∎

Now let E be an A-module, let $(M_\lambda)_{\lambda \in I}$ be a family of submodules, and let $I = \bigcup_{\mu \in J} I_\mu$ be a partition of the index set I; the linear mapping

$$f_\mu \colon \bigoplus_{\lambda \in I_\mu} M_\lambda \to E, \quad \mu \in J,$$

defined as f in Corollary 5 to Proposition 2, but only for $\lambda \in I_\mu$, has, according to this Corollary, $\sum_{\lambda \in I_\mu} M_\lambda$ as image; if $g_\mu \colon \bigoplus_{\lambda \in I_\mu} M_\lambda \to \sum_{\lambda \in I_\mu} M_\lambda$, $\mu \in J$, is the homomorphism obtained from f_μ by restricting the codomain, then g_μ is surjective, $\mu \in J$, and hence, according to Corollary 4 to Proposition 2, § 7, it follows that $\bigoplus_{\mu \in J} g_\mu$ is also surjective. Consider the diagram

$$\begin{array}{ccc} \bigoplus_{\mu \in J} (\bigoplus_{\lambda \in I_\mu} M_\lambda) & \xrightarrow[\sim]{u} & \bigoplus_{\lambda \in I} M_\lambda \\ {\scriptstyle \bigoplus_{\mu \in J} g_\mu} \downarrow & & \downarrow {\scriptstyle f} \\ \bigoplus_{\mu \in J} (\sum_{\lambda \in I_\mu} M_\lambda) & \xrightarrow{f'} & E, \end{array} \tag{3}$$

where f is the homomorphism obtained by Corollary 5, f' is the similar homomorphism for the family of submodules $(\sum_{\lambda \in I_\mu} M_\lambda)_{\mu \in J}$ of E, while u is the isomorphism given by Corollary 3 to Proposition 1, § 7, for the family $(M_\lambda)_{\lambda \in I}$ of modules and for the given partition of the index set I. Diagram (3) is commutative: since if $((x_\lambda)_{\lambda \in I_\mu})_{\mu \in J} \in \bigoplus_{\mu \in J} (\bigoplus_{\lambda \in I_\mu} M_\lambda)$, then

$$\begin{aligned} &f(u(((x_\lambda)_{\lambda \in I_\mu})_{\mu \in J})) = f((x_\lambda)_{\lambda \in I}) = \\ &= \sum_{\lambda \in I} x_\lambda = \sum_{\mu \in J} (\sum_{\lambda \in I_\mu} x_\lambda) = f'((\sum_{\lambda \in I_\mu} x_\lambda)_{\mu \in J}) = \\ &= f'[(g_\mu((x_\lambda)_{\lambda \in I_\mu}))_{\mu \in J}] = f'[(\bigoplus_{\mu \in J} g_\mu)(((x_\lambda)_{\lambda \in I_\mu})_{\mu \in J})], \end{aligned} \tag{4}$$

the third equality in (4) being according to those shown in § 1.

From the commutativity of diagram (3), considering the images and taking into account that u and $\bigoplus_{\mu} g_\mu$ are surjective, it follows also that

$$\sum_{\lambda \in I} M_\lambda = \sum_{\mu \in J} \Big(\sum_{\lambda \in I_\mu} M_\lambda \Big),$$

equality which is known even in the more general case when $I = \bigcup_{\mu \in J} I_\mu$ is not necessarily a partition of I.[1]

From Corollary 5 follows

COROLLARY 6. *Let E be an A-module, and let $(M_\lambda)_{\lambda \in I}$ be a family of submodules of E directed to the right with respect to inclusion; then $\sum_{\lambda \in I} M_\lambda = \bigcup_{\lambda \in I} M_\lambda$.*

Proof. The fact that the sum contains the union follows by the definition of the sum; to prove the converse it suffices, according to Corollary 5, to show that the sum of any fs-family $(x_\lambda)_{\lambda \in I}$, with $x_\lambda \in M_\lambda$, $\lambda \in I$, is in the union: but such a sum is by definition the finite sum $\sum_{\lambda \in J} x_\lambda$, J being the support of the family; $(M_\lambda)_{\lambda \in I}$ being directed to the right, there exists $\lambda_0 \in I$ such that $M_\lambda \subseteq M_{\lambda_0}$ for any $\lambda \in J$; it follows that $x_\lambda \in M_{\lambda_0}$ for any $\lambda \in J$, hence $\sum_{\lambda \in J} x_\lambda \in M_{\lambda_0} \subseteq \bigcup_{\lambda \in I} M_\lambda$. ∎

From Corollary 6 we get

COROLLARY 7. *Let E be an A-module, and let $(M_\lambda)_{\lambda \in I}$ be a family of submodules of E directed to the right with respect to inclusion; let N be a submodule of E; then*

$$(5) \qquad \Big(\sum_{\lambda \in I} M_\lambda\Big) \cap N = \sum_{\lambda \in I} (M_\lambda \cap N).$$

Proof. According to Corollary 6 we have

$$\sum_{\lambda \in I} M_\lambda = \bigcup_{\lambda \in I} M_\lambda;$$

the family $(M_\lambda)_\lambda$ being directed to the right, the family of submodules $(M_\lambda \cap N)_{\lambda \in I}$ is also directed to the right; then, by Corollary 6,

$$\sum_{\lambda \in I} (M_\lambda \cap N) = \bigcup_{\lambda \in I} (M_\lambda \cap N);$$

relation (5) now follows from the distributivity of intersection over union in the set of subsets of E. ∎

We also have

COROLLARY 8. *Let E be an A-module, and let $(M_\lambda)_{\lambda \in I}$ be a family of submodules of it; if S_λ is a set of generators of M_λ, $\lambda \in I$, then $\bigcup_{\lambda \in I} S_\lambda$ is a set of generators of $\sum_{\lambda \in I} M_\lambda$.*

---

[1] We need the commutativity of diagram (3) later on: see the proof of Proposition 4, § 9.

Proof. If $x \in \sum_{\lambda \in I} M_\lambda$, then, according to Corollary 5, x may be written as $x = \sum_{\lambda \in I} x_\lambda$, where $(x_\lambda)_{\lambda \in I}$ is an fs-family of elements of E such that $x_\lambda \in M_\lambda$, $\lambda \in I$; but then, if the support of $(x_\lambda)_{\lambda \in I}$ is J, we have $x = \sum_{\lambda \in J} x_\lambda$; for any $\lambda \in J$, x_λ is, according to Proposition 2, a linear combination of elements of S_λ, hence of $\bigcup_{\lambda \in I} S_\lambda$, and x is also such a combination. As $S_\lambda \subseteq M_\lambda \subseteq \sum_{\lambda \in I} M_\lambda$ for any $\lambda \in I$, it follows $\bigcup_{\lambda \in I} S_\lambda \subseteq \sum_{\lambda \in I} M_\lambda$, and the Corollary is proved. ▮

It follows that *if in the A-module E a family of elements* $(x_\lambda)_{\lambda \in I}$ *is given, then* $E = \sum_{\lambda \in I} Ax_\lambda$ *if and only if* $(x_\lambda)_{\lambda \in I}$ *is a system of generators of E.*

We mention, as a particular case, that if $(M_\lambda)_{\lambda \in I}$ is the family of all submodules of finite type of an A-module E (indexed, for instance, identically by the set of these submodules as index set), then, since for any $x \in E$ we have $x \in Ax$, and since Ax is finitely generated, it follows that $E = \bigcup_{\lambda \in I} M_\lambda$; as, by the preceding considerations, we have $\bigcup_{\lambda \in I} M_\lambda \subseteq \sum_{\lambda \in I} M_\lambda$, it follows that $E = \sum_{\lambda \in I} M_\lambda$, i.e., *any A-module is the sum of the family of its finitely generated submodules.*

REMARK. According to Corollary 8, if M_λ and M_μ are finitely generated submodules of an A-module E, then $M_\lambda + M_\mu$ is also a finitely generated submodule containing M_λ and M_μ; it follows that *the family* $(M_\lambda)_{\lambda \in I}$ *of all finitely generated submodules of an A-module E is directed to the right with respect to inclusion* as a family of subsets of E (see also the corresponding particular case of the notion of inductive limit, § 13). ▮

Now let E be an A-module, and let M, N be two of its submodules; since $M \cap N \subseteq M$, $M \cap N \subseteq N$, it follows that $M \cap N$ is a submodule of M and of N; let $i_M: M \cap N \to M$, $i_N: M \cap N \to N$ be the canonical injections, and let

$$u: M \cap N \to M \times N = M \oplus N$$

be the homomorphism obtained from i_M and i_N by the universal property of the product; next, since $M \subseteq M + N$, $N \subseteq M + N$, we have the canonical injections $j_M: M \to M + N$ and $j_N: N \to M + N$; let

$$v: M \oplus N \to M + N$$

be the homomorphism obtained from j_M and $-j_N$ by the universal property of the direct sum. We have:

PROPOSITION 3. *The diagram*

$$0 \to M \cap N \xrightarrow{u} M \oplus N \xrightarrow{v} M + N \to 0$$

is an exact sequence of A-modules and A-homomorphisms.

Proof. According to Proposition 1, *a*) and *b*), § 5, it suffices to show that u is injective, v is surjective, and $\mathrm{Im}(u) = \mathrm{Ker}(v)$: for the first assertion, we recall that if $x \in M \cap N$, then $u(x) = (i_M(x), i_N(x)) = (x, x)$; but then $u(x) = u(y)$, with $x, y \in M \cap N$, implies $(x, x) = (y, y)$, hence $x = y$; for the second assertion, we recall that if $(x, y) \in M \oplus N$, then $v(x, y) = x - y$; thus, if $z \in M + N$, z takes on the form $z = x + y$, $x \in M$, $y \in N$, hence $z = x + y = v(x, -y)$, and v turns out to be surjective; finally, the fact that (x, y) is in $\mathrm{Ker}(v)$ is equivalent to $x - y = 0$, hence to $x = y \in M \cap N$, and hence to $(x, y) = u(x) \in \mathrm{Im}(u)$. ∎

Similarly, if E is an A-module and M, N are two of its submodules, then, as M and N both contain $M \cap N$ and are both contained in $M + N$, there exist the canonical surjections

$$p_M: E/(M \cap N) \to E/M, \qquad p_N: E/(M \cap N) \to E/N,$$

$$q_M: E/M \to E/(M + N), \qquad q_N: E/N \to E/(M + N);$$

the linear mappings p_M and $-p_N$ determine, by virtue of the universal property of the direct product, a linear mapping

$$f: E/(M \cap N) \to (E/M) \oplus (E/N) = (E/M) \times (E/N);$$

moreover, the linear mappings q_M and q_N determine, by the universal property of the direct sum, a linear mapping

$$g: (E/M) \oplus (E/N) \to E/(M + N).$$

We have

PROPOSITION 4. *The diagram*

$$0 \to E/(M \cap N) \xrightarrow{f} (E/M) \oplus (E/N) \xrightarrow{g} E/(M + N) \to 0$$

is an exact sequence of A-modules and homomorphisms.

The proof is left to the reader as an exercise. ∎

PROPOSITION 5 (**Second Isomorphism Theorem for modules**). *Let E be an A-module, and let M, N be two submodules of E; then there exists a canonical isomorphism of A-modules*

$$M/(M \cap N) \cong (M + N)/N.$$

Proof. Let us consider the linear mapping

$$u: M \to (M + N)/N$$

obtained as the composition of

$$M \xrightarrow{j} M + N \xrightarrow{p} (M + N)/N,$$

where j and p are, respectively, the canonical injection and surjection; u is a surjective mapping: since given an arbitrary element of $(M+N)/N$, this is of the form $\widehat{x+y}$, with $x \in M$ and $y \in N$, the sign "$\wedge$" designating the coset modulo N; but $x+y \equiv x \pmod N$ hence $\widehat{x+y} = \hat{x}$, $x \in M$; we have

$$u(x) = p(j(x)) = p(x) = \hat{x}.$$

If $x \in \mathrm{Ker}(u)$, then $\hat{x} = 0$, hence $x \in N$, and $x \in M \cap N$; conversely, if $x \in M \cap N$, then $u(x) = \hat{x} = 0$, since $x \in N$; it follows that $\mathrm{Ker}(u) = M \cap N$. The stated isomorphism is now $\hat{u}$ obtained from u according to Corollary 1 to Proposition 1, § 4, that is, the one for which

$$\hat{u}\,(\check{x}) = u(x) = \hat{x},$$

where $x \in M$ and $\check{x}$ is the coset of x modulo $M \cap N$. ∎

As mentioned above, *the partially ordered set of all submodules of an A-module E* is a complete meet-semilattice and a complete join-semilattice; thereby, it *is a complete lattice*; its first element is the zero submodule of E, while the last one is the total submodule of E.

PROPOSITION 6. *Given a ring A and an A-module E, the complete lattice of all submodules of E is a completely modular lattice.*

Proof. Let $(M_\lambda)_{\lambda\in I}$, $(N_\lambda)_{\lambda\in I}$ be two families of submodules of E with the property $M_\lambda \subseteq N_\mu$ for any $\lambda, \mu \in I$ such that $\lambda \neq \mu$; let $\lambda \in I$; we have $M_\lambda \cap N_\lambda \subseteq \subseteq M_\lambda \subseteq N_\mu$ for any $\mu \neq \lambda$; next, $M_\lambda \cap N_\lambda \subseteq N_\lambda$, hence

$$M_\lambda \cap N_\lambda \subseteq \bigcap_{\nu\in I} N_\nu;$$

since $M_\lambda \cap N_\lambda \subseteq M_\lambda \subseteq \sum_{\nu\in I} M_\nu$, it follows that

$$M_\lambda \cap N_\lambda \subseteq (\sum_{\nu\in I} M_\nu) \cap (\bigcap_{\nu\in I} N_\nu);$$

λ being arbitrary in I, we have

$$\sum_{\lambda\in I} (M_\lambda \cap N_\lambda) \subseteq (\sum_{\lambda\in I} M_\lambda) \cap (\bigcap_{\lambda\in I} N_\lambda). \tag{6}$$

Let now $x \in (\sum_{\lambda\in I} M_\lambda)(\bigcap_{\lambda\in I} \cap\, N_\lambda)$; as $x \in \sum_{\lambda\in I} M_\lambda$, according to Corollary 5 to Proposition 2 x may be written as $x = \sum_{\lambda\in I} x_\lambda$, where $(x_\lambda)_{\lambda\in I}$ is an fs-family of elements of E and $x_\lambda \in M_\lambda$, $\lambda \in I$; if $\lambda \in I$ and $\mu \neq \lambda$, then $x_\mu \in M_\mu \subseteq N_\lambda$, and taking into account that $x_\lambda = x - \sum_{\mu \neq \lambda} x_\mu$ and that $x \in N_\lambda$, it follows that $x_\lambda \in N_\lambda$; thus $x_\lambda \in M_\lambda \cap N_\lambda$ for any $\lambda \in I$, and hence $x \in \sum_{\lambda \in I} (M_\lambda \cap N_\lambda)$; hence we have equality in (6), and this completes the proof. ▮

* The notions of intersection and sum of submodules are recovered in the Theory of categories as follows: Let $\mathscr{C}$ be an abelian category; let $E \in \operatorname{Ob} \mathscr{C}$; given two subobjects of E, let (M, i) and (N, j) be some pairs representing these objects, respectively (see Definition 5, § 3); let (P, p), respectively (Q, q) be a cokernel of i, respectively j; consider the morphism

$$f : E \to P \times Q$$

obtained from the morphisms $p : E \to P$, $q : E \to Q$ by applying the definition of the direct product; a kernel (K, k) of f determines a subobject of E; this subobject does not change when replacing (M, i), respectively (N, j), by equivalent pairs, or if two other cokernels are considered for i, respectively j, and so on; this subobject uniquely determined by the equivalence *classes* of (M, i) and (N, j), and hence by the given subobjects, is called the *intersection* of the two given subobjects; as agreed in Definition 5, § 3, sometimes a pair (K, k) representing it is considered as the intersection, or even just the object component K; for this last meaning we sometimes write, of course, by an abuse of notation, $K = M \cap N$; all stated for two subobjects is valid for any *finite* family of subobjects of E.

If in the abelian category $\mathscr{C}$ Axiom AB3°) is satisfied, that is, there exist direct products indexed by arbitrary sets (see § 6), then in the same way we may define the *intersection* of an arbitrary family (indexed by a set) of subobjects of E, and we maintain the notation (this time $\bigcap_{\lambda \in I} M_\lambda$ or $(\bigcap_{\lambda \in I} M_\lambda, i)$ if $((M_\lambda, i_\lambda))_{\lambda \in I}$ is a family of representatives for the given family of subobjects).

Next, given an object E and two pairs (M, i), (M', i'), where M, M' are objects in the (not necessarily abelian) category $\mathscr{C}$, and i, respectively i', is a monomorphism of source M, respectively M', and range E, the pair (M, i) is said to *precede* (or to be *less than*) (M', i') or, alternatively, (M', i') is said to *succeed to* (or to be *greater than*) (M, i) if there exists a morphism $v : M \to M'$ making commutative the diagram

$$\begin{array}{ccc} M & & \\ & \searrow^{i} & \\ {\scriptstyle v}\downarrow & & E; \\ & \nearrow_{i'} & \\ M' & & \end{array}$$

in this case, we write $(M, i) \leqslant (M', i')$. The relation $\leqslant$ defined in this way is a preorder relation in the class of all pairs (M, i) described as above (the object E

being fixed); the equivalence relation associated with this preorder relation is exactly the one described in § 3; the relation $\leqslant$ canonically induces in the corresponding quotient class, i.e., in the class of subobjects of E (see Definition 5, § 3), a partial order relation denoted also by $\leqslant$ and for which the terminology used for pairs is maintained.

The subobject of E defined by the pair $(E, 1_E)$ (i.e. the one containing this pair) is greater than any subobject of E, hence it is the last element in the class of subobjects of E; it will be called the *total subobject* of the object E; according to the convention established in Definition 5, § 3, sometimes the pair $(E, 1_E)$ itself is called the *total subobject* of E. If we assume $\mathscr{C}$ to be an abelian category, then, as the unique morphism $0 \to E$ is a monomorphism, the pair $(0, 0)$ defines a subobject of E preceding any subobject of E, being hence the first element of the class of subobjects of E; this subobject (or the pair $(0, 0)$ itself) will be called the *zero subobject* of E.

If the category $\mathscr{C}$ is an abelian category, then the above defined intersection of two subobjects of E is the greatest lower bound of the two subobjects, in the sense of the partial order relation $\leqslant$; the same is valid for any finite family of subobjects of E; if, moreover, the abelian category $\mathscr{C}$ satisfies also Axiom AB3°), then the intersection of an arbitrary family (that is, indexed by an arbitrary set) of subobjects of the object E is the greatest common lower bound of the subobjects of the given family, in the sense of the same partial order relation:

Indeed, if $((M_\lambda, i_\lambda))_{\lambda\in I}$ is a complete system of representatives of the given family of subobjects, then let (P_λ, p_λ) be a cokernel of the morphism i_λ, $\lambda \in I$; let $f : E \to \prod_{\lambda\in I} P_\lambda$ be the morphism obtained applying the definition of the direct product to the family of morphisms $(p_\lambda)_{\lambda\in I}$; hence, denoting by π_λ the canonical projection of index λ, $\lambda \in I$, we have

$$\pi_\lambda \circ f = p_\lambda, \qquad \lambda \in I;$$

let (K, k) be a kernel of f; i_λ being a monomorphism, (M_λ, i_λ) is a kernel of p_λ, $\lambda \in I$; but for any $\lambda \in I$ we have

$$p_\lambda \circ k = \pi_\lambda \circ f \circ k = 0,$$

hence there exists $r_\lambda \in \mathrm{Hom}_{\mathscr{C}}(K, M_\lambda)$ unique with the property that

$$i_\lambda \circ r_\lambda = k;$$

it follows that the subobject of E defined by the pair (K, k) is less than any subobject of the given family; if another subobject of E represented by (X, ξ) has the same property, then there exist some morphisms $\eta_\lambda : X \to M_\lambda$, $\lambda \in I$, such that for any $\lambda \in I$ the following diagram commutes:

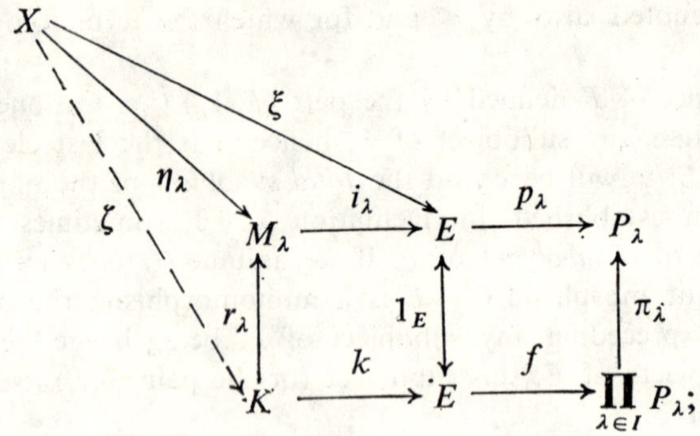

we have

$$\pi_\lambda \circ f \circ \xi = p_\lambda \circ \xi = p_\lambda \circ i_\lambda \circ \eta_\lambda = 0$$

for any $\lambda \in I$, hence, in view of the uniqueness, $f \circ \xi = 0$; it follows that there exists a unique morphism $\zeta : X \to K$ with the property $k \circ \zeta = \xi$; hence the subobject represented by (X, ξ) precedes the subobject represented by (K, k).

Consider again an abelian category $\mathcal{C}$ and an object E of it; given two subobjects of E and two pairs (M, i), (N, j) respectively representing them, consider the morphism

$$f : M \oplus N \to E$$

obtained from the morphisms $i : M \to E$, $j : N \to E$ by applying the definition of the direct sum; the image of f considered as a *class* of equivalent pairs turns out to be independent of the representatives (M, i), (N, j) chosen for the given subobjects; this class, or any representative of it, is called the *sum* of the two given subobjects; if we consider the object component of the pair chosen to represent the sum, this object is denoted, also by an abuse of notation, by $M + N$. The sum of two subobjects is the least upper bound of the two subobjects, in the sense of the partial order relation $\leqslant$; this fact is shown below in a more general setting.

If in the abelian category $\mathcal{C}$ Axiom AB3) is satisfied, that is, there exist direct sums indexed by arbitrary sets (see § 7), the *sum* of an arbitrary family (indexed by a set) of subobjects of $\mathcal{C}$ may be defined in the same way, and the notation subsists (this time $\sum_{\lambda \in I} M_\lambda$, where $((M_\lambda, i_\lambda))_{\lambda \in I}$ is a family of representatives of the given family of subobjects); in this case, the sum of the family is the least common upper bound of the subobjects of the given family, in the sense of the partial order relation defined above.

Let us demonstrate this fact: considering, in the preceding notation, the morphism

$$f : \bigoplus_{\lambda \in I} M_\lambda \to E$$

obtained from the family of morphisms $(i_\lambda)_{\lambda \in I}$ by applying the definition of the direct sum, we have

$$f \circ j_\lambda = i_\lambda, \qquad \lambda \in I,$$

where j_λ is the canonical injection $M_\lambda \to \bigoplus_{\lambda \in I} M_\lambda$, $\lambda \in I$; if (J, r) is an image of f, then r is a monomorphism, and hence (J, r) defines a subobject of E; if (C, q) is a coimage of f in the category $\mathscr{C}$, then there exists a unique isomorphism $\overline{f} : C \rightleftarrows J$ with the property that

$$r \circ \overline{f} \circ q = f;$$

we have then for any $\lambda \in I$,

$$r \circ \overline{f} \circ q \circ j_\lambda = f \circ j_\lambda = i_\lambda,$$

hence the subobject of E defined by (J, r) is greater than any subobject of the given family. If a subobject of E defined, say, by the pair (Y, η) has the same property, then for any $\lambda \in I$ there exists $\zeta_\lambda : M_\lambda \to Y$ such that the following diagram commutes:

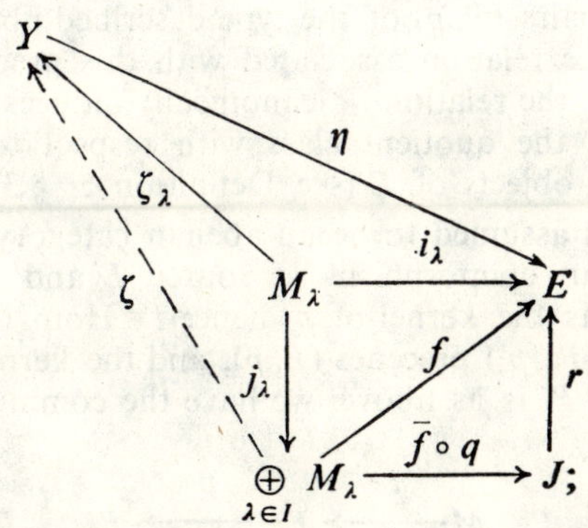

consequently, there exists a unique morphism $\zeta : \bigoplus_{\lambda \in I} M_\lambda \to Y$ with the property that

$$\zeta \circ j_\lambda = \zeta_\lambda, \qquad \lambda \in I;$$

we have

$$\eta \circ \zeta \circ j_\lambda = \eta \circ \zeta_\lambda = i_\lambda = f \circ j_\lambda$$

for any $\lambda \in I$, hence $\eta \circ \zeta = f$; denoting by (K, k) a kernel of f, we have $\eta \circ \zeta \circ k = f \circ k = 0$ or, as η is a monomorphism, $\zeta \circ k = 0$; hence ζ factors through q, since (C, q) is a cokernel of k; thus, there exists $\rho : C \to Y$ such that $\rho \circ q = \zeta$; then, as $\eta \circ \rho \circ q = \eta \circ \zeta = f = r \circ \bar{f} \circ q$, and as q is an epimorphism, it follows $\eta \circ \rho = r \circ \bar{f}$, hence

$$\eta \circ \rho \circ \bar{f}^{-1} = r;$$

consequently, the subobject of E defined by the pair (Y, η) is greater than the subobject of E defined by the pair (J, r).

As for the quotient objects of an object in a category, there exists a partial order relation dual to the partial order relation for subobjects, namely: Let $\mathcal{C}$ be a category, and let $E \in \operatorname{Ob} \mathcal{C}$; given two pairs (P, p), (P', p'), where P, P' are objects of the category, while p, respectively p', is an epimorphism of source E and range P, respectively P', the pair (P', p') is said to *precede* (or to be *less than*) (P, p) or, equivalently, (P, p) is said to *succeed to* (or to be *greater than*) (P', p') if there exists a morphism $w : P \to P'$ rendering commutative the diagram

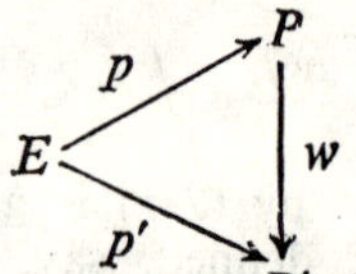

in this case, we write $(P', p') \prec (P, p)$. The relation $\prec$ thus defined is a preorder relation in the class of pairs (P, p) of the type described above (where the object E is fixed); the equivalence relation associated with this preorder relation is exactly the one described in § 3; the relation $\prec$ canonically induces a partial order relation denoted also by $\prec$ in the quotient class with respect to this equivalence, i.e., in the class of quotient objects of E (see Definition 5, § 3).

If, moreover, $\mathcal{C}$ is assumed to be an abelian category, then to a pair (P, p), with $P \in \operatorname{Ob} \mathcal{C}$ and p an epimorphism of source E and range P, it is assigned a pair (M, i) obtained as the kernel of p; hence $i \in \operatorname{Hom}_{\mathcal{C}}(M, E)$ and i is a monomorphism; if the pair (P', p') precedes (P, p), and the kernel of (P', p') is (M', i'), then, assuming $w : P \to P'$ is as above, we have the commutative diagram

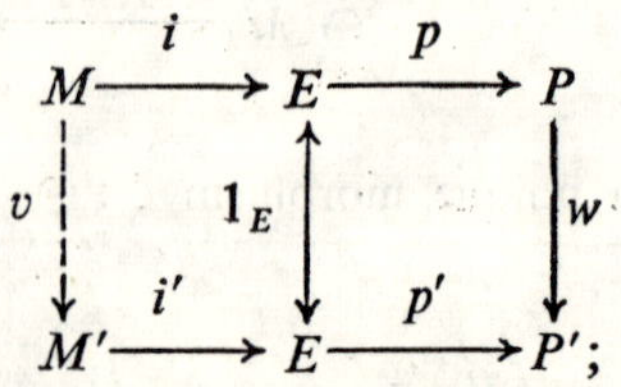

by virtue of those shown in § 4 concerning the diagram (2), it follows that there exists a unique morphism $v : M \to M'$ making the diagram commutative; this means that $(M, i) \leqslant (M', i')$; in particular, if (P, p) and (P', p') are equivalent pairs in the sense of § 3, then so are the pairs (M, i), (M', i') obtained as kernels of p and p', respectively; this mapping between pairs defines thus a mapping between equivalence classes, that is, a mapping from the class of quotient objects of E to the class of subobjects of E, and this mapping turns to be antimonotone.

Similarly, a mapping from the class of subobjects of the object E to the class of quotient objects of E may be defined by assigning to a subobject defined, say, by the pair (M, i) the quotient object defined by the pair (P, p) which is a cokernel of i. This mapping turns also to be antimonotone, and it is the inverse mapping of the one defined above; these two mappings are thus bijective and antimonotone; they will be called in the sequel the *canonical mappings* (or *bijections*) between the two classes.

Given a subobject of E defined, say, by the pair (M, i), the quotient object corresponding to it under the canonical mapping is denoted, by an abuse of notation, E/M; similarly, the subobject corresponding under the canonical mapping to the quotient object defined by the pair (P, p) is denoted, also by an abuse of notation, $E \setminus P$.[1]

It follows that in the class of all quotient objects of an object E of an abelian category (respectively of an abelian category with direct sums) any finite family (respectively any family) of elements has a greatest lower bound in the sense of the partial order relation defined above; also, if $\mathscr{C}$ is an abelian category (respectively an abelian category with direct products), then any finite family (respectively any family) of quotient objects of an object E has a least upper bound.

Now we are going to show what do become in the Theory of categories the notions of image and inverse image of submodules under a homomorphism, defined in § 3:

Let $\mathscr{C}$ be an abelian category, let $E, F \in \operatorname{Ob}\mathscr{C}$, and let $u \in \operatorname{Hom}_{\mathscr{C}}(E, F)$; given a subobject of E represented by the pair (M, i), let us consider the pair (I, r), where (I, r) is an image of $u \circ i$; if (M', i') is another pair, where $i' \in \operatorname{Hom}_{\mathscr{C}}(M', E)$ and i' is a monomorphism, and if the pair (M, i) precedes the pair (M', i'), then there exists $f : M \to M'$ such that $i' \circ f = i$; hence we have the commutative diagram

$$\begin{array}{ccccc} M & \xrightarrow{i} & E & \xrightarrow{u} & F \\ {\scriptstyle f}\downarrow & & & & \updownarrow{\scriptstyle 1_F} \\ M' & \xrightarrow{i'} & E & \xrightarrow{u} & F; \end{array}$$

consequently, denoting by (I', r') an image of $u \circ i'$, there exists a morphism $g : I \to I'$ such that $r' \circ g = r$ (see the discussion concerning diagram (2), § 4); hence $(I, r) \leqslant$

---

[1] This is not to be confounded with the notation of the complement of a subclass in a class, with which it has in common only the symbol.

$\leqslant (I', r')$; in particular, if (M', i') defines the same subobject of E as (M, i), then (I', r') defines the same subobject of F as (I, r), and we defined thus a mapping between the class of subobjects of E and the class of subobjects of F; according to the preceding considerations, this mapping is monotone; the subobject of F corresponding under this mapping to a subobject of E is called the *image under the morphism u* of this subobject; if, as we agreed, the given subobject of E is represented by (M, i), then the image under u of this subobject is, by an abuse of notation, denoted $u(M)$.

The definition of the image of an object under a morphism has its origin in the fact that, if $\mathcal{C} = \mathbf{Mod}(A)$, if M is a submodule of the module E with the inclusion mapping i, and $u : E \to F$ is an A-homomorphism, then $\operatorname{Im}(u \circ i) = u(M)$.

Analogously, under the same conditions, if a subobject of F represented by (N, j) is given, then let (Q, q) be a cokernel of j; if (K, k) is a kernel of $q \circ u$, then, using as before the results of § 4, we obtain that the equivalence class of (K, k) does not depend on the chosen representative (N, j); so, a mapping between the class of subobjects of F and the class of subobjects of E is defined, and this in addition is monotone; the subobject of E corresponding under this mapping to a subobject of F is called the *inverse image under the morphism u* of the latter subobject; the inverse image of the subobject represented by the pair (N, j) is, by an abuse of notation, denoted $u^{-1}(N)$.

The definition of the inverse image of a subobject under a morphism is the translation into categorical language of the fact that if $\mathcal{C} = \mathbf{Mod}(A)$, if N is a submodule of the module F with the canonical surjection $q : F \to F/N$, and $u : E \to F$ is an A-homomorphism, then $u^{-1}(N) = \operatorname{Ker}(q \circ u)$.

Using the above defined canonical bijections between the class of subobjects and the one of quotient objects of an object in an abelian category, we get, for a morphism $u \in \operatorname{Hom}_{\mathcal{C}}(E, F)$, a canonical monotone mapping defined on the class of quotient objects of E and taking values in the class of quotient objects of F; the quotient object of F corresponding under this mapping to a quotient object of E is called its *image under the morphism u*; in the same way, one obtains a canonical monotone mapping defined on the class of quotient objects of F and taking values in the class of quotient objects of E; the quotient object of E corresponding under this mapping to a quotient object of F is called its *inverse image under the morphism u*; we leave the details to the reader.*

Returning now to the category of modules over a ring A, we first notice that if E is an A-module, then the submodules of E together with the corresponding canonical injections constitute a complete system of representatives of the subobjects of E, in the sense of classes of equivalent pairs; it is left to the reader as an exercise to verify the fact that the class represented by the submodule M with the canonical injection i precedes, in the sense of the above definition for arbitrary categories, the class represented by the submodule M' with the canonical injection i' amounts to $M \subseteq M'$. Next, Proposition 1 and Corollary 5 to Proposition 2 show that the intersection and the sum of submodules are representatives of the intersection, respectively of the sum of the corresponding subobjects (considered as classes) in the category $\mathbf{Mod}(A)$.

* As for the extension of the results previously obtained for modules to arbitrary abelian categories (eventually satisfying Axiom AB3) or AB3°)), we say only a few words about Corollary 7 to Proposition 2; the corresponding assertion for abelian categories is stated as follows:

AB5) *In the abelian category* $\mathscr{C}$ *Axiom* AB3) *is satisfied, and for any object* E *of* $\mathscr{C}$, *any subobject* (N, j) *of* E, *and any family* $((M_\lambda, i_\lambda))_{\lambda \in I}$ *of subobjects of* E *directed to the right with respect to the relation* $\leqslant$ *we have*

$$(\sum_{\lambda \in I} M_\lambda) \cap N = \sum_{\lambda \in I} (M_\lambda \cap N).$$

Generally, Axiom AB5) is not valid for an abelian category even if it satisfies Axiom AB3).*

§ 9. Internal Direct Sum of Submodules

Let A be a ring, let E be an A-module, and let $(M_\lambda)_{\lambda \in I}$ be a family of submodules of E; if $i_\lambda : M_\lambda \to E$, $\lambda \in I$, is the canonical injection, let

$$f : \bigoplus_{\lambda \in I} M_\lambda \to E$$

be the linear mapping determined according to the universal property of the direct sum by the family of mappings $(i_\lambda)_{\lambda \in I}$.

PROPOSITION 1. *Let E be an A-module, and let $(M_\lambda)_{\lambda \in I}$ be a family of submodules of E; the following assertions concerning this family are equivalent:*

i) The above homomorphism f is an isomorphism;

ii) Any element $x \in E$ may be uniquely written as

$$x = \sum_{\lambda \in I} x_\lambda, \tag{1}$$

where $(x_\lambda)_{\lambda \in I}$ is an fs-family of elements of E such that $x_\lambda \in M_\lambda$ for any $\lambda \in I$;

iii) We have

$$E = \sum_{\lambda \in I} M_\lambda,$$

and any relation of the form

$$\sum_{\lambda \in I} x_\lambda = 0, \tag{2}$$

where $(x_\lambda)_{\lambda \in I}$ satisfies the conditions of ii), implies $x_\lambda = 0$ for any $\lambda \in I$;

iv) We have

$$E = \sum_{\lambda \in I} M_\lambda \quad \text{and} \quad M_\lambda \cap \Big(\sum_{\mu \neq \lambda} M_\mu\Big) = 0, \qquad \lambda \in I.$$

Proof. i) $\Rightarrow$ *ii)*: According to the proof or Corollary 5 to Proposition 2, § 8, if $(x_\lambda)_{\lambda \in I}$ is as stated, i.e. an element of $\bigoplus_{\lambda \in I} M_\lambda$, then $f((x_\lambda)_\lambda) = \sum_{\lambda \in I} x_\lambda$; but any $x \in E$ is, by hypothesis, the image under f of a uniquely determined family $(x_\lambda)_\lambda \in \bigoplus_{\lambda \in I} M_\lambda$; thereby *ii)* follows.

ii) $\Rightarrow$ *iii)*: By *ii)*, any $x \in E$ is in $\sum_{\lambda \in I} M_\lambda$ according to Corollary 5 to Proposition 2, § 8; on the other hand, relation (2) provides a representation of 0 as in (1), and as 0 admits also the representation

$$0 = \sum_{\lambda \in I} 0, \qquad 0 \in M_\lambda,$$

the uniqueness implies $x_\lambda = 0$ for any $\lambda \in I$.

iii) $\Rightarrow$ *iv)*: The first equality of *iv)* just repeats a part of *iii)*; next, let $\lambda \in I$, and let $x \in M_\lambda \cap (\sum_{\mu \neq \lambda} M_\mu)$; it follows that $x = \sum_{\mu \neq \lambda} x_\mu$, where $(x_\mu)_{\mu \neq \lambda}$ is an fs-family and $x_\mu \in M_\mu$; but then, putting $x_\lambda = -x \in M_\lambda$, we get $\sum_{\mu \in I} x_\mu = 0$, with $(x_\mu)_{\mu \in I}$ an fs-family and $x_\mu \in M_\mu$, $\mu \in I$; by *iii)* it follows $x_\mu = 0$ for any $\mu \in I$; in particular, $x = -x_\lambda = 0$.

iv) $\Rightarrow$ *i)*: Again applying Corollary 5 to Proposition 2, § 8, we have $\mathrm{Im}(f) = \sum_{\lambda \in I} M_\lambda = E$, hence f is surjective; if $(x_\lambda)_{\lambda \in I} \in \mathrm{Ker}(f)$, then $\sum_{\lambda \in I} x_\lambda = 0$, whence for any $\lambda \in I$,

$$x_\lambda = \sum_{\mu \neq \lambda} (-x_\mu) \in M_\lambda \cap (\sum_{\mu \neq \lambda} M_\mu) = 0;$$

hence $\mathrm{Ker}(f) = 0$, and f is injective (see Proposition 2, § 3). ▮

DEFINITION 1. Let E be an A-module, and let $(M_\lambda)_{\lambda \in I}$ be a family of submodules of E; we say that *the module E is the internal direct sum* (or that *E decomposes into the internal direct sum*) of the family of submodules $(M_\lambda)_{\lambda \in I}$ provided the equivalent conditions of Proposition 1 are satisfied. To describe this situation, we shall write $E = \dot{\sum_{\lambda \in I}} M_\lambda$ or also, if $I = \{1, \ldots, n\}$, $E = M_1 \dot{+} M_2 \dot{+} \ldots \dot{+} M_n$ or also $E = \dot{\sum_{i=1}^{n}} M_i$. In this last case, E is also said to be the *internal direct sum of* $M_1, M_2, \ldots, M_n$.

If $E = \dot{\sum_{\lambda \in I}} M_\lambda$, the uniquely determined element $x_\lambda \in M_\lambda$, $\lambda \in I$, in expression (1) for $x \in E$ is called the *component in M_λ* (or the *component of index λ*) of x with respect to the given decomposition into internal direct sum of E.

If F is a submodule of E, *the submodule F is said to be the internal direct sum of the family of submodules* $(M_\lambda)_{\lambda \in I}$ if $M_\lambda \subseteq F$, $\lambda \in I$, and if F considered as a module by itself is the internal direct sum of the family of its submodules $(M_\lambda)_{\lambda \in I}$.[1] ▮

REMARKS. 1) If the A-module E is the internal direct sum of the family $(M_\lambda)_{\lambda \in I}$ of its submodules, then, by Proposition 1, we have $E = \sum_{\lambda \in I} M_\lambda$; notice, however, that the notion of internal direct sum does not define a construction on modules, as do, for instance, the direct sum of modules or the sum of submodules:

---

[1] In this case, each M_λ, $\lambda \in I$, is a submodule of F (see § 3).

the notion of internal direct sum defines a situation, namely that in which the sum of the family $(M_\lambda)_{\lambda \in I}$ of submodules enjoys a special property.

2) By definition, *if the A-module E is the internal direct sum of the family* $(M_\lambda)_{\lambda \in I}$ *of its submodules, then E is A-isomorphic to the external direct sum of the family of A-modules* $(M_\lambda)_{\lambda \in I}$: indeed, in this case there exists the canonical isomorphism

$$f : \bigoplus_{\lambda \in I} M_\lambda \rightleftarrows E$$

given by Proposition 1, *i*); conversely, *if* $(E_\lambda)_{\lambda \in I}$ *is a family of A-modules, then* $E = \bigoplus_{\lambda \in I} E_\lambda$ *is the internal direct sum of some family* $(M_\lambda)_{\lambda \in I}$ *of submodules of it, indexed by the same index set I, and such that for any* $\lambda \in I$, M_λ *is isomorphic under a canonical isomorphism to* E_λ: indeed, denoting by j_λ the canonical injection $E_\lambda \to E$, $\lambda \in I$, by Lemma 1, § 7, any $x \in E$ may be uniquely written as

$$x = \sum_{\lambda \in I} j_\lambda(x_\lambda),$$

where $(x_\lambda)_{\lambda \in I}$ is an fs-family such that $x_\lambda \in E_\lambda$, $\lambda \in I$; but this means, by the characterization *ii*) of Proposition 1 of the internal direct sum, that E is the internal direct sum of the family $(M_\lambda)_{\lambda \in I}$ of submodules of it, where $M_\lambda = j_\lambda(E_\lambda)$, $\lambda \in I$; as j_λ is an injective homomorphism, we have an isomorphism $j'_\lambda : E_\lambda \rightleftarrows M_\lambda$, $\lambda \in I$ (see Corollary 2 to Proposition 1, § 4).

These facts show that the properties preserved under isomorphisms are transferred from the internal direct sum to the external direct sum, and vice versa; sometimes, no distinction is done between the two notions. However, from now on, whenever talking of an isomorphism between a module which is the internal direct sum of a family of submodules and the external direct sum of the corresponding family of modules, we shall mean, unless otherwise stated, the above canonical isomorphism. ▮

PROPOSITION 2. *Let* $(E_\lambda)_{\lambda \in I}$ *be a family of A-modules, let E be an A-module, and let* $f_\lambda : E_\lambda \to E$, $\lambda \in I$, *be a family of linear mappings; if*

$$f : \bigoplus_{\lambda \in I} E_\lambda \to E$$

is the linear mapping the family $(f_\lambda)_{\lambda \in I}$ *induces by virtue of the universal property of the direct sum, then, in order that f be an isomorphism it is necessary and sufficient that there exists a family of linear mappings*

$$g_\lambda : E \to E_\lambda, \qquad \lambda \in I,$$

satisfying the following conditions:

a) $g_\lambda \circ f_\lambda = 1_{E_\lambda}$ *for any* $\lambda \in I$;

b) $g_\lambda \circ f_\mu = 0$ *if* $\lambda \neq \mu$;

c) for any $x \in E$, *the family* $(g_\lambda(x))_{\lambda \in I}$ *is an fs-family, and we have*

$$x = \sum_{\lambda \in I} f_\lambda(g_\lambda(x));$$

if I is a finite set, then condition c) is equivalent to the equality

$$c')\ \sum_{\lambda \in I} f_\lambda \circ g_\lambda = 1_E.$$

Returning to an arbitrary set I, provided the above equivalent assertions are valid, the family $(g_\lambda)_{\lambda \in I}$ *is uniquely determined by conditions a) and b).*

Proof. Let us start with the necessity: If f is an isomorphism, we denote by g_λ, $\lambda \in I$, the composite mapping

$$E \xrightarrow{f^{-1}} \bigoplus_{\lambda \in I} E_\lambda \xrightarrow{k} \prod_{\lambda \in I} E_\lambda \xrightarrow{p_\lambda} E_\lambda,$$

where k is the canonical injection and p_λ the canonical projection of index λ; if $j_\mu : E_\mu \to \bigoplus_{\lambda \in I} E_\lambda$ is the canonical injection, then, for any $\lambda, \mu \in I$ we have

$$g_\lambda \circ f_\mu = p_\lambda \circ k \circ f^{-1} \circ f \circ j_\mu = p_\lambda \circ k \circ j_\mu,$$

which is 0 or 1_{E_λ} according as $\lambda \neq \mu$ or $\lambda = \mu$ by the definitions of j_μ and p_λ; thus, the mappings g_λ, $\lambda \in I$, satisfy conditions *a*) and *b*). As for condition *c*), we notice that if $x \in E$, then $f^{-1}(x)$, which is an element of $\bigoplus_{\lambda \in I} E_\lambda$, has $p_\lambda(k(f^{-1}(x))) = g_\lambda(x)$ as component of index λ, hence that $(g_\lambda(x))_{\lambda \in I} = f^{-1}(x)$, and so $(g_\lambda(x))_\lambda$ is an fs-family; next we have

$$\sum_{\lambda \in I} f_\lambda(g_\lambda(x)) = f((g_\lambda(x))_\lambda) = f(f^{-1}(x)) = x.$$

Conversely, if there exists the family of mappings $(g_\lambda)_{\lambda \in I}$ satisfying conditions *a*), *b*), *c*), then the linear mapping $E \to \prod_{\lambda \in I} E_\lambda$ induced by the family g_λ, $\lambda \in I$, by virtue of the universal property of the direct product has its image contained in $\bigoplus_{\lambda \in I} E_\lambda$, and hence induces by the universal property of the submodule a linear mapping

$$g : E \to \bigoplus_{\lambda \in I} E_\lambda;$$

if $x \in E$, we have $g(x) = (g_\lambda(x))_{\lambda \in I}$, hence $f(g(x)) = \sum_{\lambda \in I} f_\lambda(g_\lambda(x)) = x$ according to condition *c*). If $(x_\lambda)_{\lambda \in I} \in \bigoplus_{\lambda \in I} E_\lambda$, then, by conditions *a*) and *b*),

$$g(f((x_\lambda)_{\lambda \in I})) = g(\sum_{\lambda \in I} f_\lambda(x_\lambda)) = (g_\mu(\sum_{\lambda \in I} f_\lambda(x_\lambda)))_{\mu \in I} = (x_\mu)_{\mu \in I};$$

consequently, f is an isomorphism whose inverse is g.

Clearly, conditions *c*), *c'*) are equivalent provided I is finite.

For the last assertion, a procedure similar to the one used in Proposition 4, § 7, to prove the uniqueness of the family of morphisms $(p_i)_{i\in I}$ is applied: if $h_\lambda : E \to E_\lambda$, $\lambda \in I$, is a family of linear mappings satisfying conditions *a*) and *b*) when replacing g_λ, $\lambda \in I$, then for each $\lambda \in I$ we have $g_\lambda \circ f \circ j_\mu = h_\lambda \circ f \circ j_\mu$ for any $\mu \in I$, and hence, as in the aforementioned proof, $g_\lambda \circ f = h_\lambda \circ f$, $\lambda \in I$, whence $g_\lambda = h_\lambda$, $\lambda \in I$, since f is by hypothesis an isomorphism. ∎

COROLLARY 1. *Let E be an A-module, and let $(M_\lambda)_{\lambda\in I}$ be a family of submodules of it; in order that E be the internal direct sum of the family of submodules $(M_\lambda)_{\lambda\in I}$ it is necessary and sufficient that there exist a family of A-homomorphisms*

$$g_\lambda : E \to M_\lambda, \quad \lambda \in I,$$

such that, if i_λ is the canonical injection $M_\lambda \to E$, $\lambda \in I$, the following conditions are satisfied:

a) $g_\lambda \circ i_\lambda = 1_{M_\lambda}$ *for any* $\lambda \in I$;

b) $g_\lambda \circ i_\mu = 0$ *if* $\lambda \neq \mu$;

c) *for any $x \in E$, the family $(g_\lambda(x))_{\lambda\in I}$ is an fs-family, and*

$$x = \sum_{\lambda\in I} i_\lambda(g_\lambda(x)).$$

In this case, the family of linear mappings $(g_\lambda)_{\lambda\in I}$ is uniquely determined by conditions a) and b).

Proof. We need only take in Proposition 2 $E_\lambda = M_\lambda$ and $f_\lambda = i_\lambda$ for any $\lambda \in I$. ∎

REMARK. Under the conditions of Corollary 1, the mappings g_λ, $\lambda \in I$, act as follows: by Proposition 1, *ii*), any element $x \in E$ may be uniquely written as $x = \sum_{\lambda\in I} x_\lambda$, where $(x_\lambda)_{\lambda\in I}$ is an fs-family and $x_\lambda \in M_\lambda$; if f is the canonical isomorphism $\bigoplus_{\lambda\in I} M_\lambda \rightleftarrows E$, then $f((x_\lambda)_\lambda) = \sum_{\lambda\in I} x_\lambda$; hence, in the notations of Proposition 2,

$$g_\lambda(x) = p_\lambda(f^{-1}(x)) = p_\lambda((x_\mu)_{\mu\in I}) = x_\lambda, \quad \lambda \in I,$$

and g_λ sends $x \in E$ into its component in M_λ, $\lambda \in I$.

In particular, let a family $(E_\lambda)_{\lambda\in I}$ of A-modules be given, and let $E = \bigoplus_{\lambda\in I} E_\lambda$; we saw (Remark 2) following Definition 1) that E is the internal direct sum of the family of its submodules $(j_\lambda(E_\lambda))_{\lambda\in I}$, j_λ being the canonical injection $E_\lambda \to E$, $\lambda \in I$; in this case, the component of $x = (x_\lambda)_{\lambda\in I}$ in $j_\lambda(E_\lambda)$ is $j_\lambda(x_\lambda)$, hence $g_\lambda : E \to j_\lambda(E_\lambda)$, $\lambda \in I$, is the composition

$$E = \bigoplus_{\lambda\in I} E_\lambda \xrightarrow{k} \prod_{\lambda\in I} E_\lambda \xrightarrow{p_\lambda} E_\lambda \xrightarrow{j'_\lambda} j_\lambda(E_\lambda),$$

where k is the canonical injection, p_λ is the canonical projection of index λ, and j'_λ is induced by j_λ. ∎

DEFINITION 2. Given an A-module E, an endomorphism u of E is said to be a *projector* of E if $u \circ u = u$, in other words, if u is an idempotent element of the ring $\mathrm{End}_A(E)$; if $(u_\lambda)_{\lambda \in I}$ is a family of projectors of E, this family is said to be *orthogonal* if $u_\lambda \circ u_\mu = 0$ for $\lambda \neq \mu$. ▮

We also have the following characterization of the fact that a module decomposes into the internal direct sum of a family of its submodules, a characterization which actually is a transcription of Corollary 1:

COROLLARY 2. *Let E be an A-module, and let $(M_\lambda)_{\lambda \in I}$ be a family of submodules of E; then E is the internal direct sum of the family of submodules $(M_\lambda)_{\lambda \in I}$ if and only if there exists an orthogonal family of projectors $(u_\lambda)_{\lambda \in I}$ such that $M_\lambda = u_\lambda(E)$, $\lambda \in I$, and that for any $x \in E$ the family $(u_\lambda(x))_{\lambda \in I}$ has finite support and*

$$x = \sum_{\lambda \in I} u_\lambda(x).$$

For I finite, the last condition is equivalent to

$$\sum_{\lambda \in I} u_\lambda = 1_E.$$

Proof. The condition is necessary: Using the notation of Corollary 1, let u_λ, $\lambda \in I$, be the composite linear mapping

$$E \xrightarrow{g_\lambda} M_\lambda \xrightarrow{i_\lambda} E;$$

then, by conditions *a*) and *b*) of Corollary 1 it follows readily that $(u_\lambda)_{\lambda \in I}$ is an orthogonal family of projectors of E. Since for $x \in E$, $g_\lambda(x) = u_\lambda(x)$, $\lambda \in I$, the last assertion of the proposition is equivalent to *c*) of Corollary 1; finally, $u_\lambda(E) = g_\lambda(E) = M_\lambda$, $\lambda \in I$, the last equality being valid since, by *a*) of Corollary 1, g_λ is surjective (see the Corollary to Proposition 1, § 1).

The condition is sufficient: Denote by g_λ, $\lambda \in I$, the linear mapping $E \to M_\lambda$ induced by u_λ by virtue of Corollary 2 to Proposition 1, § 4; we have $i_\lambda \circ g_\lambda = u_\lambda$, $\lambda \in I$, hence for any $x \in E$, $g_\lambda(x) = u_\lambda(x)$, and condition *c*) of Corollary 1 is satisfied; it follows also that $g_\lambda(E) = M_\lambda$, hence that g_λ is surjective; if $\lambda \in I$, then from the fact that u_λ is an idempotent we have

$$i_\lambda \circ g_\lambda \circ i_\lambda \circ g_\lambda = i_\lambda \circ g_\lambda,$$

and from the fact that i_λ is a monomorphism and g_λ is an epimorphism we obtain *a*) of Corollary 1; similarly, from the orthogonality conditions follows also *b*) of Corollary 1, and this completes the proof. ▮

REMARK. The proof of Corollary 2 shows that u_λ *sends* $x \in E$ *into its component in* M_λ, $\lambda \in I$. The uniqueness of the representation $x = \sum_{\lambda \in I} x_\lambda$, with $x_\lambda \in M_\lambda$, $\lambda \in I$, implies that *the family* $(u_\lambda)_{\lambda \in I}$ with the properties given in Corollary 2 *is uniquely determined by the given internal direct sum decomposition of* E. ▮

The family $(u_\lambda)_{\lambda \in I}$ constructed in Corollary 2 to Proposition 2 will be called the orthogonal family of projectors *corresponding* to the given internal direct sum decomposition of E.

Corollary 3. *Let E be an A-module, and let u be one of its projectors; then: a) E is the internal direct sum of* $\mathrm{Im}(u)$ *and* $\mathrm{Ker}(u)$ *considered as submodules of E; b) the component of $x \in E$ in* $\mathrm{Im}(u)$ *with respect to this decomposition is* $u(x)$; *c) $1_E - u$ is a projector of E, and*

$$\mathrm{Im}(1_E - u) = \mathrm{Ker}(u), \quad \mathrm{Ker}(1_E - u) = \mathrm{Im}(u).$$

Proof. We start with assertion *c*): we have

$$(1_E - u)\circ(1_E - u) = 1_E - u - u + u^2 = 1_E - u;$$

next, $u(x) = 0$ is equivalent to $x = x - u(x) = (1_E - u)(x)$, and hence $\mathrm{Ker}(u) = \mathrm{Im}(1_E - u)$; as $u = 1_E - (1_E - u)$, the other assertion of *c*) also follows. Further, since $u\circ(1_E - u) = u - u\circ u = 0$, the family of projectors $(u, 1_E - u)$ is orthogonal; as

$$x = u(x) + (1_E - u)(x), \tag{3}$$

the family of submodules $(\mathrm{Im}(u), \mathrm{Ker}(u))$ satisfies the conditions of Corollary 2, which proves *a*). From (3) assertion *b*) also follows. ∎

Let E be an A-module, and let $(M_\lambda)_{\lambda\in I}$ be a family of submodules of E such that $E = \dot{\sum}_{\lambda\in I} M_\lambda$; let F be a submodule of E; denote by $(u_\lambda)_{\lambda\in I}$ the orthogonal family of projectors corresponding to the given internal direct sum decomposition of E.

Definition 3. The submodule $u_\lambda(F)$, $\lambda\in I$, is called the *component of F in M_λ* (or also the *component of index λ of F*) with respect to the given internal direct sum decomposition of E; the submodule F is said to be *homogeneous* (with respect to the given internal direct sum decomposition of E) if for any $\lambda \in I$ the component of index λ of F is contained in F. ∎

From the Remark following Corollary 2, the component of F in M_λ is the set of components in M_λ of the elements of F, $\lambda\in I$; hence this component is contained in M_λ for any $\lambda \in I$.

Proposition 3. *Let E be an A-module, and let $(M_\lambda)_{\lambda\in I}$ be a family of submodules of E such that $E = \dot{\sum}_{\lambda\in I} M_\lambda$; let F be a submodule of E homogeneous with respect to the given decomposition of E; then, if N_λ is the component of F in M_λ, $\lambda\in I$, we have*

$$F = \dot{\sum_{\lambda\in I}} N_\lambda;$$

moreover, there exists a canonical isomorphism of A-modules

$$E/F \rightleftarrows \bigoplus_{\lambda\in I} (M_\lambda/N_\lambda).$$

Proof. By hypothesis, $N_\lambda \subseteq F$, $\lambda \in I$; next, if $x \in F$, we have

$$x = \sum_{\lambda \in I} x_\lambda, \tag{4}$$

where x_λ is the component of x in M_λ with respect to the given decomposition; hence $x_\lambda \in N_\lambda$, $\lambda \in I$; but the representation (4) of x as a sum of elements of the submodules N_λ, $\lambda \in I$, is unique, since any other representation of this kind provides another representation of x as a sum of elements of the submodules M_λ, $\lambda \in I$.

For the second assertion, let i_λ be the canonical injection of M_λ into E, $\lambda \in I$; since $i_\lambda(N_\lambda) = N_\lambda \subseteq F$, i_λ induces by passing to the quotient an A-linear mapping

$$f_\lambda : M_\lambda/N_\lambda \to E/F,$$

$\lambda \in I$; we will show that the linear mapping

$$f : \bigoplus_{\lambda \in I} (M_\lambda/N_\lambda) \to E/F$$

obtained from the family $(f_\lambda)_{\lambda \in I}$ by the universal property of the direct sum is an isomorphism; to do this, we construct a family of linear mappings

$$g_\lambda : E/F \to M_\lambda/N_\lambda, \qquad \lambda \in I,$$

such that conditions *a*), *b*), *c*) of Proposition 2 are satisfied: Let $(u_\lambda)_{\lambda \in I}$ be the orthogonal family of projectors corresponding to the given internal direct sum decomposition of E; u_λ induces by restriction to E and M_λ a linear mapping $v_\lambda : E \to M_\lambda$, $\lambda \in I$; as $v_\lambda(F) = N_\lambda$, $\lambda \in I$, we get by passing to the quotient the linear mapping g_λ, $\lambda \in I$. If $\lambda, \mu \in I$ and $x_\mu \in M_\mu$, then

$$(g_\lambda \circ f_\mu)(x_\mu + N_\mu) = g_\lambda(x_\mu + F) = u_\lambda(x_\mu) + N_\lambda,$$

and since $u_\lambda(x_\mu)$ is 0 or x_μ according as $\lambda \neq \mu$ or $\lambda = \mu$, the mappings g_λ, $\lambda \in I$, satisfy conditions *a*) and *b*) of Proposition 2; since for any $\lambda \in I$ and any $x \in E$,

$$g_\lambda(x + F) = u_\lambda(x) + N_\lambda,$$

it follows that $(g_\lambda(x + F))_{\lambda \in I}$ is an fs-family, and we have

$$\sum_{\lambda \in I} f_\lambda(g_\lambda(x + F)) = \sum_{\lambda \in I} f_\lambda(u_\lambda(x) + N_\lambda) = \sum_{\lambda \in I} (u_\lambda(x) + F) = x + F;$$

hence the family $(g_\lambda)_{\lambda \in I}$ satisfies also condition *c*) of Proposition 2. The stated isomorphism is f^{-1}. ∎

REMARK. The isomorphism f^{-1} constructed in the proof of Proposition 3 and which is the canonical isomorphism stated by this Proposition acts as follows: if $x \in E$, and if, in the preceding notation, $x_\lambda = u_\lambda(x)$, $\lambda \in I$, then

$$f^{-1}(x + F) = (x_\lambda + N_\lambda)_{\lambda \in I};$$

indeed,

$$f((x_\lambda + N_\lambda)_{\lambda\in I}) = \sum_{\lambda\in I} f_\lambda(x_\lambda + N_\lambda) = \sum_{\lambda\in I} (x_\lambda + F) = x + F. \blacksquare$$

The property asserted by the following Proposition is called the "associativity of the internal direct sum".

PROPOSITION 4. *Let E be an A-module, and let $(M_\lambda)_{\lambda\in I}$ be a family of submodules of it; let $I = \bigcup_{\mu\in J} I_\mu$ be a partition of the index set I; then we have*

(5) $$E = \dot{\sum_{\lambda\in I}} M_\lambda$$

if and only if

(6) $$\sum_{\lambda\in I_\mu} M_\lambda = \dot{\sum_{\lambda\in I_\mu}} M_\lambda$$

for any $\mu \in J$ and, moreover,

(7) $$E = \dot{\sum_{\mu\in J}} (\sum_{\lambda\in I_\mu} M_\lambda).$$

Proof. Under condition (5), in the commutative diagram (3) of § 8, f is an isomorphism; hence, in the notation of that diagram, $f\circ u = f'\circ(\bigoplus_{\mu\in J} g_\mu)$ is also an isomorphism; but then $\bigoplus_{\mu\in J} g_\mu$ is, by the arguments of § 3, a monomorphism; when studying that diagram we showed that $\bigoplus_{\mu\in J} g_\mu$ is also surjective; thus $\bigoplus_{\mu\in J} g_\mu$ is an isomorphism, and, by Corollary 4 to Proposition 2, § 7, every g_μ is an isomorphism, $\mu \in J$; but g_μ is for $\sum_{\lambda\in I_\mu} M_\lambda$ and for the family $(M_\lambda)_{\lambda\in I_\mu}$ of its submodules what f is for E and $(M_\lambda)_{\lambda\in I}$, because of the very manner in which g_μ acts, $\mu \in J$; consequently, we have (6) for any $\mu \in J$; since $\bigoplus_{\mu\in J} g_\mu$ is an isomorphism, f' is also an isomorphism, and we have (7).

Conversely, if for any $\mu \in J$ we have (6), and if, moreover, we have (7), then, in the same notation, g_μ is an isomorphism for any $\mu \in J$ and f' is an isomorphism; but then, $\bigoplus_{\mu\in J} g_\mu$ is also an isomorphism; it follows that f is an isomorphism, and hence we have (5). ▮

DEFINITION 4. If E is an A-module and M, N are two of its submodules, then M and N are said to be *supplementary submodules* provided E decomposes into the internal direct sum of M and N; in this case, one also says that N is a *submodule supplementary to* M (in E) and vice versa or that *M has N as a supplementary submodule* (in E) and vice versa.

If a submodule M of an A-module E has a supplementary submodule, then M is said to be a *direct summand* of (or in) E. ▮

By Proposition 1, *iv*), in order that the submodules M and N of E be supplementary submodules it is necessary and sufficient that

$$E = M + N \quad \text{and} \quad M \cap N = 0;$$

or also, by Proposition 1, *ii*), it is necessary and sufficient that any element $x \in E$ may be uniquely represented as $x = x_1 + x_2$, with $x_1 \in M$, $x_2 \in N$.

If M and N are supplementary submodules of E, then by the Second Isomorphism Theorem for modules (Proposition 4, § 8), there exists a canonical isomorphism

$$M \cong M/0 = M/(M \cap N) \cong (M + N)/N = E/N$$

given by

$$x \mapsto x + N, \quad x \in M;$$

this isomorphism may also be thought of as the composition of the linear mappings

$$M \xrightarrow{i} E \xrightarrow{p} E/N,$$

where i is the canonical injection and p the canonical surjection. Herefrom it follows that *any submodule supplementary to N is isomorphic to E/N*, hence that *two submodules supplementary to N are isomorphic to each other.*

By Proposition 4, *if E is the internal direct sum of the family $(M_\lambda)_{\lambda \in I}$ of its submodules, then every M_λ, $\lambda \in I$, is a direct summand of E*: indeed, we have only to consider the partition $I = \{\lambda\} \cup (I \setminus \{\lambda\})$.

Also by Proposition 4 it follows that *if M is a direct summand of E, and if N is a direct summand of M, then N is a direct summand of E*: for if M' is a submodule supplementary to M in E and N' is a submodule supplementary to N in M, then $M' + N'$ is a submodule supplementary to N in E.

From Corollaries 2 and 3 to Proposition 2 we obtain

PROPOSITION 5. *If E is an A-module and M is a submodule of it, then M is a direct summand of E if and only if there exists a projector of E having M as image, or, equivalently, if and only if there exists a projector of E having M as kernel.*

We leave to the reader as an easy exercise to work out the details of the *proof.* ▮

Not every submodule of a module has a supplementary submodule, as shown by the following

EXAMPLE. 1) Consider the **Z**-module **Z** (i.e. the abelian group **Z**; see Example 4), § 1); take as submodule the subgroup 2**Z** of multiples of 2; if 2**Z** had a supplementary submodule N in **Z**, this would imply $2\mathbf{Z} + N = \mathbf{Z}$, hence $N \neq 0$; as N contains non-zero integers and is a submodule, it contains at least a positive integer; let n be the smallest positive integer of N; then, if $k \in N$, we may write

$$k = nq + r, \quad \text{with} \quad 0 \leqslant r < n;$$

but $r \in N$, since k and n belong to N; we cannot have $r > 0$, since this would contradict the choice of n; hence any $k \in N$ is a multiple of n, and as any multiple of n is in N, it follows that N is the submodule $n\mathbf{Z}$ of all multiples of n. On the other hand, we should have $2\mathbf{Z} \cap n\mathbf{Z} = 0$; but $2n \in 2\mathbf{Z} \cap n\mathbf{Z}$ and $2n \neq 0$, which contradicts the fact that $2\mathbf{Z}$ is a direct summand of $\mathbf{Z}$.

As an additional result, we proved that any subgroup N of $\mathbf{Z}$ is of the form $N = n\mathbf{Z}$ for some natural n. ∎

Consider the short exact sequence of A-modules and A-homomorphisms

$$0 \to E \xrightarrow{f} F \xrightarrow{g} G \to 0. \tag{8}$$

PROPOSITION 6. *The following assertions concerning the exact sequence* (8) *are equivalent:*

i) $\mathrm{Im}(f) = \mathrm{Ker}(g)$ *is a direct summand of* F;

ii) *There exists a linear mapping* $r : F \to E$ *such that* $r \circ f = 1_E$, *i.e. a linear retraction associated with* f;

iii) *There exists a linear mapping* $s : G \to F$ *such that* $g \circ s = 1_G$, *i.e. a linear section associated with* g.

When these equivalent conditions are fulfilled, the linear mapping $E \oplus G \to F$ *obtained from* f *and* s *by the universal property of the direct sum is an isomorphism.*

Proof. i) ⇒ *ii*): The linear mapping f induces an isomorphism

$$f' : E \rightleftarrows f(E)$$

(Corollary 2 to Proposition 1, § 4); by Proposition 5, there exists a projector u of F such that $u(F) = f(E)$; u induces a surjective homomorphism

$$u' : F \to f(E);$$

we define r as the composition $f'^{-1} \circ u'$; if $x \in E$, then, if $f(x) = u'(y)$, with $y \in F$, we have $u'(f(x)) = u'(u'(y)) = u'(y) = f(x)$, and hence,

$$(r \circ f)(x) = f'^{-1}(u'(f(x))) = f'^{-1}(f(x)) = x.$$

ii) ⇒ *i*): We have $f \circ r \circ f \circ r = f \circ r$, hence $f \circ r : F \to F$ is a projector of F; by *a*) of Corollary 3 to Proposition 2, $\mathrm{Im}(f \circ r)$ is a direct summand of F; but according to § 3, r is an epimorphism since $r \circ f = 1_E$, and thus $\mathrm{Im}(f \circ r) = \mathrm{Im}(f)$.

We mention that a submodule supplementary to $\mathrm{Im}(f)$ is $\mathrm{Ker}(f \circ r) = \mathrm{Ker}(r)$ (the last equality following from the injectivity of f).

i) ⇒ *iii*): By Corollary 1 to Proposition 1, § 4, g induces an isomorphism

$$\hat{g} : F/\mathrm{Ker}(g) \rightleftarrows G$$

such that $\hat{g}(y + \mathrm{Ker}\, g) = g(y)$, $y \in F$; on the other hand, if N is a submodule supplementary to $\mathrm{Ker}\, g$ in F, then there exists the isomorphism

$$h : N \rightleftarrows F/\mathrm{Ker}\, g$$

given by $h(y) = y + \operatorname{Ker} g$, $y \in N$; we define s as the composition

$$G \xrightarrow{\hat{g}^{-1}} F/\operatorname{Ker} g \xrightarrow{h^{-1}} N \xrightarrow{j} F,$$

where j is the canonical injection; if $z = g(y)$ is an element of G, with $y \in F$, then

$$g(s(z)) = g(j(h^{-1}(\hat{g}^{-1}(g(y))))) = g(j(h^{-1}(y + \operatorname{Ker} g))) = g(y) = z.$$

iii) ⇒ *i)*: The mapping $s \circ g$ is a projector of F, hence $\operatorname{Ker}(s \circ g)$ is a direct summand of F; but s being injective, we have $\operatorname{Ker}(s \circ g) = \operatorname{Ker}(g)$.

We mention that a submodule supplementary to $\operatorname{Ker}(g)$ is $\operatorname{Im}(s \circ g) = \operatorname{Im}(s)$.

For the last assertion we notice that s induces the isomorphism $s' : G \rightleftarrows s(G)$ and that the stated mapping is the composition of the mapping $\operatorname{Im}(f) \oplus \operatorname{Im}(s) \to F$ induced by the canonical injections with the isomorphism $f' \oplus s'$; but the first mapping is also an isomorphism, since $F = \operatorname{Im}(f) \dotplus \operatorname{Im}(s)$. ∎

REMARK. A categorical proof of Proposition 6 will be given below. ∎

DEFINITION 5. A short exact sequence (8) satisfying the equivalent conditions of Proposition 6 is called a *split exact sequence*; we say equivalently that the short exact sequence (8) *splits*. In this case, the extension (F, f, g) of G by E (see the Remark following Proposition 3, § 5) is said to be a *trivial* extension. ∎

If E, F are two A-modules, then the exact sequence

$$0 \to E \xrightarrow{j_E} E \oplus F \xrightarrow{p_F} F \to 0,$$

where j_E, j_F are the canonical injections and p_E, p_F are the canonical projections (see § 7), is a split exact sequence: indeed, by the Remark 2) following Definition 1, $j_E(E)$ is a direct summand of $E \oplus F$.

Let now $f : E \to F$ be a left invertible linear mapping and r a left inverse of f (see Definition 6, § 1); as $r \circ f = 1_E$, it follows that f is injective and, consequently, that we have the exact sequence

$$0 \to E \xrightarrow{f} F \xrightarrow{g} \operatorname{Coker} f \to 0, \tag{9}$$

g being the canonical surjection; but r being a linear retraction associated with f, the exact sequence (9) splits, and hence, according to Proposition 6, $f(E)$ is a direct summand of F; conversely, if $f : E \to F$ is injective and $f(E)$ is a direct summand of F, then (9) is an exact sequence, and by Proposition 6, there exists a linear retraction associated with f. Hence we proved half (the other half being proved by dual considerations) of the following

COROLLARY. *A linear mapping $f : E \to F$ is left invertible if and only if f is injective and* $\operatorname{Im}(f)$ *is a direct summand of F; a linear mapping $g : F \to G$ is right invertible if and only if g is surjective and* $\operatorname{Ker}(g)$ *is a direct summand of F.* ∎

* Though the concept of decomposition into internal direct sum of an object might be recovered in an abelian category satisfying axiom AB3) (using, for instance, the characterization *iv*) of Proposition 1 given for categories of modules, or also other characterizations, as those given by Corollary 1 or by Corollary 2 to Proposition 2 in the case of a finite family), we shall not especially insist on this question for an arbitrary abelian category, since the concept is of interest for categories of modules, related to the calculus with elements in a module. We shall insist a little only on the concept of split exact sequence:

Let $\mathcal{C}$ be an abelian category, and let the short exact sequence

$$0 \to E \xrightarrow{f} F \xrightarrow{g} G \to 0$$

be given in $\mathcal{C}$; if there exists a morphism $r : F \to E$ such that $r \circ f = 1_E$, then consider the diagram in $\mathcal{C}$

$$\begin{array}{ccccccc} 0 \longrightarrow & E & \xrightarrow{\ f\ } & F & \xrightarrow{\ g\ } & G & \longrightarrow 0; \\ & & & \Big\downarrow{\scriptstyle 1_F - f\circ r} & \swarrow{\scriptstyle s} & & \\ & & & F & & & \end{array} \tag{10}$$

we have

$$(1_F - f\circ r)\circ f = f - f\circ(r\circ f) = f - f = 0;$$

as the pair (G, g) is a cokernel of f in the sense of Definition 1, § 4 (see also Proposition 2, § 5), there exists a morphism $s : G \to F$ uniquely determined by the property of rendering diagram (10) commutative; hence we have

$$f\circ r + s\circ g = 1_F; \tag{11}$$

by composition on the left with g we get, taking into account that $g\circ f = 0$,

$$g\circ s\circ g = g = 1_G\circ g;$$

as g is an epimorphism (see § 5), it follows $g\circ s = 1_G$. The converse can be shown by dual considerations: if for the given exact sequence there exists $s : G \to F$ such that $g\circ s = 1_G$, then there exists $r : F \to E$ such that $r\circ f = 1_E$. In this way we proved that if in an abelian category $\mathcal{C}$ the short exact sequence (8) is given, then assertions *ii*) and *iii*) of Proposition 6 are still equivalent (of course, with the replacement in these assertions of "linear mapping" by "morphism").

Again, under the hypotheses we started with, and with the constructed s, we have

$$r\circ s\circ g = r\circ(1_F - f\circ r) = r - r\circ f\circ r = r - r = 0,$$

and as g is an epimorphism, it follows that $r \circ s = 0$; hence we have

$$r \circ f = 1_E, \quad g \circ s = 1_G, \quad g \circ f = 0, \quad r \circ s = 0, \tag{12}$$

which together with (11) show, by virtue of Proposition 4, § 7, that (F, r, g) is a direct product in the category $\mathcal{C}$ of the family of two objects (E, G) and that the direct sum associated with this direct product is (F, f, s); an analogous result is obtained when starting with a given morphism $s : G \to F$ for which $g \circ s = 1_G$ and we construct r by the procedure dual to the one above: (F, f, s) is a direct sum of the family (E, G), and the direct product associated with it is (F, r, g).

Conversely, assume that in the abelian category $\mathcal{C}$ the diagram

$$0 \to E \xrightarrow{f} F \xrightarrow{g} G \to 0 \tag{13}$$

is given and that there exist the morphisms $r : F \to E$ and $s : G \to F$ such that (F, f, s), respectively (F, r, g), represent a direct sum, respectively a direct product of the family (E, G), sum and product which are associated with each other; hence we have (11) and (12); from $r \circ f = 1_E$ it follows that f is a monomorphism, hence, by the considerations in § 5, diagram (13) is an exact sequence at E; similarly, from $g \circ s = 1_G$ we get that g is an epimorphism, hence that (13) is an exact sequence at G; by the results of § 4, (E, f) is an image of f; if we prove that (E, f) is a kernel of g, diagram (13) will be proved to be an exact sequence; but $g \circ f = 0$, and if $u : X \to F$ is a morphism such that $g \circ u = 0$, then

$$u = 1_F \circ u = (f \circ r + s \circ g) \circ u = f \circ r \circ u;$$

hence u factors through f; the fact that this factorization is unique is ensured by the monomorphism character of f.

Thus we proved:

PROPOSITION 7. *Given the diagram* (13) *in the abelian category* $\mathcal{C}$, *the following assertions are equivalent:*

i) Diagram (13) *is an exact sequence, and there exists a morphism* $r : F \to E$ *such that* $r \circ f = 1_E$;

ii) Diagram (13) *is an exact sequence, and there exists a morphism* $s : G \to F$ *such that* $g \circ s = 1_G$;

iii) There exist some morphisms $r : F \to E$ *and* $s : G \to F$ *satisfying relations* (11) *and* (12), *i.e.*, (F, f, s), *respectively* (F, r, g), *represents a direct sum, respectively a direct product of the family* (E, G), *sum and product which are associated with each other.* ▮

We shall maintain in an abelian category $\mathcal{C}$ the term "*split exact sequence*" employed in **Mod**(A), for a diagram (13) satisfying the equivalent conditions of Proposition 7.

REMARKS. 1) A split exact sequence of the abelian category $\mathcal{C}$ is an exact sequence of the category.

2) In case we take as $\mathcal{C}$ the category **Mod**(A), the notion of split exact sequence is the one given by Definition 5.

3) If (13) is a split exact sequence of the category $\mathcal{C}$, then considering in $\mathcal{C}^\circ$ the diagram

$$0 \to G^\circ \xrightarrow{g^\circ} F^\circ \xrightarrow{f^\circ} E^\circ \to 0,$$

this is a split exact sequence, since in the notation of Proposition 7, there exists $s^\circ : F^\circ \to G^\circ$ and $r^\circ : E^\circ \to F^\circ$ satisfying the conditions analogous to (11) and (12). ∎

As an application of Proposition 7, we shall prove

PROPOSITION 8 (**Hexagonal Lemma**). *Suppose in the abelian category $\mathcal{C}$ the following diagram is given:*

(14)

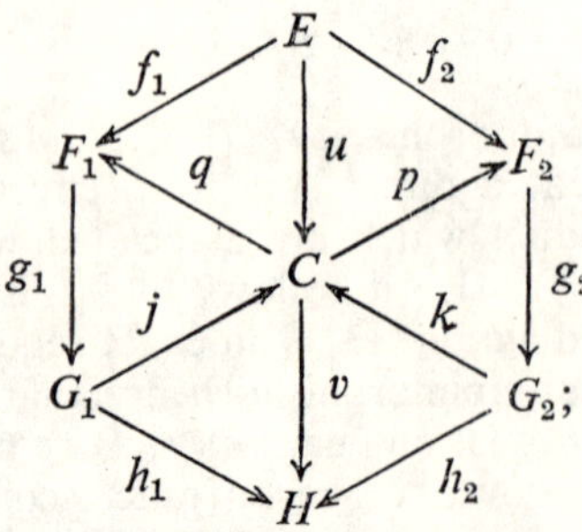

assume that the two upper and the two lower triangles are commutative, that

(15) $$g_1 \circ q \circ j = 1_{G_1}, \quad p \circ k \circ g_2 = 1_{F_2},$$

and that (j, p) and (k, q) are exact sequences; then

$$h_1 \circ g_1 \circ f_1 + h_2 \circ g_2 \circ f_2 = v \circ u.$$

Proof. Because of (15), it follows, as shown in § 3, that j is a monomorphism and that p is an epimorphism, hence that the diagram

$$0 \to G_1 \xrightarrow{j} C \xrightarrow{p} F_2 \to 0$$

is an exact sequence of the category $\mathcal{C}$; moreover, it is split because of the first relation (15); in the same way as in the proof of the implication $i) \Rightarrow iii)$ in Proposition 7, we get a morphism $s : F_2 \to C$ such that

(16) $$p \circ s = 1_{F_2}, \; g_1 \circ q \circ s = 0, \; j \circ g_1 \circ q + s \circ p = 1_C;$$

from the second relation (15) and from $g_1 \circ q \circ k \circ g_2 = 0$ we get, in view of the uniqueness of the direct sum associated with a given direct product (see Proposition 4, § 7), that $s = k \circ g_2$, hence, by (16), that

$$j \circ g_1 \circ q + k \circ g_2 \circ p = 1_C;$$

then the commutativity of the upper and lower triangles in (14) implies

$$h_1 \circ g_1 \circ f_1 + h_2 \circ g_2 \circ f_2 = v \circ j \circ g_1 \circ q \circ u + v \circ k \circ g_2 \circ p \circ u = v \circ u. \blacksquare$$

REMARK. It would be sufficient to require in Proposition 8 only $q \circ k = 0$ instead of the exactness of (k, q). ▮

Now let $\mathcal{C}, \mathcal{C}'$ be abelian categories, and let $T : \mathcal{C} \to \mathcal{C}'$ be an additive functor; *if* (13) *is a split exact sequence of the category* $\mathcal{C}$, *then the diagram*

$$(17) \qquad 0 \to T(E) \xrightarrow{T(f)} T(F) \xrightarrow{T(g)} T(G) \to 0$$

is a split exact sequence of the category $\mathcal{C}'$: for, using the characterization *iii*) of Proposition 7 for the notion of split exact sequence, we have, in the same notations as there,

$$T(f) \circ T(r) + T(s) \circ T(g) = 1_{T(F)},$$

$$T(r) \circ T(f) = 1_{T(E)}, \quad T(g) \circ T(s) = 1_{T(G)}, \quad T(g) \circ T(f) = 0, \quad T(r) \circ T(s) = 0,$$

since T is an additive functor; then, for diagram (17), condition *iii*) of Proposition 7 is satisfied with $T(r)$, respectively $T(s)$, instead of r, respectively s.*

From the above considerations and from the fact that for any A-module E the functor $h^E : \mathbf{Mod}(A) \to \mathbf{Ab}$ is an additive functor (see § 2) it follows that if

$$0 \to F' \xrightarrow{u} F \xrightarrow{v} F'' \to 0$$

is a split exact sequence of the category $\mathbf{Mod}(A)$, then

$$0 \to \mathrm{Hom}_A(E, F') \xrightarrow{\mathrm{Hom}(1_E, u)} \mathrm{Hom}_A(E, F) \xrightarrow{\mathrm{Hom}(1_E, v)} \mathrm{Hom}_A(E, F'') \to 0$$

is a split exact sequence of the category $\mathbf{Ab}$; likewise, the fact that for any A-module F the functor $h_F : (\mathbf{Mod}(A))^\circ \to \mathbf{Ab}$ is an additive functor implies that if

$$0 \to E' \xrightarrow{w} E \xrightarrow{t} E'' \to 0$$

is a split exact sequence of the category $\mathbf{Mod}(A)$, then

$$0 \to \mathrm{Hom}_A(E'', F) \xrightarrow{\mathrm{Hom}(t, 1_F)} \mathrm{Hom}_A(E, F) \xrightarrow{\mathrm{Hom}(w, 1_F)} \mathrm{Hom}_A(E', F) \to 0$$

is a split exact sequence of the category $\mathbf{Ab}$ (compare with Proposition 4, § 5).

* From the preceding considerations it can be proved that *if two abelian categories* $\mathcal{C}, \mathcal{C}'$ *and a functor* $T : \mathcal{C} \to \mathcal{C}'$ *are given such that the exactness of the sequence*

$$0 \to E \xrightarrow{u} F \xrightarrow{v} G \to 0$$

in the category $\mathcal{C}$ implies the exactness of the sequence

$$0 \to T(E) \xrightarrow{T(u)} T(F) \xrightarrow{T(v)} T(G)$$

(*respectively of the sequence*

$$T(E) \xrightarrow{T(u)} T(F) \xrightarrow{T(v)} T(G) \to 0)$$

in the category $\mathcal{C}'$, then T is an additive functor; this allows us to dispense with *a priori* requiring in Definition 2, § 5, that T is an additive functor provided $\mathcal{C}, \mathcal{C}'$ are abelian categories.

Let us justify the first assertion; notice that T with the above property carries the zero object of $\mathcal{C}$ into the zero object of $\mathcal{C}'$, and hence, as the zero morphism $E \to F$, $E, F \in \operatorname{Ob} \mathcal{C}$, is the composition

$$E \to 0 \to F,$$

T carries the zero morphism $E \to F$ into the zero morphism $T(E) \to T(F)$: indeed, as for any object E of $\mathcal{C}$ we have the exact sequence

$$0 \to 0 \to E \xrightarrow{1_E} E \to 0,$$

it follows that we have in $\mathcal{C}'$ the exact sequence

$$0 \to T(0) \to T(E) \xrightarrow{1_{T(E)}} T(E),$$

hence that $T(0)$ is a kernel of $1_{T(E)}$ (see Proposition 2, § 5), whence $T(0) = 0$.

Now let $E, F \in \operatorname{Ob} \mathcal{C}$, and let $u_1, u_2 \in \operatorname{Hom}_{\mathcal{C}}(E, F)$; let (S, i_1, i_2) be a direct sum in $\mathcal{C}$ of the finite family of objects (E, E);[1] let (S, p_1, p_2) be the direct product associated with the given direct sum; hence we have

$$p_1 \circ i_1 = 1_E, \; p_2 \circ i_2 = 1_E, \; p_1 \circ i_2 = 0, \quad p_2 \circ i_1 = 0,$$

$$i_1 \circ p_1 + i_2 \circ p_2 = 1_S;$$

let $d : E \to S$ be the diagonal morphism, i.e. the unique morphism satisfying the conditions

$$p_1 \circ d = 1_E, \qquad p_2 \circ d = 1_E;$$

let $h : S \to F$ be the unique morphism satisfying the conditions

$$h \circ i_1 = u_1, \qquad h \circ i_2 = u_2;$$

---

[1] Notice that we should say "of the family of objects (E_1, E_2), where $E_1 = E, E_2 = E$".

we have

$$h \circ d = h \circ 1_S \circ d = h \circ (i_1 \circ p_1 + i_2 \circ p_2) \circ d =$$

$$= h \circ i_1 \circ p_1 \circ d + h \circ i_2 \circ p_2 \circ d = u_1 + u_2,$$

whence

(18) $$T(u_1 + u_2) = T(h) \circ T(d).$$

According to Proposition 7 above, namely to *iii*) $\Leftrightarrow$ *i*), the diagram

$$0 \to E \xrightarrow{i_1} S \xrightarrow{p_2} E \to 0$$

is an exact sequence of the category $\mathcal{C}$; by hypothesis, the diagram

$$0 \to T(E) \xrightarrow{T(i_1)} T(S) \xrightarrow{T(p_2)} T(E)$$

is an exact sequence of the category $\mathcal{C}'$; as we have also $T(p_2) \circ T(i_2) = 1_{T(E)}$, it follows that $T(p_2)$ is an epimorphism, hence that we have the split exact sequence

$$0 \to T(E) \xrightarrow{T(i_1)} T(S) \xrightarrow{T(p_2)} T(E) \to 0;$$

again applying Proposition 7, *ii*) $\Rightarrow$ *iii*), we get the existence of a morphism $p_1'\colon T(S) \to$ $\to T(E)$ such that the conditions

$$p_1' \circ T(i_1) = 1_{T(E)},\ T(p_2) \circ T(i_2) = 1_{T(E)},\ p' \circ T(i_2) = 0,$$

$$T(p_2) \circ T(i_1) = 0,\quad T(i_1) \circ p_1' + T(i_2) \circ T(p_2) = 1_{T(S)}$$

are satisfied; hence $(T(S), T(i_1), T(i_2))$ is a direct sum in $\mathcal{C}'$ of the family of objects $(T(E), T(E))$; as we have also

$$T(p_1) \circ T(i_1) = 1_{T(E)},\quad T(p_1) \circ T(i_2) = 0,$$

it follows, by the uniqueness of the morphism defined by the family of morphisms $(1_{T(E)}, 0)$, that $p_1' = T(p_1)$. Hence

$$T(i_1) \circ T(p_1) + T(i_2) \circ T(p_2) = 1_{T(S)};$$

having also

$$T(p_1) \circ T(d) = 1_{T(E)},\ T(p_2) \circ T(d) = 1_{T(E)},\ T(h) \circ T(i_1) = T(u_1),\ T(h) \circ T(i_2) = T(u_2),$$

we get

$$T(u_1) + T(u_2) = T(h) \circ T(d),$$

which together with (18) gives

$$T(u_1 + u_2) = T(u_1) + T(u_2).^*$$

The notion of split exact sequence may be generalized to "long" exact sequences; a more general notion will be studied first.

Let

(19) $$\ldots \to E_{n-1} \xrightarrow{u_{n-1}} E_n \xrightarrow{u_n} E_{n+1} \to \ldots$$

be a diagram of A-modules and A-homomorphisms, where all the situations described when we considered diagram (2), § 5, may occur; if

(20) $$\ldots \to E'_{n-1} \xrightarrow{u'_{n-1}} E'_n \xrightarrow{u'_n} E'_{n+1} \to \ldots$$

is a similar diagram, with n running through the same index set, say I, then consider a family of A-homomorphisms $f_n : E_n \to E'_n$, $n \in I$, such that the following diagram be commutative:

$$\begin{array}{ccccccccc} \ldots & \longrightarrow & E_{n-1} & \xrightarrow{u_{n-1}} & E_n & \xrightarrow{u_n} & E_{n+1} & \longrightarrow & \ldots \\ & & \downarrow{\scriptstyle f_{n-1}} & & \downarrow{\scriptstyle f_n} & & \downarrow{\scriptstyle f_{n+1}} & & \\ \ldots & \longrightarrow & E'_{n-1} & \xrightarrow{u'_{n-1}} & E'_n & \xrightarrow{u'_n} & E'_{n+1} & \longrightarrow & \ldots; \end{array}$$

the family $(f_n)_{n \in I}$ is called a *morphism* from the diagram (19) to the diagram (20).

DEFINITION 6. If diagrams (19) and (20) as above are complexes, and if $(f_n)_n$ and $(g_n)_n$ are morphisms from the first to the second diagram, then a *homotopy between* $(f_n)_n$ *and* $(g_n)_n$ is a family of A-homomorphisms

$$s_n : E_n \to E'_{n-1},$$

where n runs through the whole index set I except perhaps the left extremity provided this exists, such that for any $n \in I$ except the extremities provided they exist, we have

(21) $$s_{n+1} \circ u_n + u'_{n-1} \circ s_n = f_n - g_n;$$

whenever a homotopy between $(f_n)_n$ and $(g_n)_n$ exists, we say that the first morphism is *homotopic* to the second, and we write $(f_n)_n \simeq (g_n)_n$. ▮

Obviously, the relation "homotopic to" on the set of all morphisms from a complex (19) to a complex (20) defined for the same index set is an equivalence relation; consequently, we may speak of morphisms $(f_n)_n$, $(g_n)_n$ as above which are *homotopic.*

By the considerations in § 5, we may see that a morphism $(f_n)_n$ from the complex (19) to the complex (20) induces homomorphisms between the homology modules

$$f_n^\natural : \operatorname{Ker} u_n / \operatorname{Im} u_{n-1} \to \operatorname{Ker} u'_n / \operatorname{Im} u'_{n-1},$$

where n runs through the index set I except perhaps the extremities.

PROPOSITION 9. *If* (19) *and* (20) *are complexes as above, and* $(f_n)_n$, $(g_n)_n$ *are homotopic morphisms from the first to the second, then the induced homomorphisms* $f_n^\natural$ *and* $g_n^\natural$ *between the homology modules coincide for any* n *for which they are defined, i.e., except perhaps for the extremal indices.*

Proof. If $x_n \in \operatorname{Ker}(u_n)$, respectively $y'_n \in \operatorname{Ker}(u'_n)$, we denote by $\hat{x}_n$, respectively $\hat{y}'_n$, the coset of x_n, respectively y'_n, modulo $\operatorname{Im}(u_{n-1})$, respectively $\operatorname{Im}(u'_{n-1})$; then

$$f_n^\natural(\hat{x}_n) = \widehat{f(x_n)},$$

and similarly for $g_n^\natural$; by (21) we get

$$f_n(x_n) - g_n(x_n) = u'_{n-1}(s_n x_n),$$

since $u_n x_n = 0$; hence $f_n^\natural(\hat{x}_n) - g_n^\natural(\hat{x}_n) = 0.$ ∎

COROLLARY. *If* (19) *is a complex, and if the family of identity homomorphisms* $(1_{E_n})_n$ *is homotopic to the family of zero homomorphisms* $E_n \to E_n$, *then* (19) *is an exact sequence.*

Proof. Obviously, the identity homomorphism 1_{E_n}, respectively the zero homomorphism $E_n \to E_n$, induces the identity homomorphism of $\operatorname{Ker}(u_n)/\operatorname{Im}(u_{n-1})$, respectively the zero homomorphism $\operatorname{Ker}(u_n)/\operatorname{Im}(u_{n-1}) \to \operatorname{Ker}(u_n)/\operatorname{Im}(u_{n-1})$; but by Proposition 9, these homomorphisms coincide; hence $\operatorname{Ker}(u_n)/\operatorname{Im}(u_{n-1}) = 0$ for any n except perhaps for the extremities, and this means that (19) is an exact sequence. ∎

DEFINITION 7. If (19) is an exact sequence of A-modules and A-homomorphisms, then it is said to be a *split exact sequence* provided the family of identity homomorphisms (1_{E_n}) is homotopic to the family of zero homomorphisms $E_n \to E_n$, i.e., if there exists a family of homomorphisms $s_n : E_n \to E_{n-1}$ defined for all n except perhaps for the left extremity so that we have for any n except perhaps for the extremities,

$$s_{n+1} \circ u_n + u_{n-1} \circ s_n = 1_{E_n}. \blacksquare$$

From Proposition 7 it follows that the notion of split exact sequence defined for short exact sequences is a particular case of the one given by Definition 7.

§ 10. Linearly Independent Families, Bases, Free Modules

Let A be a ring, and let I be a set; the module $A_s^{(I)}$ (see § 7) is the external direct sum of the family $(E_t)_{t\in I}$ of A-modules such that $E_t = A_s$ for any $t \in I$; it may be also thought of as the set of mappings $I \to A_s$ having finite support, with componentwise addition and scalar multiplication. Let $j_t : A_s \to A_s^{(I)}$ be the canonical injection of index t, $t \in I$; j_t carries an element $\alpha \in A$ into the family whose component of index t is α and whose component of index t' is 0 for $t' \neq t$. By Lemma 1, § 7, any $x \in A_s^{(I)}$ is uniquely written as

(1) $$x = \sum_{t\in I} j_t(\xi_t) = \sum_{t\in I} \xi_t j_t(1) = \sum_{t\in I} \xi_t e_t,$$

where we denoted $e_t = j_t(1)$, $t \in I$, and where $(\xi_t)_{t\in I}$ is an fs-family of elements of A (by the proof of Lemma 1, § 7, ξ_t is just the component of index t of x, $t \in I$).

On the other hand, let us consider the mapping

$$\varphi : I \to A_s^{(I)}$$

given by $\varphi(t) = e_t$, $t \in I$; if $A \neq 0$, then φ is injective; it will be called the *canonical mapping* $I \to A_s^{(I)}$; the injectivity of φ allows us to identify t with e_t, $t \in I$; this identification is sometimes supposed to be done, and by means of it expression (1) for x becomes

(2) $$x = \sum_{t\in I} \xi_t \cdot t;$$

this expression will be sometimes useful in applications; due to it, the elements of $A_s^{(I)}$ are called *formal linear combinations of the elements of* I *with coefficients in* A, the module $A_s^{(I)}$ itself being called the *module of formal linear combinations of the elements of* I *with coefficients in* A;[1] of course, $A_d^{(I)}$ will be called the *right module of formal linear combinations of the elements of* I *with coefficients in* A.

PROPOSITION 1. *Let E be an A-module, let I be a set, and let $f : I \to E$ be a mapping from I to the underlying set of the module E; then there exists a*

---

[1] The presence of the term "formal" is justified by the fact that sometimes one constructs $A_s^{(I)}$ starting with expressions of the form (2), the operations having a formal meaning.

unique linear mapping $\bar{f} : A_s^{(I)} \to E$ *rendering commutative the diagram of sets*

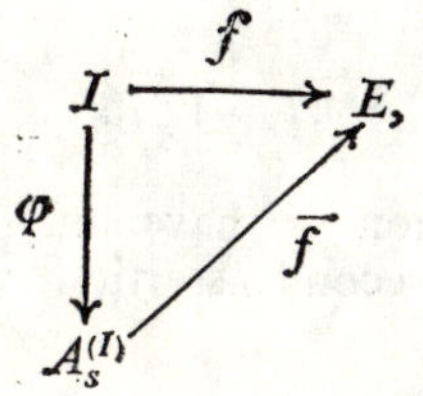

where φ *is the canonical mapping.*

Proof. If (using the preceding notations) $(\xi_t)_{t\in I} = \sum_{t\in I} \xi_t e_t$ is an element of $A_s^{(I)}$, then we define

$$\bar{f}(\sum_{t\in I} \xi_t e_t) = \sum_{t\in I} \xi_t f(t);$$

obviously, $\bar{f}$ is linear and makes the diagram commute, since in the representation of $e_t = \varphi(t)$ as $\sum_{t'\in I} \xi_{t'} e_{t'}$ we have $\xi_t = 1$ and $\xi_{t'} = 0$ if $t' \neq t$; if $g : A_s^{(I)} \to E$ is a linear mapping rendering the diagram commutative, then

$$g(\sum_{t\in I} \xi_t e_t) = \sum_{t\in I} \xi_t g(e_t) = \sum_{t\in I} \xi_t g(\varphi(t)) = \sum_{t\in I} \xi_t f(t) = \bar{f}(\sum_{t\in I} \xi_t e_t),$$

and $g = \bar{f}$. ∎

The mapping $\bar{f}$ is called the *linear continuation* of the mapping f or, equivalently, is said to be obtained from f by *linear extension.*

COROLLARY. *Let* I, J *be two sets, and let* $f : I \to J$ *be a mapping of sets; then there exists a linear mapping* $\bar{f} : A_s^{(I)} \to A_s^{(J)}$ *uniquely determined by the property of rendering commutative the diagram*

(3)

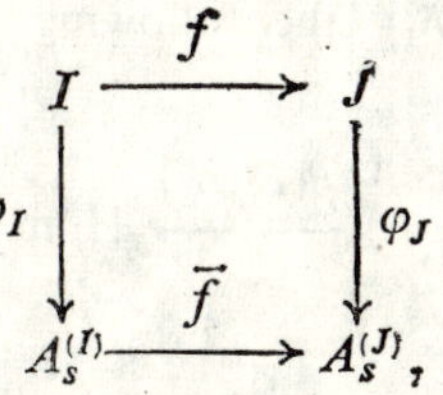

where φ_I, φ_J *are the canonical mappings; if* L *is a third set and* $g : J \to L$ *is a mapping of sets, then, denoting by* $\bar{g}$, *respectively* $\overline{g\circ f}$, *the linear mapping* $A_s^{(J)} \to A_s^{(L)}$, *respectively* $A_s^{(I)} \to A_s^{(L)}$, *corresponding by the above procedure to* g, *respectively* $g\circ f$, *we have*

(4)
$$\overline{g\circ f} = \bar{g}\circ\bar{f};$$

furthermore, if $\overline{1}_I$ *is the linear mapping* $A_s^{(I)} \to A_s^{(I)}$ *corresponding to the identity mapping of* I, *we have*

$$\overline{1}_I = 1_{A_s^{(I)}}.$$

Proof. For the first assertion we have only to replace in Proposition 1 E by $A_s^{(J)}$ and f by $\varphi_J \circ f$; for the second assertion, we have

$$\overline{g} \circ \overline{f} \circ \varphi_I = \varphi_L \circ g \circ f = \overline{g \circ f} \circ \varphi_I,$$

and (4) follows by the aforementioned uniqueness; the third assertion follows similarly, since

$$1_{A_s^{(I)}} \circ \varphi_I = \varphi_I \circ 1_I = \overline{1}_I \circ \varphi_I. \blacksquare$$

The Corollary to Proposition 1 shows that in the same notation, the mappings

$$I \mapsto A_s^{(I)}, \qquad f \mapsto \overline{f}$$

define a functor from the category **Ens** to the category **Mod**(A); this functor will be denoted in this Section by T_0; obviously, there exists an analogous functor **Ens** $\to$ **Mod**$_d$(A).

* Definition 1. Let $\mathcal{C}$, $\mathcal{C}'$ be two categories, and let $S: \mathcal{C} \to \mathcal{C}'$ be a functor; a functor $T: \mathcal{C}' \to \mathcal{C}$ is said to be a *left adjoint* (*functor*) *of the functor* S or *left adjoint to* S provided for any pair of objects $E' \in \mathrm{Ob}\,\mathcal{C}'$, $F \in \mathrm{Ob}\,\mathcal{C}$, a bijective mapping of sets

$$\theta(E', F) : \mathrm{Hom}_{\mathcal{C}}(T(E'), F) \rightrightarrows \mathrm{Hom}_{\mathcal{C}'}(E', S(F))$$

is given such that the following condition is satisfied: for any pair of objects E', $E'_1 \in \mathrm{Ob}\,\mathcal{C}'$, any pair of objects $F, F_1 \in \mathrm{Ob}\,\mathcal{C}$, and any pair of morphisms $u' \in \mathrm{Hom}_{\mathcal{C}'}(E'_1, E')$, $v \in \mathrm{Hom}_{\mathcal{C}}(F, F_1)$, the following diagram of sets and mappings of sets is commutative:

$$\text{(5)} \qquad \begin{array}{ccc} \mathrm{Hom}_{\mathcal{C}}(T(E'), F) & \xrightarrow[\sim]{\theta(E', F)} & \mathrm{Hom}_{\mathcal{C}'}(E', S(F)) \\ {\scriptstyle \mathrm{Hom}_{\mathcal{C}}(T(u'), v)} \downarrow & & \downarrow {\scriptstyle \mathrm{Hom}_{\mathcal{C}'}(u', S(v))} \\ \mathrm{Hom}_{\mathcal{C}}(T(E'_1), F_1) & \xrightarrow[\sim]{\theta(E'_1, F_1)} & \mathrm{Hom}_{\mathcal{C}'}(E'_1, S(F_1)). \end{array}$$

If T is left adjoint to S, then S is said to be a *right adjoint* (*functor*) *of* T or *right adjoint to* T.

In case the categories $\mathcal{C}$, $\mathcal{C}'$ are abelian, it is usually required that the above mappings $\theta(E', F)$ be isomorphisms of abelian groups. ▮ *

Considering A-modules again, let us denote in this Section by S_0 the forgetful functor $\mathbf{Mod}(A) \to \mathbf{Ens}$ (see § 2).

Now let I be a set, hence an object of **Ens**, and let E be an object of $\mathbf{Mod}(A)$; we define a mapping

$$\theta(I, E) : \mathrm{Hom}_A(T_0(I), E) \to \mathrm{Hom}_{\mathbf{Ens}}(I, S_0(E))$$

taking for any A-homomorphism $l : A_s^{(I)} = T_0(I) \to E$ as $\theta(I, E)(l)$ the composite mapping

$$I \xrightarrow{\varphi_I} A_s^{(I)} \xrightarrow{l} E,$$

where φ_I is the canonical mapping and l is considered as a mapping of sets from $S_0(A_s^{(I)})$ to $S_0(E)$. Proposition 1 expresses just the fact that the mapping $\theta(I, E)$ is bijective for any set I and any A-module E. We leave to the reader to verify, taking the commutativity of diagram (3) into account, the commutativity of diagram (5) for this case, i.e., with $T = T_0$, $S = S_0$, etc. Hence *the functor* $T_0 : \mathbf{Ens} \to \mathbf{Mod}(A)$ *is a left adjoint of the forgetful functor* $S_0 : \mathbf{Mod}(A) \to \mathbf{Ens}$.

Let E be an A-module, and let $(a_t)_{t \in I}$ be a family of elements of E; this means a mapping of sets $f : I \to E$ is given such that $f(t) = a_t$, $t \in I$; according to Proposition 1, there exists a linear mapping $\bar{f} : A_s^{(I)} \to E$ uniquely determined by the property that $\bar{f} \circ \varphi = f$, where φ is the canonical mapping $I \to A_s^{(I)}$; this implies that $\bar{f}(e_t) = a_t$, $t \in I$, or, identifying by means of φ the set I with a subset of $A_s^{(I)}$, that $\bar{f}$ coincides with f on the elements of I. The mapping $\bar{f}$ thus obtained is said to be the linear mapping *determined* by the family $(a_t)_{t \in I}$.

If $(\xi_t)_{t \in I}$ is an element of $A_s^{(I)}$, then, in the preceding notation, we have $(\xi_t)_{t \in I} = \sum_{t \in I} \xi_t e_t$, hence

$$\bar{f}((\xi_t)_{t \in I}) = \sum_{t \in I} \xi_t \bar{f}(e_t) = \sum_{t \in I} \xi_t a_t; \tag{6}$$

consequently, if $R = \mathrm{Ker}(\bar{f})$, then an element $(\xi_t)_{t \in I}$ of $A_s^{(I)}$ belongs to R if and only if $\sum_{t \in I} \xi_t a_t = 0$; for this reason, R is called the *submodule of linear relations among the elements of the family* $(a_t)_{t \in I}$.

Proposition 2. *Let E be an A-module, let $(a_t)_{t \in I}$ be a family of elements of E, and let $\bar{f}$ be the linear mapping $A_s^{(I)} \to E$ determined by this family; the following assertions are valid:*

a) The mapping $\bar{f}$ is surjective if and only if the family $(a_t)_{t \in I}$ is a system of generators of E;

b) The mapping $\bar{f}$ is injective if and only if any relation of the form

$$\sum_{t \in I} \xi_t a_t = 0,$$

where $(\xi_t)_{t \in I}$ is an fs-family of elements of A, implies $\xi_t = 0$ for all $t \in I$;

c) The mapping $\bar{f}$ is bijective, i.e. an A-isomorphism, if and only if the family $(a_t)_{t\in I}$ is a system of generators of E and has the property given at b), or, equivalently, if and only if any element $x \in E$ may be uniquely written as

$$x = \sum_{t\in I} \xi_t a_t,$$

where $(\xi_t)_{t\in I}$ is an fs-family of elements of A.

The *proof* is immediate, taking into account the manner (given by (6)) in which the mapping $\bar{f}$ acts. ∎

DEFINITION 2. A family $(a_t)_{t\in I}$ of elements of an A-module E is said to be *linearly independent* (over A) or also *A-free* if it satisfies the equivalent conditions given in Proposition 2, *b*); otherwise $(a_t)_{t\in I}$ is said to be *linearly dependent*. A subset M of an A-module E is said to be *linearly independent* (over A) or also *A-free* if the family indexed identically by M is linearly independent; otherwise the subset M is said to be *linearly dependent* (over A).

An element x of an A-module E is said to be *linearly independent*, respectively *linearly dependent*, if the subset $\{x\}$ is a linearly independent, respectively linearly dependent subset of E. ∎

To say that the family $(a_t)_{t\in I}$ of elements of the A-module E is linearly dependent is to say that there exists a family $(\xi_t)_{t\in I}$ of elements of A with finite non-void support (hence such that there exists at least one $t \in I$ with $\xi_t \neq 0$) having the property that $\sum_{t\in I} \xi_t a_t = 0$; hence, *if the family $(a_t)_{t\in I}$ is linearly independent and $A \neq 0$, then for $t \neq t'$ we have $a_t \neq a_{t'}$*: for otherwise, from the relation

$$1 \cdot a_t - 1 \cdot a_{t'} = 0$$

it would follow, taking $\xi_t = 1$, $\xi_{t'} = -1$ and $\xi_u = 0$ if $u \neq t$, $u \neq t'$, that the family of scalars $(\xi_u)_{u\in I}$ satisfies the condition $\sum_{u\in I} \xi_u a_u = 0$ though not all ξ_u are zero. Consequently, in the case of a linearly independent family $(a_t)_{t\in I}$ the mapping $t \mapsto a_t$ is bijective, and we could very well replace the index set I by the set of values; hence, *there is no essential distinction between the concepts of linearly independent set and linearly independent family*; conversely, given a linearly independent subset M of E, the family defined by any bijective mapping of sets $I \rightleftarrows M$, i.e., bijectively indexed, is linearly independent.

If $(a_t)_{t\in I}$ is a linearly independent family of elements of E, and $(a_t)_{t\in J}$, with $J \subseteq I$, $J \neq \emptyset$, is a subfamily of it, then the subfamily $(a_t)_{t\in J}$ is linearly independent too: for otherwise, from a family $(\xi_t)_{t\in J}$ with finite non-empty support such that $\sum_{t\in J} \xi_t a_t = 0$ one obtains the family $(\xi_t)_{t\in I}$ with $\xi_t = 0$ if $t \in I \setminus J$ and with finite non-empty support such that $\sum_{t\in I} \xi_t a_t = 0$; it follows that *if for a family $(a_t)_{t\in I}$ there exists a non-empty linearly dependent subfamily, then the given family is linearly dependent*. As for the empty family, it is linearly independent: since for $I = \emptyset$, $A_s^{(I)} = 0$ (see Definition 1, § 6), and in the notation of Proposition 2, the mapping

$\bar{f}: A_s^{(I)} \to E$ is the unique mapping $0 \to E$, hence it is injective; consequently, the above restriction $J \neq \emptyset$ may be removed. From here it follows in particular that *any finite subfamily of a linearly independent family is linearly independent*; conversely, *if any finite subfamily of a family $(a_t)_{t\in I}$ is linearly independent, then the family itself is linearly independent*: for otherwise, let $(\xi_t)_{t\in I}$ be a family with finite non-empty support J such that $\sum_{t\in I} \xi_t a_t = 0$; this actually means that $\sum_{t\in J} \xi_t a_t = 0$, with $\xi_t \neq 0$, $t \in J$, and this contradicts the linear independence of the family $(a_t)_{t\in J}$.

From the preceding considerations it follows in particular that: *any subset of a linearly independent subset is linearly independent, hence any element of such a subset is linearly independent; any subset containing a linearly dependent subset is itself linearly dependent*; hence, as $0 \in E$ is linearly dependent provided $A \neq 0$, since $1 \cdot 0 = 0$, *any subset containing 0 is linearly dependent*; also, *the empty set is linearly independent.*

If $(a_t)_{t\in I}$ is a family of elements of an A-module E, $A \neq 0$, and if a_t, for a given $t \in I$, is a linear combination of the family $(a_{t'})_{t'\neq t}$, then the family $(a_t)_{t\in I}$ is linearly dependent: for, $a_t = \sum_{t'\neq t} \xi_{t'} a_{t'}$ implies $\sum_{t'\in I} \eta_{t'} a_{t'} = 0$, with $\eta_t = 1 \neq 0$ and $\eta_{t'} = -\xi_{t'}$ if $t' \neq t$. The converse is valid in a particular case:

PROPOSITION 3. *If $(a_t)_{t\in I}$ is a family of elements of a vector space over a skew-field K, then from the linear dependence of $(a_t)_{t\in I}$ it follows that there exists some index $t \in I$ such that a_t is a linear combination of the family $(a_{t'})_{t'\neq t}$.*

Proof. There exists a family of scalars $(\xi_t)_{t\in I}$ with finite non-empty support such that

$$\sum_{t\in I} \xi_t a_t = 0; \tag{7}$$

if t is an element of the support of $(\xi_t)_{t\in I}$, as $\xi_t \neq 0$, $\xi_t \in K$, and K is a skew-field, we may multiply (7) by ξ_t^{-1} and separate a_t in one side of the equality, thus obtaining the desired result. ▮

The fact that the assertion of Proposition 3 is not valid for an arbitrary ring A is shown by the following

EXAMPLE. 1) Let $A = \mathbf{Z}$, and let the family of two elements (2, 3) in the A-module A be given; we have

$$3 \cdot 2 + (-2) \cdot 3 = 0,$$

hence the given family is linearly dependent; however, there exists no $n \in \mathbf{Z}$ such that $2 = n \cdot 3$ or $3 = n \cdot 2$. ▮

Example 1) shows also that *in an A-module E a linearly dependent subset may exist each element of which is linearly independent*: indeed, $k \cdot 2 = 0$, with $k \in \mathbf{Z}$, implies $k = 0$, and similarly for 3.

DEFINITION 3. A family $(a_t)_{t\in I}$ of elements from an A-module E is called a *basis* (over A) of the module E or, equivalently, an *A-basis* of E if it satisfies the equivalent conditions of Proposition 2, *c*); if M is a subset of the A-module

E, M is said to be a *basis* (over A) or an *A-basis* of E if the family indexed identically by M is a basis of E; a module which admits a basis is said to be a *free module*.

An abelian group is *free* if, considered as a **Z**-module, it is a free module (see Example 4), § 1).

If E is a free A-module and $(a_\iota)_{\iota\in I}$ is a basis of it, then for any element $x \in E$, the element $\xi_\iota \in A$ from the (unique) representation $x = \sum_{\iota\in I} \xi_\iota a_\iota$ is called the *coordinate* (or also the *component*) of index ι (or of order ι) of x with respect to (or relative to or in) the basis $(a_\iota)_{\iota\in I}$. ∎

Hence, a family $(a_\iota)_{\iota\in I}$ of elements of the A-module E is a basis if and only if it is both a system of generators and a linearly independent family; consequently, if $(a_\iota)_{\iota\in I}$ is a linearly independent family of elements from a (not necessarily free) module E, then the submodule F generated by this family in E is free, and $(a_\iota)_{\iota\in I}$ is a basis of F. By the arguments at the beginning of this Section, for any ring A and any set I, $A_s^{(I)}$ *is a free module*, and, in the same notation as there, $(e_\iota)_{\iota\in I}$ *is a basis of it*; it will be called the *canonical basis* of $A_s^{(I)}$. As the zero module is generated by the empty family (see § 8), it follows that 0 *is a free module having* Ø *as basis* (see also Definition 1, § 6, which implies that $A_s^{\emptyset} = 0$). The previously defined functor $T_0 : \mathbf{Ens} \to \mathbf{Mod}(A)$ for which $T_0(I) = A_s^{(I)}$ is referred to as the *functor of taking the generated free module*.

Obviously, any A-module E isomorphic to a module of the form $A_s^{(I)}$ is free, since under an isomorphism a basis is carried onto a basis; conversely, for any free A-module E having a basis $(a_\iota)_{\iota\in I}$, the linear mapping $\bar{f} : A_s^{(I)} \to E$ determined by this family is an isomorphism.

We notice also that *if E is a finitely generated free A-module, and if S is a basis of it, then S is finite*: for let M be a finite set of generators of E; each $x \in M$ is represented as a linear combination of a finite number of elements of S; denoting by S' the subset of elements of S employed to represent all $x \in M$, S' is a finite subset of S; but any element of E is represented as a linear combination of elements of M, hence of S' (see § 1); if there existed $y \in S \setminus S'$, then y would also be represented as a linear combination of elements of S', and the subset $S' \cup \{y\} \subseteq S$ would, as shown above, be linearly dependent; hence $S = S'$. From here it follows that *the finitely generated free A-modules are exactly the free A-modules having a finite basis, hence the A-modules isomorphic to some A_s^n, with $n \geqslant 0$.*

Proposition 4. *Let E be a free A-module, and let $(a_\iota)_{\iota\in I}$ be a basis of it; let F be an A-module, and let $(b_\iota)_{\iota\in I}$ be a family of elements of F indexed by the same set I; then there exists a unique linear mapping $u : E \to F$ such that $u(a_\iota) = b_\iota$ for any $\iota \in I$; the mapping u is, respectively, injective, surjective, bijective if and only if $(b_\iota)_{\iota\in I}$ is, respectively, a linearly independent family, a system of generators, a basis of F.*

Proof. Making use of the isomorphism $A_s^{(I)} \rightrightarrows E$ determined by the basis $(a_\iota)_{\iota\in I}$, the proof follows as a transcription of Proposition 2 *via* this isomorphism. ∎

If two sets I, J are given, and if there exists a bijection $f : I \rightrightarrows J$, then, as f is an isomorphism in the category **Ens**, $T_0(f) : A_s^{(I)} \to A_s^{(J)}$ turns out to be an A-

isomorphism; from here it follows that *if E and F are free A-modules having* $(a_\iota)_{\iota\in I}$, *respectively* $(b_u)_{u\in J}$ *as bases such that I and J have the same cardinality, then E and F are isomorphic*: for in this case, there exists a bijection $I \rightleftarrows J$, and E is isomorphic to $A_s^{(I)}$, F to $A_s^{(J)}$; as for the converse, i.e. the uniqueness of the cardinal number of the different bases of a free A-module, this will be discussed later (see Corollary 2 to Proposition 11, as well as Example 8)).

PROPOSITION 5. *Let K be a skew-field, and let E be a vector space over K; if L is a linearly independent subset of E and S is a set of generators of E over K such that* $L \subseteq S$, *then there exists a basis B of K such that* $L \subseteq B \subseteq S$.

Proof. Let $\mathcal{M}$ be the set of K-free subsets of E containing L and contained in S; we have $L \in \mathcal{M}$, hence $\mathcal{M} \neq \emptyset$; $\mathcal{M}$ is partially ordered by inclusion. To show that it is inductively ordered, consider a non-void subset $\mathcal{N}$ of $\mathcal{M}$, totally ordered by inclusion; if $M = \bigcup_{P\in\mathcal{N}} P$, then to show that M is an element of $\mathcal{M}$ it suffices to show that M is linearly independent; as shown above, a sufficient condition for this is the linear independence of any finite subset N of M; $\mathcal{N}$ being totally ordered by inclusion, there exists $P \in \mathcal{N}$ such that $N \subseteq P$; since P is linearly independent, so is N; now, M is an upper bound for the subsets from $\mathcal{N}$, and $M \in \mathcal{M}$.

By Zorn's Lemma, $\mathcal{M}$ has a maximal element, say B; once we show that B is a system of generators of E, as $B \in \mathcal{M}$, our Proposition is completely proved. But if $x \in S$, there are two possibilities: either $x \in B$, and then x is a linear combination of the elements of B, or $x \in S \setminus B$, and then, as $B \cup \{x\}$ strictly contains B and, moreover, $L \subseteq B \cup \{x\} \subseteq S$, $B \cup \{x\}$ cannot be linearly independent, since this would imply $B \cup \{x\} \in \mathcal{M}$, contradicting the fact that B is a maximal element of $\mathcal{M}$; hence $B \cup \{x\}$ is a linearly dependent subset of E; then there exists an fs-family $(\alpha_y)_{y\in B}$ of elements of K and $\alpha \in K$ such that at least one of the elements α, α_y, with $y \in B$, is non-zero and that

$$\alpha x + \sum_{y\in B} \alpha_y y = 0; \tag{8}$$

if $\alpha = 0$, then (8) would imply the linear dependence of B; hence $\alpha \neq 0$, and as the ring of scalars is a skew-field, by (8) x turns to be a linear combination of the elements of B; thus any element of S is a linear combination of the elements of B. Since S is a system of generators of E, by § 1 any element of E is a linear combination of the elements of B. ▮

COROLLARY 1. *Any vector space E over a skew-field K is a free K-module.*

Proof. Taking in Proposition 5 $L = \emptyset$ and $S = E$, we get that in E there always exists a basis. ▮

COROLLARY 2. *Let K be a skew-field, and let E be a K-vector space; if L is a linearly independent subset of E, then there exists a basis B of E containing L* (*or, briefly, any linearly independent subset of E may be extended to a basis*); *if S is a set of generators of E, then there exists a basis of E contained in S* (*or, briefly, from any set of generators it is possible to extract a basis*).

Proof. For the first assertion, we consider in Proposition 5 the given L and $S = E$; for the second assertion, take in Proposition 5 $L = \emptyset$ and the given S. ▮

COROLLARY 3. *If B is a subset of a vector space E over a skew-field K, then the following assertions are equivalent*:

i) B is a basis of E;

ii) B is a maximal linearly independent subset of E;

iii) B is a minimal set of generators of E.

Proof. If B is a basis of E, and if there exists a linearly independent subset L of E with $B \subseteq L$, $B \neq L$, then let $x \in L \setminus B$; x is a linear combination of the elements of B, this being a basis; hence $B \cup \{x\}$ is linearly dependent, and $B \cup \{x\} \subseteq L$; since in this way we get a contradiction, we must have *i*) $\Rightarrow$ *ii*). If B is a maximal linearly independent subset of E, then, as in the proof of Proposition 5 with $L = \emptyset$ and $S = E$, we obtain that any element of E is a linear combination of the elements of B; thus *ii*) $\Rightarrow$ *i*).

If B is a basis, then for any set of generators S such that $S \subseteq B$, we have $S = B$, since otherwise there would exist an element $x \in B \setminus S$, and the subset $S \cup \{x\} \subseteq B$ would be linearly dependent; thus *i*)$\Rightarrow$*iii*). Conversely, if B is a minimal set of generators, B must be linearly independent, since otherwise by Proposition 3 there would exist an element $x \in B$ such that x is a linear combination of the elements of $B \setminus \{x\}$, and $B \setminus \{x\}$ would be a set of generators; so *iii*) $\Rightarrow$ *i*). ∎

COROLLARY 4 (**Exchange Theorem**). *Let E be a vector space over the skew-field K; if L is a linearly independent subset of E and S is a set of generators of E, then there exists a subset S' of S such that $L \cup S'$ is a basis of E and $L \cap S' = \emptyset$.*

Proof. Apply Proposition 5 to the linearly independent subset L and to the set of generators $L \cup S$; we have $L \subseteq L \cup S$, and so, there exists a basis B of E such that $L \subseteq B \subseteq L \cup S$; put $S' = B \setminus L$. Since any element of S' is in $L \cup S$ but not in L, we have $S' \subseteq S$. ∎

If, however, A is an arbitrary ring, not every A-module is free, as shown by the following

EXAMPLE. 2) Let $A = \mathbf{Z}$, and let E be the $\mathbf{Z}$-module $\mathbf{Z}/n\mathbf{Z}$, $n > 1$; if $x \in \mathbf{Z}/n\mathbf{Z}$, we have $nx = 0$, with $n \neq 0$, hence x is linearly dependent, and thus cannot be an element of a basis; but neither can $\emptyset$ be a basis for $\mathbf{Z}/n\mathbf{Z}$, since the empty set generates the zero $\mathbf{Z}$-module and $\mathbf{Z}/n\mathbf{Z} \neq 0$. ∎

A free module may have non-zero linearly dependent elements, and also, *each non-zero element of a module may be linearly independent without implying the module to be free*; these facts are proved respectively by the following two

EXAMPLES. 3) Let $A = \mathbf{Z}/6\mathbf{Z}$, i.e. the quotient ring of the integers modulo 6; let $E = A$ considered as an A-module; E is a free A-module, since $E = A^1$; but denoting by $\hat{n}$ the coset in E of $n \in \mathbf{Z}$, we have $\hat{2} \neq 0$ and $\hat{3}.\hat{2} = 0$, with $\hat{3} \neq 0$.

4) Let $A = \mathbf{Z}$, and let $E = \mathbf{Q}$; if E were free, then a basis of it could have at most one element: for if r_1/s_1 and r_2/s_2 are two non-zero rational numbers, then $r_1, r_2, s_1, s_2 \neq 0$ in $\mathbf{Z}$ and

$$r_2 s_1 (r_1/s_1) + (-r_1 s_2)(r_2/s_2) = 0,$$

with $r_2s_1 \neq 0$; if a basis consisting of a single element $a \in \mathbf{Q}$ existed, then $a \neq 0$, and we would have $a/2 = na$, with $n \in \mathbf{Z}$, hence $1/2 = n$, which is impossible; as $\mathbf{Q} \neq 0$, $\mathbf{Q}$ cannot have the empty basis either. ▮

PROPOSITION 6. *Any A-module E is isomorphic to a quotient module of a free A-module.*

Proof. Let M be a set of generators of E; the linear mapping $\bar{f}: A_s^{(M)} \to E$ determined by M considered as a family is surjective by Proposition 2, *a*); but then $E \approx A_s^{(M)}/\mathrm{Ker}(\bar{f})$ (see Corollary 1 to Proposition 1, § 4). ▮

REMARK. If in the proof of Proposition 6 we take, in particular, $M = E$, then for any A-module E we get a surjective homomorphism $A_s^{(E)} \to E$; this sends the element $\sum_{x \in E} \xi_x e_x$ of $A_s^{(E)}$, $(e_x)_{x \in E}$ being the canonical basis of $A_s^{(E)}$, into the element $\sum_{x \in E} \xi_x x$ of E; this linear mapping will be called the *canonical linear mapping* $A_s^{(E)} \to E$. ▮

As an additional result to Proposition 6, we notice that if E is a finitely generated A-module having a set of generators consisting of, say, n elements, $n \geqslant 0$, then the linear mapping $A_s^n \to E$ determined by the family of these n generators (indexed by the natural numbers $1, \ldots, n$) is surjective; conversely, if there exists a surjective homomorphism $A_s^n \to E$ for some $n \geqslant 0$, then by Corollary 2 to Proposition 2, § 8, E is finitely generated; hence, *the finitely generated A-modules are the A-modules isomorphic to a quotient module of some module* A_s^n, *with* $n \geqslant 0$.

In particular, if E is a cyclic A-module and a is a generator of E, taking $n = 1$ in the preceding considerations, we get the surjective homomorphism $\bar{f}: A_s^1 = A_s \to E$ sending $1 \in A$ into a; then $\bar{f}(\alpha) = \alpha\bar{f}(1) = \alpha a$, $\alpha \in A$; the kernel of $\bar{f}$ consists of those $\alpha \in A$ for which $\alpha a = 0$.

DEFINITION 4. If A is a ring, E is an A-module, and M is a subset of E, then an element $\alpha \in A$ is said to *annihilate* M provided $\alpha x = 0$ for any $x \in M$; the set of all $\alpha \in A$ anihilating M is called the *annihilator* of M, and is denoted $\mathrm{Ann}_A(M)$ or $\mathrm{Ann}(M)$.

If $M = \{x\}$, then instead of $\mathrm{Ann}(\{x\})$ one writes $\mathrm{Ann}(x)$. ▮

It follows from the definition that *for any subset M of an A-module E,* $\mathrm{Ann}(M)$ *is a left ideal of E*: for, $0 \in \mathrm{Ann}(M)$, and if, for instance, $\alpha \in A$, $\beta \in \mathrm{Ann}(M)$, then for any $x \in M$ we have $(\alpha\beta)x = \alpha(\beta x) = 0$, hence $\alpha\beta \in \mathrm{Ann}(M)$; *if*, however, *M is a submodule of E, then* $\mathrm{Ann}(M)$ *is a two-sided ideal of A*: for if $\alpha \in A$, $\beta \in \mathrm{Ann}(M)$, then for any $x \in M$ we have $\alpha x \in M$, hence $(\beta\alpha)x = \beta(\alpha x) = 0$, and $\beta\alpha \in \mathrm{Ann}(M)$.

Notice that *the element* $x \in E$ *is linearly independent if and only if* $\mathrm{Ann}(x) = 0$.

Of course, the concept of annihilator may be defined also for a subset M of a right A-module E; *in this case*, using the same notation, $\mathrm{Ann}(M)$ *is*, in general, *a right ideal of A*, and it is a two-sided ideal of A provided M is a submodule.

Considering again a (left) A-module E, we have $\mathrm{Ann}(0) = A$, and *if* $\mathrm{Ann}(x) = A$, *then* $x = 0$: the first assertion is obvious, while for the second notice that since $1 \in \mathrm{Ann}(x)$, it follows $x = 1 \cdot x = 0$. Moreover, *if* M_1, M_2 *are subsets of E such that* $M_1 \subseteq M_2$, *then* $\mathrm{Ann}(M_1) \supseteq \mathrm{Ann}(M_2)$, which is obvious.

Next, *if* $(M_\lambda)_{\lambda\in I}$ *is a family of subsets of the A-module E, then*

$$\mathrm{Ann}(\bigcup_{\lambda\in I} M_\lambda) = \bigcap_{\lambda\in I} \mathrm{Ann}(M_\lambda):$$

for, $\bigcup_{\lambda\in I} M_\lambda \supseteq M_\lambda$, hence $\mathrm{Ann}(\bigcup_{\lambda\in I} M_\lambda) \subseteq \mathrm{Ann}(M_\lambda)$ for any $\lambda \in I$; conversely, if $\alpha \in \bigcap_{\lambda\in I}\mathrm{Ann}(M_\lambda)$, and if $x \in \bigcup_{\lambda\in I} M_\lambda$, then $x \in M_\lambda$ for some $\lambda \in I$, and as $\alpha \in \mathrm{Ann}(M_\lambda)$, it follows that $\alpha x = 0$; hence $\alpha \in \mathrm{Ann}\,(\bigcup_{\lambda\in I} M_\lambda)$; in particular, *for any subset M of E we have*

$$\mathrm{Ann}(M) = \bigcup_{x\in M} \mathrm{Ann}(x).$$

We mention also that *if* $(M_\lambda)_{\lambda\in I}$ *is a family of submodules of E, then*

$$\mathrm{Ann}(\sum_{\lambda\in I} M_\lambda) = \bigcap_{\lambda\in I} \mathrm{Ann}(M_\lambda):$$

since, by Corollary 5 to Proposition 2, § 8, for any $x \in \sum_{\lambda\in I} M_\lambda$ there exists a family of elements $x_{\lambda_i} \in M_{\lambda_i}$, $i = 1,\ldots, n$, such that $x = \sum_{i=1}^{n} x_{\lambda_i}$; if $\alpha \in \bigcap_{\lambda\in I} \mathrm{Ann}(M_\lambda)$, then $\alpha x_{\lambda_i} = 0$ for any $i = 1,\ldots, n$, hence $\alpha x = 0$, and $\alpha \in \mathrm{Ann}(\sum_{\lambda\in I} M_\lambda)$; the converse follows from the preceding considerations. Finally, *for any element x of the A-module E and any* $\alpha \in A$, $\mathrm{Ann}(\alpha x)$ *consists of all* $\beta \in A$ *for which* $\beta\alpha \in \mathrm{Ann}(x)$.

The annihilator of the A-module E may be also characterized as follows: defining the mapping

$$\varphi\colon A \to \mathrm{End}_{\mathbf{Z}}(E)$$

by $\varphi(\alpha) = h_\alpha$, with $\alpha \in A$ and h_α the homothety of ratio α in E (see the Remark following Example 10), § 1), we get a homomorphism of rings with unit element: for instance,

$$h_{\alpha\beta}(x) = (\alpha\beta)\, x = \alpha(\beta x) = (h_\alpha \circ h_\beta)\,(x)$$

for any $x \in E$, hence $h_{\alpha\beta} = h_\alpha \circ h_\beta$, and

$$h_1(x) = 1 \cdot x = x = 1_E(x)$$

for any $x \in E$, hence $h_1 = 1_E$; then

$$\mathrm{Ann}(E) = \varphi^{-1}(0),$$

0 of the right hand side being the zero endomorphism: indeed, we have $\alpha \in \mathrm{Ann}(E)$ if and only if $\alpha x = 0$ for any $x \in E$, hence if and only if $\varphi(\alpha) = h_\alpha = 0$.

The A-module E is said to be a *faithful module* if $\mathrm{Ann}(E) = 0$; in this case φ is injective, since φ is also a homomorphism of abelian groups between the underlying additive abelian groups of A and $\mathrm{End}_{\mathbf{Z}}(E)$. With a given A-module E a faithful module over a certain ring canonically constructed may be associated: indeed, if $\mathfrak{a} = \mathrm{Ann}_A(E)$, consider the quotient ring $A/\mathfrak{a} = B$; on the underlying additive abelian group of E define a B-module structure as follows: let $x \in E$, and let $\hat{\alpha}$ be the coset of $\alpha \in A$ with respect to the two-sided ideal $\mathfrak{a}$; then, if $\alpha' \in A$ is in the same coset, it follows that $\alpha' - \alpha \in \mathrm{Ann}(E)$, hence $0 = (\alpha' - \alpha)x = \alpha'x - \alpha x$, and the element αx does not depend on the chosen representative α in the coset $\hat{\alpha}$; define

$$\hat{\alpha} x = \alpha x$$

for any of the representatives α of the coset $\hat{\alpha}$; in this way a B-module structure is obtained on E: for instance, if $\hat{\alpha}, \hat{\beta} \in B$, $x \in E$, then

$$(\hat{\alpha} + \hat{\beta})x = \widehat{(\alpha + \beta)} \cdot x = (\alpha + \beta)x = \alpha x + \beta x = \hat{\alpha}x + \hat{\beta}x.$$

If $\hat{\alpha}$ is an element of the annihilator of E, with $\alpha \in A$, where E is considered as a B-module, then for any $x \in E$,

$$\alpha x = \hat{\alpha} x = 0,$$

and $\alpha \in \mathrm{Ann}_A(E) = \mathfrak{a}$, hence $\hat{\alpha} = 0$; in other words, $\mathrm{Ann}_B(E) = 0$, and E is a faithful B-module; the module constructed in this way is called the *faithful module associated with E*.

Considering again the case of a cyclic A-module E having a as generator, we have seen above that the canonical homomorphism $\bar{f}: A_s \to E$ has $\mathrm{Ann}(a)$ as kernel; by the First Isomorphism Theorem for modules, there exists an isomorphism

$$A_s/\mathrm{Ann}(a) \xrightarrow{\sim} E$$

under which the coset $\hat{\alpha}$ of $\alpha \in A$ with respect to $\mathrm{Ann}(a)$ is sent into αa.

Conversely, if E is isomorphic to a quotient module of A_s, then as A_s is cyclic, it follows from the considerations following Corollary 2 to Proposition 2, § 8, that E is also cyclic.

Thus we have proved:

PROPOSITION 7. *An A-module E is cyclic if and only if it is isomorphic to a quotient module of A_s.* ∎

REMARK. If the ring A is commutative and the A-module E is cyclic, then, whatever the generator a of E, we have $\mathrm{Ann}(E) = \mathrm{Ann}(a)$: indeed, in this case, if $\beta \in \mathrm{Ann}(a)$, then $\beta(\alpha a) = \alpha(\beta a) = 0$ for any $\alpha \in A$. ∎

PROPOSITION 8. *If A is an integral domain, then in any A-module E the subset of all elements having a non-zero annihilator is a submodule of E.*

Proof. We have seen above that $0 \in E$ is a linearly dependent element, hence an element whose annihilator is different from 0; next, if $x, y \in E$ have non-zero annihilators, let $\alpha \in \mathrm{Ann}(x)$, $\beta \in \mathrm{Ann}(y)$, $\alpha, \beta \neq 0$; we have

$$(\alpha\beta)(x + y) = \alpha\beta x + \alpha\beta y = 0;$$

furthermore, $\alpha\beta \neq 0$, since A is an integral domain; consequently, $x + y$ is a linearly dependent element; if x is a linearly dependent element and $\alpha x = 0$, $\alpha \neq 0$, then for any $\beta \in A$ we have $\alpha(\beta x) = 0$, hence βx has a non-zero annihilator. ∎

DEFINITION 5. If A is an integral domain and E is an A-module, an element $x \in E$ is said to be a *torsion element* if $\mathrm{Ann}(x) \neq 0$ or, equivalently, if x is a linearly dependent element; the submodule consisting of all torsion elements of E is called the *torsion submodule* of E, and is denoted $t(E)$. If $E = t(E)$, the module E is said to be a *torsion module.* If $t(E) = 0$ or, equivalently, if any non-zero element of E is linearly independent, the A-module E is said to be a *torsion-free module.* ∎

In an A-module *the set of all elements having zero annihilator is not, generally, a submodule* even when the ring A is an integral domain, and even if the zero element is joined to the set mentioned; moreover, *if the ring is not an integral domain, the set of all elements having a non-zero annihilator is not, generally, a submodule*; these facts are respectively shown by the following two

EXAMPLES. 5) Let $A = \mathbf{Z}$, and let $E = \mathbf{R}/\mathbf{Z}$; denote by $\hat{\alpha}$ the coset of $\alpha \in \mathbf{R}$ modulo $\mathbf{Z}$; if α is an irrational real number, then each of the elements $\hat{\alpha}$ and $\widehat{1/2-\alpha}$ is linearly independent; however, $\hat{\alpha} + (\widehat{1/2-\alpha}) = \widehat{1/2}$ is a non-zero linearly dependent element, since $2 \cdot \widehat{1/2} = \hat{0}$.

6) Let A and E be as in Example 3) above; in the same notation, the elements $\hat{3}$ and $\hat{4}$ have non-zero annihilators, since $\hat{2} \cdot \hat{3} = 0$, with $\hat{2} \neq 0$, and $\hat{3} \cdot \hat{4} = 0$, with $\hat{3} \neq 0$; however, $\hat{3} + \hat{4} = \hat{1}$, whose annihilator is zero. ∎

If A is an integral domain, if E, F are A-modules, and if $u : E \to F$ is an A-homomorphism, then $u(t(E)) \subseteq t(F)$: for let $x \in t(E)$, and let $\alpha \in \mathrm{Ann}(x)$, $\alpha \neq 0$; we have $\alpha(ux) = u(\alpha x) = 0$, hence $\alpha \in \mathrm{Ann}(ux)$. It follows that u induces by restriction to $t(E)$ and $t(F)$ an A-homomorphism $t(u) : t(E) \to t(F)$. It is straightforward that the mappings

$$E \mapsto t(E), \quad u \mapsto t(u),$$

where E runs over all objects and u over all morphisms of $\mathbf{Mod}(A)$, define an additive functor

$$t : \mathbf{Mod}(A) \to \mathbf{Mod}(A),$$

which will be referred to as the *torsion functor.*

The torsion functor is left exact: since let

$$0 \to E \xrightarrow{u} F \xrightarrow{v} G \tag{9}$$

be an exact sequence of A-modules, where A is an integral domain; by Proposition 3, § 5, the assertion will result if we show that the diagram

$$0 \to t(E) \xrightarrow{t(u)} t(F) \xrightarrow{t(v)} t(G)$$

is an exact sequence of A-modules; but, firstly, $t(u)$ is injective, because it acts in the same way as u; next, $t(v) \circ t(u) = 0$, since $v \circ u = 0$; if $y \in \mathrm{Ker}(t(v))$, we have $vy = 0$, and by the exactness of (9), there exists $x \in E$ such that $u(x) = y$; but $x \in t(E)$, since if $\alpha \in \mathrm{Ann}(y)$, $\alpha \neq 0$, we have $u(\alpha x) = \alpha y = 0$, and as u is injective, $\alpha x = 0$.

If A is an integral domain and $(E_\lambda)_{\lambda \in I}$ is a family of A-modules, then[1]

$$t(\bigoplus_{\lambda \in I} E_\lambda) = \bigoplus_{\lambda \in I} t(E_\lambda): \tag{10}$$

for let $(x_\lambda)_{\lambda \in I} \in t(\bigoplus_{\lambda \in I} E_\lambda)$; if $\alpha \in A$, $\alpha \neq 0$, is such that $\alpha(x_\lambda)_\lambda = 0$, then $\alpha x_\lambda = 0$ for any $\lambda \in I$, and $x_\lambda \in t(E_\lambda)$; hence $t(\bigoplus_{\lambda \in I} E_\lambda) \subseteq \bigoplus_{\lambda \in I} t(E_\lambda)$; conversely, let $(x_\lambda)_{\lambda \in I} \in \bigoplus_{\lambda \in I} t(E_\lambda)$; the family $(x_\lambda)_\lambda$ has a finite support, say J, and for any $\lambda \in J$ we can find a non-zero element α_λ of $\mathrm{Ann}(x_\lambda)$; we have[2] $\prod_{\lambda \in J} \alpha_\lambda \neq 0$, and, since for $\lambda \in I \setminus J$, $x_\lambda = 0$, we get

$$(\prod_{\lambda \in J} \alpha_\lambda)(x_\lambda)_{\lambda \in I} = 0;$$

thus $(x_\lambda)_{\lambda \in I} \in t(\bigoplus_{\lambda \in I} E_\lambda)$, and (10) is completely proved.

In particular, *the direct sum of a family of torsion modules over an integral domain is again a torsion module.* The following example shows that the preceding result cannot be extended to direct products:

EXAMPLE. 7) Let $A = \mathbf{Z}$, and let $E_n = \mathbf{Z}/n\mathbf{Z}$ for any natural number $n > 1$; for any $n > 1$, E_n is a torsion $\mathbf{Z}$-module, but $E = \prod_{n>1} (\mathbf{Z}/n\mathbf{Z})$ is not such a module: indeed, the family $(1 + n\mathbf{Z})_{n>1} \in E$ of cosets[3] of $1 \in \mathbf{Z}$ is a linearly independent element, since, for instance, $m(1 + (m+1)\mathbf{Z}) \neq 0$ for any $m \geqslant 1$. ∎

---

[1] See the considerations following Corollary 1 to Proposition 2, § 7.
[2] Since A is a commutative ring, the ordering of factors in the product is unimportant.
[3] See the Remark following Corollary 5 to Proposition 2, § 8.

Proposition 9. *If A is an integral domain and E is an A-module, then the quotient module $E/t(E)$ is a torsion-free module.*

Proof. Let us denote by $\hat{x}$ the coset of $x \in E$ modulo $t(E)$. If $\hat{x} \in E/t(E)$, with $x \in E$, has the property that $\alpha\hat{x} = 0$ for some $\alpha \in A$, $\alpha \neq 0$, then $\alpha x \in t(E)$; consequently, there exists some $\beta \in A$, $\beta \neq 0$, such that $\beta\alpha x = 0$; but $\beta\alpha \neq 0$, and so $x \in t(E)$, hence $\hat{x} = 0$. ∎

We consider again an arbitrary ring A (with unit element).

Proposition 10. *Let E be an A-module, and let $(M_\lambda)_{\lambda \in I}$ be a family of submodules of E such that $E = \dot{\sum\limits_{\lambda \in I}} M_\lambda$; let S_λ be a set of generators, respectively a linearly independent subset, respectively a basis of M_λ for any $\lambda \in I$; then $S = \bigcup\limits_{\lambda \in I} S_\lambda$ is a set of generators, respectively a linearly independent subset, respectively a basis of E.*

Conversely, let S be a basis of E, and let $S = \bigcup\limits_{\lambda \in I} S_\lambda$ be a partition of S; if we denote by M_λ the submodule generated in E by S_λ, $\lambda \in I$, then $E = \dot{\sum\limits_{\lambda \in I}} M_\lambda$.

Proof. To start with, we consider the direct assertions: the first of these is a particular case of Corollary 8 to Proposition 2, § 8. As for the second assertion, since $S_\lambda \subseteq M_\lambda$ for any $\lambda \in I$, it follows that $S_\lambda \cap S_\mu = \emptyset$ for $\lambda \neq \mu$: indeed, if there exists $x \in S_\lambda \cap S_\mu$, then $x \in M_\lambda \cap M_\mu$; by the characterization *iv*) in Proposition 1, § 9, for the internal direct sum decomposition, $M_\lambda \cap M_\mu = 0$; but 0 cannot belong to a linearly independent subset. Consequently, the representation $S = \bigcup\limits_{\lambda \in I} S_\lambda$ is a partition of S; hence any relation of the form

$$\sum_{x \in S} \alpha_x \cdot x = 0,$$

where $(\alpha_x)_{x \in S}$ is an fs-family of elements of A, may be written as

$$\sum_{\lambda \in I} \sum_{x \in S_\lambda} \alpha_x \cdot x = 0, \tag{11}$$

where $(\sum\limits_{x \in S_\lambda} \alpha_x \cdot x)_{\lambda \in I}$ is an fs-family (see § 1) and $\sum\limits_{x \in S_\lambda} \alpha_x \cdot x \in M_\lambda$, $\lambda \in I$; by the characterization *iii*) in Proposition 1, § 9, of the internal direct sum decomposition, we get from (11)

$$\sum_{x \in S_\lambda} \alpha_x \cdot x = 0$$

for any $\lambda \in I$; since S_λ is a linearly independent subset, we have $\alpha_x = 0$ for any $x \in S_\lambda$ and any $\lambda \in I$, hence for any $x \in S$. The third assertion is a consequence of the first two ones.

As for the converse assertion, we make use again of Corollary 8 to Proposition 2, § 8: since the submodule generated by $S = \bigcup\limits_{\lambda \in I} S_\lambda$ is by this Corollary

$\sum_{\lambda \in I} M_\lambda$, and since, on the other hand, by hypothesis this submodule is E, it follows $E = \sum_{\lambda \in I} M_\lambda$; given the relation

$$\sum_{\lambda \in I} x_\lambda = 0, \tag{12}$$

with $x_\lambda \in M_\lambda$, $\lambda \in I$, and $(x_\lambda)_{\lambda \in I}$ an fs-family, then, assuming its support, say J, to be non-void, we have for any $\lambda \in J$,

$$x_\lambda = \sum_{y \in S_\lambda} \alpha_y \cdot y,$$

with $(\alpha_y)_{y \in S_\lambda}$ an fs-family of elements of A; setting $\alpha_y = 0$ for $y \in S_\lambda$, $\lambda \in I \setminus J$, the family $(\alpha_y)_{y \in S}$ is an fs-family, and by (12),

$$\sum_{y \in S} \alpha_y \cdot y = 0;$$

the set S being linearly independent, $\alpha_y = 0$ for any $y \in S$ and hence also for any $y \in S_\lambda$, $\lambda \in J$; hence $x_\lambda = 0$ for any $\lambda \in J$, which is absurd; the support of the family $(x_\lambda)_{\lambda \in I}$ turns to be empty, and hence $x_\lambda = 0$ for any $\lambda \in I$; by Proposition 1, *iii*), § 9, we get the required result. ∎

COROLLARY. *Let E be an A-module, and let $(a_\lambda)_{\lambda \in I}$ be a family of elements of E; in order that $(a_\lambda)_{\lambda \in I}$ be a basis of E it is necessary and sufficient that E be the internal direct sum of the family of cyclic submodules $(Aa_\lambda)_{\lambda \in I}$ and a_λ be a linearly independent element of E for any $\lambda \in I$.*

Proof. If $E = \dot{\sum_{\lambda \in I}} (Aa_\lambda)$ and a_λ is linearly independent, $\lambda \in I$, then a_λ is a basis of Aa_λ; by Proposition 10, $\bigcup_{\lambda \in I} \{a_\lambda\}$ is a basis of E; but the family $(a_\lambda)_{\lambda \in I}$ is bijectively indexed by I according to an argument similar to the one used in the proof of Proposition 10 when showing that the sets S_λ, $\lambda \in I$, considered there are mutually disjoint; hence $(a_\lambda)_{\lambda \in I}$ is a basis of E.

Conversely, if $(a_\lambda)_{\lambda \in I}$ is a basis of E, then if $S = \bigcup_{\lambda \in I} \{a_\lambda\}$, this representation of S is a partition, since $(a_\lambda)_\lambda$ is bijectively indexed by I; but then $E = \dot{\sum_{\lambda \in I}} Aa_\lambda$; the fact that a_λ is a linearly independent element for any $\lambda \in I$ follows from the fact that it was picked up from a linearly independent subset. ∎

REMARK. Now it is possible to point out that Proposition 1 or the part of Proposition 4 concerning the existence of the linear mapping u follow by virtue of the universal property of the direct sum once they are proved for a set I consisting of a single element: indeed, considering a basis $(a_\lambda)_{\lambda \in I}$ of the given free A-module, this module is, by the Corollary to Proposition 10, isomorphic to $\bigoplus_{\lambda \in I} Aa_\lambda$, and we need only apply the universal property of the direct sum to the already defined mappings from each Aa_λ, $\lambda \in I$, to the second module. ∎

Now, we are going to consider the question of the uniqueness of the cardinality for the different bases of a free A-module.

PROPOSITION 11. *Let E be an A-module which is the internal direct sum of an infinite family $(M_\lambda)_{\lambda\in I}$ of cyclic non-zero submodules of it; then for any set of generators S of E we have* $\mathrm{Card}(S) \geqslant \mathrm{Card}(I)$.

Proof. Let $(u_\lambda)_{\lambda\in I}$ be the orthogonal family of projectors corresponding to the given decomposition of E into internal direct sum; by Corollary 2 to Proposition 2, § 9, for any $x \in E$ the family $(u_\lambda(x))_{\lambda\in I}$ is an fs-family of elements of E; denote for any $x \in E$ by I_x the support of this family; notice that $I_x = \emptyset$ implies $x = 0$; we claim that

$$\bigcup_{x\in S} I_x = I: \tag{13}$$

indeed, let $\lambda \in I$; if a_λ is a generator of M_λ, then a_λ may be represented as $a_\lambda = \sum_{i=1}^{n} \alpha_i x_i$, with $\alpha_i \in A$ and $x_i \in S$, $i = 1,\ldots, n$, since S is a set of generators of E; if for any $i = 1,\ldots, n$ we had $u_\lambda(x_i) = 0$, it would follow that $u_\lambda(a_\lambda) = 0$; but, on the other hand, as $a_\lambda \in M_\lambda$, we have $u_\lambda(a_\lambda) = a_\lambda$; hence we would have $a_\lambda = 0$, which would imply $M_\lambda = 0$, contradicting the hypothesis; consequently, there exists $i \in [1, n]$ such that $u_\lambda(x_i) \neq 0$, and as $x_i \in S$, $\lambda \in \bigcup_{x\in S} I_x$.

By (13) we have

$$\mathrm{Card}(I) \leqslant \sum_{x\in S} \mathrm{Card}(I_x);$$

but, as every set I_x, $x \in S$, is finite, if S were finite, I would be finite too, contradicting the hypothesis; hence S is an infinite set, and as I_x, $x \in S$, are finite sets and all except perhaps I_0 are non-void, it follows

$$\sum_{x\in S} \mathrm{Card}(I_x) = \mathrm{Card}(S). \blacksquare$$

COROLLARY 1. *Let E be an A-module; let $(M_\lambda)_{\lambda\in I}$, $(N_\mu)_{\mu\in J}$ be two families of cyclic non-zero submodules of E such that $E = \sum_{\lambda\in I} M_\lambda = \sum_{\mu\in J} N_\mu$; if I is infinite, then* $\mathrm{Card}(I) = \mathrm{Card}(J)$.

Proof. Let b_μ be a generator of N_μ for any $\mu \in J$; then the subset S consisting of all b_μ, $\mu \in J$, is by Proposition 10 a set of generators of E; by Proposition 11 we have

$$\mathrm{Card}(S) \geqslant \mathrm{Card}(I);$$

but $\mathrm{Card}(S) = \mathrm{Card}(J)$, since if $\mu \neq \mu'$, then $b_\mu \neq b_{\mu'}$: indeed, $N_\mu \cap N_{\mu'} = 0$, and $b_\mu = b_{\mu'}$ would imply $b_\mu = 0$, hence $N_\mu = 0$, contradicting the hypothesis; it follows that

$$\mathrm{Card}(J) \geqslant \mathrm{Card}(I);$$

hence J is an infinite set, and interchanging the roles of the two families we get the required result. ▮

COROLLARY 2. *If E is a free A-module having an infinite basis B, then for any basis B' of E we have* $\text{Card}(B') = \text{Card}(B)$, *and for any set of generators S of E,* $\text{Card}(S) \geqslant \text{Card}(B)$; *in particular, under the given hypothesis, any basis and any set of generators of E are infinite.*

Proof. We start with the first assertion: by the Corollary to Proposition 10, $E = \dot{\sum\limits_{x \in B}} Ax = \dot{\sum\limits_{y \in B'}} Ay$; since $x \in B$, respectively $y \in B'$, implies $x \neq 0$, respectively $y \neq 0$, it follows that $Ax \neq 0$ for any $x \in B$ and $Ay \neq 0$ for any $y \in B'$; as B is infinite, we are under the conditions of Corollary 1, and therefore $\text{Card}(B) = \text{Card}(B')$; for the second assertion we make use only of the decomposition $E = \dot{\sum\limits_{x \in B}} Ax$ of E into internal direct sum; by Proposition 11, $\text{Card}(S) \geqslant \text{Card}(B)$. ▮

Corollary 2 to Proposition 11 may be extended to arbitrary, not necessarily infinite bases provided the ring A is a skew-field:

PROPOSITION 12. *Any two bases of a vector space E over a skew-field K have the same cardinality.*

Proof. If E has an infinite basis, the result is a consequence of Corollary 2 to Proposition 11. Assume that any basis of E is finite, and let B be such a basis having, say, n elements; we shall prove that if B' is another basis of E, then B' has at most n elements, and by interchanging the roles of B and B' the result will be proved.

We shall do it by induction on n: if $n = 0$, we have $E = 0$, and any basis has 0 elements; suppose $n \geqslant 1$ and the assertion valid for any vector space having a basis of less then n elements; as above, it is not possible to have $B' = \emptyset$, since that would imply $E = 0$. Let $a \in B'$; applying the Exchange Theorem (Corollary 4 to Proposition 5) to the linearly independent subset $\{a\}$ and to the set of generators B, it follows that a subset C of E, $C \subseteq B$, exists such that $\{a\} \cup C$ is a basis of E and $\{a\} \cap C = \emptyset$; we cannot have $C = B$, since this would imply that $\{a\} \cup B$ strictly contains B, and B would no longer be a maximal linearly independent subset (see Corollary 3 to Proposition 5); hence C has at most $n - 1$ elements.

Let F, respectively F', be the subspace generated in E by C, respectively $B' \setminus \{a\}$; according to Proposition 10 we have

$$E = F \dotplus Ka = F' \dotplus Ka,$$

and so, by § 9, there exists a K-isomorphism $u : F \rightrightarrows F'$. Since C is a basis of F, $u(C)$ is a basis of F', and has at most $n - 1$ elements; as $B' \setminus \{a\}$ is another basis of F, making use of the induction hypothesis we obtain that $B' \setminus \{a\}$ has at most $n - 1$ elements; hence B' has at most n elements. ▮

DEFINITION 6. Given a skew-field K and a K-vector space E, the *dimension of E over K*, denoted $\dim_K E$ or $\dim E$, is the cardinal number of any of its bases[1]; if this cardinal number is a natural number, respectively the natural number n, then E is said to be a *finite-dimensional*, respectively an *n-dimensional vector space* over K; if this cardinal number is transfinite, E is said to be an *infinite-dimensional vector space.* ∎

The fact that Corollary 2 to Proposition 11 cannot, in general, be extended to any free module having a finite basis, in other words, the fact that *there exists free modules over certain rings such that the number of elements of a basis may differ from basis to basis* is shown by the following

EXAMPLE. 8) Let C be an arbitrary ring, and let A be the ring of endomorphisms of the C-module $C_s^{(\mathbf{N})}$; A_s is a free A-module having a basis consisting of one element, namely $1_{C_s^{(\mathbf{N})}}$, the unit element of A; we shall construct another basis of A consisting of two elements, and this will settle the question:

Let u_1, u_2 be the endomorphisms of $C_s^{(\mathbf{N})}$ defined on the canonical basis $(e_k)_{k\in\mathbf{N}}$ as follows

$$u_1(e_{2n}) = e_n,\ u_1(e_{2n+1}) = 0,$$

$$u_2(e_{2n+1}) = e_n,\ u_2(e_{2n}) = 0$$

for any $n \in \mathbf{N}$; by virtue of Proposition 1 and of the identification of the elements of $\mathbf{N}$ with the corresponding elements of the canonical basis, the endomorphisms u_1 and u_2 are in this way uniquely defined on the whole $C_s^{(\mathbf{N})}$. The endomorphisms u_1 and u_2 form a basis over A of A_s: indeed, if $u \in A_s$, let $v_1, v_2 \in A$ be defined on the basis $(e_k)_{k\in\mathbf{N}}$ by

$$v_1(e_n) = u(e_{2n}), \quad v_2(e_n) = u(e_{2n+1});$$

it is easy to check that $u = v_1 \circ u_1 + v_2 \circ u_2$; next, if $w_1, w_2 \in A$ have the property that

$$w_1 \circ u_1 + w_2 \circ u_2 = 0,$$

then

$$w_1(e_n) = w_1(u_1(e_{2n})) + w_2(u_2(e_{2n})) = 0,$$

hence $w_1 = 0$, and similarly $w_2 = 0$. ∎

* Let $\mathcal{C}, \mathcal{C}'$ be two categories, and let T and U be two functors $\mathcal{C} \to \mathcal{C}'$.

DEFINITION 7. A *natural transformation* (or also a *morphism of functors*) Φ *of T into U* or *from T to U* or *between T and U*, denoted $\Phi : T \to U$, consists of the assignment to each object E of $\mathcal{C}$ of a morphism of $\mathcal{C}'$

$$\Phi(E) : T(E) \to U(E)$$

[1] It exists and is independent of the considered basis by virtue of Corollary 1 to Proposition 5 and by Proposition 12.

in such a way that for any morphism $u \in \mathrm{Hom}_{\mathcal{C}}(E, F)$ the following diagram in $\mathcal{C}'$ is commutative:

$$
\tag{14}
\begin{array}{ccc}
T(E) & \xrightarrow{T(u)} & T(F) \\
\Phi(E)\downarrow & & \downarrow\Phi(F) \\
U(E) & \xrightarrow{U(u)} & U(F).
\end{array}
$$

The morphism $\Phi(E)$ is sometimes called the *component at the object E* of the natural transformation Φ. Of course, two natural transformations $\Phi_1, \Phi_2 : T \to U$ are considered identical if for any object E of $\mathcal{C}$ we have $\Phi_1(E) = \Phi_2(E)$ as morphisms of $\mathcal{C}'$. ▮

If V is another functor $\mathcal{C} \to \mathcal{C}'$, and if $\Psi : U \to V$ is a natural transformation, then it follows easily that assigning to each $E \in \mathrm{Ob}\,\mathcal{C}$ the morphism

$$\Theta(E) = \Psi(E)\circ\Phi(E) : T(E) \to V(E),$$

a natural transformation $\Theta : T \to V$ is obtained; this natural transformation is called the *composition* (or the *composite natural transformation*) of Ψ and Φ (or of Ψ with Φ) and is denoted $\Psi\circ\Phi$. The operation of composition is obviously associative. Moreover, for any functor $T : \mathcal{C} \to \mathcal{C}'$ a natural transformation $1_T : T \to T$ is obtained defining for any $E \in \mathrm{Ob}\,\mathcal{C}$,

$$1_T(E) = 1_{T(E)};$$

the natural transformation 1_T is called the *identity natural transformation* of T. Obviously, the identity natural transformation of T is a two-sided neutral element for all natural transformations whose composition with 1_T is defined.

REMARK. Now we could have all the necessary data for defining a new category: as objects we could take the functors $\mathcal{C} \to \mathcal{C}'$, and for any two such functors T, U we could take as $\mathrm{Hom}(T, U)$ the class of all natural transformations $T \to U$, the composition and the identity natural transformation being those defined above. However, the natural transformations $T \to U$ do not, in general, form a set (see Definition 4, § 1); it is necessary to make additional hypotheses on $\mathcal{C}$ or $\mathcal{C}'$ (for instance, $\mathcal{C}$ should be a small category), and then we indeed get such a category, called the *category of functors from $\mathcal{C}$ to $\mathcal{C}'$* and denoted **Fonct**$(\mathcal{C}, \mathcal{C}')$ or **Hom**$(\mathcal{C}, \mathcal{C}')$ or, sometimes, $\mathcal{C}'^{\mathcal{C}}$; when this is the case, we shall also proceed in this way (see § 12 and § 20). ▮

In the same notation as above, if $\Phi : T \to U$ is a natural transformation, Φ is said to be a *natural equivalence between T and U* or *from T to U* (or also, an *isomorphism of functors from T to U*), and is written as $\Phi : T \xrightarrow{\sim} U$, if there exists a natural transformation $\Psi : U \to T$ such that

(15) $$\Psi \circ \Phi = 1_T, \qquad \Phi \circ \Psi = 1_U;$$

this is equivalent to saying that for any $E \in \mathrm{Ob}\, \mathcal{C}$, $\Phi(E)$ is an isomorphism of the category $\mathcal{C}'$. If there exists an isomorphism $\Phi : T \xrightarrow{\sim} U$, then the functor T is said to be *naturally equivalent* (or also *isomorphic*) to the functor U; to describe this situation, we write $T \approx U$. The natural transformation Ψ is uniquely determined by relations (15), and is a natural equivalence; it is called the *natural equivalence inverse to* Φ or also the *inverse of* Φ, and is denoted Φ^{-1}.In general, the same conventions of terminology and notation used for categories (see § 1) will be adopted (see the Remark following Definition 7).*

Let A be a ring, let I and J be two sets; the mappings

$$(((E_\lambda)_{\lambda \in I})^\circ, (F_\mu)_{\mu \in J}) \mapsto \mathrm{Hom}_A(\bigoplus_{\lambda \in I} E_\lambda, \prod_{\mu \in J} F_\mu),$$

$$(((u_\lambda)_{\lambda \in I})^\circ, (v_\mu)_{\mu \in J}) \mapsto \mathrm{Hom}(\bigoplus_{\lambda \in I} u_\lambda, \prod_{\mu \in J} v_\mu)$$

(see § 7) define a functor $((\mathbf{Mod}(A))^I)^\circ \times (\mathbf{Mod}(A))^J \to \mathbf{Ab}$; in the same way, the mappings

$$(((E_\lambda)_{\lambda \in I})^\circ, (F_\mu)_{\mu \in J}) \mapsto \prod_{(\lambda, \mu) \in I \times J} \mathrm{Hom}_A(E_\lambda, F_\mu),$$

$$(((u_\lambda)_{\lambda \in I})^\circ, (v_\mu)_{\mu \in J}) \mapsto \prod_{(\lambda, \mu) \in I \times J} \mathrm{Hom}(u_\lambda, v_\mu)$$

define a functor $((\mathbf{Mod}(A))^I)^\circ \times (\mathbf{Mod}(A))^J \to \mathbf{Ab}$; we leave to the reader to verify these facts; the isomorphism defined by Proposition 3, § 7, is the component at the object $(((E_\lambda)_{\lambda \in I})^\circ, (F_\mu)_{\mu \in J})$ of the category $((\mathbf{Mod}(A))^I)^\circ \times (\mathbf{Mod}(A))^J$ of a natural equivalence between these two functors, as shown by the commutativity of diagram (11), § 7.

* If $\mathcal{C}, \mathcal{C}'$ are two categories and $S : \mathcal{C} \to \mathcal{C}'$, $T : \mathcal{C}' \to \mathcal{C}$ are two functors, then the mappings

$$(E'^\circ, F) \mapsto \mathrm{Hom}\,(T(E'), F),\ (u'^\circ, v) \mapsto \mathrm{Hom}(T(u'), v),$$

with $E' \in \mathrm{Ob}\, \mathcal{C}'$, $F \in \mathrm{Ob}\, \mathcal{C}$, $u' \in \mathrm{Hom}_{\mathcal{C}'}(E'_1, E')$, $v \in \mathrm{Hom}_{\mathcal{C}}(F, F_1)$, define a functor

$$\mathrm{Hom}(T(-), -) : \mathcal{C}'^\circ \times \mathcal{C} \to \mathbf{Ens};$$

also, with the same meaning of the symbols, the mappings

$$(E'^\circ, F) \mapsto \mathrm{Hom}_{\mathcal{C}'}(E', S(F)),\ (u'^\circ, v) \mapsto \mathrm{Hom}_{\mathcal{C}'}(u', S(v))$$

define a functor

$$\mathrm{Hom}(-, S(-)) : \mathcal{C}'^{\circ} \times \mathcal{C} \to \mathbf{Ens}.$$

If, moreover, T is a left adjoint of S, then the bijective mappings $\theta(E', F)$ given in Definition 1 define a natural equivalence from $\mathrm{Hom}(T(-), -)$ to $\mathrm{Hom}(-, S(-))$, as proved by the commutativity of diagram (5).

Let $\mathcal{C}$, $\mathcal{C}'$ be two categories and U, V two functors defined on $\mathcal{C}$ and taking values in $\mathcal{C}'$; let $\Phi : U \to V$ be a natural transformation; if $\mathcal{C}''$ is another category and $T : \mathcal{C}' \to \mathcal{C}''$ is a functor, then, taking for any object E of $\mathcal{C}$ the morphism $T(\Phi(E)) : (T \circ U)(E) \to (T \circ V)(E)$, the mapping

(16) $$E \mapsto T(\Phi(E)), \qquad E \in \mathrm{Ob}\, \mathcal{C},$$

defines a natural transformation $T \circ U \to T \circ V$: since if $u \in \mathrm{Hom}_{\mathcal{C}}(E, F)$, then, Φ being a natural transformation, the diagram (14) above in which T, U are respectively replaced by U, V is commutative; applying the functor T on this diagram, we get the commutative diagram

$$\begin{array}{ccc} (T \circ U)(E) & \xrightarrow{(T \circ U)(u)} & (T \circ U)(F) \\ {\scriptstyle T(\Phi(E))}\downarrow & & \downarrow{\scriptstyle T(\Phi(F))} \\ (T \circ V)(E) & \xrightarrow{(T \circ V)(u)} & (T \circ V)(F), \end{array}$$

and this proves our assertion; the natural transformation defined by (16) will be, by an abuse of notation, denoted $T(\Phi)$.

If the categories $\mathcal{C}$, $\mathcal{C}'$, the functors U, V, and the natural transformation Φ are given as above, then for any $u \in \mathrm{Hom}_{\mathcal{C}}(E, F)$, from diagram (14) with the same replacements as above we get the commutative diagram in $(\mathcal{C}')^{\circ}$

$$\begin{array}{ccc} V^{\circ}(F^{\circ}) & \xrightarrow{V^{\circ}(u^{\circ})} & V^{\circ}(E^{\circ}) \\ {\scriptstyle (\Phi(F))^{\circ}}\downarrow & & \downarrow{\scriptstyle (\Phi(E))^{\circ}} \\ U^{\circ}(F^{\circ}) & \xrightarrow{U^{\circ}(u^{\circ})} & U^{\circ}(E^{\circ}), \end{array}$$

where U°, respectively V°, are the functors dual to U, respectively V. This means that if we assign to each object E° of $\mathcal{C}^{\circ}$ the morphism $(\Phi(E))^{\circ} : V^{\circ}(E^{\circ}) \to U^{\circ}(E^{\circ})$ of the category $(\mathcal{C}')^{\circ}$, we obtain a natural transformation from V° to U°, denoted Φ° and called the *natural transformation dual to* Φ. Further, if T is a functor defined on $(\mathcal{C}')^{\circ}$ and taking values in a category $\mathcal{C}''$, it follows from the preceeding considerations that $T(\Phi^{\circ})$ is a natural transformation from $T \circ V^{\circ}$ to $T \circ U^{\circ}$.

Now let $\mathcal{C}, \mathcal{C}', \mathcal{C}''$ be three categories, let $T : \mathcal{C} \to \mathcal{C}'$ be a functor, let $U, V : \mathcal{C}' \to \mathcal{C}''$ be two functors, and let $\Phi : U \to V$ be a natural transformation; for any object E of $\mathcal{C}$ let us consider the morphism $\Phi(T(E)) \in \in \mathrm{Hom}_{\mathcal{C}''}((U \circ T)(E), (V \circ T)(E))$; the mapping

$$E \mapsto \Phi(T(E)), \qquad E \in \mathrm{Ob}\, \mathcal{C}, \tag{17}$$

defines a natural transformation $U \circ T \to V \circ T$; to prove this, let $u \in \mathrm{Hom}_{\mathcal{C}}(E, F)$; then as $T(u) \in \mathrm{Hom}_{\mathcal{C}'}(T(E), T(F))$, and as Φ is a natural transformation, we have the commutative diagram

$$\begin{array}{ccc} U(T(E)) & \xrightarrow{U(T(u))} & U(T(F)) \\ \Phi(T(E)) \downarrow & & \downarrow \Phi(T(F)) \\ V(T(E)) & \xrightarrow{V(T(u))} & V(T(F)), \end{array}$$

which proves our assertion; the natural transformation defined by (17) will be, by an abuse of notation, denoted $\Phi(T)$.

If $\mathcal{C}, \mathcal{C}', \mathcal{C}''$ and T are given as above, but U, V are functors defined on the category $(\mathcal{C}')^\circ$ and taking values in the category $\mathcal{C}''$, while $\Phi : U \to V$ is a natural transformation, then T°, i.e. the dual of the functor T, is a functor $\mathcal{C}^\circ \to (\mathcal{C}')^\circ$, and we get in particular the natural transformation $\Phi(T^\circ) : U \circ T^\circ \to V \circ T^\circ$ assigning to each object E° of $\mathcal{C}^\circ$ the morphism $\Phi(T^\circ(E^\circ)) \in \mathrm{Hom}_{\mathcal{C}''}(U(T^\circ(E^\circ)), V(T^\circ(E^\circ)))$.

Now let $\mathcal{C}, \mathcal{C}'$ be two categories, let $S, S_1 : \mathcal{C} \to \mathcal{C}'$ be two functors, and let $T, T_1 : \mathcal{C}' \to \mathcal{C}$ be two functors such that T, respectively T_1, is left adjoint to S, respectively to S_1; let $\Xi : S \to S_1$ be a natural transformation; then, denoting by θ, respectively θ_1, the natural equivalences $\mathrm{Hom}(T(-), -) \rightleftarrows \mathrm{Hom}(-, S(-))$, respectively $\mathrm{Hom}(T_1(-), -) \rightleftarrows \mathrm{Hom}(-, S_1(-))$, given by the adjointness situation, we define a natural transformation

$$\Omega : T_1 \to T$$

setting for any $E' \in \mathrm{Ob}\, \mathcal{C}'$,

$$\Omega(E') = \theta_1^{-1}(E', T(E'))\ [\Xi(T(E')) \circ \{\theta(E', T(E'))(1_{T(E')})\}].$$

For any $E' \in \mathrm{Ob}\, \mathcal{C}'$ and $F \in \mathrm{Ob}\, \mathcal{C}$, the following diagram turns to be commutative:

$$\begin{array}{ccc} \mathrm{Hom}_{\mathcal{C}}(T(E'), F) & \xrightarrow[\sim]{\theta(E', F)} & \mathrm{Hom}_{\mathcal{C}'}(E', S(F)) \\ \mathrm{Hom}\,(\Omega(E'), 1_F) \downarrow & & \downarrow \mathrm{Hom}\,(1_{E'}, \Xi(F)) \\ \mathrm{Hom}_{\mathcal{C}}(T_1(E'), F) & \xrightarrow[\sim]{\theta_1(E', F)} & \mathrm{Hom}_{\mathcal{C}'}(E', S_1(F)); \end{array} \tag{18}$$

Ω is the unique natural transformation $T_1 \to T$ making diagram (18) commute. The assignment $\Xi \mapsto \Omega$ is "functor-like": this means that if, in addition, the functor $S_2 : \mathcal{C} \to \mathcal{C}'$ is given, and if $T_2 : \mathcal{C}' \to \mathcal{C}$ is a left adjoint of S_2, while $\Xi_1 : S_1 \to S_2$ is a natural transformation, and $\Omega_1 : T_2 \to T_1$ corresponds to Ξ_1 in the above sense, then the natural transformation $T_2 \to T$ corresponding to $\Xi_1 \circ \Xi$ as above is $\Omega \circ \Omega_1$; moreover, the identity morphism of T corresponds to the identity morphism 1_S. From here it follows that *if Ξ is a natural equivalence, then the corresponding Ω is also a natural equivalence.* In particular, taking $S_1 = S$ and $\Xi = 1_S$, a natural equivalence $\Omega : T_1 \rightrightarrows T$ is assigned to 1_S; hence, *if S is a functor $\mathcal{C} \to \mathcal{C}'$ and T, T_1 are two left adjoints of S, then T is isomorphic to T_1: that is, a left adjoint of S is uniquely determined up to a natural equivalence.*

Again let $S : \mathcal{C} \to \mathcal{C}'$ be a functor, and let $T : \mathcal{C}' \to \mathcal{C}$ be a left adjoint of S, with the natural equivalence $\theta : \mathrm{Hom}(T(-), -) \rightrightarrows \mathrm{Hom}(-, S(-))$; defining for any object E' of $\mathcal{C}'$, a morphism

$$\Phi(E') : E' \to S(T(E'))$$

by

$$\Phi(E') = \theta(E', T(E'))(1_{T(E')}),$$

the morphisms $\Phi(E')$, with $E' \in \mathrm{Ob}\, \mathcal{C}'$, define a natural transformation

$$\Phi : 1_{\mathcal{C}'} \to S \circ T; \tag{19}$$

likewise, defining for any object F of $\mathcal{C}$, the morphism

$$\Psi(F) : T(S(F)) \to F$$

by

$$\Psi(F) = (\theta(S(F), F))^{-1}(1_{S(F)}),$$

the morphisms $\Psi(F)$, with $F \in \mathrm{Ob}\, \mathcal{C}$, define a natural transformation

$$\Psi : T \circ S = 1_{\mathcal{C}}. \tag{20}$$

It may be shown that the composite natural transformations

$$\begin{aligned} &T \xrightarrow{T(\Phi)} T \circ S \circ T \xrightarrow{\Psi(T)} T, \\ &S \xrightarrow{\Phi(S)} S \circ T \circ S \xrightarrow{S(\Psi)} S \end{aligned} \tag{21}$$

are the identity natural transformations.

Conversely, if $S : \mathcal{C} \to \mathcal{C}'$ and $T : \mathcal{C}' \to \mathcal{C}$ aie two functors, and if there exist natural transformations Φ and Ψ given as in (19) and (20) and such that the composite natural transformations (21) are the identity natural transformations, then, defining for any pair of objects $E' \in \mathrm{Ob}\,\mathcal{C}'$, $F \in \mathrm{Ob}\,\mathcal{C}$, the mapping of sets

$$\theta(E', F) : \mathrm{Hom}_{\mathcal{C}}(T(E'), F) \to \mathrm{Hom}_{\mathcal{C}'}(E', S(F))$$

by

$$\theta(E', F)(f) = S(f)\circ\Phi(E'),$$

where $f \in \mathrm{Hom}_{\mathcal{C}}(T(E'), F)$, the mappings $\theta(E', F)$ turn out to be bijective and, under the conditions of Definition 1, the corresponding diagram (5) is commutative; hence T is a left adjoint of S; we mention that, in the above notation,

$$\theta^{-1}(E', F)(g') = \Psi(F)\circ T(g'),$$

where $g' \in \mathrm{Hom}_{\mathcal{C}'}(E', S(F))$.

Combining the two aspects for two functors $S : \mathcal{C} \to \mathcal{C}'$ and $T : \mathcal{C}' \to \mathcal{C}$, the mapping

$$\theta \mapsto (\Phi, \Psi),$$

where we used the same notations as above, defines a bijection between the *class* of all natural equivalences

$$\mathrm{Hom}(T(-), -) \rightleftarrows \mathrm{Hom}(-, S(-))$$

and the class of those pairs of natural transformations

$$\Phi : 1_{\mathcal{C}'} \to S\circ T, \qquad \Psi : T\circ S \to 1_{\mathcal{C}}$$

for which the composite natural transformations (21) are the corresponding identity natural transformations.*

Consider again the forgetful functor $S_0 : \mathbf{Mod}(A) \to \mathbf{Ens}$ and the functor $T_0 : \mathbf{Ens} \to \mathbf{Mod}(A)$ of taking the generated free module; we saw that T_0 is a left adjoint of S_0; let us calculate for any set I and any A-module E, the mapping of sets $\Phi(I) : I \to A_s^{(I)}$ and the homomorphism of modules $\Psi(E) : A_s^{(E)} \to E$, where Φ and Ψ are the natural transformations associated by the above general theory with the natural equivalence θ defined for this particular case at the beginning of this Section: we have

$$\Phi(I) = \theta(I, T_0(I))(1_{T_0(I)}) = 1_{T_0(I)}\circ\varphi_I = \varphi_I,$$

where φ_I is the canonical mapping $I \to A_s^{(I)}$; similarly, if p_E is the canonical linear mapping $A_s^{(E)} \to E$ (see the Remark following Proposition 6), then if φ_E is the canonical mapping $E \to A_s^{(E)}$, we have

$$\theta(S_0(E), E)(p_E) = p_E\circ\varphi_E,$$

and for any $x \in E$ we have, in the preceding notation,

$$(p_E \circ \varphi_E)(x) = p_E(e_x) = x;$$

hence $p_E \circ \varphi_E = 1_{S_0(E)}$; whence,

$$\Psi(E) = (\theta(S_0(E), E))^{-1}(1_{S_0(E)}) = p_E.$$

Hence $\Phi(I)$ and $\Psi(E)$ are exactly the canonical mappings previously defined.
* Concerning the adjoint functors we also have the following result:

PROPOSITION 13. *Let $\mathcal{C}$, $\mathcal{C}'$ be two categories, and let $S : \mathcal{C} \to \mathcal{C}'$ be a functor; assume that to any object E' of $\mathcal{C}'$ an object denoted $T(E')$ of the category $\mathcal{C}$ is assigned, and that for any pair of objects, E' of $\mathcal{C}'$ and F of $\mathcal{C}$, a bijection of sets*

$$\theta(E', F) : \mathrm{Hom}_{\mathcal{C}}(T(E'), F) \xrightarrow{\sim} \mathrm{Hom}_{\mathcal{C}'}(E', S(F))$$

is given such that for any $E' \in \mathrm{Ob}\, \mathcal{C}'$, any $F, F_1 \in \mathrm{Ob}\, \mathcal{C}$, and any $v \in \mathrm{Hom}_{\mathcal{C}}(F, F_1)$, the following diagram commutes:

$$\begin{array}{ccc} \mathrm{Hom}_{\mathcal{C}}(T(E'), F) & \xrightarrow{\theta(E', F)} & \mathrm{Hom}_{\mathcal{C}'}(E', S(F)) \\ {\scriptstyle \mathrm{Hom}\,(1_{TE'}, v)} \downarrow & & \downarrow {\scriptstyle \mathrm{Hom}\,(1_{E'}, S(v))} \\ \mathrm{Hom}_{\mathcal{C}}(T(E'), F_1) & \xrightarrow{\theta(E', F_1)} & \mathrm{Hom}_{\mathcal{C}'}(E', S(F_1)); \end{array} \tag{22}$$

then there exists a functor $T : \mathcal{C}' \to \mathcal{C}$ which coincides on the objects of $\mathcal{C}'$ with the given mapping $E' \mapsto T(E')$ and which is a left adjoint of S, with the natural equivalence $\mathrm{Hom}(T(-), -) \xrightarrow{\sim} \mathrm{Hom}(-, S(-))$ given by the above bijections $\theta(E', F)$.

Proof. To define T on morphisms, let $u' \in \mathrm{Hom}_{\mathcal{C}'}(E'_1, E')$, with E', $E'_1 \in \mathrm{Ob}\, \mathcal{C}'$; set

$$T(u') = \theta(E'_1, T(E'))^{-1} [\theta(E', T(E'))\,(1_{TE'}) \circ u']. \tag{23}$$

Let us show that, in the preceding notation, for any object F of $\mathcal{C}$, the following diagram is commutative:

$$\begin{array}{ccc} \mathrm{Hom}_{\mathcal{C}}(T(E'), F) & \xrightarrow[\sim]{\theta(E', F)} & \mathrm{Hom}_{\mathcal{C}'}(E', S(F)) \\ {\scriptstyle \mathrm{Hom}\,(T(u'), 1_F)} \downarrow & & \downarrow {\scriptstyle \mathrm{Hom}\,(u', 1_{SF})} \\ \mathrm{Hom}_{\mathcal{C}}(T(E'_1), F) & \xrightarrow[\sim]{\theta(E'_1, F)} & \mathrm{Hom}_{\mathcal{C}'}(E'_1, S(F)); \end{array} \tag{24}$$

let then $f \in \mathrm{Hom}_{\mathscr{C}}(T(E'), F)$; in view of the commutativity of the diagram

$$
(25) \qquad \begin{array}{ccc}
\mathrm{Hom}_{\mathscr{C}}(T(E_1'), T(E')) & \xrightarrow[\sim]{\theta(E_1', T(E'))} & \mathrm{Hom}_{\mathscr{C}'}(E_1', S(T(E'))) \\
\Big\downarrow \mathrm{Hom}\,(1_{TE_1'}, f) & & \Big\downarrow \mathrm{Hom}\,(1_{E_1'}, S(f)) \\
\mathrm{Hom}_{\mathscr{C}}(T(E_1'), F) & \xrightarrow[\sim]{\theta(E_1', F)} & \mathrm{Hom}_{\mathscr{C}'}(E_1', S(F))
\end{array}
$$

(which is only a particular case of (22)), we have

$$(26) \qquad \theta(E_1', F)(f \circ T(u')) = S(f) \circ [\theta(E_1', TE')(T(u'))];$$

by (23) we have

$$(27) \qquad \theta(E_1', T(E'))(T(u')) = \theta(E', T(E'))(1_{TE'}) \circ u';$$

because of the commutativity of diagram (25) in which E_1' is replaced by E', we have

$$(28) \qquad S(f) \circ [\theta(E', T(E'))(1_{TE'})] = \theta(E', F)(f);$$

from (26), (27), and (28) it follows

$$\theta(E_1', F)(f \circ T(u')) = \theta(E', F)(f) \circ u';$$

this means that diagram (24) is commutative.

Further, let $E_2' \in \mathrm{Ob}\,\mathscr{C}'$, and let $v' \in \mathrm{Hom}_{\mathscr{C}'}(E_2', E_1')$; by definition we have

$$(29) \qquad T(u' \circ v') = \theta(E_2', T(E'))^{-1}\,[\theta(E', T(E'))(1_{TE'}) \circ u' \circ v'] =$$

$$= [\theta(E_2', T(E'))^{-1} \circ \mathrm{Hom}(v', 1_{STE'}) \circ \theta(E_1', T(E'))][\theta(E_1', T(E'))^{-1}\{\theta(E', T(E'))(1_{TE'}) \circ u'\}] =$$

$$= [(\theta(E_2', T(E')))^{-1} \circ \mathrm{Hom}(v', 1_{STE'}) \circ \theta(E_1', T(E'))](T(u')).$$

The last expression in (29) is $T(u') \circ T(v')$, as follows from the commutativity of the diagram obtained from (24) when replacing E' by E_1', E_1' by E_2', u' by v', and F by $T(E')$; hence we get

$$T(u' \circ v') = T(u') \circ T(v').$$

The relation

$$T(1_{E'}) = 1_{TE'}$$

is a straightforward consequence of the definition (23).

Thus, T is a functor $\mathcal{C}' \to \mathcal{C}$ which coincides on the objects with the given mapping $E' \mapsto T(E')$.

To complete the proof, we have to show that given E', $E'_1 \in \text{Ob}\,\mathcal{C}'$, F, $F_1 \in \text{Ob}\,\mathcal{C}$, $u' \in \text{Hom}_{\mathcal{C}'}(E'_1, E')$, $v \in \text{Hom}_{\mathcal{C}}(F, F_1)$, diagram (5) in which the mappings θ are those given is commutative: this follows from the commutativity of diagram (22) and of diagram (24) in which F is replaced by F_1. ∎

Remark. The universal properties we were hitherto concerned with (kernel, cokernel, direct product, free module $A_s^{(I)}$, etc.), as well as those we are going to give from now on (inverse limit, direct limit; see §§ 12, 13, etc.) express in fact the existence of some (left or right) adjoint functors of certain given functors; this explains why after constructing the respective objects we could, by standard procedures, give the corresponding constructions for morphisms (see the Corollary to Proposition 1 of this Section, Corollary 1 to Proposition 1, § 6, Corollary 1 to Proposition 1, § 7, as well as, further on, Corollary 1 to Proposition 1, § 12, Corollary 1 to Proposition 1, § 13, etc.): these last constructions are only particular cases of the construction given in Proposition 13. ∎

Proposition 14. *Let $\mathcal{C}, \mathcal{C}', \mathcal{C}''$ be three categories, let $S: \mathcal{C} \to \mathcal{C}'$, $S': \mathcal{C}' \to \mathcal{C}''$ be two functors; if $T: \mathcal{C}' \to \mathcal{C}$ and $T': \mathcal{C}'' \to \mathcal{C}'$ are left adjoints of S and S', respectively, then $T \circ T'$ is a left adjoint of $S' \circ S$.*

Proof. Let $\theta : \text{Hom}(T(-), -) \rightrightarrows \text{Hom}(-, S(-))$ and $\theta' : \text{Hom}(T'(-), -) \rightrightarrows \text{Hom}(-, S'(-))$ be natural equivalences given by the adjointness relation. Then, for any $E'' \in \text{Ob}\,\mathcal{C}''$, $F \in \text{Ob}\,\mathcal{C}$, let us denote by $\theta''(E'', F)$ the composition of the bijections

$$\text{Hom}_{\mathcal{C}}((T \circ T')(E''), F) \xrightarrow[\sim]{\theta(T'(E''), F)} \text{Hom}_{\mathcal{C}'}(T'(E''), S(F)) \xrightarrow[\sim]{\theta'(E'', S(F))}$$
$$\longrightarrow \text{Hom}_{\mathcal{C}''}(E'', (S' \circ S)(F));$$

the fact that the mappings

$$(E'', F) \mapsto \theta''(E'', F), \qquad E'' \in \text{Ob}\,\mathcal{C}'', \quad F \in \text{Ob}\,\mathcal{C},$$

define a natural transformation is an immediate consequence of the fact that θ and θ' are natural transformations. ∎*

§ 11. Modules of Finite Length

DEFINITION 1. Let A be a ring, and let E be an A-module; a *composition series* of the A-module E is a finite family $(E_i)_{0 \leqslant i \leqslant n}$ of submodules of E such that $E_0 = E$, $E_n = 0$, and such that for any $i = 0, \ldots, n-1, E_{i+1} \subseteq E_i$; the number $n+1$ will be called the *number of terms* of the given composition series; the quotient modules E_i/E_{i+1}, $i = 0, \ldots, n-1$, are called the *factors* of the composition series $(E_i)_i$; if $(F_j)_{0 \leqslant j \leqslant m}$ is another composition series of E, then $(F_j)_j$ is said to be a *refinement* of $(E_i)_i$ provided there exists an injection $\varphi : [0, n] \to [0, m]$ such that for any $i = 0, \ldots, n$, $F_{\varphi(i)} = E_i$.

Let E and F be A-modules, and let $(E_i)_{0 \leqslant i \leqslant m}$, respectively $(F_j)_{0 \leqslant j \leqslant n}$, be a composition series of E, respectively F; the two composition series are said to be *equivalent* if $m = n$ and if there exists a bijection $\psi : [0, m-1] \rightleftarrows [0, m-1]$ such that for any $i = 0, \ldots, m-1$,

$$F_{\psi(i)}/F_{\psi(i)+1} \approx E_i/E_{i+1}.$$

A *Jordan-Hölder series* of the A-module E is a composition series $(E_i)_{0 \leqslant i \leqslant n}$ of E which has the property that E_{i+1} is strictly contained in E_i for any $= 0, \ldots, n-1$, and which admits no other refinement with this property but itself. ▮

LEMMA 1 (**Zassenhaus Lemma for modules**). *Let E be an A-module, and let M, N, M', N' be submodules of it with the property that $M' \subseteq M$ and $N' \subseteq N$; then we have a canonical isomorphism of A-modules*

$$\frac{M' + (M \cap N)}{M' + (M \cap N')} \underset{\sim}{\rightleftarrows} \frac{N' + (N \cap M)}{N' + (N \cap M')}.$$

REMARK. In view of the monotonicity of the sum and intersection with respect to inclusion (see § 8), the stated quotient modules make sense. ▮

Proof of Lemma 1. Considering the submodules $M \cap N$ and M' of E, we have, by the Second Isomorphism Theorem for modules (Proposition 5, § 8), a canonical isomorphism

$$f : (M \cap N)/(M' \cap N) \rightleftarrows (M' + (M \cap N))/M'.$$

Since

$$M \cap N \supseteq (M' \cap N) + (M \cap N') \supseteq M' \cap N,$$

it follows that $((M'\cap N)+(M\cap N'))/(M'\cap N)$ is a submodule of $(M\cap N)/(M'\cap N)$, and taking into account the way f acts, we have

$$f[((M'\cap N)+(M\cap N'))/(M'\cap N)]=(M'+(M\cap N'))/M',$$

where $(M'+(M\cap N'))/M'$ is regarded as a submodule of $(M'+(M\cap N))/M'$.

Applying the Third Isomorphism Theorem for modules (the Corollary to Proposition 2, § 4), we get an isomorphism

$$\text{(1)}\qquad \frac{M\cap N}{(M'\cap N)+(M\cap N')}\stackrel{\sim}{\rightarrow}\frac{(M\cap N)/(M'\cap N)}{((M'\cap N)+(M\cap N'))/(M'\cap N)}.$$

By the same theorem applied to the homomorphism f, we get an isomorphism

$$\text{(2)}\qquad \frac{(M\cap N)/(M'\cap N)}{((M'\cap N)+(M\cap N'))/(M'\cap N)}\stackrel{\sim}{\rightarrow}\frac{f[(M\cap N)/(M'\cap N)]}{f[((M'\cap N)+(M\cap N'))/(M'\cap N)]}=$$

$$=\frac{(M'+(M\cap N))/M'}{(M'+(M\cap N'))/M'}.$$

Finally, applying again the Corollary to Proposition 2, § 4, we get an isomorphism

$$\text{(3)}\qquad \frac{(M'+(M\cap N))/M'}{(M'+(M\cap N'))/M'}\stackrel{\sim}{\rightarrow}\frac{M'+(M\cap N)}{M'+(M\cap N')}.$$

By the composition of the isomorphisms (3), (2), (1), we obtain an isomorphism

$$\text{(4)}\qquad \frac{M\cap N}{(M'\cap N)+(M\cap N')}\stackrel{\sim}{\rightarrow}\frac{M'+(M\cap N)}{M'+(M\cap N')}.$$

In the same way, interchanging the roles of M and N, we get the isomorphism

$$\text{(5)}\qquad \frac{N'+(N\cap M)}{N'+(N\cap M')}\stackrel{\sim}{\rightarrow}\frac{N\cap M}{(N'\cap M)+(N\cap M')}.$$

Taking the composite isomorphism of (4) and (5), and then the inverse of this composite isomorphism, we get the required isomorphism. ∎

PROPOSITION 1 (**Schreier's Refinement Theorem for modules**). *Given two composition series of an A-module E, for each of these series there exists a refinement such that the two refinements are equivalent.*

Proof. Let $(E_i)_{0\leqslant i\leqslant m}$, $(F_j)_{0\leqslant j\leqslant n}$ be the given composition series; set

$$E'_{in+j} = E_{i+1} + (E_i \cap F_j), \qquad F'_{jm+i} = F_{j+1} + (F_j \cap E_i) \tag{6}$$

for all pairs (i, j) with i and j independently running through $[0, m-1]$ and $[0, n-1]$, respectively; furthermore, set

$$E'_{mn} = F'_{mn} = 0.$$

We notice that for the indicated values of i and j, given $in + j$, the pair (i, j) is uniquely determined, since it consists of the quotient and respectively the reduced remainder (i.e. the one between 0 and $n - 1$) obtained upon dividing $in + j$ by n; moreover, for the same values of i and j we have $0 \leqslant in + j \leqslant mn - 1$; thus, the notations introduced for the modules E'_k, $k = 0,\ldots, mn$, are not ambiguous, that is, no risk that E'_k represent different modules for the same value of the index k may arise; the same is true for the modules F'_l, $l = 0,\ldots, nm$.

Let us show that the families of submodules $(E'_k)_{0\leqslant k\leqslant mn}$ and $(F'_l)_{0\leqslant l\leqslant mn}$ are composition series of E: we have, by (6),

$$E'_{in} = E_{i+1} + (E_i \cap F_0) = E_i$$

for any $i = 0,\ldots, m - 1$, since $F_0 = E$; in particular, $E'_0 = E_0 = E$; next, for any $i = 0,\ldots, m - 1$ and $j = 0,\ldots, n - 2$ we have

$$E'_{in+j} \supseteq E'_{in+j+1}$$

by virtue of the monotonicity of the sum and intersection with respect to inclusion; but for any $i = 0,\ldots, m - 2$ we have

$$E'_{(i+1)n} = E_{i+1} \subseteq E_{i+1} + (E_i \cap F_{n-1}) = E'_{in+(n-1)};$$

as

$$E'_{mn} = 0 \subseteq E'_{mn-1},$$

it follows that $(E'_k)_{0\leqslant k\leqslant mn}$ is indeed a composition series of E; similarly,

$$F'_{jm} = F_j$$

for any $j = 0,\ldots, n - 1$, and hence $(F'_l)_{0\leqslant l\leqslant nm}$ is a composition series of E.

Defining the mapping

$$\varphi_E : [0, m] \to [0, mn]$$

by

$$\varphi_E(i) = in, \quad i \in [0, m],$$

we obviously get an injective mapping[1]; for $i=0,\ldots,m-1$ we have $E'_{\varphi_E(i)}=E'_{in}= = E_i$; moreover, $E'_{\varphi_E(m)} = E'_{mn} = 0 = E_m$; hence $(E'_k)_k$ is a refinement of $(E_i)_i$; similarly, using the injection

$$\varphi_F : [0, n] \to [0, nm]$$

given by

$$\varphi_F(j) = jm, \quad j \in [0, n],$$

we obtain that $(F'_l)_{0\leqslant l\leqslant nm}$ is a refinement of $(F_j)_{0\leqslant j\leqslant n}$.

Now define a mapping

$$\psi : [0, mn-1] \to [0, mn-1]$$

by

$$\psi(in+j) = jm + i,$$

where $i = 0,\ldots, m-1$, $j = 0,\ldots, n-1$; the mapping ψ is injective, since, as previously noticed, $jm + i = j'm + i'$, with i, i', j, j' taking the above values, implies $j = j'$ and $i = i'$; it follows that ψ is bijective, its domain and range being the finite set $[0, mn-1]$. Now let $0 \leqslant i < m-1$, $0 \leqslant j < n-1$; by the Zassenhaus Lemma,

$$\frac{F'_{\psi(in+j)}}{F'_{\psi(in+j)+1}} = \frac{F'_{jm+i}}{F'_{jm+i+1}} = \frac{F_{j+1} + (F_j \cap E_i)}{F_{j+1} + (F_j \cap E_{i+1})} \approx \frac{E_{i+1} + (E_i \cap F_j)}{E_{i+1} + (E_i \cap F_{j+1})} = \frac{E'_{in+j}}{E'_{in+j+1}};$$

if $i = m-1$ and $0 \leqslant j < n-1$, then, by the Second Isomorphism Theorem for modules (Proposition 5, § 8) we have

$$\frac{F'_{\psi((m-1)n+j)}}{F'_{\psi((m-1)n+j)+1}} = \frac{F'_{jm+(m-1)}}{F'_{(j+1)m}} = \frac{F_{j+1} + (F_j \cap E_{m-1})}{F_{j+1}} \approx \frac{F_j \cap E_{m-1}}{F_{j+1} \cap E_{m-1}} =$$

$$= \frac{E_m + (E_{m-1} \cap F_j)}{E_m + (E_{m-1} \cap F_{j+1})} = \frac{E'_{(m-1)n+j}}{E'_{(m-1)n+j+1}},$$

since $E_m = 0$; in the same way, we get for $0 \leqslant i < m-1$, $j = n-1$,

$$\frac{F'_{\psi(in+(n-1))}}{F'_{\psi(in+(n-1))+1}} \approx \frac{E'_{in+(n-1)}}{E'_{(i+1)n}};$$

---

[1] Here $m \neq 0$ and $n \neq 0$, since, for instance, $m = 0$, $E_0 = E$, $E_m = 0$ would imply $E = 0$, case which is excluded because of its triviality.

finally, for $i = m - 1$, $j = n - 1$ we have

$$\frac{F'_{\psi((m-1)n+n-1)}}{F'_{\psi((m-1)n+n-1)+1}} = \frac{F'_{(n-1)m+(m-1)}}{F'_{nm}} \approx F_n + (F_{n-1} \cap E_{m-1}) = F_{n-1} \cap E_{m-1} =$$

$$= E_m + (E_{m-1} \cap F_{n-1}) \approx \frac{E'_{(m-1)n+(n-1)}}{E'_{mn}}.$$

It follows that $(E'_k)_k$ and $(F'_l)_l$ are equivalent composition series. ▮

DEFINITION 2. The A-module E is said to be *simple* or *irreducible* if $E \neq 0$ and E has no submodules different from 0 and E, that is, it has no proper submodules (see Example 1), § 3). ▮

If E is a simple A-module and a is a non-zero element of it, then the submodule Aa generated by a in E is not the zero submodule, hence $E = Aa$, and *E is a cyclic A-module*; any non-zero element of E is a generator of it.

If K is a skew-field, and if E is a one-dimensional K-vector space (see Definition 6, § 10), *then E is a simple K-module*: indeed, E has a basis consisting of one element, say x, hence $E \neq 0$; next, if F is a subspace of E, $F \neq 0$, and $y \in F$, $y \neq 0$, then $y = \alpha x$ for some $\alpha \in K$, $\alpha \neq 0$; hence $x = \alpha^{-1}y$; consequently, any element of E, being of the form $\beta x = \beta\alpha^{-1}y$, with $\beta \in K$, belongs to F; thus $F = E$.

PROPOSITION 2. *Let E and F be two A-modules, and let $u: E \to F$ be a linear mapping different from the zero mapping; if E, respectively F, is a simple A-module, then u is injective, respectively surjective; if E and F are both simple A-modules, then u is an isomorphism.*

Proof. If E is a simple A-module, then $\mathrm{Ker}(u)$ is 0 or E; the second possibility is excluded, since $u \neq 0$; hence $\mathrm{Ker}(u) = 0$, and u is injective by Proposition 2, § 3; similarly, if F is a simple A-module, as we have $\mathrm{Im}(u) \neq 0$, it follows that $\mathrm{Im}(u) = F$, and u is surjective; the last assertion of the Proposition is a consequence of the first two ones. ▮

Proposition 2 has the following

COROLLARY **(Schur's Lemma)**. *Let E be a simple A-module; then the ring*[1] $\mathrm{End}_A(E)$ *is a skew-field.*

Proof. As $E \neq 0$, we have $1_E \neq 0$, hence $\mathrm{End}_A(E)$ has at least two elements; if u is a non-zero A-endomorphism, as E is simple, by Proposition 2 u is an isomorphism, hence an invertible element of the ring $\mathrm{End}_A(E)$. ▮

PROPOSITION 3. *Let E be an A-module, and let $(E_i)_{0 \leqslant i \leqslant n}$ be a composition series of E with the property that E_{i+1} is strictly contained in E_i for $i = 0, \ldots, n-1$; in order that $(E_i)_i$ be a Jordan-Hölder series of E it is necessary and sufficient that all its factors be simple modules.*

---

[1] See § 2.

Proof. The condition is necessary: Let E_i/E_{i+1}, with $i \in [0, n-1]$, be a factor which is not a simple module of the Jordan-Hölder series $(E_i)_i$; there exists a submodule F of E_i/E_{i+1} distinct from E_i/E_{i+1} and from the zero submodule; by the Third Isomorphism Theorem for modules (Proposition 2, § 4) applied to the canonical surjection $E_i \to E_i/E_{i+1}$, the submodule F is of the form E'/E_{i+1}, where E' is a submodule of E such that $E_i \supseteq E' \supseteq E_{i+1}$, $E' \neq E_i$, $E' \neq E_{i+1}$; the family $(E'_j)_{0 \leqslant j \leqslant n+1}$ defined by

$$E'_j = \begin{cases} E_j & \text{if } j \leqslant i, \\ E' & \text{if } j = i+1, \\ E_{j-1} & \text{if } j > i+1 \end{cases}$$

obviously is a composition series of E and a refinement of $(E_i)_i$ distinct from $(E_i)_i$; however, all the inclusions in $(E'_j)_j$ are strict; this contradicts the fact that $(E_i)_i$ is a Jordan-Hölder series.

The condition is sufficient: If, under the given conditions, $(E_i)_i$ is not a Jordan-Hölder series, then there exists a refinement of it, say $(E'_j)_{0 \leqslant j \leqslant m}$, distinct from $(E_i)_i$ and such that E'_{j+1} is strictly contained in E'_j for $j = 0, \ldots, m-1$; but this means that $m > n$, and hence that there exists $j_0 \in [0, m]$ such that $E'_{j_0} \neq E_i$ for any $i = 0, \ldots, n$; we cannot have $j_0 = m$, since $E'_m = 0 = E_n$; consequently, there exists $i_0 \in [0, n-1]$ such that $E_{i_0} \supseteq E'_{j_0} \supseteq E_{i_0+1}$ and such that the inclusions are strict; by the Third Isomorphism Theorem for modules, the factor E_{i_0}/E_{i_0+1} of the given composition series contains the proper submodule E'_{j_0}/E_{i_0+1}, hence it is not simple. ▌

PROPOSITION 4 (**Jordan-Hölder Theorem for modules**). *Let E be an A-module; any two Jordan-Hölder series of E are equivalent (considered as composition series of E).*

Proof. Let $(E_i)_{0 \leqslant i \leqslant m}$, $(F_j)_{0 \leqslant j \leqslant n}$ be two Jordan-Hölder series of E; by Schreier's Refinement Theorem, there exist some refinements $(E'_k)_{0 \leqslant k \leqslant mn}$ and $(F'_l)_{0 \leqslant l \leqslant mn}$ of $(E_i)_i$, respectively of $(F_j)_j$, such that $(E'_k)_k$ and $(F'_l)_l$ are equivalent composition series. But, for instance, $(E_i)_i$ admits no refinement except itself having only strict inclusions; so that, if $(E'_k)_k$ is different from $(E_i)_i$, then at least one of the inclusions $E'_{k+1} \subseteq E'_k$, $k \in [0, mn-1]$, is an equality, i.e., $(E'_k)_k$ contains a so-called "repetition"; suppressing all these repetitions, we get a refinement of $(E_i)_i$, since $(E_i)_i$ has only strict inclusions; but this new refinement must be, by the above arguments, exactly $(E_i)_i$; by suppressing the repetitions in $(E'_k)_k$, we suppress some zero factors corresponding to some zero factors of the equivalent series $(F'_l)_l$; by suppressing in $(F'_l)_l$ the repetitions corresponding to these zero factors, we get, on the one hand, a composition series equivalent to $(E_i)_i$, hence with only non-zero factors, on the other hand, a refinement of $(F_j)_j$, since we suppressed only zero factors of $(F'_l)_l$; therefore, what we get is just $(F_j)_j$, which is equivalent to $(E_i)_i$. ▌

COROLLARY. *Let E be an A-module having a Jordan-Hölder series; then any composition series $(F_j)_{0 \leqslant j \leqslant n}$ of E, with F_{j+1} strictly contained in F_j for any $j = 0, \ldots, n-1$, has a refinement which is a Jordan-Hölder series.*

Proof. Let $(E_i)_{0 \leqslant i \leqslant m}$ be a Jordan-Hölder series of E; making use of the notations and arguments in the proof of Proposition 4, we get at the end instead of $(F_j)_j$ itself, a refinement of $(F_j)_j$, because $(F_j)_j$ has only strict inclusions; but this refinement of $(F_j)_j$ is equivalent to $(E_i)_i$; hence, it has all inclusions strict and all factors simple, since its factors are, in a certain order, isomorphic to those of $(E_i)_i$, which is a Jordan-Hölder series; by Proposition 3 we obtain that this refinement of $(F_j)_j$ is a Jordan-Hölder series. ∎

In view of Proposition 4, if the A-module E has a Jordan-Hölder series, then any two Jordan-Hölder series have the same number of factors.

DEFINITION 3. The A-module E is said to be a *module of finite length* if E has at least one Jordan-Hölder series; in this case the common number of factors of all Jordan-Hölder series is called the *length* of E, and is denoted $\text{long}_A(E)$ or $\text{long}(E)$. By convention, the zero A-module is of finite length, and we define $\text{long}_A(0) = 0$. ∎

If E is a simple A-module, then the family $(E_i)_{i=0,1}$, where $E_0 = E$, $E_1 = 0$, is, by Proposition 3, a Jordan-Hölder series of E, and hence, in this case, $\text{long}_A(E)=1$; conversely, if E is a module of finite length and $\text{long}_A(E) = 1$, then E is a simple module. If E is an A-module of finite length and $E \neq 0$, then $\text{long}_A(E) > 0$. Also, if E is a module of finite length and F is a module such that $F \approx E$, then, obviously, F is a module of finite length, and $\text{long}(F) = \text{long}(E)$.

PROPOSITION 5. *If K is a skew-field, and if E is a K-vector space, then E is finite-dimensional over K if and only if it is a K-module of finite length, and in this case*

$$\dim_K E = \text{long}_K(E). \tag{7}$$

Proof. If $E = 0$, then $\dim_K E = \text{long}_K(E) = 0$. Let $E \neq 0$ be a finite-dimensional K-vector space, and let $\dim_K E = n$; let $(a_1,\ldots, a_n)$ be a basis of E, let E_n be the zero subspace of E, and let E_i be the subspace generated in E by the family $(a_1,\ldots, a_{n-i})$, $i = 0,\ldots, n-1$; the family of submodules $(E_i)_{0 \leqslant i \leqslant n}$ is a composition series of E, and as a_{n-i} cannot be a linear combination of the elements of the family $(a_1,\ldots, a_{n-i-1})$, it follows that E_{i+1} is strictly contained in E_i, $i = 0, 1,\ldots, n-1$; moreover, in E_i/E_{i+1} the element $a_{n-i} + E_{i+1}$ is both a generator (see Corollary 1 to Proposition 2, § 8, applied to the canonical surjection $E_i \to E_i/E_{i+1}$) and a linear independent element, hence $\dim_K(E_i/E_{i+1}) = 1$; by the preceding arguments, E_i/E_{i+1} is a simple K-module; hence, by Proposition 3, $(E_i)_{0 \leqslant i \leqslant n}$ is a Jordan-Hölder series of E; relation (7) follows now from the very construction.

Conversely, if $E \neq 0$ is a K-module of finite length, let $(E_i)_{0 \leqslant i \leqslant n}$ be a Jordan-Hölder series of it; the factors E_i/E_{i+1}, $i = 0,\ldots, n-1$, being simple modules, they are non-zero cyclic K-modules; let $a_i + E_{i+1}$ be a generator of E_i/E_{i+1}; then $(a_0,\ldots, a_{n-1})$ is a set of generators of E; by Corollary 2 to Proposition 5, § 10, E is finite-dimensional over K. ∎

REMARK. Proposition 5 above supplies us with a new proof of the invariance of the number of elements of a basis in a vector space over a skew-field provided we know that all bases of the given vector space are finite. ∎

By virtue of Proposition 5, all subsequent results concerning modules of finite length and, respectively, the length of such a module are to be automatically applied to finite-dimensional vector spaces, respectively to the dimension of such a vector space.

We return to an arbitrary ring A (with unit element).

Not any A-module is of finite length, as shown by the following

EXAMPLE. 1) Let $A = \mathbf{Z}$ and $E = \mathbf{Z}$; if $\mathbf{Z}$ were of finite length and its length were n, then any composition series of it with strict inclusions would have at most $n + 1$ terms, as follows from the Corollary to the Jordan-Hölder Theorem; but the family $(E_i)_{0 \leqslant i \leqslant n+1}$, where $E_i = 2^i\mathbf{Z}$ is the submodule consisting of the multiples of 2^i for $i = 0, \ldots, n$, and $E_{n+1} = 0$, is a composition series with strict inclusions and with $n + 2$ terms. ▮

We also give the following

EXAMPLE. 2) Let $A = \mathbf{Z}$, and let $p \in \mathbf{Z}$, $p > 0$, be a prime number; the $\mathbf{Z}$-module $\mathbf{Z}/p\mathbf{Z}$ is simple, hence of finite length; indeed, if $\mathbf{Z}/p\mathbf{Z}$ had a proper submodule, then this would be of the form $F/p\mathbf{Z}$, where F is a submodule of $\mathbf{Z}$ strictly contained in $\mathbf{Z}$ and strictly containing $p\mathbf{Z}$; if $F = n\mathbf{Z}$ (see Example 1), § 9), we would have $n|p$; but $n = 1$ would imply $F = \mathbf{Z}$, while $n = p$ would imply $F = p\mathbf{Z}$; both situations are however excluded. ▮

PROPOSITION 6. *Let E be an A-module, and let M be one of its submodules; E is a module of finite length if and only if both M and E/M are modules of finite length; in this case we have*

$$\text{(8)} \qquad \operatorname{long}(E) = \operatorname{long}(M) + \operatorname{long}(E/M).$$

Proof. Assume that E is of finite length; we consider the case where $M \neq E$ and $M \neq 0$, otherwise the Proposition is trivial; in this case the family $(E_i)_{i=0,1,2}$, with $E_0 = E$, $E_1 = M$, $E_2 = 0$, is a composition series of E, and hence, by the Corollary to Proposition 4, there exists a Jordan-Hölder series $(F_j)_{0 \leqslant j \leqslant n}$ of E such that $F_k = M$ for some $k \in [0, n]$; but then $(F_j)_{k \leqslant j \leqslant n}$ is a Jordan-Hölder series of M, while $(F_j/M)_{0 \leqslant j \leqslant k}$ is a Jordan-Hölder series of E/M: indeed, by the Third Isomorphism Theorem for modules (the Corollary to Proposition 2, § 4), we have a canonical isomorphism

$$F_j/F_{j+1} \rightleftarrows \frac{F_j/M}{F_{j+1}/M}, \qquad j = 0, \ldots, k-1;$$

it turns out that M and E/M are modules of finite length and that we have (8).

Conversely, if M and E/M are modules of finite length, let $(F_j)_{0 \leqslant j \leqslant m}$, respectively $(G_k)_{0 \leqslant k \leqslant n}$ be a Jordan-Hölder series of M, respectively of E/M; let H_k, $k = 0, \ldots, n$, be the inverse image of G_k under the canonical surjection $E \to E/M$; by the Third Isomorphism Theorem for modules, we have $M \subseteq H_k$ and

$G_k = H_k/M$, $k = 0,\ldots, n$; the family $(L_l)_{0\leqslant l\leqslant m+n}$ such that

$$L_l = \begin{cases} H_l & \text{for } l = 0,\ldots, n, \\ F_{l-n} & \text{for } l = n,\ldots, m+n \end{cases}$$

is a Jordan-Hölder series of E. ▮

COROLLARY 1. *Given the exact sequence of A-modules and A-homomorphisms*

$$0 \to E \xrightarrow{u} F \xrightarrow{v} G \to 0,$$

F is a module of finite length if and only if E and G are modules of finite length; in this case we have

$$\operatorname{long}(F) = \operatorname{long}(E) + \operatorname{long}(G).$$

Proof. Indeed, in this case we have $E \approx u(E)$ and $G \approx F/u(E)$ (Corollaries to Proposition 1, § 4). ▮

COROLLARY 2. *Given a finite family* $(E_1,\ldots, E_n)$ *of A-modules,* $E_1 \oplus \ldots \oplus E_n$ *is a module of finite length if and only if each of* $E_1,\ldots, E_n$ *is a module of finite length, and in this case we have*

$$\operatorname{long}(E_1 \oplus \ldots \oplus E_n) = \sum_{i=1}^{n} \operatorname{long}(E_i).$$

Proof. For $n = 2$, denoting $E_1 = E$, $E_2 = F$, we have, as shown in § 7, the exact sequence

$$0 \to E \xrightarrow{j_E} E \oplus F \xrightarrow{p_F} F \to 0,$$

where j_E is the canonical injection and p_F the canonical projection; for $n > 2$ we proceed by induction on n, since, in view of the associativity of the direct sum, we have

$$E_1 \oplus \ldots \oplus E_n \approx (E_1 \oplus \ldots \oplus E_{n-1}) \oplus E_n. \text{ ▮}$$

COROLLARY 3. *Let E be an A-module of finite length, and let M be a submodule of it; then M is a module of finite length, and we have*

(9) $$\operatorname{long}(M) \leqslant \operatorname{long}(E),$$

the equality holding if and only if $M = E$; *under the same hypotheses, the quotient module* E/M *is of finite length, and we have*

(10) $$\operatorname{long}(E/M) \leqslant \operatorname{long}(E),$$

the equality holding if and only if $M = 0$.

Proof. By Proposition 6, M and E/M are of finite length; since the length of a module is a non-negative integer, inequalities (9) and (10) follow from (8); finally, if we have equality in (9), respectively (10), then from (8) we obtain $\operatorname{long}(E/M) = 0$, respectively $\operatorname{long}(M) = 0$, hence $E/M = 0$, respectively $M = 0$. ∎

COROLLARY 4. *Let E, F be two A-modules, and let $u : E \to F$ be a linear mapping; if E is a module of finite length, then* Ker(u) *and* Im(u) *are modules of finite length, and we have*

$$\operatorname{long}(\operatorname{Ker} u) + \operatorname{long}(\operatorname{Im} u) = \operatorname{long}(E);$$

if F is a module of finite length, then Im(u) *and* Coker(u) *are modules of finite length, and we have*

$$\operatorname{long}(\operatorname{Im} u) + \operatorname{long}(\operatorname{Coker} u) = \operatorname{long}(F).$$

Proof. We have, indeed, the exact sequences

$$0 \to \operatorname{Ker}(u) \to E \to \operatorname{Im}(u) \to 0$$

and

$$0 \to \operatorname{Im}(u) \to F \to \operatorname{Coker}(u) \to 0. ∎$$

COROLLARY 5. *Let E be an A-module, and let M, N be two of its submodules of finite length; then $M + N$ and $M \cap N$ are modules of finite length, and we have*

$$\operatorname{long}(M + N) + \operatorname{long}(M \cap N) = \operatorname{long}(M) + \operatorname{long}(N).$$

If $(M_1, \ldots, M_n)$ is a finite family of submodules of finite length of E, then $M_1 + \ldots + M_n$ is a module of finite length, and we have

$$\operatorname{long}(M_1 + \ldots + M_n) \leqslant \sum_{i=1}^{n} \operatorname{long}(M_i), \tag{11}$$

the equality holding if and only if $\sum_{i=1}^{n} M_i = \dot{\sum_{i=1}^{n}} M_i$.

Proof. For the first assertion, we consider the exact sequence

$$0 \to M \cap N \to M \oplus N \to M + N \to 0$$

given by Proposition 3, § 8, and we apply the above Corollaries 1 and 2; for the second assertion, we either argue by induction or we consider the exact sequence

$$0 \to \operatorname{Ker}(f') \to \bigoplus_{i=1}^{n} M_i \xrightarrow{f'} \sum_{i=1}^{n} M_i \to 0,$$

where f' is obtained from f given in Corollary 5 to Proposition 2, § 8, by restricting the codomain, taking into account that $\operatorname{Im}(f)=\sum_{i=1}^{n} M_i$; finally, the equality in (11) is, by Corollaries 1 and 2, equivalent to the injectivity of f', hence to $\sum_{i=1}^{n} M_i = \dot{\sum_{i=1}^{n}} M_i$. ▮

COROLLARY 6. *Let E be a module, and let M, N be two submodules of E such that E/M and E/N be modules of finite length; then $E/(M \cap N)$ and $E/(M+N)$ are modules of finite length, and we have*

$$\operatorname{long}(E/(M \cap N)) + \operatorname{long}(E/(M+N)) = \operatorname{long}(E/M) + \operatorname{long}(E/N).$$

Proof. By Proposition 4, § 8, we have an exact sequence of A-modules and A-homomorphisms

$$0 \to E/(M \cap N) \to (E/M) \oplus (E/N) \to E/(M+N) \to 0,$$

and we need only apply Corollaries 1 and 2. ▮

COROLLARY 7. *Let E be an A-module, and let $(E_i)_{0 \leqslant i \leqslant n}$ be a composition series of E; if each of the factors E_i/E_{i+1}, $i = 0, \ldots, n-1$, is a module of finite length, then E is a module of finite length, and we have*

$$\operatorname{long}(E) = \sum_{i=0}^{n-1} \operatorname{long}(E_i/E_{i+1}).$$

Proof. If $n = 1$, then everything is clear; if $n = 2$, then, by Proposition 6, we have

$$\operatorname{long}(E) = \operatorname{long}(E/E_1) + \operatorname{long}(E_1/0),$$

and the Corollary is proved; for $n > 2$, we apply induction on n: assume the assertion true for a composition series with n terms; but then E_1 is a module of finite length, and

$$\operatorname{long}(E_1) = \sum_{i=1}^{n-1} \operatorname{long}(E_i/E_{i+1});$$

now we again apply Proposition 6. ▮

COROLLARY 8. *Let E be an A-module, and let $(M_1, \ldots, M_n)$ be a finite family of submodules of E such that E/M_i is a module of finite length for $i = 1, \ldots, n$; then, for any $i = 2, \ldots, n$, the module $(M_1 \cap \ldots \cap M_{i-1})/(M_1 \cap \ldots \cap M_{i-1} \cap M_i)$ is of finite length, and we have*

$$\operatorname{long}\left(\frac{M_1 \cap \ldots \cap M_{i-1}}{M_1 \cap \ldots \cap M_i}\right) \leqslant \operatorname{long}(E/M_i);$$

moreover, $E\Big/\Big(\bigcap_{i=1}^{n} M_i\Big)$ *is a module of finite length, and we have*

$$\operatorname{long}\left(\frac{E}{\bigcap_{i=1}^{n} M_i}\right) \leqslant \sum_{i=1}^{n} \operatorname{long}(E/M_i).$$

Proof. For the first assertion, we notice that, by The Second Isomorphism Theorem for modules (Proposition 5, § 8), we have

$$\frac{M_1 \cap \ldots \cap M_{i-1}}{M_1 \cap \ldots \cap M_{i-1} \cap M_i} \approx \frac{(M_1 \cap \ldots \cap M_{i-1}) + M_i}{M_i},$$

and the second quotient module may be considered as a submodule of E/M_i(see § 3); for the second assertion, we notice that $(F_i)_{0 \leqslant i \leqslant n}$, where

$$F_i = \frac{M_1 \cap \ldots \cap M_i}{M_1 \cap \ldots \cap M_n},$$

$i = 0, \ldots, n$, hence where $F_0 = E/(M_1 \cap \ldots \cap M_n)$, is a composition series of $E/(M_1 \cap \ldots \cap M_n)$; since by the Third Isomorphism Theorem for modules we have

$$F_{i-1}/F_i \approx (M_1 \cap \ldots \cap M_{i-1})/(M_1 \cap \ldots \cap M_i),$$

$i = 1, \ldots, n$, it follows by the first assertion and by Corollary 7 that

$$\operatorname{long}\left(\frac{E}{M_1 \cap \ldots \cap M_n}\right) = \sum_{i=1}^{n} \operatorname{long}(F_{i-1}/F_i) \leqslant \sum_{i=1}^{n} \operatorname{long}(E/M_i).$$

For the second assertion we may also argue as follows: the exact sequence

$$0 \to \bigcap_{i=1}^{n} M_i \to E \to \prod_{i=1}^{n} (E/M_i) = \bigoplus_{i=1}^{n} (E/M_i)$$

given by Proposition 1, § 8, for this case is considered; then Corollary 1 to Proposition 1, § 4, is applied, and we obtain an injective homomorphism

$$E\Big/\Big(\bigcap_{i=1}^{n} M_i\Big) \to \bigoplus_{i=1}^{n} (E/M_i). \blacksquare$$

COROLLARY 9. *Given the exact sequence of A-module and A-homomorphisms*

$$0 \to E_0 \xrightarrow{u_0} E_1 \xrightarrow{u_1} E_2 \xrightarrow{u_2} \ldots \to E_{n-1} \xrightarrow{u_{n-1}} E_n \to 0,$$

where $n > 0$ *and* E_i, $i = 0, \ldots, n$, *are modules of finite length, we have*

$$\sum_{i=0}^{n} (-1)^i \operatorname{long}(E_i) = 0.$$

Proof. We argue by induction: if $n = 1$, then $E_0 \approx E_1$, and

$$\operatorname{long}(E_0) - \operatorname{long}(E_1) = 0;$$

if $n = 2$, we have by Corollary 1,

$$\operatorname{long}(E_0) - \operatorname{long}(E_1) + \operatorname{long}(E_2) = 0;$$

suppose $n > 2$ and the assertion true for sequences with at most $n - 1$ homomorphisms; the diagram

$$0 \to E_0 \xrightarrow{u_0} E_1 \xrightarrow{u_1} E_2 \xrightarrow{u_2} \cdots \xrightarrow{u_{n-3}} E_{n-2} \xrightarrow{p} \operatorname{Im}(u_{n-2}) \to 0,$$

where p is the canonical surjection, is an exact sequence, hence, by the induction hypothesis,

$$\sum_{i=0}^{n-2} (-1)^i \operatorname{long}(E_i) + (-1)^{n-1} \operatorname{long}(\operatorname{Im}(u_{n-2})) = 0;$$

on the other hand, we have the exact sequence

$$0 \to \operatorname{Im}(u_{n-2}) \xrightarrow{j} E_{n-1} \xrightarrow{u_{n-1}} E_n \to 0,$$

where j is the canonical injection; hence, by Corollary 1 we have

$$\operatorname{long}(\operatorname{Im}(u_{n-2})) = \operatorname{long}(E_{n-1}) - \operatorname{long}(E_n),$$

and the Corollary is proved. ▌

* Definition 4. Let $\mathcal{C}$ be an abelian category, and let $\mathcal{S}$ be a subclass of the class $\operatorname{Ob} \mathcal{C}$; $\mathcal{S}$ is said to be a *Serre class* of the category $\mathcal{C}$ if the zero object 0 of the category $\mathcal{C}$ belongs to $\mathcal{S}$ and if for any exact sequence of $\mathcal{C}$ of the form

(12) $$0 \to E \to F \to G \to 0,$$

the object F belongs to $\mathcal{S}$ if and only if E and G belong to $\mathcal{S}$.

Let $\mathscr{C}$ be an abelian category, and let $\mathscr{S}$ be a Serre class of $\mathscr{C}$; let Γ be an abelian group additively written, and let the mapping

$$\varepsilon : \mathscr{S} \to \Gamma$$

be such that

1) $\varepsilon(0) = 0$,

2) if (12) is an exact sequence of the category $\mathscr{C}$, with $E, F, G \in \mathscr{S}$, then

$$\varepsilon(F) = \varepsilon(E) + \varepsilon(G);$$

in this case ε is called an *Euler-Poincaré mapping* defined on $\mathscr{S}$ and taking values in Γ. ▮

From the definition it follows that if $\mathscr{S}$ is a Serre class of the abelian category $\mathscr{C}$ and $E \in \mathscr{S}$, then any object F of $\mathscr{C}$ isomorphic to E belongs to $\mathscr{S}$: since if $u : E \to F$ is an isomorphism, we have in $\mathscr{C}$ the exact sequence (see § 5)

$$0 \to E \xrightarrow{u} F \to 0 \to 0,$$

and 0 and E belong to $\mathscr{S}$. Moreover, if ε is an Euler-Poincaré mapping defined on $\mathscr{S}$, then by conditions 1) and 2) of Definition 4 we also have

$$\varepsilon(F) = \varepsilon(E).$$

Next, if $\mathscr{S}$ is a Serre class of the abelian category $\mathscr{C}$ and $E \in \mathscr{S}$, and if (M, i), (P, p) are, respectively, a subobject and a quotient object of E, then M and P belong to $\mathscr{S}$: since arguing, for instance, for M, we have only to consider the exact sequence

$$0 \to M \xrightarrow{i} E \to \mathrm{Coker}(i) \to 0.$$

It is possible to show also that given in $\mathscr{C}$ the exact sequence

$$E \to F \to G,$$

with E, G belonging to the Serre class $\mathscr{S}$, F belongs to $\mathscr{S}$ too.*

By Definition 3 and Corollary 1 to Proposition 6 we obtain that in **Mod**(A) the class of all modules of finite length is a Serre class, and the mapping

$$E \mapsto \mathrm{long}_A(E)$$

is an Euler-Poincaré mapping defined on this class and taking values in the abelian group **Z** of integers.

* If $\mathcal{C}$ is an abelian category, if $\mathcal{S}$ is a Serre class of it, and if ε is an Euler-Poincaré mapping defined on $\mathcal{S}$ and taking values in the abelian group Γ, then Corollaries 2, 4, the first assertion in Corollary 5, and Corollary 6 remain valid when replacing "module of finite length" by "object of $\mathcal{S}$" and the mapping "long" by the mapping ε (notice that the contents of Corollary 1 coincides with the definition): for instance, Corollary 2 follows by induction, the step for $n = 2$ being based on Proposition 7, § 9; as regards Corollary 7, this holds under the same conditions provided a definition replacing the notion of composition series in the category **Mod**(A) by a suitable notion in the category $\mathcal{C}$ is given; this may be done by considering the partial order relation among subobjects (see § 8) instead of the inclusion relation among submodules; the other Corollaries to Proposition 6 cannot be proved in general; however, under additional conditions, this could be done: for instance, taking $\Gamma = \mathbf{Z}$, the additive group of integers, and assuming that for any $E \in \mathcal{S}$ we have $\varepsilon(E) \geqslant 0$.

Let now $\mathcal{C}$ be an abelian category, and let $\mathcal{S}$ be a Serre class of $\mathcal{C}$; denote by $\hat{\mathcal{S}}$ the quotient class of $\mathcal{S}$ with respect to the equivalence relation given by the existence of an isomorphism of $\mathcal{C}$ between two objects of $\mathcal{S}$; *let us assume that $\hat{\mathcal{S}}$ is a set.*

Let $\mathbf{Z}^{(\hat{\mathcal{S}})}$ be the module of formal linear combinations of the elements of $\hat{\mathcal{S}}$ with coefficients in the ring $\mathbf{Z}$; let $\varphi : \hat{\mathcal{S}} \to \mathbf{Z}^{(\hat{\mathcal{S}})}$ be the canonical mapping; if $E \in S$, denote by $\hat{E}$ the equivalence class in $\hat{\mathcal{S}}$ containing E; let B be the $\mathbf{Z}$-submodule of $\mathbf{Z}^{(\hat{\mathcal{S}})}$ generated by all elements

$$\varphi(\hat{F}) - \varphi(\hat{E}) - \varphi(\hat{G})$$

for which there exists in $\mathcal{C}$ an exact sequence of the form (12) with $E, F, G \in \mathcal{S}$.

DEFINITION 5. Under the above conditions, the abelian group $\mathbf{Z}^{(\hat{\mathcal{S}})}/B$ is called the *Grothendieck group* of the class $\mathcal{S}$, and is denoted $K(\mathcal{S})$. ▮

We have:

PROPOSITION 7. *Under the same conditions as above and in the same notation, the mapping*

$$\gamma : \mathcal{S} \to K(\mathcal{S})$$

obtained as the composite mapping

$$\mathcal{S} \xrightarrow{\pi} \hat{\mathcal{S}} \xrightarrow{\varphi} \mathbf{Z}^{(\hat{\mathcal{S}})} \xrightarrow{p} \mathbf{Z}^{(\hat{\mathcal{S}})}/B = K(\mathcal{S}),$$

where π and p are the canonical surjections, is an Euler-Poincaré mapping defined on $\mathcal{S}$ and taking values in $K(\mathcal{S})$; furthermore, $K(\mathcal{S})$ and γ are universal with this property, i.e., for any Euler-Poincaré mapping ε defined on $\mathcal{S}$ and taking values in some abelian group Γ there exists a homomorphism of abelian groups

$$f : K(\mathcal{S}) \to \Gamma$$

uniquely determined by the property of rendering commutative the diagram (of classes and mappings of classes) [1]

$$\begin{array}{ccc} \mathcal{S} & \xrightarrow{\gamma} & K(\mathcal{S}) . \\ {\scriptstyle \varepsilon}\big\downarrow & \swarrow {\scriptstyle f} & \\ \Gamma & & \end{array}$$

Proof. If 0 is the zero object of $\mathcal{C}$, then we have the exact sequence

$$0 \to 0 \to 0 \to 0 \to 0,$$

hence $\varphi(\hat{0}) - \varphi(\hat{0}) - \varphi(\hat{0}) = - \varphi(\hat{0}) \in B$, and, consequently,

$$\gamma(0) = p(\varphi(\hat{0})) = 0 \in K(\mathcal{S});$$

further, if (12) is an exact sequence of the category $\mathcal{C}$ with $E, F, G \in \mathcal{S}$, then, by definition, we have $\varphi(\hat{F}) - \varphi(\hat{E}) - \varphi(\hat{G}) \in B$, hence

$$\gamma(F) - \gamma(E) - \gamma(G) = p(\varphi(\hat{F})) - p(\varphi(\hat{E})) - p(\varphi(\hat{G})) = 0.$$

If ε is an Euler-Poincaré mapping defined on $\mathcal{S}$ and taking values in Γ, then, from $E \approx F$ in $\mathcal{C}$, with $E, F \in \mathcal{S}$, it follows that $\varepsilon(E) = \varepsilon(F)$; hence the mapping ε is compatible with the equivalence relation defined on $\mathcal{S}$ by putting in the same equivalence class all objects mutually isomorphic; consequently, ε induces a unique mapping of sets

$$\hat{\varepsilon} : \hat{\mathcal{S}} \to \Gamma$$

with the property that $\hat{\varepsilon} \circ \pi = \varepsilon$; by Proposition 1, § 10, there exists a unique **Z**-linear mapping

$$\theta : \mathbf{Z}^{(\hat{\mathcal{S}})} \to \Gamma$$

with the property that $\theta \circ \varphi = \hat{\varepsilon}$; if (12) is an exact sequence of the category $\mathcal{C}$ with $E, F, G \in \mathcal{S}$, then

$$\theta[\varphi(\hat{F}) - \varphi(\hat{E}) - \varphi(\hat{G})] = \hat{\varepsilon}(\hat{F}) - \hat{\varepsilon}(\hat{E}) - \hat{\varepsilon}(\hat{G}) = \varepsilon(F) - \varepsilon(E) - \varepsilon(G) = 0;$$

---

[1] The notion of diagram is here extended to classes.

by Corollary 3 to Proposition 2, § 8, θ vanishes on all B; consequently, by the universal property of the cokernel (Proposition 1, § 3), there exists a unique **Z**-linear mapping

$$f : K(\mathcal{S}) \to \Gamma$$

with the property that $f \circ p = \theta$; f is the required homomorphism of abelian groups. ▮*

By the Jordan-Hölder theorem for modules, if A is a ring, and if ε' is a mapping defined on the class of all simple A-modules taking values in an abelian group Γ such that if F, F' are simple modules and $F \approx F'$ we have $\varepsilon'(F) = \varepsilon'(F')$, then we may define an Euler-Poincaré mapping ε on the Serre class of all A-modules of finite length and taking values in Γ and such that ε extends ε': let E be an A-module of finite length, and let $(E_i)_{0 \leqslant i \leqslant n}$ be a Jordan-Hölder series of the module E; then, by definition,

$$\varepsilon(E) = \sum_{i=0}^{n-1} \varepsilon'(E_i/E_{i+1});$$

the Jordan-Hölder theorem ensures the independence of $\varepsilon(E)$, under the stated condition, of the chosen Jordan-Hölder series.

Let L be a skew-field; considering the Serre class $\mathcal{S}$ of all finite-dimensional L-vector spaces and using the above notation, the class $\hat{\mathcal{S}}$ is a set, since any n-dimensional vector space over L, $n \geqslant 0$, is isomorphic to L_s^n; moreover, in this case, *the Grothendieck group* $K(\mathcal{S})$ *is isomorphic to* **Z**: indeed, by Proposition 7, it suffices to show that **Z** together with a suitable $\gamma : \mathcal{S} \to \mathbf{Z}$ have the universal property stated by that Proposition; taking for γ the mapping given by $\gamma(E) = \dim_L E$, with $E \in \mathcal{S}$, we define for any abelian group Γ and any Euler-Poincaré mapping $\varepsilon : \mathcal{S} \to \Gamma$,

$$f(n) = n \cdot \varepsilon(L_s), \qquad n \in \mathbf{Z},$$

and the desired result follows.

§ 12. Projective Limits

Let I be a small category; let A be a ring, and let $T : I^\circ \to \mathbf{Mod}(A)$ be a functor from the dual of I to the category of A-modules; taking into account the definition of the dual of a category, this amounts to assigning an A-module E_λ to each object λ of the category I, and an A-homomorphism

$$u_f : E_\mu \to E_\lambda$$

to each morphism $f \in \mathrm{Hom}_I(\lambda, \mu)$ such that the following conditions be satisfied:

a) if the diagram in I

$$\lambda \xrightarrow{f} \mu \xrightarrow{g} \nu$$

is given, then we have the equality of homomorphisms

$$u_{g \circ f} = u_f \circ u_g;$$

b) for any $\lambda \in \mathrm{Ob}\, I$,

$$u_{1_\lambda} = 1_{E_\lambda}.$$

In the particular case when I is the category associated with a preordered set denoted also I (see Example 13), § 1), to give the functor $T : I^\circ \to \mathbf{Mod}(A)$ is to give for any element $\lambda \in I$ an A-module E_λ, and for any pair λ, μ for which $\lambda \leqslant \mu$ in I an A-homomorphism

$$u_{\lambda\mu} : E_\mu \to E_\lambda$$

such that if $\lambda \leqslant \mu \leqslant \nu$ in I, the diagram

$$\begin{array}{ccc} E_\nu & \xrightarrow{u_{\mu\nu}} & E_\mu \\ & \searrow^{u_{\lambda\nu}} & \downarrow u_{\lambda\mu} \\ & & E_\lambda \end{array}$$

commutes, i.e., $u_{\lambda\mu} \circ u_{\mu\nu} = u_{\lambda\nu}$, and, moreover, $u_{\lambda\lambda} = 1_{E_\lambda}$ for any $\lambda \in I$.

DEFINITION 1. A functor T as described above is called a *projective system* (or *an inverse system*) *of A-modules* indexed by the category I (of indices); the above projective system T will sometimes also be denoted $(E_\lambda, u_f)_{\lambda,f}$ or $(E_\lambda, u_f)_I$; in the particular case when I is the category associated with the preordered set I, the above projective system of A-modules indexed by I is also denoted $(E_\lambda, u_{\lambda\mu})_{\lambda\in I}$ or $(E_\lambda, u_{\lambda\mu})_I$ or, simply, $(E_\lambda, u_{\lambda\mu})$. ▮

REMARK. Of course, in the same way the notion of *projective system of right A-modules* may be defined, considering a functor defined on the dual of a small category of indices and taking values in the category $\mathbf{Mod}_d(A)$. ▮

Consider again the above projective system of A-modules; in the direct product[1] $\prod_{\lambda\in I} E_\lambda$ let us consider the subset E consisting of all families $(x_\lambda)_{\lambda\in I}$ which satisfy the condition: for any λ, $\mu\in I$ and any $f\in \mathrm{Hom}_I(\lambda, \mu)$,

$$u_f(x_\mu) = x_\lambda;$$

obviously, $E \neq \emptyset$, since, the mappings u_f being homomorphisms, E contains the zero element of $\prod_{\lambda\in I} E_\lambda$; moreover, *$E$ is a submodule of* $\prod_{\lambda\in I} E_\lambda$: let us verify, for instance, that if $x = (x_\lambda)_{\lambda\in I}$ and $y = (y_\lambda)_{\lambda\in I}$ are in E, then $x + y$ performed in $\prod_{\lambda\in I} E_\lambda$ is an element of E; indeed, if λ, μ and f are as above, then

$$u_f(x_\mu + y_\mu) = u_f(x_\mu) + u_f(y_\mu) = x_\lambda + y_\lambda.$$

E may be also characterized as the intersection over all $f\in \prod_{(\lambda,\mu)\in I\times I} \mathrm{Hom}_I(\lambda, \mu)$ of the kernels of the linear mappings $\prod_{\lambda\in I} E_\lambda \to E_\lambda$ given by $(x_\lambda)_\lambda \mapsto u_f(x_\mu) - x_\lambda$ (here $f\in \mathrm{Hom}_I(\lambda, \mu)$).

DEFINITION 2. Given a projective system T as above, the constructed module E is called the *projective limit* or the *inverse limit* or, simply, the *limit* or the *left root* of the given projective system, and is denoted $\varprojlim_I T$ or also $\varprojlim_I (E_\lambda, u_f)$ or even $\varprojlim_{\lambda\in I} E_\lambda$ or $\varprojlim_I E_\lambda$ or $\varprojlim E_\lambda$; if I is the category associated with the preordered set I, then instead of the general notation introduced one also writes $\varprojlim_I (E_\lambda, u_{\lambda\mu})$; the linear mapping $E \to E_\lambda$, $\lambda\in I$, obtained as the composition of the canonical projection $\prod_{\lambda\in I} E_\lambda \to E_\lambda$ with the canonical injection $E \to \prod_{\lambda\in I} E_\lambda$ is called the *canonical mapping* of index λ, $\lambda\in I$. ▮

The projective limit has the following universal property:

---

[1] Here and further on, in order to simplify the writing, we shall, by an abuse of notation, replace "$\lambda\in \mathrm{Ob}\, I$" by "$\lambda\in I$".

PROPOSITION 1. *Let* $E = \varprojlim_I T$ *be the projective limit of the projective system* $T = (E_\lambda, u_f)_{\lambda, f}$; *denote by* p_λ *the canonical mapping of index* λ, $\lambda \in I$; *then for any* $\lambda, \mu \in I$ *and any* $f \in \mathrm{Hom}_I(\lambda, \mu)$ *we have*

$$(1) \qquad u_f \circ p_\mu = p_\lambda;$$

the system $(E, (p_\lambda)_{\lambda \in I})$ *is universal with this property, that is, for any* A*-module* X *and any family of linear mappings* $(r_\lambda)_{\lambda \in I}$, *where* $r_\lambda : X \to E_\lambda$, $\lambda \in I$, *are such that*

$$u_f \circ r_\mu = r_\lambda$$

for any $\lambda, \mu \in I$ *and* $f \in \mathrm{Hom}_I(\lambda, \mu)$, *there exists a homomorphism* $r : X \to E$ *uniquely determined by the property*

$$(2) \qquad p_\lambda \circ r = r_\lambda$$

for any $\lambda \in I$.

This is displayed on the following diagram:

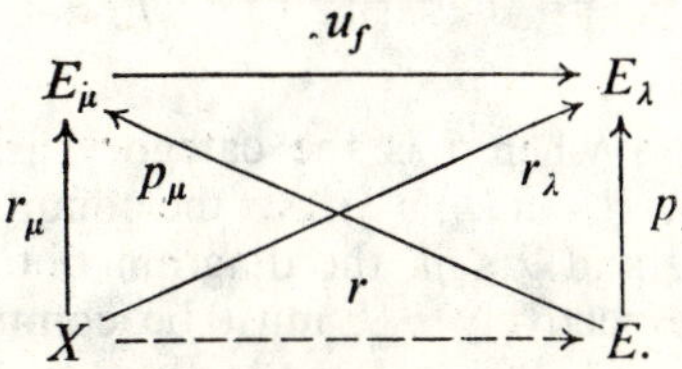

Proof. To prove relation (1), let $(x_\lambda)_{\lambda \in I} \in E$; we have by the definition of E,

$$(u_f \circ p_\mu)((x_\lambda)_{\lambda \in I}) = u_f(x_\mu) = x_\lambda = p_\lambda((x_\lambda)_{\lambda \in I}).$$

For the second assertion, let us consider, under the stated conditions, the linear mapping $r' : X \to \prod_{\lambda \in I} E_\lambda$ obtained from the family of linear mappings $(r_\lambda)_{\lambda \in I}$ by the universal property of the direct product, i.e. such that $p'_\lambda \circ r' = r_\lambda$, $\lambda \in I$, where p'_λ is the canonical projection $\prod_\lambda E_\lambda \to E_\lambda$, $\lambda \in I$; we have $r'(X) \subseteq E$: for if $x \in X$, then $r'(x) = (r_\lambda(x))_{\lambda \in I}$, and if $f \in \mathrm{Hom}_I(\lambda, \mu)$, then, by hypothesis, $u_f(r_\mu(x)) = r_\lambda(x)$; by the universal property of the submodule (see § 3), there exists a uniquely determined linear mapping $r : X \to E$ for which, denoting by i the canonical injection of E into $\prod_\lambda E_\lambda$, we have $i \circ r = r'$; but then, for any $x \in X$ we have $(p_\lambda \circ r)(x) = (p'_\lambda \circ r')(x) = r_\lambda(x)$, $\lambda \in I$.

If there existed $r^* : X \to E$ satisfying condition (2) when put instead of r, then, denoting $r'^* = i \circ r^*$, it would follow that

$$r_\lambda = p'_\lambda \circ i \circ r^* = p'_\lambda \circ r'^*$$

for any $\lambda \in I$; hence, by virtue of the uniqueness of r', $r'^* = r'$, and since i is injective, $r^* = r$. ∎

Consider now two projective systems of A-modules $T = (E_\lambda, u_f)_{\lambda, f}$, $U = (F_\lambda, v_f)_{\lambda, f}$ indexed by the same category I; T and U being functors $I^\circ \to \mathbf{Mod}(A)$, assume a natural transformation from T to U is given; this amounts to giving for any $\lambda \in I$ a homomorphism

$$l_\lambda : E_\lambda \to F_\lambda$$

such that, if $\lambda, \mu \in I$, $f \in \mathrm{Hom}_I(\lambda, \mu)$, the following diagram commutes:

$$(3) \qquad \begin{array}{ccc} E_\mu & \xrightarrow{u_f} & E_\lambda \\ {\scriptstyle l_\mu}\downarrow & & \downarrow{\scriptstyle l_\lambda} \\ F_\mu & \xrightarrow{v_f} & F_\lambda. \end{array}$$

In the particular case when I is the category associated with a preordered set, and $T = (E_\lambda, u_{\lambda\mu})_{\lambda \in I}$, $U = (F_\lambda, v_{\lambda\mu})_{\lambda \in I}$, the condition required for the family $(l_\lambda)_{\lambda \in I}$ becomes: if $\lambda, \mu \in I$ and $\lambda \leqslant \mu$, the diagram obtained from (3) replacing u_f, respectively v_f, by $u_{\lambda\mu}$, respectively $v_{\lambda\mu}$, must be commutative.

Definition 3. If T and U are projective systems of A-modules indexed by the same category I of indices, a natural transformation $T \to U$ is called a *projective* (or *inverse*) *system of linear mappings* or also a *projective* (or *inverse*) *morphism* or, alternatively, a *morphism of projective* (or *inverse*) *systems* (of A-modules) *from T to U* or *between T and U*; the above described projective morphism will be also denoted $(l_\lambda)_{\lambda \in I}$. ∎

Now, notice that if T and U are projective systems of A-modules indexed by the same (small) category I, any projective morphism $(l_\lambda)_{\lambda \in I}$ from T to U is, using the preceding notations for T and U, an element of the direct product $\prod_{\lambda \in I} \mathrm{Hom}_A(E_\lambda, F_\lambda)$; consequently, the class of projective morphisms from T to U is a set, and hence we may speak of the *category of projective systems of A-modules* (A fixed) *indexed by the (small) category I of indices* (I fixed); it is the category denoted $\mathbf{Fonct}(I^\circ, \mathbf{Mod}(A))$ or $(\mathbf{Mod}(A))^{I^\circ}$ constructed in our case according to the Remark following Definition 7, § 10. It is not difficult to check that $(\mathbf{Mod}(A))^{I^\circ}$ is an abelian category: for instance, the sum of two projective morphisms $(l_\lambda)_{\lambda \in I}$, $(l'_\lambda)_{\lambda \in I}$ from T to U is the projective morphism $(l_\lambda + l'_\lambda)_{\lambda \in I}$.

REMARK. The notation $(\mathbf{Mod}(A))^{I^\circ}$ now introduced is not at all in disagreement with the notation $\mathcal{C}^I$ introduced in Definition 5, § 2, for a category $\mathcal{C}$ and a set I, being in fact a generalization of it: indeed, in case I is a set, it may be regarded as a small category in which the objects are the elements of the set, and if $\lambda, \mu \in I$, then $\mathrm{Hom}_I(\lambda, \mu)$ is the void set or the set consisting if a single element, namely the identity morphism of λ, according as $\lambda \neq \mu$ or $\lambda = \mu$ (such a category is called a *discrete category* even in the case when I is an arbitrary class); obviously, in this case, we may identify I° with I according to $\lambda \mapsto \lambda^\circ$, $\lambda \in \mathrm{Ob}\, I$; further, for a discrete category I, an object of $(\mathbf{Mod}(A))^I$, that is, a functor $I \to \mathbf{Mod}(A)$, is determined by the "values" it takes on the objects of I; hence, it is a mapping $I \to \mathrm{Ob}(\mathbf{Mod}(A))$, that is, a family $(E_\lambda)_{\lambda \in I}$ of A-modules indexed by I; a morphism $(E_\lambda)_{\lambda \in I} \to (F_\lambda)_{\lambda \in I}$ is a natural transformation, that is, simply, a family of linear mappings $u_\lambda : E_\lambda \to F_\lambda$, $\lambda \in I$, since the condition required for natural transformations to render commutative certain diagrams is here automatically satisfied, as in I there are no other morphisms then the identity ones; further, the composition is done componentwise, and in this way we recover the definition of $(\mathbf{Mod}(A))^I$ given by Definition 5, § 2.

* This Remark is valid for an arbitrary category instead of $\mathbf{Mod}(A)$.* ▮

We have the following Corollaries to Proposition 1:

COROLLARY 1. *Let $T = (E_\lambda, u_f)_{\lambda,f}$, $U = (F_\lambda, v_f)_{\lambda,f}$ be two projective systems of A-modules indexed by the same category I, and let $(l_\lambda)_{\lambda \in I}$ be a projective morphism from T to U; then there exists an A-homomorphism $l : \varprojlim_I E_\lambda \to \varprojlim_I F_\lambda$ uniquely determined by the property that for any $\lambda \in I$ it renders commutative the diagram*

$$\begin{array}{ccc} \varprojlim_I E_\lambda & \xrightarrow{\;l\;} & \varprojlim_I F_\lambda \\ \downarrow{\scriptstyle p_\lambda} & & \downarrow{\scriptstyle q_\lambda} \\ E_\lambda & \xrightarrow{\;l_\lambda\;} & F_\lambda, \end{array}$$

p_λ and q_λ being the canonical mappings of index λ.

Proof. If $\lambda, \mu \in I$ and $f \in \mathrm{Hom}_I(\lambda, \mu)$, then, by the definition of the projective morphisms we have

$$v_f \circ l_\mu \circ p_\mu = l_\lambda \circ u_f \circ p_\mu = l_\lambda \circ p_\lambda;$$

we now apply Proposition 1 taking $(F_\lambda, v_f)_{\lambda,f}$ instead of $(E_\lambda, u_f)_{\lambda,f}$, $\varprojlim_I E_\lambda$ instead of X, and $(l_\lambda \circ p_\lambda)_{\lambda \in I}$ instead of $(r_\lambda)_{\lambda \in I}$. ▮

REMARK. If $(x_\lambda)_{\lambda \in I} \in \varprojlim E_\lambda$, then $l((x_\lambda)_{\lambda \in I}) = (l_\lambda(x_\lambda))_{\lambda \in I}$, as one can see applying the canonical mapping for each index. ▮

DEFINITION 4. In the notation of Corollary 1, l is called the *projective limit* or the *inverse limit* of the given projective morphism $(l_\lambda)_{\lambda \in I}$, and is denoted $\varprojlim_{\lambda \in I} l_\lambda$ or $\varprojlim_I l_\lambda$ or, provided no confusion may arise, $\varprojlim l_\lambda$. ▮

COROLLARY 2. *Let three projective systems of A-modules* $T = (E_\lambda, u_f)_{\lambda, f}$, $U = (F_\lambda, v_f)_{\lambda, f}$, $V = (G_\lambda, w_f)_{\lambda, f}$ *indexed by the same category I be given; let* $(l_\lambda)_{\lambda \in I}$, $(m_\lambda)_{\lambda \in I}$ *be two morphisms of projective systems from T to U, respectively from U to V; then the projective limit of the composite projective morphism* $(m_\lambda \circ l_\lambda)_{\lambda \in I}$ *from T to V is the composition* $(\varprojlim_{\lambda \in I} m_\lambda) \circ (\varprojlim_{\lambda \in I} l_\lambda)$; *also, the projective limit of the identity projective morphism* $(1_{E_\lambda})_{\lambda \in I}$ *of* $(E_\lambda, u_f)_{\lambda, f}$ *is the identity mapping of* $\varprojlim_{\lambda \in I} E_\lambda$; *finally, if another projective morphism* $(l'_\lambda)_{\lambda \in I}$ *from T to U is given, then the projective limit of the sum* $(l_\lambda + l'_\lambda)_{\lambda \in I}$ *is the sum of the projective limits* $\varprojlim l_\lambda + \varprojlim l'_\lambda$.

The *proof* is analogous to the one of Corollary 2 to Proposition 1, § 6. ▮

Corollaries 1 and 2 show that, fixing the category of indices I and the ring A, the mappings

$$(E_\lambda, u_f)_I \mapsto \varprojlim_I E_\lambda, \quad (l_\lambda)_{\lambda \in I} \mapsto \varprojlim_I l_\lambda$$

define an additive functor from the category $(\mathbf{Mod}(A))^{I^\circ}$ to $\mathbf{Mod}(A)$.

DEFINITION 5. The additive functor $(\mathbf{Mod}(A))^{I^\circ} \to \mathbf{Mod}(A)$ defined above is called the *projective* (or *inverse*) *limit functor*, and is denoted $\varprojlim_I$ or, if no confusion may arise, $\varprojlim$. ▮

Consider now another small category J and the product category $I \times J$; we saw (§ 2) that $(I \times J)^\circ$ is canonically identified with $I^\circ \times J^\circ$; consider a projective system of A-modules indexed by the category $I \times J$; denote by E_λ^ρ the A-module corresponding in this projective system to the object $(\lambda, \rho) \in \mathrm{Ob}(I \times J)$; also, if $f \in \mathrm{Hom}_I(\lambda, \mu)$, $g \in \mathrm{Hom}_J(\rho, \sigma)$, denote by u_f^g the linear mapping $E_\mu^\sigma \to E_\lambda^\rho$ which is assigned under the given projective system to the morphism (f, g); consequently, if, moreover, $f' \in \mathrm{Hom}_I(\mu, \nu)$, $g' \in \mathrm{Hom}_J(\sigma, \tau)$ are given, we have

$$u_{f' \circ f}^{g' \circ g} = u_f^g \circ u_{f'}^{g'} \text{ and } u_{1_\lambda}^{1_\rho} = 1_{E_\lambda^\rho}. \tag{4}$$

In case I and J are the categories associated with the preordered sets I, respectively J, the product category $I \times J$ may be identified in an obvious manner with the category associated with the preordered product set $I \times J$; then, denoting by $u_{\lambda\mu}^{\rho\sigma}$ the A-homomorphism $E_\mu^\sigma \to E_\lambda^\rho$ corresponding to the relation $(\lambda, \rho) \leqslant (\mu, \sigma)$ in $I \times J$, and if $\lambda \leqslant \mu \leqslant \nu$ in I and $\rho \leqslant \sigma \leqslant \tau$ in J, we have

$$u_{\lambda\nu}^{\rho\tau} = u_{\lambda\mu}^{\rho\sigma} \circ u_{\mu\nu}^{\sigma\tau} \quad \text{and} \quad u_{\lambda\lambda}^{\rho\rho} = 1_{E_\lambda^\rho}.$$

Returning to the general case, let us denote $E = \varprojlim_{I \times J} (E_\lambda^\rho, u_f^g)$, with the canonical mappings $p_\lambda^\rho : E \to E_\lambda^\rho$, $\lambda \in I$, $\rho \in J$. Fix an index $\rho \in J$, and consider for any $\lambda \in I$ the module E_λ^ρ and for any $f \in \mathrm{Hom}_I(\lambda, \mu)$ the linear mapping $u_f^{1_\rho} : E_\mu^\rho \to E_\lambda^\rho$; since for $f' \in \mathrm{Hom}_I(\mu, \nu)$ we have

$$u_{f' \circ f}^{1_\rho} = u_f^{1_\rho} \circ u_{f'}^{1_\rho} \text{ and } u_{1_\lambda}^{1_\rho} = 1_{E_\lambda^\rho},$$

it turns out that $(E_\lambda^\rho, u_f^{1_\rho})_{\lambda, f}$ is a projective system of A-modules indexed by the category I of indices; denote

$$F^\rho = \varprojlim_{I} (E_\lambda^\rho, u_f^{1_\rho}),$$

with the canonical mappings $q_\lambda^\rho : F^\rho \to E_\lambda^\rho$, $\lambda \in I$; if σ is another fixed index in J and $g \in \mathrm{Hom}_J(\rho, \sigma)$ is fixed, then the family of linear mappings $u_{1_\lambda}^g : E_\lambda^\sigma \to E_\lambda^\rho$, $\lambda \in I$, is a morphism of projective systems

$$(E_\lambda^\sigma, u_f^{1_\sigma})_{\lambda, f} \to (E_\lambda^\rho, u_f^{1_\rho})_{\lambda, f};$$

indeed, if $f \in \mathrm{Hom}_I(\lambda, \mu)$, then, by (4), the following diagram is commutative:

(5)
$$\begin{array}{ccc} E_\mu^\sigma & \xrightarrow{u_f^{1_\sigma}} & E_\lambda^\sigma \\ \downarrow{\scriptstyle u_{1_\mu}^g} & & \downarrow{\scriptstyle u_{1_\lambda}^g} \\ E_\mu^\rho & \xrightarrow{u_f^{1_\rho}} & E_\lambda^\rho ; \end{array}$$

denote

$$v^g = \varprojlim_{\lambda \in I} u_{1_\lambda}^g ;$$

we have $v^g : F^\sigma \to F^\rho$.

Doing the same thing for any $\rho \in J$ and any $g \in \mathrm{Hom}_J(\rho, \sigma)$, $\rho, \sigma \in J$, we get the projective system of A-modules $(F^\rho, v^g)_{\rho, g}$ indexed by the category J of indices: indeed, if besides the above data the object $\tau \in J$ and the morphism $g' \in \mathrm{Hom}_J(\sigma, \tau)$ are given, then, by Corollary 2 to Proposition 1, we have

$$v^{g' \circ g} = \varprojlim_{I} u_{1_\lambda}^{g' \circ g} = \varprojlim_{I} (u_{1_\lambda}^g \circ u_{1_\lambda}^{g'}) = \varprojlim_{I} u_{1_\lambda}^g \circ \varprojlim_{I} u_{1_\lambda}^{g'} = v^g \circ v^{g'},$$

and, likewise,

$$v^{1_\rho} = 1_{F_\rho}.$$

Denote

$$F = \varprojlim_{J} (F^{\rho},\ v^{g}),$$

with the canonical mappings $r^{\rho}: F \to F^{\rho}$, $\rho \in J$.

Again fixing ρ in J, we have the family of linear mappings

$$p^{\rho}_{\lambda}: E \to E^{\rho}_{\lambda}, \quad \lambda \in I;$$

since for any $f \in \mathrm{Hom}_I(\lambda, \mu)$ we have the commutative diagram

(6)

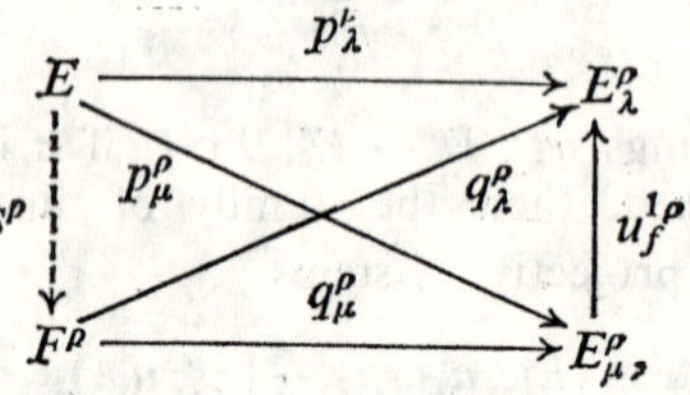

it follows, by the universal property of the projective limit (Proposition 1), that there exists a unique linear mapping $s^{\rho}: E \to F^{\rho}$ such that

(7) $$q^{\rho}_{\lambda} \circ s^{\rho} = p^{\rho}_{\lambda}$$

for any $\lambda \in I$; doing that for all $\rho \in J$, we get the family of linear mappings $s^{\rho}: E \to F^{\rho}$, $\rho \in J$; but if $g \in \mathrm{Hom}_J(\rho, \sigma)$, then

(8) $$v^{g} \circ s^{\sigma} = s^{\rho}:$$

indeed, by the uniqueness assertion in the universal property of the projective limit, it suffices to show that for any $\lambda \in I$,

(9) $$q^{\rho}_{\lambda} \circ v^{g} \circ s^{\sigma} = q^{\rho}_{\lambda} \circ s^{\rho};$$

but (9) follows from the fact that: by the definition of v^{g} we have $q^{\rho}_{\lambda} \circ v^{g} = u^{g}_{1_\lambda} \circ q^{\sigma}_{\lambda}$, by relation (7) we have $q^{\sigma}_{\lambda} \circ s^{\sigma} = p^{\sigma}_{\lambda}$, and by Proposition 1 we have $u^{g}_{1_\lambda} \circ p^{\sigma}_{\lambda} = p^{\rho}_{\lambda}$. By (8), applying again the universal property of the projective limit, we get the existence of a unique linear mapping

$$\varphi: E \to F$$

with the property of rendering commutative the diagram

(10)

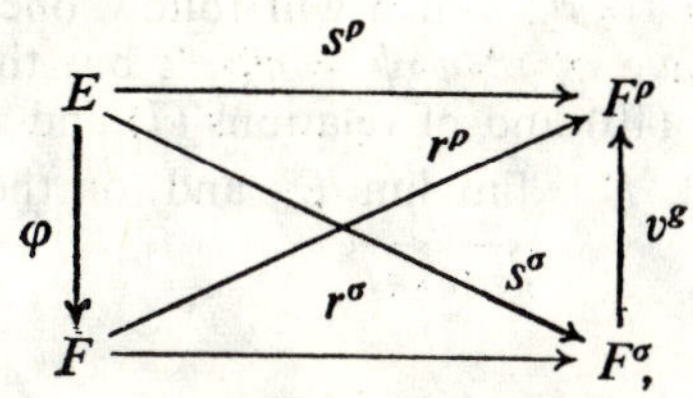

hence that $r^\rho \circ \varphi = s^\rho$ for any $\rho \in J$ (here, for the sake of clarity, F^σ and v^g have also been represented).

On the other hand, for any $\lambda \in I$ and $\rho \in J$ we have the composite linear mapping $q^\rho_\lambda \circ r^\rho \colon F \to E^\rho_\lambda$; if $f \in \mathrm{Hom}_I(\lambda, \mu)$ and $g \in \mathrm{Hom}_J(\rho, \sigma)$, then the following diagram is commutative:

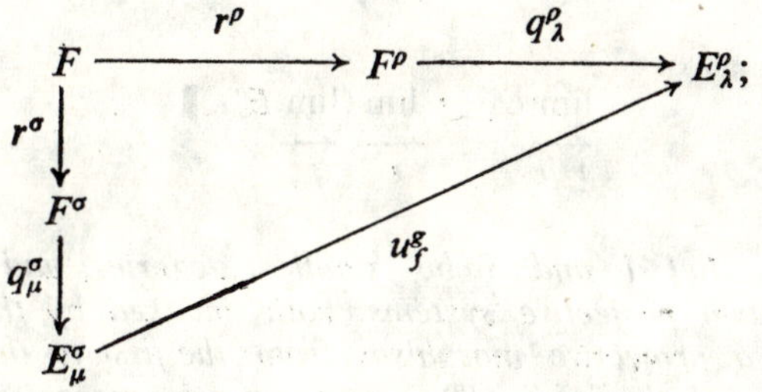

since, applying successively relations (4), the fact that diagram (6) is commutative (with σ instead of ρ), the definition of v^g, and the fact that diagram (10) is commutative, we get

$$u^g_f \circ q^\sigma_\mu \circ r^\sigma = u^g_{1_\lambda} \circ u^{1_\sigma}_f \circ q^\sigma_\mu \circ r^\sigma = u^g_{1_\lambda} \circ q^\sigma_\lambda \circ r^\sigma = q^\rho_\lambda \circ v^g \circ r^\sigma = q^\rho_\lambda \circ r^\rho;$$

by the universal property of the projective limit, there exists a unique linear mapping

$$\psi \colon F \to E$$

with the property that for any $\lambda \in I$ and $\rho \in J$ we have

(11) $$p^\rho_\lambda \circ \psi = q^\rho_\lambda \circ r^\rho.$$

Let us show that φ and ψ are isomorphisms inverse to each other: to show that $\psi \circ \varphi = 1_E$ it suffices to show that for any $\lambda \in I$

(12) $$p^\rho_\lambda \circ \psi \circ \varphi = p^\rho_\lambda;$$

but relations (12) are obtained from (11), from the commutativity of diagram (10), and from (7). In the same way, to prove that $\varphi\circ\psi = 1_F$ it suffices to show that for any $\rho \in J$ we have $r^\rho\circ\varphi\circ\psi = r^\rho$, which will follow once we have shown that for any $\lambda \in I$ and $\rho \in J$ we have $q_\lambda^\rho\circ r^\rho\circ\varphi\circ\psi = q_\lambda^\rho\circ r^\rho$; but this is a consequence of the commutativity of diagram (10) and of relations (7) and (11).

As, on the one hand, $F = \varprojlim_{\rho\in J} \varprojlim_{\lambda\in I} E_\lambda^\rho$, and, on the other hand, I and J may be interchanged, we get:

Corollary 3. *Let I and J be small categories, and let $(E_\lambda^\rho, u_f^g)_{\lambda,\rho,f,g}$ be a projective system of A-modules indexed by the product category $I\times J$; then, with the previously established meaning for the first double limit and with an analogous meaning for the second double limit, there exist canonical A-isomorphisms*

$$\varprojlim_{I\times J} E_\lambda^\rho \rightleftarrows \varprojlim_J(\varprojlim_I E_\lambda^\rho),$$

$$\varprojlim_{I\times J} E_\lambda^\rho \rightleftarrows \varprojlim_I(\varprojlim_J E_\lambda^\rho). \blacksquare$$

Corollary 4. *Let I and J be small categories, and let $(E_\lambda^\rho, u_f^g)_{\lambda,\rho,f,g}$, $(E'^\rho_\lambda, u'^g_f)_{\lambda,\rho,f,g}$ be two projective systems both indexed by the category $I\times J$; let $(l_\lambda^\rho)_{(\lambda,\rho)\in I\times J}$ be a projective morphism from the first to the second projective system; then, fixing ρ in J, the family $(l_\lambda^\rho)_{\lambda\in I}$ is a projective morphism from $(E_\lambda^\rho, u_f^{1\rho})_{\lambda,f}$ to $(E'^\rho_\lambda, u'^{1\rho}_f)_{\lambda,f}$; let $k^\rho = \varprojlim_{\lambda\in I} l_\lambda^\rho$; the family $(k^\rho)_{\rho\in J}$ is, in the notation of the proof of Corollary 3 and in the corresponding notation for $(E'^\rho_\lambda, u'^g_f)_{\lambda,\rho,f,g}$, a projective morphism from $(F^\rho, v^g)_{\rho,g}$ to $(F'^\rho, v'^g)_{\rho,g}$, and the following diagram is commutative:*

$$\begin{array}{ccc} \varprojlim_{I\times J} E_\lambda^\rho & \xrightarrow{\sim} & \varprojlim_J(\varprojlim_I E_\lambda^\rho) \\ {\scriptstyle \varprojlim_{I\times J} l_\lambda^\rho}\Big\downarrow & & \Big\downarrow{\scriptstyle \varprojlim_J(\varprojlim_I l_\lambda^\rho)} \\ \varprojlim_{I\times J} E'^\rho_\lambda & \xrightarrow{\sim} & \varprojlim_J(\varprojlim_I E'^\rho_\lambda), \end{array}$$

where the horizontal arrows represent the canonical isomorphisms; also, interchanging the roles of I and J, we get with an analogous meaning a similar commutative diagram.

The proof is left to the reader as an exercise. ▮

The property that there exists the canonical isomorphism

$$\varprojlim_{I} (\varprojlim_{J} E_{\lambda}^{\varrho}) \rightleftarrows \varprojlim_{J} (\varprojlim_{I} E_{\lambda}^{\varrho})$$

obtained as the composition of the first isomorphism in Corollary 3 with the inverse of the second one together with the fact that the corresponding diagram analogous to the one in Corollary 4 is commutative will be called the "property of projective limits of commuting with projective limits".

* DEFINITION 6. If $\mathcal{C}$ and $\mathcal{D}$ are categories, $\mathcal{D}$ is said to be a *subcategory* of $\mathcal{C}$ if the following conditions are satisfied:

a) Ob $\mathcal{D}$ is a subclass of Ob $\mathcal{C}$, hence any object of $\mathcal{D}$ is also an object of $\mathcal{C}$;

b) if $E, F \in \mathrm{Ob}\,\mathcal{D}$, then $\mathrm{Hom}_{\mathcal{D}}(E, F) \subseteq \mathrm{Hom}_{\mathcal{C}}(E, F)$;

c) for $E, F, G \in \mathrm{Ob}\,\mathcal{D}$, the mapping

$$\mathrm{Hom}_{\mathcal{D}}(E, F) \times \mathrm{Hom}_{\mathcal{D}}(F, G) \to \mathrm{Hom}_{\mathcal{D}}(E, G)$$

given by the composition of morphisms is induced by the analogous mapping of $\mathcal{C}$, in other words, the composition of any two morphisms in $\mathcal{D}$ is their composition in $\mathcal{C}$;

d) if $E \in \mathrm{Ob}\,\mathcal{D}$, then the identity morphism of E in the category $\mathcal{D}$ coincides with the identity morphism of E in $\mathcal{C}$.

The mappings

$$E \mapsto E, \qquad u \mapsto u,$$

where $E \in \mathrm{Ob}\,\mathcal{D}$, $u \in \mathrm{Hom}_{\mathcal{D}}(E, F)$, define a functor $\mathcal{D} \to \mathcal{C}$, called the *imbedding functor* or the *immersion functor* or the *inclusion functor* of the subcategory $\mathcal{D}$ into the category $\mathcal{C}$.

A subcategory $\mathcal{D}$ of a category $\mathcal{C}$ is said to be a *full subcategory* if for any pair of objects $E, F \in \mathrm{Ob}\,\mathcal{D}$ we have

$$\mathrm{Hom}_{\mathcal{D}}(E, F) = \mathrm{Hom}_{\mathcal{C}}(E, F).$$

A subcategory $\mathcal{D}$ of a category $\mathcal{C}$ is said to be a *cofinal subcategory* of $\mathcal{C}$ if for any object E of $\mathcal{C}$ there exists an object F of $\mathcal{D}$ such that the set $\mathrm{Hom}_{\mathcal{C}}(E, F)$ is not empty, hence such that there exists at least one morphism $u : E \to F$ of the category $\mathcal{C}$.

If $T : \mathcal{C} \to \mathcal{C}'$ is a functor and $\mathcal{D}$ is a subcategory of $\mathcal{C}$, then the mappings

$$E \mapsto T(E), \qquad u \mapsto T(u),$$

where $E \in \mathrm{Ob}\,\mathcal{D}$, $u \in \mathrm{Hom}_{\mathcal{D}}(E, F)$, define a functor $U : \mathcal{D} \to \mathcal{C}'$, called the *functor obtained by the restriction of T to $\mathcal{D}$*. ▮

We notice that the functor $U : \mathcal{D} \to \mathcal{C}'$ obtained by the restriction of $T : \mathcal{C} \to \mathcal{C}'$ to the subcategory $\mathcal{D}$ of $\mathcal{C}$ is actually the composition of T with the imbedding functor of $\mathcal{D}$ into $\mathcal{C}$. We also notice that if $\mathcal{D}$ is a subcategory of the category $\mathcal{C}$, then the dual $\mathcal{D}^\circ$ of $\mathcal{D}$ is identified in an obvious mannei with a subcategory of the dual $\mathcal{C}^\circ$ of $\mathcal{C}$. Further, any full subcategory of a category is uniquely determined by the class of its objects.

DEFINITION 7. A category $\mathcal{C}$ is said to be a (*right*) *pseudofiltered category* provided it satisfies the following conditions:

a) for any diagram in $\mathcal{C}$ of the form

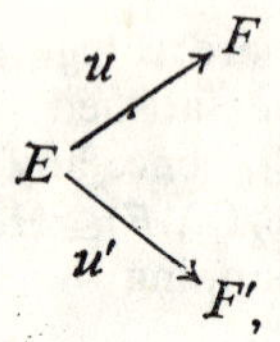

there exists an object G of $\mathcal{C}$ and two morphisms $v : F \to G$, $v' : F' \to G$ such that the following diagram commutes:

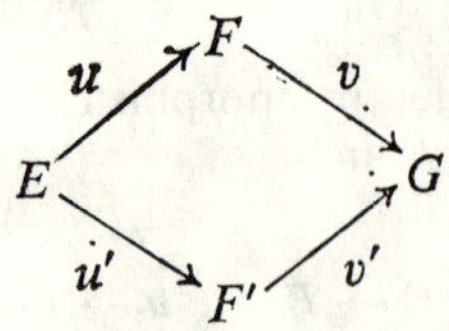

b) for any pair of morphisms u, u' having the same source E and the same range F there exists an object G and a morphism $v : F \to G$ such that

$$v \circ u = v \circ u'.$$

A category $\mathcal{C}$ is said to be a *connected category* if for any pair of objects E, F of the category there exists a finite sequence of objects $E_0, \ldots, E_n$, $n \geqslant 1$, such that $E_0 = E$, $E_n = F$ and such that for each $i = 0, \ldots, n-1$ at least one of the sets $\mathrm{Hom}_{\mathcal{C}}(E_i, E_{i+1})$ and $\mathrm{Hom}_{\mathcal{C}}(E_{i+1}, E_i)$ is not empty. ∎*

If I is a preordered set directed to the right, the category associated with I is a pseudofiltered category: indeed, in such a category two morphisms with the same source and the same range are equal. Next, if J is a cofinal subset of the preordered set I, then the category associated with the preordered set J, with the ordering induced by I, is a cofinal subcategory of the category associated with the preordered set I.

Proposition 1 has also the following Corollary:

COROLLARY 5. *Let I be a small pseudofiltered category, and let J be a full cofinal subcategory of it; let A be a ring, let $T = (E_\lambda, u_f)_{\lambda, f}$ be a projective* *system of A-modules indexed by the category I, and let* $U = (E_\rho, u_g)_{\rho, g}$ *be the projective system of A-modules obtained by the restriction of T to the subcategory* J° *of* I°*; then there exists a canonical A-isomorphism*

$$\varphi : \varprojlim_I E_\lambda \cong \varprojlim_J E_\rho$$

such that, if $p_\lambda : \varprojlim_I E_\lambda \to E_\lambda$, $\lambda \in I$, *and* $q_\rho : \varprojlim_J F_\rho \to F_\rho$, $\rho \in J$, *are the canonical mappings,* $q_\rho \circ \varphi = p_\rho$ *for any* $\rho \in J$*; if, moreover,* $T' = (E'_\lambda, u'_f)_{\lambda, f}$ *is another projective system of A-modules indexed by the category I, if* $U' = (E'_\rho, u'_g)_{\rho, g}$ *is the projective system obtained from T′ by restriction to* J°*, and if* $(l_\lambda)_{\lambda \in I}$ *is a projective morphism from T to T′, then* $(l_\rho)_{\rho \in J}$ *is a projective morphism from U to U′, and we have the commutative diagram*

$$\begin{array}{ccc} \varprojlim_I E_\lambda & \xrightarrow{\sim} & \varprojlim_J E_\rho \\ {\scriptstyle \varprojlim_I l_\lambda}\downarrow & & \downarrow{\scriptstyle \varprojlim_J l_\rho} \\ \varprojlim_I E'_\lambda & \xrightarrow{\sim} & \varprojlim_J E'_\rho, \end{array}$$

the horizontal arrows representing the canonical isomorphisms.

Proof. Let us denote $E = \varprojlim_I E_\lambda$, $F = \varprojlim_J E_\rho$; the family of mappings $(p_\rho)_{\rho \in J}$ obviously satisfies the condition $u_g \circ p_\sigma = p_\rho$ for any $\rho, \sigma \in J$ and $g \in \operatorname{Hom}_J(\rho, \sigma)$; by Proposition 1 there exists an A-isomorphism $\varphi : E \to F$ uniquely determined by the property that

$$q_\rho \circ \varphi = p_\rho$$

for any $\rho \in J$. Now, let $\lambda \in I$; J being by hypothesis a cofinal subcategory of I, there exist $\rho \in J$ and $h \in \operatorname{Hom}_I(\lambda, \rho)$; we claim that *the composite morphism*

$$F \xrightarrow{q_\rho} E_\rho \xrightarrow{u_h} E_\lambda$$

does not depend on the chosen ρ *and* h *but only on* λ: indeed, let (h', ρ') be another pair with the properties of (h, ρ); I being a pseudofiltered category, there exist $\mu \in I$ and $k \in \operatorname{Hom}_I(\rho, \mu)$, $k' \in \operatorname{Hom}_I(\rho', \mu)$ such that $k \circ h = k' \circ h'$; next, as J is cofinal, there exist $\sigma \in J$ and $j \in \mathrm{Hom}_I(\mu, \sigma)$; if $g = j \circ k$, $g' = j \circ k'$, then we have the commutative diagram

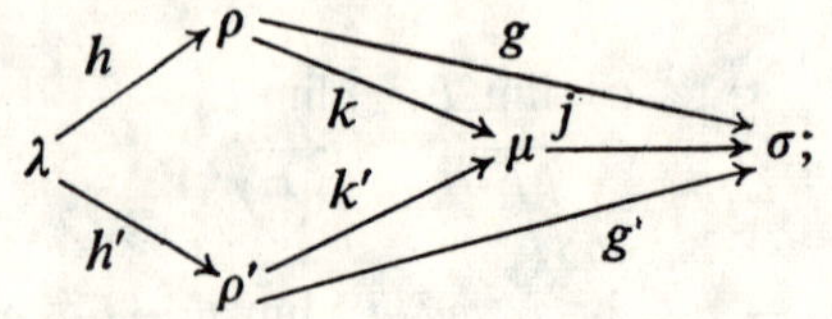

as $\rho, \sigma \in J$ and J is a full subcategory of I, $g \in \mathrm{Hom}_J(\rho, \sigma)$; likewise, $g' \in \mathrm{Hom}_J(\rho', \sigma)$; consequently, we have

$$u_h \circ q_\rho = u_h \circ u_g \circ q_\sigma = u_{h'} \circ u_{g'} \circ q_\sigma = u_{h'} \circ q_{\rho'},$$

which proves our assertion. Let us denote

$$v_\lambda = u_h \circ q_\rho;$$

the family of mappings $v_\lambda : F \to E_\lambda$, $\lambda \in I$, satisfies the condition that if $\lambda, \mu \in I$ and $f \in \mathrm{Hom}_I(\lambda, \mu)$, then $u_f \circ v_\mu = v_\lambda$: indeed, let ρ, respectively ρ', be objects of J such that $h \in \mathrm{Hom}_I(\lambda, \rho)$, $h' \in \mathrm{Hom}_I(\mu, \rho')$; I being a pseudofiltered category and J a cofinal full subcategory of I, there exist $\sigma \in J$ and $g \in \mathrm{Hom}_J(\rho, \sigma)$, $g' \in \mathrm{Hom}_J(\rho', \sigma)$ such that the following diagram is commutative:

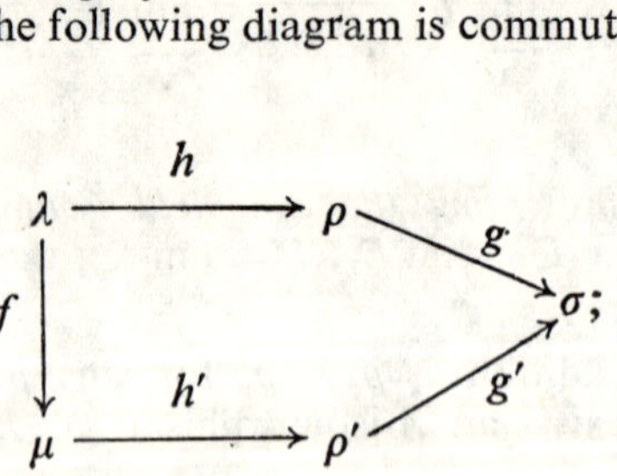

we have

$$u_f \circ v_\mu = u_f \circ u_{h'} \circ q_{\rho'} = u_f \circ u_{h'} \circ u_{g'} \circ q_\sigma = u_h \circ u_g \circ q_\sigma = u_h \circ q_\rho = v_\lambda;$$

consequently, by Proposition 1, there exists $\psi : F \to E$ uniquely determined by the property that

$$p_\lambda \circ \psi = v_\lambda$$

for any $\lambda \in I$.

But then, using the preceding notation, we have for any $\lambda \in I$,

$$p_\lambda \circ \psi \circ \varphi = v_\lambda \circ \varphi = u_h \circ q_\rho \circ \varphi = u_h \circ p_\rho = p_\lambda,$$

and, by virtue of the uniqueness,

$$\psi\circ\varphi = 1_E;$$

likewise,

$$\varphi\circ\psi = 1_F,$$

since $q_\rho\circ\varphi\circ\psi = p_\rho\circ\psi = v_\rho = q_\rho$ for any $\rho \in J$; hence φ is an isomorphism, and ψ is its inverse.

For the second part, one shows that the composition of the canonical mapping $\varprojlim_{\rho\in J} E'_\rho \to E'_\rho$ with the two homomorphisms of the diagram which should be proved to be equal gives the same homomorphism $\varprojlim_I E_\lambda \to E'_\rho$ for any $\rho \in J$; we leave the details to the reader as an exercise. ▮

PROPOSITION 2. *a) Let I be a small connected category; let A be a ring, and let $(E_\lambda, u_f)_{\lambda, f}$ be a projective system of A-modules indexed by the category I; if for any $\lambda, \mu \in I$ and any $f \in \mathrm{Hom}_I(\lambda, \mu)$ the homomorphism u_f is injective, then, denoting $E = \varprojlim E_\lambda$ and by p_λ the canonical mapping $E \to E_\lambda$, $\lambda \in I$, the linear mapping p_λ is injective for any $\lambda \in I$.*

b) Let I be a preordered set directed to the right; let $(E_\lambda, u_{\lambda\mu})_{\lambda\in I}$ be a projective system of A-modules indexed by the category I associated with the preordered set I; denoting $E = \varprojlim E_\lambda$ and by p_λ the canonical mapping $E \to E_\lambda$, $\lambda \in I$, if $u_{\lambda\mu}$ is an isomorphism for any λ, μ for which $\lambda \leqslant \mu$, then p_λ is an isomorphism for any $\lambda \in I$.

Proof. a) Let $\lambda \in I$; let $x = (x_\rho)_\rho$, $y = (y_\rho)_\rho$ be elements of E for which $p_\lambda(x) = p_\lambda(y)$, hence for which $x_\lambda = y_\lambda$; if $\mu \in I$ enjoys the property that there exists $f : \lambda \to \mu$ in I, then

$$u_f(x_\mu) = x_\lambda = y_\lambda = u_f(y_\mu),$$

and as u_f is injective, $x_\mu = y_\mu$; if $\nu \in I$ enjoys the property that there exists $g : \nu \to \lambda$ in I, then

$$x_\nu = u_g(x_\lambda) = u_g(y_\lambda) = y_\nu;$$

the case of an arbitrary $\rho \in I$ instead of μ and ν reduces to these two cases, since I is a connected category; hence $x_\rho = y_\rho$ for any $\rho = I$, and $x = y$.

b) First we notice that if I is a preordered set directed to the right, then the (small) category associated with I is connected; hence using *a*), it follows from the hypotheses that p_λ is injective for any $\lambda \in I$; thus, we have only to prove the surjectivity: let $\lambda \in I$ and $x_\lambda \in E_\lambda$; for any $\mu \in I$ choose a $\nu \in I$ such that $\lambda \leqslant \nu$, $\mu \leqslant \nu$, and define the element $x_\mu \in E_\mu$ by

$$x_\mu = (u_{\mu\nu}\circ u_{\lambda\nu}^{-1})(x_\lambda);$$

once shown that the family $x=(x_\mu)_{\mu\in I}\in\prod\limits_{\mu\in I}E_\mu$ is an element of E, everything is proved, since in that case, $p_\lambda(x)=x_\lambda$. Let then $\mu,\ \rho\in I,\ \mu\leqslant\rho$; assume that for μ the above ν is chosen, while for ρ say $\pi\in I$ is chosen such that $\lambda\leqslant\pi,\ \rho\leqslant\pi$; hence we have $x_\rho=(u_{\rho\pi}\circ u_{\lambda\pi}^{-1})(x_\lambda)$; let $\sigma\in I$ such that $\nu\leqslant\sigma,\ \pi\leqslant\sigma$; we have the diagram of A modules and A-homomorphisms

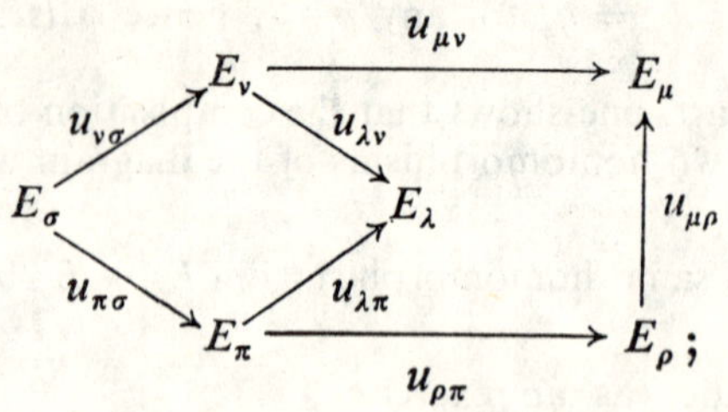

since $(E_\lambda, u_{\lambda\mu})$ is a projective system, we have $u_{\lambda\pi}\circ u_{\pi\sigma}=u_{\lambda\sigma}=u_{\lambda\nu}\circ u_{\nu\sigma}$, hence $u_{\lambda\pi}^{-1}=$ $=u_{\pi\sigma}\circ u_{\nu\sigma}^{-1}\circ u_{\lambda\nu}^{-1}$; also, $u_{\mu\nu}\circ u_{\nu\sigma}=u_{\mu\sigma}$, hence $u_{\mu\nu}=u_{\mu\sigma}\circ u_{\nu\sigma}^{-1}$; consequently,

$$u_{\mu\rho}(x_\rho)=(u_{\mu\rho}\circ u_{\rho\pi}\circ u_{\lambda\pi}^{-1})(x_\lambda)=(u_{\mu\pi}\circ u_{\pi\sigma}\circ u_{\nu\sigma}^{-1}\circ u_{\lambda\nu}^{-1})(x_\lambda)=(u_{\mu\sigma}\circ u_{\nu\sigma}^{-1}\circ u_{\lambda\nu}^{-1})(x_\lambda)=$$

$$=(u_{\mu\nu}\circ u_{\lambda\nu}^{-1})(x_\lambda)=x_\mu. \ \blacksquare$$

Some particular cases are now studied:

As a first particular case consider a ring A and an arbitrary small category I; let X be a fixed A-module, and let T be the projective system of A-modules (indexed by the category I) associating with every $\lambda\in I$ the A-module X, and with every morphism $f\in\mathrm{Hom}_I(\lambda,\mu)$ the linear mapping 1_X; this projective system is called the *constant projective* (or *inverse*) *system of value* X indexed by the category I of indices; if $U=(E_\lambda, u_f)_{\lambda, f}$ is a projective system of A-modules indexed by the same category I, then a projective morphism from T to U is a family of linear mappings $r_\lambda: X\to E_\lambda,\ \lambda\in I$, such that for any $\lambda,\mu\in I$ and $f\in\mathrm{Hom}_I(\lambda,\mu)$ we have $u_f\circ r_\mu=r_\lambda$; thus, the family of canonical mappings $p_\lambda\colon \underset{\overleftarrow{I}}{\lim}\, U\to E_\lambda,\ \lambda\in I$, is a projective morphism from the constant projective system of value $\underset{\longleftarrow}{\lim}\, U$ to U; likewise, the family of linear mappings $(r_\lambda)_{\lambda\in I}$ considered in Proposition 1 is a projective morphism from the constant projective system of value X to $(E_\lambda, u_f)_{\lambda, f}$.

If X, Y are A-modules and $u\in\mathrm{Hom}_A(X, Y)$, then the family of mappings $(u_\lambda)_{\lambda\in I}$, where $u_\lambda=u$ for any $\lambda\in I$, is a projective morphism from the constant projective system of value X to the constant projective system of value Y; it will be called the *constant projective* (or *inverse*) *morphism of value u* (indexed by the category I).

It is clear that the mappings

$X \mapsto$ the constant projective system of value X indexed by I,

$u \mapsto$ the constant projective morphism of value u indexed by I

define an additive functor

$$S\colon \mathbf{Mod}(A) \to (\mathbf{Mod}(A))^{I^\circ};$$

S will be called the *immersion functor* or the *imbedding functor* $\mathbf{Mod}(A) \to (\mathbf{Mod}(A))^{I^\circ}$.

PROPOSITION 3. *If A is a ring and I is a small category, the immersion functor* $S\colon \mathbf{Mod}(A) \to (\mathbf{Mod}(A))^{I^\circ}$ *is a left adjoint of the functor* $\varprojlim_I : (\mathbf{Mod}(A))^{I^\circ} \to \mathbf{Mod}(A)$.

Proof. Let $T = (E_\lambda, u_f)_{\lambda, f}$ be a projective system of A-modules indexed by the small category I, and let X be an A-module; define a mapping

$$\theta(X, T)\colon \mathrm{Hom}_A(X, \varprojlim_I T) \to \mathrm{Hom}_{(\mathbf{Mod}(A))^{I^\circ}}(SX, T)$$

as follows: if $r \in \mathrm{Hom}_A(X, \varprojlim_I T)$, then the family of mappings $(p_\lambda \circ r)_{\lambda \in I}$, $p_\lambda\colon \varprojlim T \to E_\lambda$ being the canonical mapping of index λ, $\lambda \in I$, is a projective morphism $SX \to T$, since $(p_\lambda)_{\lambda \in I}$ is a projective morphism $S(\varprojlim T) \to T$; set

$$\theta(X, T)(r) = (p_\lambda \circ r)_{\lambda \in I};$$

the mapping $\theta(X, T)$ is bijective, since if $(r_\lambda)_{\lambda \in I} \in \mathrm{Hom}_{(\mathbf{Mod}(A))^{I^\circ}}(SX, T)$, then, by Proposition 1, there exists a unique $r\colon X \to \varprojlim T$ with the property that $p_\lambda \circ r = r_\lambda$ for any $\lambda \in I$.

To complete the proof, we should prove the commutativity of the diagram corresponding to diagram (5), § 10. This is left to the reader. ▮

Let I be a connected category, let T be the constant projective system of value X indexed by the category I, and let X' be the projective limit of T; if $(x_\lambda)_{\lambda \in I} \in X'$, then for any $\lambda, \mu \in I$ we have $x_\lambda = x_\mu$: this is clear if there exists $f \in \mathrm{Hom}_I(\lambda, \mu)$, and the general case reduces to this one using the fact that I is a connected category; denote by x the element equal to all x_λ, $\lambda \in I$; the linear mapping $X \to X'$ obtained from the family of identity mappings $X \to X$ for any $\lambda \in I$ by the universal property of the projective limit, and sending an $x \in X$ into the family $(x_\lambda)_\lambda \in X'$ for which $x_\lambda = x$ for any $\lambda \in I$ is bijective; we notice that the inverse of this bijection is any of the canonical mappings $X' \to X$; *hence, in this case, the projective limit X' may be identified with X using this isomorphism.*

This is the case, for instance, when I is the category associated with a preordered set directed to the right.

Returning to the case of an arbitrary connected category I, and considering a homomorphism $u : X \to Y$ of A-modules, denote by X', respectively Y', respectively u', the projective limit of the constant projective system of value X, respectively of the constant projective system of value Y, respectively of the constant projective morphism of value u indexed by the category I; we have the commutative diagram

$$\begin{array}{ccc} X' & \xrightarrow{\ \sim\ } & X \\ {\scriptstyle u'}\downarrow & & \downarrow{\scriptstyle u} \\ Y' & \xrightarrow{\ \sim\ } & Y, \end{array}$$

where the horizontal isomorphisms are the canonical mappings, hence the isomorphisms used to identify X' with X and Y' with Y; in turns out that, under this identification, *u' is identified with u.*

Up to this identification of X' with X, the projective limit of the projective morphism $(r_\lambda)_{\lambda \in I}$ from the statement of Proposition 1 is, for this case, the morphism $r : X \to E$ given in the aforementioned Proposition. Therefore, the homomorphism r constructed in Proposition 1 is sometimes denoted, in the notation used there, $\varprojlim_{\lambda \in I} r_\lambda$, this notation being, by an abuse, extended to the case where I is an arbitrary category not necessarily connected.

The second particular case considered will be the one in which the category I of indices is discrete (see the Remark preceding Corollary 1 to Proposition 1); in this case, if $(E_\lambda, u_f)_{\lambda, f}$ is a projective system of A-modules indexed by the category I, then, as easily seen, any element of $\prod_{\lambda \in I} E_\lambda$ belongs to $\varprojlim_{\lambda \in I} E_\lambda$; hence, in this case, the projective limit coincides with the product of the family $(E_\lambda)_{\lambda \in I}$ of A-modules; certainly, the canonical mapping of index λ, $\lambda \in I$, is in this case the canonical projection $\prod_{\lambda \in I} E_\lambda \to E_\lambda$; consequently, *the direct product of a nonvoid family is a particular case of projective limit.*

The direct product may also be viewed in another way as a projective limit: Let I be a set; let $(E_\lambda)_{\lambda \in I}$ be a family of A-modules; let J be the set partially ordered by inclusion of all finite subsets of I, and let for any $M \in J$,

$$F_M = \prod_{\lambda \in M} E_\lambda;$$

if $M, N \in J$ and $M \subseteq N$, then let u_{MN} be the linear mapping $F_N \to F_M$ obtained from the family of canonical projections $F_N \to E_\lambda$, $\lambda \in M$, by the universal property of the direct product; (F_M, u_{MN}) is a projective system of A-modules indexed by the category J associated with the partially ordered set J; one may show that there exists a canonical isomorphism

$$\varphi\colon \prod_{\lambda \in I} E_\lambda \xrightarrow{\sim} \varprojlim_{M \in J} F_M,$$

the details being given for the dual situation of inductive limits (see § 13).

As a third particular case let us consider a category I with two objects λ and μ; we take by definition

$$\mathrm{Hom}_I(\lambda, \lambda) = \{1_\lambda\}, \quad \mathrm{Hom}_I(\mu, \mu) = \{1_\mu\}, \quad \mathrm{Hom}_I(\mu, \lambda) = \emptyset,$$

and $\mathrm{Hom}_I(\lambda, \mu)$ to be a set consisting of two arrows denoted, say, by f and g; the composition of morphisms is defined in an obvious manner; let us consider the following projective system T of A-modules indexed by the category I: with μ, respectively λ, we associate an arbitrary but fixed module E, respectively F, with f we associate an arbitrary but fixed homomorphism $u : E \to F$, and with g we associate the zero homomorphism $E \to F$; the projective limit of this system consists of those pairs $(y, x) \in F \times E$ for which

$$u(x) = y \quad \text{and} \quad 0(x) = y,$$

hence of those pairs of the form $(0, x) \in F \times E$ for which $u(x) = 0$; it may be seen that the composition

$$\mathrm{Ker}(u) \xrightarrow{i} E \xrightarrow{j_E} F \oplus E = F \times E,$$

where i and j_E are the canonical injections, induces by the procedure described in Corollary 2 to Proposition 1, § 4, an isomorphism $k : \mathrm{Ker}(u) \xrightarrow{\sim} \varprojlim_I T$; if p_E is the canonical mapping $\varprojlim_I T \to E$, we have $p_E \circ k = i$; consequently, $\mathrm{Ker}(u)$ may be identified with $\varprojlim_I T$, under this identification the injection $\mathrm{Ker}(u) \to E$ being identified with one of the canonical mappings; up to this identification, we may say that *the kernel of a linear mapping is a particular case of projective limit.*

It is possible to show that the isomorphism k is "natural", in the following sense: if T' is another projective system indexed by the same category I, defined by the diagram $E' \overset{u'}{\underset{0}{\rightrightarrows}} F'$ with obvious meaning, and if (v, w) is a projective morphism $T \to T'$, hence a pair of linear mappings rendering commutative the diagram

$$\begin{array}{ccc} E & \xrightarrow{u} & F \\ {\scriptstyle v}\downarrow & & \downarrow{\scriptstyle w} \\ E' & \xrightarrow{u'} & F', \end{array}$$

then the following diagram is commutative:

$$\begin{array}{ccc}
\operatorname{Ker}(u) & \xrightarrow[\sim]{k} & \varprojlim_{I} T \\
\downarrow & & \downarrow \\
\operatorname{Ker}(u') & \xrightarrow[\sim]{k'} & \varprojlim_{I} T';
\end{array}$$

here k and k' are the isomorphisms constructed above, the left vertical arrow represents the linear mapping induced by v (see diagram (3), § 4), while the right one represents the projective limit of the given projective morphism $T \to T'$.

As a fourth particular case we consider the following situation: let E be an A-module, and let $(M_\lambda)_{\lambda\in I}$, $I \neq \emptyset$, be a family of submodules of it which as a family of subsets of E is directed to the left; let us define on I a preorder relation denoted $\leqslant$ as follows: if $\lambda, \mu \in I$, then, by definition,

$$\lambda \leqslant \mu \text{ in } I \text{ if and only if } M_\lambda \supseteq M_\mu;$$

in this way I becomes a preordered set directed to the right; we now define the following projective system of A-modules $(E_\lambda, u_{\lambda\mu})_{\lambda\in I}$ indexed by the category I associated with the preordered set I: for $\lambda \in I$, set $E_\lambda = M_\lambda$; if $\lambda \leqslant \mu$, take for $u_{\lambda\mu}$ the canonical injection $M_\mu \to M_\lambda$; let then $(x_\lambda)_{\lambda\in I} \in \varprojlim_{I} (E_\lambda, u_{\lambda\mu})$, and let $\lambda, \nu \in I$; there exists $\mu \in I$ such that $\lambda \leqslant \mu$, $\nu \leqslant \mu$; we have

$$x_\mu = u_{\lambda\mu}(x_\mu) = x_\lambda, \qquad x_\mu = u_{\nu\mu}(x_\mu) = x_\nu,$$

implying that all x_λ, $\lambda \in I$, are equal to a certain element $x \in E$; as $x = x_\lambda \in M_\lambda$ for any $\lambda \in I$, it follows that $x \in \bigcap_{\lambda\in I} M_\lambda$; this means that the linear mapping [1]

$$r: \bigcap_{\lambda\in I} M_\lambda \to \varprojlim_{\lambda\in I} M_\lambda$$

obtained from the family of inclusion mappings $r_\lambda: \bigcap_{\lambda\in I} M_\lambda \to M_\lambda$, $\lambda \in I$ (which satisfy $u_{\lambda\mu} \circ r_\mu = r_\lambda$ for $\lambda \leqslant \mu$) by the universal property of the projective limit is surjective; by virtue of the way in which it acts, it is also injective; hence r turns

[1] Here we replaced, by an abuse of notation, $\varprojlim E_\lambda$ by $\varprojlim M_\lambda$, taking into account that $E_\lambda = M_\lambda$, $\lambda \in I$.

out to be an isomorphism; if $p_\lambda\colon \varprojlim_I M_\lambda \to M_\lambda$ is the canonical mapping, then $p_\lambda \circ r = r_\lambda$ for any $\lambda \in I$; consequently, identifying $\bigcap_\lambda M_\lambda$ with $\varprojlim_\lambda M_\lambda$ by means of the isomorphism r, the canonical mappings are identified with the inclusion mappings; up to the isomorphism r, we can say that *the intersection of a non-void directed to the left family of submodules of a module is a particular case of projective limit.*

The fifth particular case considered is the following: let I be a preordered set consisting of three elements λ, μ, ν, on which the preorder relation $\leqslant$ is defined apart from $\lambda \leqslant \lambda$, etc. by

$$\nu \leqslant \lambda, \quad \nu \leqslant \mu;$$

we consider the category associated with the preordered set I, and the projective system of A-modules $(E_\lambda, E_\mu, E_\nu, u_{\nu\lambda}, u_{\nu\mu})$ indexed by the category I ; to simplify the writing, we put

$$E_\lambda = E,\ E_\mu = F,\ E_\nu = G,\ u_{\nu\lambda} = u, \quad u_{\nu\mu} = v;$$

thus we have the diagram

(13)
$$\begin{array}{ccc} & & E \\ & & \downarrow u \\ F & \xrightarrow{v} & G; \end{array}$$

let Q be the projective limit of this system, with the canonical mappings

$$q_E\colon Q \to E, \quad q_F\colon Q \to F, \quad q_G\colon Q \to G.$$

Let us now consider in the direct product $E \times F$ the subset P of all pairs (x, y) , $x \in E, y \in F$, for which $u(x) = v(y)$; P is the kernel of the homomorphism

$$u \circ p_E - v \circ p_F\colon E \times F \to G,$$

where p_E and p_F are the canonical projections.

DEFINITION 8. Given the diagram (13) of A-modules and A-homomorphisms, the constructed module P is called the *fibered* (or the *amalgamated*) *product of E and F with respect to* (or *relative to*) *this diagram* (or, if there is no danger of confusion, *with respect to G* or *relative to G*), and is denoted, again if no confusion may arise, by $E \prod_G F$ or by $E \times_G F$. ▮

Let us now denote by i the canonical injection $P \to E \times F$, and let $u' = p_E \circ i$, $v' = p_F \circ i$; obviously, the following diagram is commutative:

$$(14)\qquad \begin{array}{ccc} P & \xrightarrow{u'} & E \\ {\scriptstyle v'}\downarrow & & \downarrow{\scriptstyle u} \\ F & \xrightarrow{v} & G; \end{array}$$

moreover, the system (P, u', v') is universal with this property: in other words, given the system (X, r, s), where X is an A-module and $r : X \to E$, $s : X \to F$ are such that the following diagram commutes

$$(15)\qquad \begin{array}{ccc} X & \xrightarrow{r} & E \\ {\scriptstyle s}\downarrow & & \downarrow{\scriptstyle u} \\ F & \xrightarrow{v} & G, \end{array}$$

there exists a unique homomorphism $\varphi : X \to P$ with the property that

$$(16)\qquad u' \circ \varphi = r, \qquad v' \circ \varphi = s;$$

indeed, let $t : X \to E \times F$ be the unique mapping (given by the universal property of the direct product) such that $p_E \circ t = r$, $p_F \circ t = s$; we have $(u \circ p_E - v \circ p_F) \circ t = 0$, and φ is now defined to be the unique homomorphism $X \to P$ (obtained by the universal property of the kernel) such that $i \circ \varphi = t$; if $\psi : X \to P$ is a mapping satisfying relations (16) when put instead of φ, then, by the definition of t, it follows that $i \circ \psi = t$, and hence $\psi = \varphi$.

Let us now consider the family of mappings u', v', w', where $w' : P \to G$ is defined by

$$w' = u \circ u' = v \circ v';$$

by virtue of Proposition 1, there exists a linear mapping $\rho : P \to Q$ uniquely determined by the relations

$$(17)\qquad q_E \circ \rho = u', \quad q_F \circ \rho = v', \quad q_G \circ \rho = w';$$

on the other hand we have, again by Proposition 1,

$$u \circ q_E = q_G = v \circ q_F;$$

hence, by the universal property of (P, u', v') of rendering commutative diagram (14), there exists $\theta : Q \to P$ unique with the property that

$$u' \circ \theta = q_E, \quad v' \circ \theta = q_F. \tag{18}$$

Relations (17) and (18) imply that

$$q_E \circ \rho \circ \theta = q_E, \quad q_F \circ \rho \circ \theta = q_F, \quad q_G \circ \rho \circ \theta = q_G,$$

so that, from the uniqueness of the mapping $Q \to Q$ corresponding by virtue of the universal property of the projective limit to the family of mappings q_E, q_F, q_G it follows

$$\rho \circ \theta = 1_Q;$$

likewise, as from (17) and (18) it follows also $u' \circ \theta \circ \rho = u'$, $v' \circ \theta \circ \rho = v'$, by an analogous argument this time for the fibered product, we get

$$\theta \circ \rho = 1_P;$$

consequently, θ and ρ are isomorphisms inverse to each other; using, for instance, the isomorphism $\rho : P \xrightarrow{\sim} Q$, one may identify the fibered product of E and F with respect to the diagram (13) with the projective limit of the given projective system; by means of this identification u', v' are, respectively, identified with the canonical mappings q_E, q_F; up to this identification, we can say that *the fibered product is a particular case of projective limit.*

As an application of these considerations, we give the following Proposition:

PROPOSITION 4. *Let I be a small category, and let*

$$T = (E_\lambda, k_f)_{\lambda, f}, \quad U = (F_\lambda, l_f)_{\lambda, f}, \quad V = (G_\lambda, m_f)_{\lambda, f}$$

be three projective systems of A-modules indexed by the category I; let $(u_\lambda)_\lambda : T \to$ $\to U$, $(v_\lambda)_\lambda : U \to V$ be projective systems of linear mappings such that for any $\lambda \in I$ the diagram

$$0 \to E_\lambda \xrightarrow{u_\lambda} F_\lambda \xrightarrow{v_\lambda} G_\lambda \tag{19}$$

is an exact sequence; then, denoting $u = \varprojlim_I u_\lambda$, $v = \varprojlim_I v_\lambda$, the diagram

$$0 \to \varprojlim_I E_\lambda \xrightarrow{u} \varprojlim_I F_\lambda \xrightarrow{v} \varprojlim_I G_\lambda$$

is an exact sequence.

REMARK. The fact that for any $\lambda \in I$ diagram (19) is an exact sequence is equivalent to the fact that in the abelian category $(\mathbf{Mod}(A))^{I^\circ}$ the diagram

$$0 \to T \to U \to V,$$

where the arrows represent exactly the given projective morphisms, is an exact sequence; Proposition 4 then states that the projective limit functor is left exact (see Proposition 3, § 5, the equivalence *iv*) ⇔ *v*)). ∎

We are not going to give in detail the *proof of Proposition* 4; the basic idea is the following: as in the above third particular case, consider a category J with two objects ρ, σ and two non-identity morphisms $f, g : \rho \to \sigma$; next, consider the product category $I \times J$ and the projective system of A-modules $(H_\lambda^\rho, H_\lambda^\sigma, w_h^f, w_h^g)$ indexed by the category $I \times J$, where $H_\lambda^\rho = G_\lambda$, $H_\lambda^\sigma = F_\lambda$, $\lambda \in I$, and if $h \in \mathrm{Hom}_I(\lambda, \mu)$, $w_h^f = v_\lambda \circ l_h = m_h \circ v_\mu$, $w_h^g = 0$; taking into account the property of projective limits of commuting with projective limits (Corollaries 3 and 4 to Proposition 1) and the fact that we have a canonical isomorphism $E_\lambda \rightleftarrows \varprojlim_J (H_\lambda^\rho, H_\lambda^\sigma, w_{1_\lambda}^f, w_{1_\lambda}^g)$, we get a canonical isomorphism $\varprojlim_I E_\lambda \rightleftarrows \varprojlim_J (\varprojlim_I (H_\lambda^\rho, H_\lambda^\sigma, w_h^f, w_h^g))$; in other words, $\varprojlim_I E_\lambda$ is identified with the kernel of v, the canonical homomorphism $\mathrm{Ker}\, v \to \varprojlim_I F_\lambda$ becoming u under this identification; the details are left to the reader. ∎

COROLLARY. *Let I be a small connected category, and let $T = (E_\lambda, u_f)_I$ be a projective system of A-modules indexed by the category I; if X and $(r_\lambda)_{\lambda \in I}$ are given as in Proposition* 1, *and if r_λ is a monomorphism for any $\lambda \in I$, then r constructed in Proposition* 1 *is a monomorphism.*

Proof. We saw before that the family $(r_\lambda)_{\lambda \in I}$ we are concerned with is a projective morphism from the constant projective system of value X to $(E_\lambda, u_f)_I$; since all r_λ, $\lambda \in I$, are monomorphisms, by Proposition 4 $\varprojlim_{\lambda \in I} r_\lambda$ is a monomorphism too; but I being a connected category, we know that r constructed in Proposition 1 is identified with $\varprojlim_{\lambda \in I} r_\lambda$, i.e., is obtained from this last one by composition with isomorphisms, whence the conclusion. ∎

* DEFINITION 9. If I is a small category while $\mathcal{C}$ is an arbitrary one, a functor $T : I^\circ \to \mathcal{C}$ is called a *projective* (or *inverse*) *system with values in the category* $\mathcal{C}$ indexed by the category I (of indices).

If $T(\lambda^\circ) = E_\lambda$, with $\lambda \in I$, and $T(f^\circ) = u_f$, with $f \in \mathrm{Hom}_I(\lambda, \mu)$, then the projective system T is denoted also $(E_\lambda, u_f)_{\lambda, f}$ or $(E_\lambda, u_f)_I$.

A system $(E, (p_\lambda)_{\lambda \in I})$, where, in the preceding notation, $E \in \mathrm{Ob}\, \mathcal{C}$ and $p_\lambda : E \to E_\lambda$, $\lambda \in I$, is said to be a *projective limit* or an *inverse limit* or, simply, a *limit* or also a *left root* in the category $\mathcal{C}$ of the projective system T if

$$u_f \circ p_\mu = p_\lambda$$

for any $\lambda, \mu \in I$ and any $f \in \operatorname{Hom}_I(\lambda, \mu)$ and if, moreover, $(E, (p_\lambda)_{\lambda \in I})$ is universal with this property; that is, for any system $(X, (r_\lambda)_{\lambda \in I})$, with $X \in \operatorname{Ob} \mathscr{C}$ and $r_\lambda \in \operatorname{Hom}_{\mathscr{C}}(X, E_\lambda)$, $\lambda \in I$, such that

$$u_f \circ r_\mu = r_\lambda, \qquad \lambda, \mu \in I,\ f \in \operatorname{Hom}_I(\lambda, \mu),$$

there exists a morphism $r : X \to E$ uniquely determined by the property that

$$p_\lambda \circ r = r_\lambda$$

for any $\lambda \in I$ (see the diagram given for modules in the statement of Proposition 1). ▮*

From Proposition 1 it follows that if A is a ring and $(E_\lambda, u_f)_{\lambda, f}$ is a projective system of A-modules indexed by the small category I, then the system $(\varprojlim_I E_\lambda, (p_\lambda)_{\lambda \in I})$ constructed before Proposition 1 is a projective limit in the category **Mod** (A) of the given projective system, in the sense of Definition 9.

Another example of projective limit in a category is the one in the category **Ann** of rings (with unit element); given a small category I and a projective system $(A_\lambda, u_f)_I$ with values in the category **Ann** indexed by the category I, we consider the subring A of $\prod_{\lambda \in I} A_\lambda$ (containing the unit element) consisting of all those families $(\alpha_\lambda)_{\lambda \in I}$, with $\alpha_\lambda \in A_\lambda$, $\lambda \in I$, for which, whatever $f \in \operatorname{Hom}_I(\lambda, \mu)$, $u_f(\alpha_\mu) = \alpha_\lambda$; denote by p_λ the homomorphism of rings $A \to A_\lambda$, $\lambda \in I$, obtained as the composition of the canonical projection $\prod_{\lambda \in I} A_\lambda \to A_\lambda$ with the inclusion homomorphism $A \to \prod_{\lambda \in I} A_\lambda$; the system $(A, (p_\lambda)_{\lambda \in I})$ is a projective limit in the category **Ann** of the projective system $(A_\lambda, u_f)_I$, in the sense of Definition 9.

* Returning to an arbitrary category $\mathscr{C}$, we shall use for the projective systems with values in the category $\mathscr{C}$ the same notations as for projective systems of A-modules; in particular, when I is the category associated with a preordered set.

From the definition it follows using a standard proof (see § 6) that if $(E, (p_\lambda)_{\lambda \in I})$ and $(E', (p'_\lambda)_{\lambda \in I})$ are projective limits of the projective system T with values in $\mathscr{C}$ and indexed by I, then there exists a morphism $\varphi : E \to E'$ of the category $\mathscr{C}$ uniquely determined by the property of rendering commutative the diagram

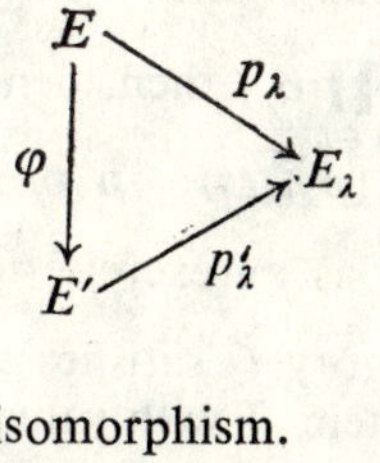

for any $\lambda \in I$; moreover, φ is an isomorphism.

Taking this into account, if for the projective system $T = (E_\lambda, u_f)_{\lambda, f}$ with values in $\mathcal{C}$ and indexed by I there exists at least one projective limit, then, from all these projective limits, one, say $(E, (p_\lambda)_{\lambda \in I})$, is chosen and called *the projective limit* of the projective system $(E_\lambda, u_f)_{\lambda, f}$; the object E is denoted in this case $\varprojlim_I T$ or also $\varprojlim_I (E_\lambda, u_f)$ or even $\varprojlim_{\lambda \in I} E_\lambda$ or $\varprojlim_I E_\lambda$ or $\varprojlim E_\lambda$, and it is this object which is sometimes, by an abuse of language, called the *projective limit* of the given projective system; the morphism p_λ, $\lambda \in I$, corresponding to this choice is called the *canonical morphism* (of index λ) from $\varprojlim_I E_\lambda$ to E_λ.

As for the existence of the projective limit, we have the following result:

Proposition 5. *Let $\mathcal{C}$ be an abelian category satisfying Axiom* AB3°), *that is, in which there exists a direct product for any family of objects; then for any small category I and any projective system T with values in the category $\mathcal{C}$ indexed by the category I there exists a projective limit in the category $\mathcal{C}$ of the projective system T.*

Proof. Let $T = (E_\lambda, u_f)_{\lambda, f}$, and let $(\prod_{\lambda \in I} E_\lambda, (p'_\lambda)_{\lambda \in I})$ be a direct product of the family of objects $(E_\lambda)_{\lambda \in I}$; for any $f \in \prod_{(\lambda, \mu) \in I \times I} \mathrm{Hom}_I(\lambda, \mu) = M$ let us consider, if $f \in \mathrm{Hom}_I(\lambda, \mu)$, the direct product $E_\mu \times E_\lambda$ and the morphism $s_f : E_\mu \times E_\lambda \to E_\lambda$ obtained as follows: by Proposition 4, § 7, there exist some morphisms $j^\mu_\lambda : E_\lambda \to E_\mu \times E_\lambda$, $j^\lambda_\mu : E_\mu \to E_\mu \times E_\lambda$ such that the system $(E_\mu \times E_\lambda, j^\lambda_\mu, j^\mu_\lambda)$ is a direct sum of the family of objects (E_μ, E_λ); then s_f is the morphism obtained from the family of morphisms

$$u_f : E_\mu \to E_\lambda, \quad -1_{E_\lambda} : E_\lambda \to E_\lambda$$

by the definition of the direct sum. Next, let $t_f : \prod_{\nu \in I} E_\nu \to E_\mu \times E_\lambda$ be the morphism obtained from the family of morphisms

$$p'_\mu : \prod_\nu E_\nu \to E_\mu, \quad p'_\lambda : \prod_\nu E_\nu \to E_\lambda,$$

by the definition of the direct product; let $w_f = s_f \circ t_f$; since the abelian category $\mathcal{C}$ satisfies Axiom AB3°), we may speak of the intersection $(\bigcap_{f \in M} \mathrm{Ker}\, w_f, i)$ of the family of subobjects $(\mathrm{Ker}\, w_f, i_f)_{f \in M}$ of $\prod_{\lambda \in I} E_\lambda$; then, denoting by p_λ the composition $p'_\lambda \circ i$, $\lambda \in I$, the system $(\bigcap_{f \in M} \mathrm{Ker}\, w_f, (p_\lambda)_{\lambda \in I})$ is a projective limit of the projective system T; we leave it to the reader as an exercise to verify this assertion. ∎

Definition 10. If the category $\mathcal{C}$ satisfies the condition that for any small category I and any projective system T with values in $\mathcal{C}$ indexed by I there exists a projective limit of T in $\mathcal{C}$, then $\mathcal{C}$ is said to be a *category with projective limits* (or *with inverse limits* or *with limits*) or also a *complete category* or a *left complete category*; if in the above statement only finite categories I are considered, $\mathcal{C}$ is said to be a *category with finite projective limits* or also a *finitely* (*left*) *complete category*. ∎

In these terms, Proposition 5 may be reformulated by saying that *an abelian category with products is complete*; from the proof of Proposition 5 it may be seen that, without assuming the validity of Axiom AB3°), *any abelian category is finitely complete.*

If $\mathcal{C}$ is an abelian[1] complete category, the results given for modules by Corollaries 1, 2, 3, 4, 5 to Proposition 1 are still valid for the category $\mathcal{C}$ with the appropriate replacements; this is because in the proofs given for modules we did not use elements but only the universal property of the projective limit, involved also in Definition 9; of course, in any of these cases it is assumed that a determined projective limit is chosen, and it is this one we are dealing with; it is also assumed that in the appropriate place the adequate definition was introduced: thus, one introduces the notion of *projective morphism* or *projective system of morphisms* from one projective system to another, as well as the notion of *projective limit* of a projective morphism; also, the notion of *projective limit functor* $\mathcal{C}^{I^\circ} \to \mathcal{C}$, and so on.

We notice here that *if M is a set partially ordered by the relation $\leqslant$, then it is a meet-semilattice if and only if the category associated with the partially ordered set M is a category with finite* (*non-empty*[2]) *direct products, or if and only if it is a category with finite projective limits*: indeed, the meet of a finite family (together with the unique morphisms given by the relation $\leqslant$) is a direct product and a projective limit of this family in the category associated with the partially ordered set M; also, *the set M partially ordered by the relation $\leqslant$ is a complete meet-semilattice if and only if the associated category is a category with arbitrary* (*non-empty*) *direct products, or if and only if it is a category with projective limits.*

As for the first particular case, one may define in the same way as for modules for any $X \in \mathrm{Ob}\,\mathcal{C}$ the *constant projective system of value X* indexed by the (small) category I of indices; if I is a connected category, then $(X, (p_\lambda)_{\lambda \in I})$, where $p_\lambda = 1_X$ for any $\lambda \in I$, is a projective limit of this projective system; for any small category I, the *immersion functor* $\mathcal{C} \to \mathcal{C}^{I^\circ}$ may be defined, and this is, in the case of a category $\mathcal{C}$ with projective limits, a left adjoint of the projective limit functor.

Using the notion of constant projective system, another interpretation of the projective limit of a projective system with values in a category may be given, namely the following:

Let $\mathcal{C}$, $\mathcal{D}$ be categories; for each $X \in \mathrm{Ob}\,\mathcal{C}$ the *constant functor $\mathcal{D} \to \mathcal{C}$ of value X*, denoted C_X, may be defined in the same way as a constant projective system, namely assigning to any object of $\mathcal{D}$ the object X and to any morphism of the category $\mathcal{D}$ the morphism 1_X. Now, if $T : \mathcal{D} \to \mathcal{C}$ is a functor, we define

---

[1] The hypothesis that $\mathcal{C}$ is an abelian category is not essential for all the established results.
[2] That is, direct products of families indexed by finite non-void sets of indices.

in the following manner a new category [1], denoted $\mathcal{C}/T$: as objects of $\mathcal{C}/T$ the natural transformations $C_X \to T$ are taken for all $X \in \mathrm{Ob}\,\mathcal{C}$; in other words, an object of $\mathcal{C}/T$ is a system $(X, (w_E)_{E \in \mathrm{Ob}\,\mathcal{D}})$, where $X \in \mathrm{Ob}\,\mathcal{C}$ and w_E is a morphism $X \to T(E)$, $E \in \mathrm{Ob}\,\mathcal{D}$, such that if $u \in \mathrm{Hom}_{\mathcal{D}}(E, F)$, the following diagram is commutative:

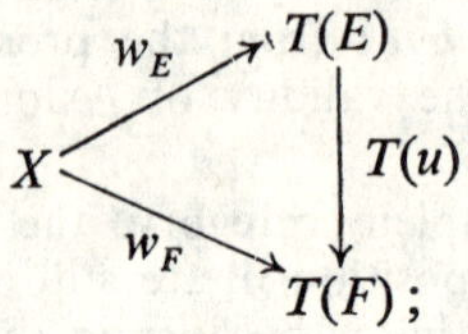

next, if $(X, (w_E)_E)$ and $(Y, (t_E)_E)$ are two such objects, a morphism $(X, (w_E)_E) \to$ $\to (Y, (t_E)_E)$ is a morphism $f \in \mathrm{Hom}_{\mathcal{C}}(X, Y)$ with the property that for any $E \in \mathrm{Ob}\,\mathcal{D}$ the following diagram is commutative:

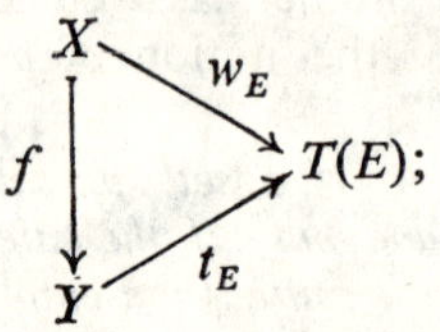

the composition of morphisms in $\mathcal{C}/T$ is their composition in $\mathcal{C}$; likewise for the identity morphisms.

In the particular case where T is the constant functor of value Z, where $Z \in \mathrm{Ob}\,\mathcal{C}$, the category $\mathcal{C}/T$ is denoted by $\mathcal{C}/Z$; of course, in this case, the category $\mathcal{D}$ may be replaced by any other category, in particular by $\mathcal{C}$, without "practically" changing the category $\mathcal{C}/Z$; in other words, the different categories C/Z obtained for the different categories $\mathcal{D}$ may be identified in an obvious manner.

The dual notion is the category denoted $T/\mathcal{C}$, respectively $Z/\mathcal{C}$; the objects of $T/\mathcal{C}$ are thus the natural transformations $T \to C_X$ for all $X \in \mathrm{Ob}\,\mathcal{C}$.

DEFINITION 11. Let $\mathcal{C}$ be a category; a *final* (or *terminal*) *object* of the category $\mathcal{C}$ is an object X with the property that for any $E \in \mathrm{Ob}\,\mathcal{C}$, the set $\mathrm{Hom}_{\mathcal{C}}(E, X)$ consists of exactly one element; an *initial object* of the category $\mathcal{C}$ is an object Y of $\mathcal{C}$ having the property that for any object E of $\mathcal{C}$, the set $\mathrm{Hom}_{\mathcal{C}}(Y, E)$ consists of exactly one element. ∎

If $\mathcal{C}$ is a preadditive category, then, by the considerations following Definition 6, § 2, an object X of $\mathcal{C}$ is an initial object if and only if it is a terminal object, or if and only if it is a zero object.

---

[1] The existence of this category is conditioned by the existence of at least one of its objects; the same remark is also valid for other constructions of this type.

Considering again an arbitrary category $\mathcal{C}$, it may be seen in the same way as for the zero objects of a preadditive category (§ 2) that *any two final objects of $\mathcal{C}$ are isomorphic*; likewise, *any two initial objects are isomorphic.*

Now let $T : I^\circ \to \mathcal{C}$ be a projective system with values in the category $\mathcal{C}$ indexed by the small category I; by Definition 9, a projective limit of T in the category $\mathcal{C}$ is exactly a final object of the category $\mathcal{C}/T$. Consequently, *the existence of a projective limit of the projective system T amounts to the existence of a final object of $\mathcal{C}/T$.*

As for the second, the third, and the fourth particular cases of projective systems with values in a category $\mathcal{C}$, things are similar to those for modules if $\mathcal{C}$ is an abelian category satisfying AB3°) (this last Axiom should not necessarily be satisfied for the third case); the only difference is that for an abelian category $\mathcal{C}$ with direct products, the *identifications* with the product, with the kernel, and with the intersection of the corresponding projective limits are useless, these constructions *being* projective limits of the considered projective systems; the second way of regarding the direct product as a projective limit is maintained also in this general case. In particular, Proposition 4 and the Corollary to it are valid also in this framework.

We shall insist a little more on the fifth particular case:

DEFINITION 12. Let $\mathcal{C}$ be a category, and let the diagram (13) be given, with E, F, G, u, v objects, respectively morphisms of the category $\mathcal{C}$; a *fibered product* (or an *amalgamated product*) in the category $\mathcal{C}$ *of E and F with respect to* (or *relative to*) *this diagram* (or, if there is no danger of confusion, *with respect to G* or *relative to G*) is a system (P, u', v'), where P is an object of $\mathcal{C}$ and $u' \in \mathrm{Hom}_{\mathcal{C}}(P, E)$, $v' \in \mathrm{Hom}_{\mathcal{C}}(P, F)$, making diagram (14), with the present meaning of vertices and arrows, commute and which is universal with this property: i.e., whatever the system (X, r, s), with $X \in \mathrm{Ob}\,\mathcal{C}$, $r : X \to E$, $s : X \to F$, such that diagram (15) commutes, it follows the existence of a unique morphism $\varphi : X \to P$ satisfying conditions (16).

A diagram of type (14) of the category $\mathcal{C}$ is called a *cartesian square* or a *cartesian diagram* or a *pullback* provided (P, u', v') is a fibered product of E and F with respect to the diagram (13) obtained from (14) by dropping P, u', v'. ▮*

We notice that the module $P = E \times_G F$ together with the linear mappings u', v' constructed for modules constitute an amalgamated product of E and F in the category **Mod**(A) with respect to the diagram (13) of A-modules, in the sense of Definition 12.

* Returning to an arbitrary category $\mathcal{C}$, it may be shown using the standard procedure that if (P, u', v') and (P_1, u_1', v_1') are fibered products of E and F in the category $\mathcal{C}$ with respect to the same diagram, say (13), then there exists an isomorphism $\theta : P \xrightarrow{\sim} P_1$ of the category $\mathcal{C}$ such that

$$u_1' \circ \theta = u', \quad v_1' \circ \theta = v';$$

then, as usual, when for diagram (13) at least a fibered product exists, one of them, say (P, u', v'), is chosen, and it is called *the fibered* (or *the amalgamated*) *product* of E and F with respect to the given diagram (13); the object P is denoted, if there is no danger of confusion, by $E \times F$ or by $E \prod_G F$, and sometimes it is, by an abuse of language, called the respective *fibered product*.

Concerning the existence, we have the following Proposition:

PROPOSITION 6. *If $\mathcal{C}$ is an abelian category, then for any diagram of the type* (13) *there exists a fibered product of E and F relative to this diagram, or, equivalently, any diagram of the type* (13) *may be enlarged to a cartesian square.*

Proof. If (H, p_E, p_F) is a direct product of the family of objects (E, F), let (P, i) be a kernel of the morphism

$$u \circ p_E - v \circ p_F : H \to G;$$

then, denoting $u' = p_E \circ i$, $v' = p_F \circ i$, the system (P, u', v') is a fibered product of E and F with respect to diagram (13). ▮

It may be shown that if (P', u', v') is a fibered product of E and F with respect to diagram (13) of objects and morphisms of the category $\mathcal{C}$, then (P, u', v', w'), where $w' = u \circ u' = v \circ v'$, is a projective limit in $\mathcal{C}$ of the projective system $(E_\lambda, E_\mu, E_\nu, u_{\nu\lambda}, u_{\nu\mu})$ defined as in the analogous case for modules; conversely, if (P, u', v', w') is a projective limit of this projective system, then (P, u', v') is a fibered product of E and F with respect to diagram (13).

We mention the following property of the fibered product:

PROPOSITION 7. *Let $\mathcal{C}$ be a category, and let* (14) *be a cartesian square of this category; if u is a monomorphism, respectively an isomorphism of the category $\mathcal{C}$, then v' is also a monomorphism, respectively an isomorphism; likewise for v and u'.*

Proof. Let u be a monomorphism, let $X \in \text{Ob}\,\mathcal{C}$, and $X \underset{w_2}{\overset{w_1}{\rightrightarrows}} P$ be two morphisms such that

$$v' \circ w_1 = v' \circ w_2; \tag{20}$$

it follows

$$u \circ u' \circ w_1 = v \circ v' \circ w_1 = v \circ v' \circ w_2 = u \circ u' \circ w_2,$$

and since u is a monomorphism of the category $\mathcal{C}$,

$$u' \circ w_1 = u' \circ w_2; \tag{21}$$

since P is a fibered product of E and F relative to G, from (20) and (21) it follows, by the uniqueness property in the definition of the fibered product, that $w_1 = w_2$; consequently, v' is a monomorphism.

If u is an isomorphism, let us consider the object F and the morphisms $u^{-1}\circ v : F \to E$ and $1_F : F \to F$; we have the diagram

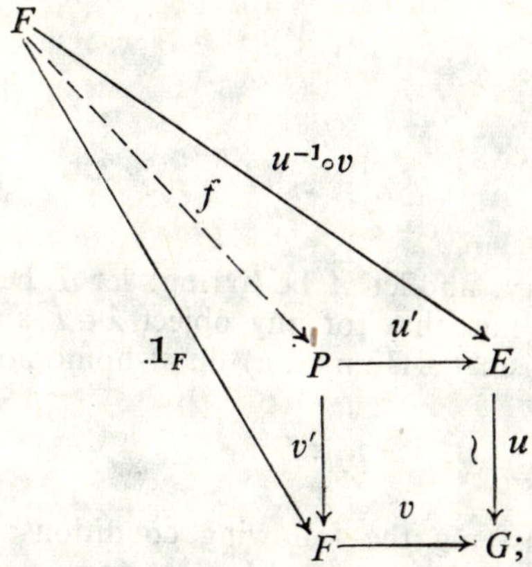

since trivially $u\circ u^{-1}\circ v = v\circ 1_F$, and since we started from a cartesian square, there exists $f : F \to P$ such that

$$v'\circ f = 1_F;$$

hence $v'\circ f\circ v' = v'$; since u is a monomorphism, it follows from the first part that v' is a monomorphism; hence

$$f\circ v' = 1_P,$$

and v' is an isomorphism. ▮*

§ 13. Inductive Limits

Let I be a small category, and let A be a ring; let T be a functor $I \to \mathbf{Mod}(A)$; this is equivalent to the fact that for any object $\lambda \in I$ an A-module E_λ is given, and that for any morphism $f \in \mathrm{Hom}_I(\lambda, \mu)$ an A-homomorphism

$$u_f : E_\lambda \to E_\mu$$

is given, these data satisfying the following conditions:

a) for any diagram in the category I of the form

$$\lambda \xrightarrow{f} \mu \xrightarrow{g} \nu$$

we have

$$u_{g \circ f} = u_g \circ u_f;$$

b) for any $\lambda \in I$ we have

$$u_{1_\lambda} = 1_{E_\lambda}.$$

In the particular case where I is the category associated with a preordered set denoted also by I, to give the functor T as above amounts to giving for any $\lambda \in I$ an A-module E_λ, and for any pair λ, μ for which $\lambda \leqslant \mu$ in I an A-homomorphism

$$u_{\lambda\mu} : E_\lambda \to E_\mu$$

such that if $\lambda \leqslant \mu \leqslant \nu$ in I, then $u_{\mu\nu} \circ u_{\lambda\mu} = u_{\lambda\nu}$, and, moreover, $u_{\lambda\lambda} = 1_{E_\lambda}$ for any $\lambda \in I$.

DEFINITION 1. A functor T as the one described above is called an *inductive* (or *direct*) *system of A-modules* indexed by the category I (of indices); if T is the inductive system given above, then it will be sometimes denoted $(E_\lambda, u_f)_{\lambda, f}$ or $(E_\lambda, u_f)_I$, and in the particular case when I is the category associated with the preordered set I, it will be denoted $(E_\lambda, u_{\lambda\mu})_{\lambda \in I}$ or $(E_\lambda, u_{\lambda\mu})$. ∎

REMARK. If a functor $T : I \to \mathbf{Mod}_d(A)$ is given, then we have to do with an *inductive system of right A-modules*. ∎

Let us consider the inductive system $T = (E_\lambda, u_f)_{\lambda, f}$ of left A-modules indexed by the small category I; in the direct sum $\bigoplus_{\lambda \in I} E_\lambda$, with the canonical injections $j'_\lambda : E_\lambda \to \bigoplus_\lambda E_\lambda$, let us consider for any $\lambda, \mu \in I$, for any $f \in \mathrm{Hom}_I(\lambda, \mu)$, and for any $x_\lambda \in E_\lambda$, the element

$$j'_\mu(u_f(x_\lambda)) - j'_\lambda(x_\lambda); \tag{1}$$

let R be the submodule of $\bigoplus_{\lambda \in I} E_\lambda$ generated by all the elements of the form (1) with the given properties, and let $E = (\bigoplus_{\lambda \in I} E_\lambda)/R$. Notice that R may be characterized as being the sum over all $f \in \coprod_{(\lambda, \mu) \in I \times I} \mathrm{Hom}_I(\lambda, \mu)$ of the images of the linear mappings $E_\lambda \to \bigoplus_\lambda E_\lambda$ given by $x_\lambda \mapsto j'_\mu(u_f(x_\lambda)) - j'_\lambda(x_\lambda)$ (here $f \in \mathrm{Hom}_I(\lambda, \mu)$).

DEFINITION 2. Given an inductive system T as above, the constructed module E is called the *inductive limit* or the *direct limit* or the *colimit* or the *right root* of the given inductive system, and is denoted $\varinjlim_I T$ or also $\varinjlim_I (E_\lambda, u_f)$ or also $\varinjlim_{\lambda \in I} E_\lambda$ or $\varinjlim_I E_\lambda$ or $\varinjlim E_\lambda$; if I is the category associated with the preordered set I, then instead of the general notation introduced one also writes $\varinjlim_{\lambda \in I} (E_\lambda, u_{\lambda\mu})$; the linear mapping $E_\lambda \to E$, $\lambda \in I$, obtained as the composition of the canonical surjection $p : \bigoplus_{\lambda \in I} E_\lambda \to (\bigoplus_{\lambda \in I} E_\lambda)/R$ with j'_λ is called the *canonical mapping* of index λ, $\lambda \in I$. ∎

The inductive limit has the following universal property:

PROPOSITION 1. *Let $E = \varinjlim_I T$ be the inductive limit of the inductive system $T = (E_\lambda, u_f)_{\lambda, f}$ of A-modules indexed by the small category I; denoting by j_λ, $\lambda \in I$, the canonical mapping of index λ, we have for any $\lambda, \mu \in I$ and any $f \in \mathrm{Hom}_I(\lambda, \mu)$,*

$$j_\mu \circ u_f = j_\lambda; \tag{2}$$

the system $(E, (j_\lambda)_{\lambda \in I})$ is universal with this property, that is, whatever the A-module Y and the family of linear mappings $(s_\lambda)_{\lambda \in I}$, with $s_\lambda : E_\lambda \to Y$, $\lambda \in I$, such that

$$s_\mu \circ u_f = s_\lambda \tag{3}$$

for any $\lambda, \mu \in I$ and $f \in \mathrm{Hom}_I(\lambda, \mu)$, there exists a homomorphism $s : E \to Y$ uniquely determined by the property that

$$s \circ j_\lambda = s_\lambda \tag{4}$$

for any $\lambda \in I$.

Proof. To prove relation (2), let $x_\lambda \in E_\lambda$; using the preceding notations, any element of the form (1) is in R; hence

$$(j_\mu \circ u_f)(x_\lambda) - j_\lambda(x_\lambda) = p((j'_\mu \circ u_f)(x_\lambda) - j'_\lambda(x_\lambda)) = 0.$$

To prove the second assertion of the Proposition, let us consider the homomorphism

$$s' : \bigoplus_{\lambda \in I} E_\lambda \to Y$$

obtained from the family of homomorphisms $(s_\lambda)_{\lambda \in I}$ by virtue of the universal property of the direct sum; but s' vanishes on R: by Corollary 3 to Proposition 2, § 8, it suffices to show that s' vanishes at any element of the form (1); this follows from relations (3) and from the fact that $s' \circ j'_\lambda = s_\lambda$ for any $\lambda \in I$; then, by the universal property of the quotient module (see § 3), it follows the existence of the unique linear mapping $s : E \to Y$ with the property that $s \circ p = s'$; by composition with j'_λ, $\lambda \in I$, we get relations (4); the uniqueness of s satisfying these relations is a consequence of the uniqueness of the mappings obtained, respectively, by the universal properties of the direct sum and of the quotient module. ▌

* DEFINITION 3. Let I be a small category, and let $\mathcal{C}$ be an arbitrary category; an *inductive* (or *direct*) *system with values in the category* $\mathcal{C}$ indexed by the category I (of indices) is a functor $T : I \to \mathcal{C}$.

In this case, denoting $T(\lambda) = E_\lambda$, $\lambda \in I$, and $T(f) = u_f$, $f \in \mathrm{Hom}_I(\lambda, \mu)$, the inductive system T is also denoted $(E_\lambda, u_f)_{\lambda, f}$ or $(E_\lambda, u_f)_I$, and if I is the category associated with the preordered set I, then T is denoted $(E_\lambda, u_{\lambda\mu})_{\lambda \in I}$ or $(E_\lambda, u_{\lambda\mu})_I$, where $u_{\lambda\mu}$ is the morphism in $\mathcal{C}$ obtained by applying the functor T (in case $\lambda \leqslant \mu$) to the unique morphism $\lambda \to \mu$.

A system $(E, (j_\lambda)_{\lambda \in I})$, where, using the preceding notation, $E \in \mathrm{Ob}\,\mathcal{C}$ and $j_\lambda : E_\lambda \to E$, $\lambda \in I$, is said to be an *inductive limit* or a *direct limit* or also a *colimit* or, alternatively, a *right root* in the category $\mathcal{C}$ of the inductive system T if for any $\lambda, \mu \in I$ and any $f \in \mathrm{Hom}_I(\lambda, \mu)$ we have

$$j_\mu \circ u_f = j_\lambda$$

and if, moreover, the system $(E, (j_\lambda)_{\lambda \in I})$ is universal with this property, that is, whatever the system $(Y, (s_\lambda)_{\lambda \in I})$, with $Y \in \mathrm{Ob}\,\mathcal{C}$ and $s_\lambda \in \mathrm{Hom}_{\mathcal{C}}(E_\lambda, Y)$, $\lambda \in I$, such that

$$s_\mu \circ u_f = s_\lambda$$

for any $\lambda, \mu \in I$ and $f \in \mathrm{Hom}_I(\lambda, \mu)$, there exists a morphism $s : E \to Y$ uniquely determined by the property of satisfying the relations

$$s \circ j_\lambda = s_\lambda$$

for any $\lambda \in I$. ▌

REMARK. The notion of inductive limit turns out to be dual to that of projective limit: more precisely, if $T : I \to \mathcal{C}$ is an inductive system with values in the category $\mathcal{C}$ indexed by the (small) category I, then the dual functor $T^\circ : I^\circ \to \mathcal{C}^\circ$ is a projective system with values in the category $\mathcal{C}^\circ$ indexed by the category I; next, if $(E, (j_\lambda)_{\lambda \in I})$ is an inductive limit in the category $\mathcal{C}$ of the inductive system T, then $(E^\circ, (j_\lambda^\circ)_{\lambda \in I})$ is a projective limit in the category $\mathcal{C}^\circ$ of the projective system T°; likewise, if $T : I^\circ \to \mathcal{C}$ is a projective system with values in the category $\mathcal{C}$ indexed by the category I, then the dual functor $T^\circ : (I^\circ)^\circ = I \to \mathcal{C}^\circ$ is an inductive system with values in the category $\mathcal{C}^\circ$ indexed by I, and the remark concerning the limits is transposed also to this case. ▮

From the definition, using the standard proof (see § 6), it follows that if $(E, (j_\lambda)_{\lambda \in I})$ and $(E', (j'_\lambda)_{\lambda \in I})$ are inductive limits of the inductive system T as above, then there exists an isomorphism $\varphi : E \xrightarrow{\sim} E'$ of the category $\mathcal{C}$ such that $\varphi \circ j_\lambda = j'_\lambda$ for any $\lambda \in I$; next, the usual procedure is applied: if for the inductive system $T = (E_\lambda, u_f)_{\lambda, f}$ there exists at least one inductive limit, one of these limits, say $(E, (j_\lambda)_{\lambda \in I})$, is chosen, and this is called *the inductive limit* of the inductive system T; the object E is denoted in this case $\varinjlim_I T$ or also $\varinjlim_I (E_\lambda, u_f)$ or $\varinjlim_I E_\lambda$ or $\varinjlim E_\lambda$, and it is sometimes, by an abuse of language, called the *inductive limit* of the given inductive system; the morphism j_λ, $\lambda \in I$, corresponding to this choice is called the *canonical morphism* (of index λ) from E_λ to $\varinjlim_I E_\lambda$.*

By Proposition 1, if A is a ring and $(E_\lambda, u_f)_{\lambda, f}$ is an inductive system of A-modules indexed by the small category I, then the system $(\varinjlim_I E_\lambda, (j_\lambda)_{\lambda \in I})$ constructed before Proposition 1 is an inductive limit in the category **Mod**(A) of the given inductive system, in the sense of Definition 3.

Let $T = (E_\lambda, u_f)_{\lambda, f}$, $U = (f_\lambda, v_f)_{\lambda, f}$ be two inductive systems of A-modules indexed by the same category I; to give a natural transformation $T \to U$ is to give for any $\lambda \in I$ a homomorphism

$$l_\lambda : E_\lambda \to F_\lambda$$

such that, if $f \in \mathrm{Hom}_I(\lambda, \mu)$, with $\lambda, \mu \in I$, the following diagram is commutative:

$$\begin{array}{ccc} E_\lambda & \xrightarrow{u_f} & E_\mu \\ {\scriptstyle l_\lambda}\downarrow & & \downarrow{\scriptstyle l_\mu} \\ F_\lambda & \xrightarrow{v_f} & F_\mu . \end{array}$$

DEFINITION 4. If T and U are inductive systems of A-modules indexed by the same category I, a natural transformation $T \to U$ is called an *inductive* (or *direct*) *system of linear mappings* or also an *inductive* (or *direct*) *morphism* or a *morphism of inductive* (or *direct*) *systems from T to U* or *between T and U*.

* If T and U are inductive systems with values in the category $\mathcal{C}$, the same definition is given for the notion of *inductive* (or *direct*) *system of morphisms* or *inductive* (or *direct*) *morphism from T to U*.

An inductive morphism from T to U assigning to the object $\lambda \in I$ the morphism l_λ of the category $\mathcal{C}$ is also denoted $(l_\lambda)_{\lambda \in I}$.* ▌

As in the case of projective systems, if I is a small category and A a ring, taking the inductive systems of A-modules indexed by the category I as objects, and the inductive morphisms between these systems as morphisms, we get an abelian category, namely the category **Fonct** $(I, \mathbf{Mod}(A))$ or $(\mathbf{Mod}(A))^I$ (see § 10).

* The same except for the character of abelian category may be said if $\mathbf{Mod}(A)$ is replaced by an arbitrary category $\mathcal{C}$; in this way we get the category **Fonct** $(I, \mathcal{C})$ or $\mathcal{C}^I$; if $\mathcal{C}$ is an abelian category, then so is $\mathcal{C}^I$ with the obvious definitions for the additional structures.*

The following Corollaries to Proposition 1 are proved by considerations dual to those for projective systems:

COROLLARY 1. *Let $T = (E_\lambda, u_f)_{\lambda, f}$, $U = (F_\lambda, v_f)_{\lambda, f}$ be two inductive systems of A-modules indexed by the category I, and let $(l_\lambda)_{\lambda \in I}$ be an inductive morphism from T to U; then there exists an A-homomorphism $l : \varinjlim_I E_\lambda \to \varinjlim_I F_\lambda$ uniquely determined by the property that for any $\lambda \in I$, the following diagram, where j_λ and k_λ are the canonical mappings of index λ, is commutative:*

$$\begin{array}{ccc} E_\lambda & \xrightarrow{\ l_\lambda\ } & F_\lambda \\ {\scriptstyle j_\lambda}\downarrow & & \downarrow{\scriptstyle k_\lambda} \\ \varinjlim E_\lambda & \xrightarrow{\ \ l\ \ } & \varinjlim F_\lambda \end{array} .$$ ▌

DEFINITION 5. In the notation of Corollary 1, l is called the *inductive* (or the *direct*) *limit* of the inductive morphism $(l_\lambda)_{\lambda \in I}$, and it is denoted $\varinjlim_{\lambda \in I} l_\lambda$ or $\varinjlim_I l_\lambda$ or also, when no confusion might arise, $\varinjlim l_\lambda$. ▌

COROLLARY 2. *Given three inductive systems of A-modules $T = (E_\lambda, u_f)_{\lambda, f}$, $U = (F_\lambda, v_f)_{\lambda, f}$, $V = (G_\lambda, w_f)_{\lambda, f}$ indexed by the same small category I, and given the inductive morphisms $(l_\lambda)_{\lambda \in I}: T \to U$ and $(m_\lambda)_{\lambda \in I}: U \to V$, we have*

$$\varinjlim_{\lambda \in I}(m_\lambda \circ l_\lambda) = (\varinjlim_{\lambda \in I} m_\lambda) \circ (\varinjlim_{\lambda \in I} l_\lambda);$$

also,

$$\varinjlim_{\lambda \in I} 1_{E_\lambda} = 1_{\varinjlim E_\lambda};$$

if, moreover, the inductive morphism $(l'_\lambda)_{\lambda \in I} : T \to U$ *is also given, then*

$$\varinjlim_{\lambda \in I} (l_\lambda + l'_\lambda) = \varinjlim_{\lambda \in I} l_\lambda + \varinjlim_{\lambda \in I} l'_\lambda. \blacksquare$$

Corollaries 1 and 2 show that for a fixed small category I of indices and for a ring A, the mappings

$$T \mapsto \varinjlim_I T, \quad (l_\lambda)_{\lambda \in I} \mapsto \varinjlim_I l_\lambda$$

define an additive functor from the category $(\mathbf{Mod}(A))^I$ to $\mathbf{Mod}(A)$.

DEFINITION 6. The above defined functor $(\mathbf{Mod}(A))^I \to \mathbf{Mod}(A)$ is called the *inductive* (or *direct*) *limit functor*, and is denoted $\varinjlim_I$ or even $\varinjlim$. $\blacksquare$

COROLLARY 3. *Let I and J be two small categories, and let* $(E^\rho_\lambda, u^g_f)_{\lambda, \rho, f, g}$ *be an inductive system of A-modules indexed by the product category $I \times J$; let* $E = \varinjlim_{I \times J} (E^\rho_\lambda, u^g_f)$; *then, for any* $\rho \in J$, $(E^\rho_\lambda, u_f^{1^\rho})_{\lambda, f}$ *is an inductive system of A-modules indexed by the category I; let F^ρ be its inductive limit; if* $g \in \coprod_{(\rho, \sigma) \in J \times J} \mathrm{Hom}_J(\rho, \sigma)$ *is fixed such that* $g \in \mathrm{Hom}_J(\rho, \sigma)$, *then the family of linear mappings* $(u^g_{1_\lambda})_{\lambda \in I}$ *is an inductive morphism from* $(E^\rho_\lambda, u_f^{1^\rho})_{\lambda, f}$ *to* $(E^\sigma_\lambda, u_f^{1^\sigma})_{\lambda, f}$; *let* $v^g = \varinjlim_{\lambda \in I} u^g_{1_\lambda}$; *then* $(F^\rho, v^g)_{\rho, g}$ *is an inductive system of A-modules indexed by the category J;* let $F = \varinjlim_J (F^\rho, v^g)$; *then there exists a canonical isomorphism* $\psi : E \xrightarrow{\sim} F$, *hence a canonical isomorphism*

$$\varinjlim_{I \times J} E^\rho_\lambda \xrightarrow{\sim} \varinjlim_J (\varinjlim_I E^\rho_\lambda);$$

likewise, there exists a canonical isomorphism

$$\varinjlim_{I \times J} E^\rho_\lambda \xrightarrow{\sim} \varinjlim_I (\varinjlim_J E^\rho_\lambda).$$

Let $(E'^{\rho}_{\lambda}, u'^{g}_{f})_{\lambda, \rho, f, g}$ *be another inductive system of A-modules indexed by the category* $I \times J$, *and let* $(l^{\rho}_{\lambda})_{(\lambda, \rho) \in I \times J}$ *be an inductive system of linear mappings from* $(E^{\rho}_{\lambda}, u^{g}_{f})_{\lambda, \rho, f, g}$ *to* $(E'^{\rho}_{\lambda}, u'^{g}_{f})_{\lambda, \rho, f, g}$; *then, for a fixed* $\rho \in J$, *the family* $(l^{\rho}_{\lambda})_{\lambda \in I}$ *is an inductive morphism from* $(E^{\rho}_{\lambda}, u_f^{1_\rho})_{\lambda, f}$ *to* $(E'^{\rho}_{\lambda}, u'^{1_\rho}_f)_{\lambda, f}$; *let* $k^{\rho} = \varinjlim_{I} l^{\rho}_{\lambda}$; *then, using the above notations for the first inductive system and the corresponding ones for the second system,* $(k^{\rho})_{\rho \in J}$ *is an inductive morphism from* $(F^{\rho}, v^{g})_{\rho, g}$ *to* $(F'^{\rho}, v'^{g})_{\rho, g}$, *and the following diagram is commutative:*

$$\begin{array}{ccc}
\varinjlim_{I \times J} E^{\rho}_{\lambda} & \xrightarrow{\ \sim\ } & \varinjlim_{J} (\varinjlim_{I} E^{\rho}_{\lambda}) \\
\Big\downarrow \varinjlim_{I \times J} l^{\rho}_{\lambda} & & \Big\downarrow \varinjlim_{J} k^{\rho} = \varinjlim_{J} (\varinjlim_{I} l^{\rho}_{\lambda}) \\
\varinjlim_{I \times J} E'^{\rho}_{\lambda} & \xrightarrow{\ \sim\ } & \varinjlim_{J} (\varinjlim_{I} E'^{\rho}_{\lambda});
\end{array}$$

here, the horizontal arrows represent the canonical isomorphisms given in the first part of the Corollary; a similar commutative diagram is obtained by interchanging the roles of I and J. ∎

The existence of the canonical isomorphism

$$\varinjlim_{I} (\varinjlim_{J} E^{\rho}_{\lambda}) \xrightarrow{\sim} \varinjlim_{J} (\varinjlim_{I} E^{\rho}_{\lambda})$$

obtained by combining the two canonical isomorphisms constructed in Corollary 3, as well as the commutativity of the corresponding diagram for a given inductive morphism express the so-called "property of inductive limits of commuting with inductive limits".

COROLLARY 4. *Let I be a small pseudofiltered category, and let J be a full cofinal subcategory of it; let A be a ring, let* $T = (E_{\lambda}, u_f)_{\lambda, f}$ *be an inductive system of A-modules indexed by the category I, and let* $U = (E_{\rho}, u_g)_{\rho, g}$ *be the inductive system of A-modules obtained by the restriction of T to the subcategory J; in this case there exists a canonical A-isomorphism*

$$\varphi : \varinjlim_{I} E_{\lambda} \xrightarrow{\sim} \varinjlim_{J} E_{\rho}$$

such that, if $j_{\lambda}: E_{\lambda} \to \varinjlim_{I} E_{\lambda}$, $\lambda \in I$, *and* $k_{\rho}: E_{\rho} \to \varinjlim_{J} E_{\rho}$, $\rho \in J$, *are the canonical mappings, then* $\varphi \circ j_{\rho} = k_{\rho}$ *for any* $\rho \in J$; *if, moreover,* $T' = (E'_{\lambda}, u'_f)_{\lambda, f}$ *is another inductive system* *of A-modules indexed by the category I, if $U' = (E'_\rho, u'_g)_{\rho, g}$ is the inductive system obtained by the restriction of T' to J, and if $(l_\lambda)_{\lambda \in I}$ is an inductive morphism from T to T', then $(l_\rho)_{\rho \in J}$ is an inductive morphism from U to U', and we have the commutative diagram*

$$\begin{array}{ccc} \varinjlim_I E_\lambda & \xrightarrow{\sim} & \varinjlim_J E_\rho \\ {\scriptstyle \varinjlim_I l_\lambda}\downarrow & & \downarrow{\scriptstyle \varinjlim_J l_\rho} \\ \varinjlim_I E'_\lambda & \xrightarrow{\sim} & \varinjlim_J E'_\rho, \end{array}$$

the horizontal arrows representing the canonical isomorphisms. ▮

*Let $\mathcal{C}$ be a category; concerning the existence of inductive limits in the category $\mathcal{C}$, we have the following result:

PROPOSITION 2. *If $\mathcal{C}$ is an abelian category satisfying Axiom* AB3), *i.e. a category with direct sums, then for any small category I and any inductive system T with values in the category $\mathcal{C}$ indexed by the category I there exists an inductive limit in $\mathcal{C}$ of the inductive system T.*

The *proof* follows by considerations dual to those in the proof of Proposition 5, § 12. ▮

DEFINITION 7. A category $\mathcal{C}$ satisfying the condition that for any small category I and any inductive system with values in $\mathcal{C}$ indexed by I there exists an inductive limit of T in $\mathcal{C}$ is said to be a *category with inductive limits* (or *with direct limits* or *with colimits*) or also a *cocomplete category* or a *right complete category*; if in the above statement only finite categories I are considered, then $\mathcal{C}$ is said to be a *category with finite inductive* (or *direct*) *limits* or also a *finitely cocomplete category.* ▮

Proposition 2 above asserts that *an abelian category with coproducts is cocomplete*; likewise, *if $\mathcal{C}$ is an arbitrary abelian category* not necessarily assumed to satisfy Axiom AB3), *then $\mathcal{C}$ is finitely cocomplete.* In an abelian category $\mathcal{C}$ with inductive limits the results stated by Corollaries 1, 2, 3, 4 to Proposition 1 are valid with the appropriate replacements and with the choice in each case of a determined inductive limit; it is also assumed that at the proper moment the notion necessary to state the subsequent results is introduced: so, once the validity for the category $\mathcal{C}$ of the results stated by Corollary 1 is proved, we have to introduce the notion of *inductive limit* of an inductive morphism in the category $\mathcal{C}$. Also, for a given small category I, the *inductive* (or *direct*) *limit functor* $\varinjlim_I : \mathcal{C}^I \to \mathcal{C}$ may be introduced.

By the Remark following Definition 3, *the dual of a category with projective limits is a category with inductive limits, and conversely.*

We remark also that *for a partially ordered set M with the partial order relation denoted $\leqslant$, the fact that M is a join-semilattice (respectively a complete join-semilattice) is equivalent to the fact that the category associated with this partially ordered set is a category with finite non-empty direct sums (respectively a category with arbitrary non-empty direct sums), and to the fact that this category is with finite inductive limits (respectively a category with inductive limits).**

Returning to a ring A, we have the following results:

PROPOSITION 3. *a) If I is a small pseudofiltered connected category and $T = (E_\lambda, u_f)_{\lambda, f}$ is an inductive system of A-modules indexed by the category I, and if $E = \varinjlim_{I} E_\lambda$, with the canonical mappings j_λ, $\lambda \in I$, then for any $x \in E$ there exist some $\lambda \in I$ and some $x_\lambda \in E_\lambda$ such that $j_\lambda(x_\lambda) = x$.*

b) If I is the category associated with the preordered set I directed to the right and $(E_\lambda, u_{\lambda\mu})_{\lambda \in I}$ is an inductive system of A-modules indexed by the category I, then from the fact that for a certain $\lambda \in I$ and a certain $x_\lambda \in E_\lambda$ we have $j_\lambda(x_\lambda) = 0$, where $j_\mu : E_\mu \to \varinjlim_{\mu \in I} E_\mu = E$ is the canonical mapping of index μ, $\mu \in I$, it follows the existence of some $\rho \in I$, with $\lambda \leqslant \rho$ in I, such that $u_{\lambda\rho}(x_\lambda) = 0$.

c) If I, $(E_\lambda, u_{\lambda\mu})_{\lambda \in I}$, and j_λ, $\lambda \in I$, have the same meaning as at b), then, for $\lambda, \mu \in I$ and $x_\lambda \in E_\lambda$, $x_\mu \in E_\mu$, we have $j_\lambda(x_\lambda) = j_\mu(x_\mu)$ in $\varinjlim E_\lambda$ if and only if there exists some $\rho \in I$, with $\lambda \leqslant \rho$, $\mu \leqslant \rho$, such that $u_{\lambda\rho}(x_\lambda) = u_{\mu\rho}(x_\mu)$ in E_ρ.

d) If I, $T = (E_\lambda, u_f)_{\lambda, f}$, E, *and* j_λ, $\lambda \in I$, *are as at a), then, from the fact that for any $f \in \coprod_{(\lambda, \mu) \in I \times I} \mathrm{Hom}_I(\lambda, \mu)$ the linear mapping u_f is surjective it follows that the linear mapping j_λ is also surjective for any $\lambda \in I$.*

e) If I is a category as at b) and $(E_\lambda, u_{\lambda\mu})_{\lambda \in I}$ is an inductive system of A-modules indexed by the category I, then, from the fact that for any $\lambda, \mu \in I$ such that $\lambda \leqslant \mu$, $u_{\lambda\mu}$ is an injective, respectively bijective mapping it follows that the canonical mapping $j_\lambda : E_\lambda \to \varinjlim_{I} E_\lambda$ is injective, respectively bijective for any $\lambda \in I$.

f) If I, $T = (E_\lambda, u_f)_{\lambda, f}$, E, and j_λ, $\lambda \in I$, have the same meaning as at a), and if for each $\lambda \in I$, S_λ is a system of generators of E_λ, then $S = \bigcup_{\lambda \in I} j_\lambda(S_\lambda)$ is a system of generators of E.

g) If I, $(E_\lambda, u_{\lambda\mu})_{\lambda \in I}$, E, and j_λ, $\lambda \in I$, have the same meaning as at b), and if for each $\lambda \in I$, S_λ is a linearly independent subset, respectively a basis of E_λ, these subsets having the property that for $\lambda \leqslant \mu$ we have $u_{\lambda\mu}(S_\lambda) \subseteq S_\mu$, then $S = \bigcup_{\lambda \in I} j_\lambda(S_\lambda)$ is a linearly independent subset, respectively a basis of E.

Proof. a) We first notice that if I is a pseudofiltered connected category, then, for any pair λ, μ of objects of I there exists some $\nu \in I$ such that $\mathrm{Hom}_I(\lambda, \nu) \neq \emptyset$ and $\mathrm{Hom}_I(\mu, \nu) \neq \emptyset$: for, I being connected, there exist $\lambda_0, \ldots, \lambda_n \in I$, $n = 2m$, $m \in \mathbf{N}$, such that $\lambda_0 = \lambda$, $\lambda_n = \mu$ and that there exist $f_k \in \mathrm{Hom}_I(\lambda_{2k}, \lambda_{2k+1})$,

$k = 0, 1, \ldots, m-1$, $g_l \in \mathrm{Hom}_I(\lambda_{2l}, \lambda_{2l-1})$, $l = 1, \ldots, m$; since I is pseudofiltered, there exists ρ_1 and some morphisms h_1, j_1 such that the following diagram is commutative:

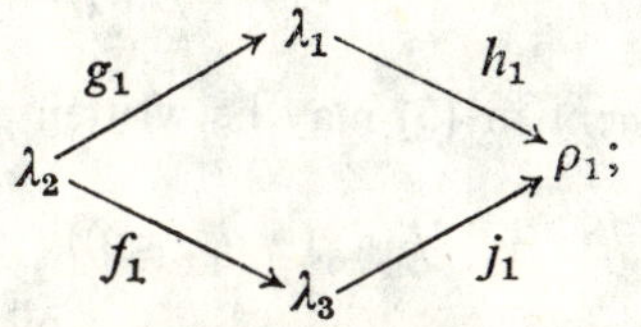

this last procedure is successively applied to $\lambda_4, \ldots, \lambda_{n-2}$ instead of λ_2, then to $\lambda_3, \lambda_5, \ldots, \lambda_{n-1}$, then to ρ_1 and to the similar elements, and so on; after a finite number of steps we get a ν and some morphisms $\lambda \to \nu$ and $\mu \to \nu$ obtained by composition.

We are now going to prove *a*): given $x \in E$, then there exists $(x_\mu)_{\mu \in I} \in \bigoplus_{\mu \in I} E_\mu$ such that if $p : \bigoplus_\mu E_\mu \to E$ is the canonical surjection, we have $p((x_\mu)_\mu) = x$; if the support of the family $(x_\mu)_{\mu \in I}$ is $\{\lambda_1, \ldots, \lambda_n\}$, then, generalizing by induction to n objects the foregoing remark for two objects of I, there exist $\lambda \in I$ and $f_i \in \mathrm{Hom}_I(\lambda_i, \lambda)$, $i = 1, \ldots, n$. Denoting by j'_μ, $\mu \in I$, the canonical injection $E_\mu \to \bigoplus_{\mu \in I} E_\mu$, we have according to Lemma 1, § 7,

$$(x_\mu)_{\mu \in I} = \sum_{\mu \in I} j'_\mu(x_\mu) = \sum_{i=1}^{n} j'_{\lambda_i}(x_{\lambda_i});$$

denoting by R the kernel of p, we have by the definition of the inductive limit that

$$\sum_{i=1}^{n} j'_{\lambda_i}(x_{\lambda_i}) - \sum_{i=1}^{n} j'_\lambda(u_{f_i}(x_{\lambda_i})) \in R;$$

consequently, applying p, we get

$$x = p((x_\mu)_\mu) = p\left(\sum_{i=1}^{n} j'_{\lambda_i}(x_{\lambda_i})\right) = p\left(\sum_{i=1}^{n} j'_\lambda(u_{f_i}(x_{\lambda_i}))\right) = j_\lambda\left(\sum_{i=1}^{n} u_{f_i}(x_{\lambda_i})\right);$$

taking $x_\lambda = \sum_{i=1}^{n} u_{f_i}(x_{\lambda_i})$, we get the required result.

b) Denote by p the canonical surjection $\bigoplus_{\lambda \in I} E_\lambda \to E$, and by R the kernel of p; if j'_λ is the canonical injection $E_\lambda \to \bigoplus_{\lambda \in I} E_\lambda$, $\lambda \in I$, then from $j_\lambda(x_\lambda) = 0$, with $x_\lambda \in E_\lambda$, it follows that $j'_\lambda(x_\lambda) \in R$; consequently, $j'_\lambda(x_\lambda)$ is a finite linear combination with coefficients in A of elements of the form

(5) $$j'_\nu(u_{\mu\nu}(x_\mu)) - j'_\mu(x_\mu),$$

with $x_\mu \in E_\mu$ and $\mu \leqslant \nu$; let $\rho \in I$ be a common upper bound of λ and of all μ and ν occurring in the given expression of $j'_\lambda(x_\lambda)$; we have

$$j'_\rho(u_{\lambda\rho}(x_\lambda)) = (j'_\rho(u_{\lambda\rho}(x_\lambda)) - j'_\lambda(x_\lambda)) + j'_\lambda(x_\lambda);$$

on the other hand, the element in (5) may be written as

$$j'_\rho(u_{\mu\rho}(x_\mu)) - j'_\mu(x_\mu) + j'_\rho(u_{\nu\rho}(-u_{\mu\nu}(x_\mu))) - j'_\nu(-u_{\mu\nu}(x_\mu)),$$

so that $j'_\rho(u_{\lambda\rho}(x_\lambda))$ may be considered as a finite linear combination of elements of the form (5) in which all ν are equal to ρ; putting together all terms corresponding to the same μ, we get for $j'_\rho(u_{\lambda\rho}(x_\lambda))$ an expression of the form

$$j'_\rho(u_{\lambda\rho}(x_\lambda)) = \sum_{i=1}^{n} (j'_\rho(u_{\mu_i\rho}(y_{\mu_i})) - j'_{\mu_i}(y_{\mu_i})), \tag{6}$$

with $\mu_i \leqslant \rho$, $y_{\mu_i} \in E_{\mu_i}$, $i = 1, \ldots, n$, and $\mu_i \neq \mu_j$ for $i \neq j$; since equality (6) is between elements of $\bigoplus_{\lambda \in I} E_\lambda$, for $\mu_i \neq \rho$ we get that $j'_{\mu_i}(y_{\mu_i}) = 0$, hence, as j'_{μ_i} are injective mappings, the term corresponding to μ_i in the right hand side of (6) is zero; but if, eventually, there exists $\mu_i = \rho$, then, as $u_{\rho\rho} = 1_{E_\rho}$, the term corresponding to $\mu_i = \rho$ in the right hand side of (6) is also zero; it follows that $j'_\rho(u_{\lambda\rho}(x_\lambda)) = 0$, hence that $u_{\lambda\rho}(x_\lambda) = 0$.

c) If there exists $\rho \in I$ such that $\lambda \leqslant \rho$, $\mu \leqslant \rho$, and $u_{\lambda\rho}(x_\lambda) = u_{\mu\rho}(x_\mu)$, then, applying j_ρ to this equality, we get $j_\lambda(x_\lambda) = j_\mu(x_\mu)$; conversely, if we have this last equality, let $\nu \in I$ with the property that $\lambda \leqslant \nu$, $\mu \leqslant \nu$; hence we have

$$j_\nu(u_{\lambda\nu}(x_\lambda) - u_{\mu\nu}(x_\mu)) = 0;$$

according to *b*) there exists $\rho \in I$, with $\nu \leqslant \rho$, such that

$$u_{\nu\rho}(u_{\lambda\nu}(x_\lambda) - u_{\mu\nu}(x_\mu)) = 0.$$

d) Let, under the hypotheses stated, $x \in E$; according to *a*), whose hypotheses are satisfied here, there exist some $\mu \in I$ and $x_\mu \in E_\mu$ such that $x = j_\mu(x_\mu)$; if $\lambda \in I$, then, by the remark at the beginning of *a*), there exist $\nu \in I$ and $f \in \mathrm{Hom}_I(\lambda, \nu)$, $g \in \mathrm{Hom}_I(\mu, \nu)$; by virtue of the surjectivity of u_f, there exists $x_\lambda \in E_\lambda$ such that in E_ν we have $u_f(x_\lambda) = u_g(x_\mu)$; but then

$$x = j_\mu(x_\mu) = j_\nu(u_g(x_\mu)) = j_\nu(u_f(x_\lambda)) = j_\lambda(x_\lambda).$$

e) For the assertion concerning the injectivity, let $\lambda \in I$ and $x_\lambda \in E_\lambda$ such that $j_\lambda(x_\lambda) = 0$; as shown at *b*), there exists some $\rho \in I$, $\lambda \leqslant \rho$, such that $u_{\lambda\rho}(x_\lambda) = 0$; since, by hypothesis, $u_{\lambda\rho}$ is injective, it follows $x_\lambda = 0$; hence j_λ is injective.

For the assertion concerning the bijectivity, if I is as stated, then the category I is pseudofiltered and connected; consequently, we have only to apply the assertion concerning the injectivity, and *d*).

f) Let $x \in E$; by *a*) there exist some $\lambda \in I$ and $x_\lambda \in E_\lambda$ such that $j_\lambda(x_\lambda) = x$; by hypothesis, there exists a finite family $x_{\lambda_1}, \ldots, x_{\lambda_n}$ of elements of S_λ such that $x_\lambda = \sum_{i=1}^{n} \alpha_i x_{\lambda_i}$, with some $\alpha_i \in A$, $i = 1, \ldots, n$; it follows

$$x = j_\lambda(x_\lambda) = \sum_{i=1}^{n} \alpha_i j_\lambda(x_{\lambda_i}),$$

where $j_\lambda(x_{\lambda_i}) \in j_\lambda(S_\lambda) \subseteq S$, $i = 1, \ldots, n$.

g) Let us prove the first assertion: as shown in § 10, it will be sufficient to prove that any *finite* subset of S is linearly independent; then let $x_1, \ldots, x_n$ be distinct elements of $\bigcup_{\lambda \in I} j_\lambda(S_\lambda)$; hence there exist some indices $\lambda_1, \ldots, \lambda_n$ and some elements $x_{\lambda_i} \in S_{\lambda_i}$, $i = 1, \ldots, n$, such that $j_{\lambda_i}(x_{\lambda_i}) = x_i$, $i = 1, \ldots, n$; if $\lambda \geqslant \lambda_i$ for any $i = 1, \ldots, n$, then $x_i = j_\lambda(u_{\lambda_i \lambda}(x_{\lambda_i}))$, $i = 1, \ldots, n$; assume that there exist $\alpha_1, \ldots, \alpha_n \in A$ such that

$$\sum_{i=1}^{n} \alpha_i x_i = 0;$$

it follows that $j_\lambda\left(\sum_{i=1}^{n} \alpha_i u_{\lambda_i \lambda}(x_{\lambda_i})\right) = 0$, and according to *b*) there exists some $\mu \in I$, $\mu \geqslant \lambda$, such that

$$0 = \sum_{i=1}^{n} \alpha_i u_{\lambda\mu}(u_{\lambda_i \lambda}(x_{\lambda_i})) = \sum_{i=1}^{n} \alpha_i u_{\lambda_i \mu}(x_{\lambda_i});$$

but $x_i = j_\mu(u_{\lambda_i \mu}(x_{\lambda_i}))$, $i = 1, \ldots, n$, and hence the elements $u_{\lambda_1 \mu}(x_{\lambda_1}), \ldots, u_{\lambda_n \mu}(x_{\lambda_n})$ are distinct; moreover, by hypothesis, $u_{\lambda_i \mu}(x_{\lambda_i}) \in u_{\lambda_i \mu}(S_{\lambda_i}) \subseteq S_\mu$, and hence $\alpha_1 = \ldots \ldots = \alpha_n = 0$.

To prove the second assertion we need only use the first assertion and *f*), taking into account the fact that a basis is a set of generators which is also a linearly independent subset. ∎

In an important particular case the inductive limit may also be constructed in a different, sometimes useful manner, namely;

Let I be the category associated with the preordered set I directed to the right; let A be a ring, and let $(E_\lambda, u_{\lambda\mu})_{\lambda \in I}$ be an inductive system of A-modules indexed by I; consider the set $F = \coprod_{\lambda \in I} E_\lambda$, i.e. the disjoint union of the family of underlying sets; denote by k_λ the canonical injection of E_λ into F, $\lambda \in I$; we introduce on F the following binary relation, denoted R : if $k_\lambda(x_\lambda)$, $k_\mu(x_\mu)$ are in F, where $x_\lambda \in E_\lambda$, $x_\mu \in E_\mu$, then, by definition, $k_\lambda(x_\lambda) \, R \, k_\mu(x_\mu)$ if and only if there exists some $\nu \in I$ with the property $\lambda \leqslant \nu$. $\mu \leqslant \nu$ and such that in E_ν

$$u_{\lambda\nu}(x_\lambda) = u_{\mu\nu}(x_\mu):$$

it is not difficult to see that R is an equivalence relation on F; let E be the quotient set F/R, and let $p: F \to F/R$ be the canonical surjection. On E an A-module structure is introduced as follows: let $x, y \in E$; there exist some $\lambda, \mu \in I$ and $x_\lambda \in E_\lambda$, $y_\mu \in E_\mu$ such that

$$x = p(k_\lambda(x_\lambda)), \qquad y = p(k_\mu(y_\mu));$$

if ν has the property $\lambda \leqslant \nu$, $\mu \leqslant \nu$, then $k_\lambda(x_\lambda) R\, k_\nu(u_{\lambda\nu}(x_\lambda))$ and $k_\mu(y_\mu)\ R\ k_\nu(u_{\mu\nu}\ (y_\mu))$; consequently,

$$x = p(k_\nu(u_{\lambda\nu}(x_\lambda))), \qquad y = p(k_\nu(u_{\mu\nu}(y_\mu)));$$

we consider the sum $u_{\lambda\nu}(x_\lambda) + u_{\mu\nu}(y_\mu)$ in E_ν, and we define

$$x + y = p(k_\nu(u_{\lambda\nu}(x_\lambda) + u_{\mu\nu}(y_\mu))); \tag{7}$$

next, if $\alpha \in A$ and x is as above, we define

$$\alpha x = p(k_\lambda(\alpha x_\lambda)); \tag{8}$$

it is not difficult to see that in (7) and (8) the results of the two operations are independent of the chosen indices λ, μ, ν, as well as of the elements x_λ, y_μ chosen with the respective properties; moreover, under the laws of composition (7) and (8), $E = F/R$ is an A-module, and the composite mapping $p \circ k_\lambda = j_\lambda$: $E_\lambda \to E$, $\lambda \in I$, is an A-homomorphism.

We will show that *the system* $(E, (j_\lambda)_{\lambda \in I})$ *is an inductive limit in the category* **Mod**(A) *of the inductive system* $(E_\lambda, u_{\lambda\mu})$, *in the sense of Definition* 3; then, according to the general theory of inductive limits in a category, it will turn out that E is isomorphic to the inductive limit of the inductive system $(E_\lambda, u_{\lambda\mu})$, in the sense of Definition 2; hence, instead of that inductive limit it will be possible to take by definition, in the particular case considered, as inductive limit of the given inductive system the module E with the linear mappings j_λ, $\lambda \in I$, now constructed:

If $\lambda \leqslant \mu$ in I, then for any $x_\lambda \in E_\lambda$ we obviously have $k_\lambda(x_\lambda)\ R\, k_\mu(u_{\lambda\mu}(x_\lambda))$; it follows that

$$j_\lambda(x_\lambda) = p(k_\lambda(x_\lambda)) = p(k_\mu(u_{\lambda\mu}(x_\lambda))) = j_\mu(u_{\lambda\mu}(x_\lambda));$$

further, if Y is an A-module and $s_\lambda : E_\lambda \to Y$ is an A-homomorphism for any $\lambda \in I$ such that $s_\mu \circ u_{\lambda\mu} = s_\lambda$ whenever $\lambda \leqslant \mu$, then the mappings s_λ, $\lambda \in I$, determine a mapping $t : F \to Y$, defined as follows: if an element $y \in F$ is given, there exists only one $\lambda \in I$ and only one $x_\lambda \in E_\lambda$ such that $y = k_\lambda(x_\lambda)$; then $t(y)$ is defined to be $s_\lambda(x_\lambda)$; as t is compatible with the equivalence relation R, it defines, by virtue of the universal property of the quotient set with respect to an equivalence relation (see § 0), a unique mapping $s : E \to Y$ with the property that $s \circ p = t$; s is an A-homomorphism, and it is unique with the property that $s \circ j_\lambda = s_\lambda$ for any $\lambda \in I$.

Now we shall study the particular cases dual to the ones considered for projective limits:

If A is a ring and I is an arbitrary small category, while Y is a fixed A-module, the *constant inductive* (or *direct*) *system of value* Y indexed by the category I

is defined to be the inductive system $T = (E_\lambda, u_f)_{\lambda, f}$ for which $E_\lambda = Y$ for any $\lambda \in I$ and $u_f = 1_Y$ for any $f \in \coprod_{(\lambda,\mu) \in I \times I} \mathrm{Hom}_I(\lambda, \mu)$; if $U = (F_\lambda, v_f)_{\lambda, f}$ is another inductive system of A-modules indexed by I, then an inductive morphism $U \to T$ is a family $(s_\lambda)_{\lambda \in I}$ of linear mappings $s_\lambda : F_\lambda \to Y$ such that for any $\lambda, \mu \in I$ and any $f \in \mathrm{Hom}_I(\lambda, \mu)$ we have $s_\mu \circ u_f = s_\lambda$; in particular, the canonical mappings $j_\lambda : F_\lambda \to \varinjlim_I F_\lambda$ form an inductive morphism from U to the constant inductive system of value $\varinjlim F_\lambda$. If Y' is another A-module and $u \in \mathrm{Hom}_A(Y, Y')$, then the family of mappings $(l_\lambda)_{\lambda \in I}$, where $l_\lambda = u$ for any $\lambda \in I$, is an inductive morphism between the constant inductive systems of values Y and Y', respectively; this morphism is called the *constant inductive* (or *direct*) *morphism of value u.*

The mappings

$Y \mapsto$ the constant inductive system of value Y,

$u \mapsto$ the constant inductive morphism of value u

define an additive functor

$$S : \mathbf{Mod}(A) \to (\mathbf{Mod}(A))^I,$$

called the *immersion* (or the *imbedding*) *functor* $\mathbf{Mod}(A) \to (\mathbf{Mod}(A))^I$.

PROPOSITION 4. *For any ring A and for any small category I, the immersion functor* $S : \mathbf{Mod}(A) \to (\mathbf{Mod}(A))^I$ *is a right adjoint of the functor* $\varinjlim_I : (\mathbf{Mod}(A))^I \to \to \mathbf{Mod}(A)$.

The *proof* is made by considerations dual to those made for Proposition 3, § 12. ∎

If I is a small connected category, if Y is an A-module, and if U is the constant inductive system of value Y indexed by I, then the family $(1_Y)_\lambda$ has the property of the family $(s_\lambda)_{\lambda \in I}$ in Proposition 1; consequently, there exists a homomorphism $\varphi : \varinjlim U \to Y$ uniquely determined by the relations $\varphi \circ k_\lambda = 1_Y$ for any $\lambda \in I$, k_λ being the canonical mapping of index λ, $\lambda \in I$; on the other hand, if $\lambda, \mu \in I$, and if there exists $f \in \mathrm{Hom}_I(\lambda, \mu)$, then, as $U(f) = 1_Y$, it follows $k_\lambda = k_\mu$; if λ, μ are arbitrary objects of I, this relation is still valid, which may be seen making use of the fact that I is connected; let $\psi = k_\lambda : Y \to \varinjlim U$; we have

$$\psi \circ \varphi \circ k_\lambda = \psi = k_\lambda$$

for any $\lambda \in I$, which implies that φ and ψ are isomorphisms inverse to each other; *hence in this case we may identify* $\varinjlim U$ *with* Y, k_λ *being identified with* $\varphi \circ k_\lambda = 1_Y$; this being done, considering an arbitrary inductive system $T = (E_\lambda, u_f)_{\lambda, f}$ indexed by the category I, a family $(s_\lambda)_{\lambda \in I}$ of linear mappings as in Proposition 1 is, as we have seen, an inductive morphism $T \to U$; the inductive limit of this inductive morphism, which is a homomorphism $\varinjlim T \to \varinjlim U$, is identified with the homomorphism s of Proposition 1; for this reason, s of Proposition 1 is also denoted $\varinjlim s_\lambda$, and this is done, of course by an abuse of notation, even when I is not a connected category.

Another particular case is the one in which I is a non-void set, and the discrete category (denoted also by I) whose objects are the elements of I is considered (see the second particular case considered for projective limits, § 12); in this case, if $(E_\lambda, u_f)_{\lambda, f}$ is an inductive system of A-modules indexed by the category I, then, with the notation used in the construction of the inductive limit for modules, the submodule R of $\bigoplus_{\lambda \in I} E_\lambda$ is the zero submodule, since all elements of the form (1) are equal to zero ; hence $\varinjlim_I E_\lambda$ is identified with $\bigoplus_{\lambda \in I} E_\lambda$ by means of the canonical isomorphism $\bigoplus_\lambda E_\lambda \to (\bigoplus_\lambda E_\lambda)/0$; by this identification the canonical mappings $E_\lambda \to \varinjlim_{\lambda \in I} E_\lambda$, $\lambda \in I$, are identified with the canonical injections $E_\lambda \to \bigoplus_{\lambda \in I} E_\lambda$. Consequently, in this sense, *the direct sum of a non-void family is a particular case of inductive limit.*

As in the case of products and projective limits, the direct sum may be also viewed in another way as an inductive limit: Let I be a set, and let $(E_\lambda)_{\lambda \in I}$ be a family of A-modules; let J be the set partially ordered by inclusion of finite subsets of I, and let, for any $M \in J$,

$$F_M = \bigoplus_{\lambda \in M} E_\lambda;$$

if $M, N \in J$ and $M \subseteq N$, then let u_{MN} be the linear mapping $F_M \to F_N$ obtained by the universal property of the direct sum from the canonical injections $j_\lambda^N : E_\lambda \to F_N$, $\lambda \in M$; hence we have $u_{MN} \circ j_\lambda^M = j_\lambda^N$, $\lambda \in M$, and u_{MN} carries a family $(x_\lambda)_{\lambda \in M} \in F_M$ into the family of F_N obtained from $(x_\lambda)_{\lambda \in M}$ by completing with zeros; the mapping assigning to $M \in J$ the A-module F_M and to the inclusion $M \subseteq N$ the A-homomorphism u_{MN} is an inductive system of A-modules indexed by the category J associated with the partially ordered set J; this inductive system will be denoted by (F_M, u_{MN}); let $u_M : F_M \to \bigoplus_{\lambda \in I} E_\lambda$ be the linear mapping obtained from the canonical injections $j_\lambda : E_\lambda \to \bigoplus_{\lambda \in I} E_\lambda$, $\lambda \in M$, by the universal property of the direct sum; hence we have $u_M \circ j_\lambda^M = j_\lambda$, $\lambda \in M$; notice that for any $\lambda \in I$, E_λ may be identified with $F_{\{\lambda\}}$ using [1] the isomorphism $j_\lambda^{\{\lambda\}}$, and that by this identification j_λ is identified with $u_{\{\lambda\}}$; moreover, if $\lambda \in M$, $u_{\{\lambda\}M}$ is identified with j_λ^M; denote for any $M \in J$ by k_M the canonical mapping $F_M \to \varinjlim_{M \in J} F_M$.

---

[1] We denoted by $\{\lambda\}$ the subset of I containing only the element λ.

The family of linear mappings $(u_M)_{M\in J}$ has the property that for $M\subseteq N$ in J we have $u_N\circ u_{MN} = u_M$, since $u_N\circ u_{MN}\circ j_\lambda^M = u_N\circ j_\lambda^N = j_\lambda = u_M\circ j_\lambda^M$ for any $\lambda\in M$; consequently, this family determines by virtue of Proposition 1 a unique linear mapping

$$\varphi : \varinjlim_{M\in J} F_M \to \bigoplus_{\lambda\in I} E_\lambda$$

with the property that $\varphi\circ k_M = u_M$ for any $M\in J$; on the other hand, the family of mappings $k_{\{\lambda\}} : E_\lambda = F_{\{\lambda\}} \to \varinjlim_J F_M$, $\lambda\in I$, determines by virtue of the universal property of the direct sum a unique linear mapping

$$\psi : \bigoplus_{\lambda\in I} E_\lambda \to \varinjlim_{M\in J} F_M$$

with the property that $\psi\circ j_\lambda = k_{\{\lambda\}}$, $\lambda\in I$. We have for any $\lambda\in I$,

$$\varphi\circ\psi\circ j_\lambda = \varphi\circ k_{\{\lambda\}} = u_{\{\lambda\}} = j_\lambda,$$

hence, by virtue of the uniqueness, $\varphi\circ\psi = 1_{\oplus E_\lambda}$; further, for any $M\in J$ and any $\lambda\in M$ we have

$$\psi\circ u_M\circ j_\lambda^M = \psi\circ j_\lambda = k_{\{\lambda\}} = k_M\circ u_{\{\lambda\}M} = k_M\circ j_\lambda^M,$$

whence, by virtue of the uniqueness, $\psi\circ u_M = k_M$; herefrom it follows for any $M\in J$,

$$\psi\circ\varphi\circ k_M = \psi\circ u_M = k_M,$$

whence, again by virtue of the uniqueness, $\psi\circ\varphi = 1_{\varinjlim F_M}$; hence φ and ψ are isomorphisms inverse to each other, and using them we may identify $\bigoplus_{\lambda\in I} E_\lambda$ with $\varinjlim_{M\in J} F_M$, the canonical injections j_λ being thus identified with the mappings $k_{\{\lambda\}}$, $\lambda\in I$; this is the second way of considering the direct sum as an inductive limit.

The next particular case considered is the one in which I is a category with two objects λ, μ and with two non-identity morphisms $f, g\in \mathrm{Hom}_I(\lambda, \mu)$; let T be an inductive system of A-modules for which, say, $T(\lambda) = E$, $T(\mu) = F$, $T(f) = u$, where E, F are A-modules and $u\in \mathrm{Hom}_A(E, F)$; take the zero homomorphism $E\to F$ as $T(g)$; if P is the cokernel of u, with the canonical surjection p, then the system $(P, 0, p)$, where 0 is the zero homomorphism $E\to P$, is an inductive limit of T in the category **Mod** (A), in the sense of Definition 3: for, $p\circ u = p\circ 0 = 0$, and if Y is an A-module while $r : E\to Y$, $s : F\to Y$ are homomorphisms such that $s\circ u = s\circ 0 = r$, it follows that $r = 0$ and that there exists a homomorphism $t : P\to Y$ uniquely determined by the property $t\circ p = s$; we have trivially $t\circ 0 = r$, and this proves our assertion. Conversely, if we consider the inductive limit $G = \varinjlim_{I} T$, with the canonical mappings j_E and j_F, then $j_F \circ u = j_F \circ 0 = j_E$, hence $j_F \circ u = 0$; if Y is an A-module and $s : F \to Y$ has the property $s \circ u = 0$, then the family of mappings s and $0 : E \to Y$ determine by virtue of the universal property of the inductive limit a unique homomorphism $t : G \to Y$ with the property that $t \circ j_F = s$; hence (G, j_F) is a cokernel of the linear mapping u, in the sense of Definition 1, § 4; with this meaning, we may say that *the cokernel is a particular case of inductive limit.*

Given another inductive system of A-modules $T' = (E', F', u', 0)$ indexed by the same category I, and an inductive morphism $T \to T'$, hence two homomorphisms $v : E \to E'$, $w : F \to F'$ rendering commutative the diagram

$$\begin{array}{ccc} E & \xrightarrow{\;u\;} & F \\ {\scriptstyle v}\downarrow & & \downarrow{\scriptstyle w} \\ E' & \xrightarrow{\;u'\;} & F', \end{array}$$

then, considering also the cokernel P' of u', with the canonical surjection p', the mapping $P \to P'$ induced by w (see diagram (3), § 4) is the inductive limit of the given inductive morphism for *this* choice of the inductive limits; conversely, taking for T and T' the inductive limits constructed at the beginning of this Section, denoted, for instance, G and G', the limit of the given morphism is the homomorphism $G \to G'$ constructed in § 4 for this choice of the cokernels.

REMARK. Here we could also have used the (dualized) procedure used to recognize the kernel as a projective limit (see § 12) as we could have used there the (dualized) procedure employed here; we made use on purpose of two different procedures, one for the first case, the other for the second. ∎

Another particular case is the following: let E be an A-module, and let $(M_\lambda)_{\lambda \in I}$, $I \neq \emptyset$, be a family of submodules of E directed to the right as a family of subsets; suppose on I the partial order relation denoted $\leqslant$ is defined by: $\lambda \leqslant \mu$, $\lambda, \mu \in I$, if and only if $M_\lambda \subseteq M_\mu$; then I is also directed to the right; assigning M_λ to $\lambda \in I$, and the inclusion mapping $M_\lambda \to M_\mu$, denoted $u_{\lambda\mu}$, to the relation $\lambda \leqslant \mu$ in I, we get an inductive system T of A-modules indexed by the category I associated with the partially ordered set I; on the other hand, the sum M of the family $(M_\lambda)_{\lambda \in I}$ of submodules is equal, by Corollary 6 to Proposition 2, § 8, to the union $\bigcup_{\lambda \in I} M_\lambda$; denote by j_λ the canonical injection $M_\lambda \to M$, $\lambda \in I$; let us show that the system $(M, (j_\lambda)_{\lambda \in I})$ is an inductive limit of the inductive system T, in the sense of Definition 3: it is clear that if $\lambda \leqslant \mu$ in I, then $j_\mu \circ u_{\lambda\mu} = j_\lambda$; further, if Y is an A-module and $s_\lambda : M_\lambda \to Y$, $\lambda \in I$, is a family of homomorphisms such that for $\lambda \leqslant \mu$, $s_\mu \circ u_{\lambda\mu} = s_\lambda$, then we define a mapping $s : M \to Y$ as follows: if $x \in M$, then there exists some $\lambda \in I$ such that $x \in M_\lambda$; if $\mu \in I$ has the same property as λ, then let $\nu \in I$ be a common upper bound of λ and μ; we have

$$s_\lambda(x) = u_{\lambda\nu}(s_\lambda(x)) = s_\nu(x) = u_{\mu\nu}(s_\mu(x)) = s_\mu(x),$$

since $u_{\lambda\nu}$ and $u_{\mu\nu}$ are inclusion mappings; consequently, $s_\lambda(x)$ does not depend on the chosen index λ for which $x \in M_\lambda$; we then define

$$s(x) = s_\lambda(x);$$

we have $s \circ j_\lambda = s_\lambda$ for any $\lambda \in I$, and if $s' : M \to Y$ has the same property, then, under the same conventions, $s'(x) = s_\lambda(x) = s(x)$; as x is arbitrary in M, it follows $s' = s$.

Consequently, the sum (or, what is the same, the union) of a non-void directed to the right family of submodules may be identified with the inductive limit of this family, by this identification the inclusion mappings into the sum being identified with the canonical mappings; in this sense, *the sum of a non-void directed to the right family of submodules is a particular case of inductive limit.*

In particular, if we consider as family of submodules $(M_\lambda)_{\lambda \in I}$ of the A-module E the family of finitely generated submodules indexed identically by the set of these submodules, then, as $E = \sum_{\lambda \in I} M_\lambda$ (see the considerations following Corollary 8 to Proposition 2, § 8), it turns out that *E may be identified with the inductive limit of the inductive system formed by its finitely generated submodules, the mappings $u_{\lambda\mu}$ being the inclusion mappings.*

Next, let us also consider the following particular case: let the diagram of A-modules and A-homomorphisms

$$\begin{array}{ccc} G & \xrightarrow{\ u\ } & E \\ {\scriptstyle v}\downarrow & & \\ F & & \end{array} \tag{9}$$

be given; in the direct sum $E \oplus F$ consider the set R consisting of those pairs (x, y), with $x \in E$, $y \in F$, for which there exists $z \in G$ such that $x = u(z)$, $y = -v(z)$; R is the image of the mapping $j_E \circ u - j_F \circ v$, where j_E and j_F are the canonical mappings of E, respectively F, into $E \oplus F$; consequently, R is a submodule of $E \oplus F$.

Definition 8. Given the diagram (9) of A-modules and A-homomorphisms, the quotient module $S = (E \oplus F)/R$, where R is the above constructed submodule of $E \oplus F$, is called the *fibered* (or *amalgamated*) *sum of E and F with respect to* (or *relative to*) *this diagram* (or, if no confusion may arise, *with respect to G* or *relative to G*), and is denoted, provided there is no danger of confusion, by $E \coprod_G F$ or also by $E \oplus_G F$. ▮

If p is the canonical surjection $E \oplus F \to S$, then, denoting $u' = p \circ j_E$, $v' = p \circ j_F$, the following diagram is commutative:

$$\begin{array}{ccc} G & \xrightarrow{\ u\ } & E \\ {\scriptstyle v}\downarrow & & \downarrow{\scriptstyle u'} \\ F & \xrightarrow{\ v'\ } & S; \end{array} \tag{10}$$

moreover, the system (S, u', v') is universal with this property: that is, for any system (Y, r, s), where Y is an A-module, $r : E \to Y$, and $s : F \to Y$, such that the following diagram be commutative

(11)
$$\begin{array}{ccc} G & \xrightarrow{u} & E \\ {\scriptstyle v}\downarrow & & \downarrow{\scriptstyle r} \\ F & \xrightarrow{s} & Y, \end{array}$$

here exists a unique homomorphism $\psi : S \to Y$ such that

(12)
$$\psi \circ u' = r, \qquad \psi \circ v' = s;$$

this may be seen by considerations dual to those for fibered products.

Let now I be a preordered set with three elements λ, μ, ν on which the preorder relation $\leqslant$ is defined (besides $\lambda \leqslant \lambda$, $\mu \leqslant \mu$, $\nu \leqslant \nu$) by $\nu \leqslant \lambda$, $\nu \leqslant \mu$; consider the inductive system T of A-modules indexed by the category I associated with the preordered set I in which, with the above meaning, $T(\lambda) = E$, $T(\mu) = F$, $T(\nu) = G$, $T(\nu \leqslant \lambda) = u$, $T(\nu \leqslant \mu) = v$;[1] we get diagram (9); if (S, u', v') have the above meaning, it may be shown that the system $(S, u', v', u' \circ u)$ is an inductive limit in the category **Mod**(A) of the inductive system T, in the sense of Definition 3; we leave to the reader as an exercise to verify this assertion; in this sense, one may say that *the fibered sum is a particular case of inductive limit.*

Finally, we also mention the following particular case, which is nothing but a generalization of the direct sum and of the manner of considering it as an inductive limit: let $(E_\lambda)_{\lambda \in I}$ be a family of A-modules indexed by a set I, and assume that for any $\lambda \in I$ a submodule F_λ of E_λ is given; a family of pairs $(E_\lambda, F_\lambda)_{\lambda \in I}$ as described is called a *restricted system* of A-modules indexed by I; if $(E_\lambda, F_\lambda)_{\lambda \in I}$ and $(E'_\lambda, F'_\lambda)_{\lambda \in I}$ are two restricted systems of A-modules indexed by the same set I, a *morphism of restricted systems* from $(E_\lambda, F_\lambda)_\lambda$ to $(E'_\lambda, F'_\lambda)_\lambda$ is a family of linear mappings $(v_\lambda)_{\lambda \in I}$, where $v_\lambda : E_\lambda \to E'_\lambda$, $\lambda \in I$, and such that $v_\lambda(F_\lambda) \subseteq F'_\lambda$ for any $\lambda \in I$.

Let $(E_\lambda, F_\lambda)_{\lambda \in I}$ be a restricted system of A-modules indexed by the set I; let J be the set partially ordered by inclusion of finite subsets of I; for any $M \in J$ define

$$P_M = \Big(\prod_{\lambda \in M} E_\lambda\Big) \times \Big(\prod_{\lambda \notin M} F_\lambda\Big);$$

because of the associativity of the direct product (see § 6), P_M is identified with the product of the family of A-modules $(G_\lambda)_{\lambda \in I}$ for which $G_\lambda = E_\lambda$ for $\lambda \in M$ and $G_\lambda = F_\lambda$ for $\lambda \in I \setminus M$; let $M, N \in J$ be such that $M \subseteq N$; define a linear mapping

$$u_{MN} : P_M \to P_N$$

[1] The abuse of notation should be noted!

as follows: u_{MN} is the direct product of the family of mappings $(i_\lambda)_{\lambda \in I}$, where

$$i_\lambda = \begin{cases} 1_{E_\lambda} & \text{for } \lambda \in M; \\ \text{the canonical injection } F_\lambda \to E_\lambda & \text{for } \lambda \in N \setminus M; \\ 1_{F_\lambda} & \text{for } \lambda \in I \setminus N. \end{cases}$$

The system (P_M, u_{MN}) is an inductive system of A-modules indexed by the category J associated with the partially ordered set J.

If $(E'_\lambda, F'_\lambda)_{\lambda \in I}$ is another restricted system of A-modules indexed by the same set I, and if $(v_\lambda)_{\lambda \in I}$ is a morphism of restricted systems from $(E_\lambda, F_\lambda)_\lambda$ to $(E'_\lambda, F'_\lambda)_\lambda$, then, denoting by (P'_M, u'_{MN}) the inductive system obtained from $(E'_\lambda, F'_\lambda)_\lambda$ in the same way as (P_M, u_{MN}) from $(E_\lambda, F_\lambda)_\lambda$, $(v_\lambda)_\lambda$ yields an inductive morphism $(w_M)_{M \in J}: (P_M, u_{MN}) \to (P'_M, u'_{MN})$ as follows: w_M, $M \in J$, is defined by

$$w_M = \prod_{\lambda \in I} t_\lambda,$$

where $t_\lambda = v_\lambda$ for $\lambda \in M$, and t_λ is the homomorphism induced by v_λ by restriction to F_λ and F'_λ for $\lambda \in I \setminus M$.

DEFINITION 9. If $(E_\lambda, F_\lambda)_{\lambda \in I}$ is a restricted system of A-modules indexed by the set I, then, in the above notation, $\varinjlim_{M \in J} (P_M, u_{MN})$ is called the *restricted product* of the given restricted system, and is denoted by $\prod_{\lambda \in I} E_\lambda : F_\lambda$.

If $(v_\lambda)_{\lambda \in I}$ is a morphism of restricted systems from the restricted system $(E_\lambda, F_\lambda)_{\lambda \in I}$ to the restricted system $(E'_\lambda, F'_\lambda)_{\lambda \in I}$, then, in the above notation, $\varinjlim_{M \in J} w_M$ is called the *restricted product* of the given morphism of restricted systems, and is denoted $\prod_{\lambda \in I} v_\lambda$. ∎

The restricted product $\prod_{\lambda \in I} E_\lambda : F_\lambda$ may be identified with the following submodule of $\prod_{\lambda \in I} E_\lambda$: consider in $\prod_{\lambda \in I} E_\lambda$ the subset P consisting of those families $(x_\lambda)_{\lambda \in I}$ for which only a finite number of components x_λ belong to the corresponding $E_\lambda \setminus F_\lambda$; it is clear that P is a submodule of $\prod_{\lambda \in I} E_\lambda$; moreover, if for any finite subset M of I the above meaning for P_M is maintained, then let $r_M : P_M \to P$ be the (linear) mapping carrying a family $(x_\lambda)_{\lambda \in I} \in P_M$ into the same family considered in P; it is easily seen that the system $(P, (r_M)_{M \in J})$ is an inductive limit in the category **Mod**(A) of the above defined inductive system (P_M, u_{MN}); hence P may be identified with the restricted product $\prod_{\lambda \in I} E_\lambda : F_\lambda$. In particular, if $F_\lambda = 0$ for any $\lambda \in I$, the restricted product $\prod_{\lambda \in I} E_\lambda : 0$ is identified with the direct sum $\bigoplus_{\lambda \in I} E_\lambda$.

In this last case it is possible to proceed also as follows: if $(E_\lambda)_{\lambda\in I}$ is a family of A-modules, then for any finite subset M of I we have the canonical isomorphism

$$\varphi_M : L_M = \bigoplus_{\lambda\in M} E_\lambda \xrightarrow{\sim} \prod_{\lambda\in M} E_\lambda \times \prod_{\lambda\notin M} 0 = P_M$$

defined in an obvious manner; if $u_{MN} : L_M \to L_N$ is (for $M \subseteq N$) the mapping carrying a family $(x_\lambda)_{\lambda\in M} \in L_M$ into the family of L_N obtained by completing with zeros (see above, the particular case of the direct sum), and if u'_{MN} is the mapping $P_M \to P_N$ carrying a family of P_M into itself considered in P_N, then the following diagram is commutative:

$$\begin{array}{ccc} L_M & \xrightarrow{u_{MN}} & L_N \\ {\scriptstyle\varphi_M}\big\downarrow\wr & & {\scriptstyle\varphi_N}\big\downarrow\wr \\ P_M & \xrightarrow{u'_{MN}} & P_N ; \end{array}$$

hence, by virtue of the functor character of the inductive limit, we have $\varinjlim L_M \approx \approx \varinjlim P_M$; but the first limit is identified with $\bigoplus_{\lambda\in I} E_\lambda$, and the second is $\prod_{\lambda\in I} E_\lambda : 0$, which allows the identification of the two modules.

By considerations dual to those used to prove Proposition 4, § 12, one may prove the following Proposition:

PROPOSITION 5. *Let I be a small category, let $T = (E_\lambda, u_f)_{\lambda,f}$, $U = (F_\lambda, v_f)_{\lambda,f}$, $V = (G_\lambda, w_f)_{\lambda,f}$ be inductive systems of A-modules indexed by the category I; let $(k_\lambda)_{\lambda\in I} : T \to U$, $(l_\lambda)_{\lambda\in I} : U \to V$ be inductive systems of linear mappings such that for any $\lambda \in I$ the diagram*

$$E_\lambda \xrightarrow{k_\lambda} F_\lambda \xrightarrow{l_\lambda} G_\lambda \longrightarrow 0$$

be an exact sequence; then, if $\varinjlim_{\lambda\in I} k_\lambda = k$, $\varinjlim_{\lambda\in I} l_\lambda = l$, *the diagram*

$$\varinjlim_{\lambda\in I} E_\lambda \xrightarrow{k} \varinjlim_{\lambda\in I} F_\lambda \xrightarrow{l} \varinjlim_{\lambda\in I} G_\lambda \to 0$$

is an exact sequence; in other words, the inductive limit functor is right exact. ▌

REMARKS. 1) It will be proved later (see the Corollary to Proposition 5, § 14) that, under certain conditions on the category I, the inductive limit functor $\varinjlim_I$ is exact.

2) In fact, for the last assertion of Proposition 5 it was necessary to show that the diagram $T \to U \to V$ is an exact sequence in the category $(\mathbf{Mod}(A))^I$ if and only if for any $\lambda \in I$ the diagram $E_\lambda \to F_\lambda \to G_\lambda$ is an exact sequence in the category $\mathbf{Mod}(A)$; we leave this to the reader (see also the Remark following the statement of Proposition 4, § 12); we shall prove later independently (see the Corollary to Proposition 3, § 14) the left, respectively right exactness of the projective, respectively inductive limit functor in an abelian complete, respectively cocomplete category. ▮

COROLLARY. *Let I be a small connected category, and let $T = (E_\lambda, u_f)_I$ be an inductive system of A-modules indexed by the category I; if Y and $(s_\lambda)_{\lambda \in I}$ are as in Proposition 1, and if s_λ is an epimorphism for any $\lambda \in I$, then s constructed as in Proposition 1 is an epimorphism.* ▮

* If $\mathcal{C}$ is an arbitrary category and I is a small category, then we may define in an obvious manner for any object Y of $\mathcal{C}$ the *constant inductive system of value Y* indexed by the category I, as well as the *constant inductive morphism of value u* for any morphism u of the category $\mathcal{C}$; we get in this way the *immersion functor* $\mathcal{C} \to \mathcal{C}^I$, which, when $\mathcal{C}$ is a category with inductive limits, is a right adjoint of the inductive limit functor; moreover, if I is a small connected category, then the system $(Y, (j_\lambda)_{\lambda \in I})$, where $j_\lambda = 1_Y$ for any $\lambda \in I$, is an inductive limit, in the sense of Definition 3, of the constant inductive system of value Y indexed by the category I.

In a manner dual to what happens for projective limits, if $T : I \to \mathcal{C}$ is an inductive system with values in the category $\mathcal{C}$ indexed by the small category I, then from Definition 3 it follows that an inductive limit of T in the category $\mathcal{C}$ is an initial object of the category $T/\mathcal{C}$ (see § 12), and conversely; hence the existence of an inductive limit of T amounts to the existence of an initial object of $T/\mathcal{C}$.

Concerning the non-void direct sum of objects, the cokernel of a morphism, and the sum of a directed to the right family of subobjects of an object in a category $\mathcal{C}$, these are particular cases of inductive limits, as in the case of modules; of course, we have to ensure their existence, which is possible assuming, for instance, that $\mathcal{C}$ is an abelian category with direct sums; in this case, Proposition 5 and its Corollary are valid, of course with the necessary modifications.

Now let $\mathcal{C}$ be an arbitrary category, and let diagram (9) be given, with E, F, G, respectively u, v, objects, respectively morphisms of $\mathcal{C}$.

DEFINITION 10. *A fibered* (or *amalgamated*) *sum in the category $\mathcal{C}$ of E and F with respect to* (or *relative to*) *diagram* (9) (or, if no confusion may arise, *with respect to G* or *relative to G*) is a system (S, u', v'), where $S \in \operatorname{Ob} \mathcal{C}$, $u' \in \operatorname{Hom}_{\mathcal{C}}(E, S)$, $v' \in \operatorname{Hom}_{\mathcal{C}}(F, S)$, such that the corresponding diagram (10) is commutative and such that, moreover, we have the following universal property: for any system (Y, r, s), with $Y \in \operatorname{Ob} \mathcal{C}$, $r \in \operatorname{Hom}_{\mathcal{C}}(E, Y)$, $s \in \operatorname{Hom}_{\mathcal{C}}(F, Y)$, for which the corresponding diagram (11) is commutative there exists a unique morphism $\psi : S \to Y$ satisfying conditions (12).

A diagram of the type (10) in the category $\mathcal{C}$ is called a *cocartesian diagram* or a *cocartesian square* or also a *pushout* provided (S, u', v') is a fibered sum of E and F with respect to the corresponding diagram (9). ▮*

By the preceding considerations, if S, u', v' are those constructed above for the diagram (9) given as diagram of A-modules and A-homomorphisms, then the system (S, u', v') is, in the sense of Definition 10, a fibered sum in the category **Mod**(A) of E and F with respect to diagram (9).

* Returning to an arbitrary category $\mathcal{C}$, it is possible to see by the standard method that if (S, u', v') and (S_1, u'_1, v'_1) are fibered sums in the category $\mathcal{C}$ of E and F with respect to the same diagram, say (9), then there exists an isomorphism $\theta : S \xrightarrow{\sim} S_1$ such that $\theta \circ u' = u'_1$, $\theta \circ v' = v'_1$; then from all fibered sums one is chosen, say (S, u', v'), called *the fibered* (or *the amalgamated*) *sum* of E and F with respect to the given diagram (9), and the corresponding object S is denoted, when no confusion may arise, $E \coprod_G F$ or $E \oplus_G F$; sometimes this object is called, by an abuse of language, the *fibered sum* of E and F with respect to diagram (9). *If $\mathcal{C}$ is an abelian category, then, whatever the diagram* (9) *in $\mathcal{C}$, there exists a fibered sum in $\mathcal{C}$ of E and F with respect to this diagram.*

Similarly as for modules and in the same sense, the fibered sum may be regarded as an inductive limit.

By considerations dual to those for fibered products the following Proposition may be proved:

Proposition 6. *Let the cocartesian square* (10) *be given in the category $\mathcal{C}$; if u is an epimorphism, respectively an isomorphism of the category $\mathcal{C}$, then v' is an epimorphism, respectively an isomorphism; likewise for v and u'.* ∎

In order to prove in an abelian category also the result obtained from Proposition 6 by replacing "epimorphism" by "monomorphism", as well as the dual result, we need the following Lemma:

Lemma 1. *Let $\mathcal{C}$ be an abelian category, and let the diagram*

$$\begin{array}{ccccccccc} & & & & 0 & & & & \\ & & & & \downarrow & & & & \\ & & & & E & & & & \\ & & & & \downarrow{\scriptstyle u} & & & & \\ 0 & \longrightarrow & F' & \xrightarrow{f} & F & \xrightarrow{g} & F'' & \longrightarrow & 0, \\ & & & & \downarrow{\scriptstyle v} & & & & \\ & & & & G & & & & \\ & & & & \downarrow & & & & \\ & & & & 0 & & & & \end{array}$$

in which the row and the column are exact sequences, be given; then $g \circ u$ is, respectively, an epimorphism, a monomorphism, an isomorphism if and only if $v \circ f$ is, respectively, an epimorphism, a monomorphism, an isomorphism.

Proof. Consider the diagram

$$
\text{(13)}\qquad
\begin{array}{ccccccccc}
 & & 0 & & 0 & & 0 & & \\
 & & \downarrow & & \downarrow & & \downarrow & & \\
0 & \longrightarrow & P & \xrightarrow{u'} & E & \xrightarrow{p} & I & \longrightarrow & 0 \\
 & & {\scriptstyle f'}\downarrow & & {\scriptstyle u}\downarrow & & \downarrow{\scriptstyle i} & & \\
0 & \longrightarrow & F' & \xrightarrow{f} & F & \xrightarrow{g} & F'' & \longrightarrow & 0 \\
 & & {\scriptstyle q}\downarrow & & {\scriptstyle v}\downarrow & & \downarrow{\scriptstyle g'} & & \\
0 & \longrightarrow & J & \xrightarrow{j} & G & \xrightarrow{v'} & S & \longrightarrow & 0, \\
 & & \downarrow & & \downarrow & & \downarrow & & \\
 & & 0 & & 0 & & 0 & &
\end{array}
$$

where the left upper square is a pullback, the right lower square is a pushout, (I, p) and (J, q) are coimages of $g \circ u$ and $v \circ f$, respectively, (I, i) and (J, j) are, respectively, images of $g \circ u$ and $v \circ f$ chosen such that diagram (13) be commutative. We claim that the rows and columns in diagram (13) are exact sequences: it suffices to prove it for the top row, the left column being in the same situation, while for the bottom row and the right column the result being obtained by duality.

Since p is an epimorphism, by Proposition 2, *a*), § 5, it suffices to show that (P, u') is a kernel of p; we shall show that (P, u') is a kernel of $g \circ u = i \circ p$, since i being a monomorphism, the fact that for some $l : X \to E$ we have $i \circ p \circ l = 0$ is equivalent to the fact that $p \circ l = 0$; we have $g \circ u \circ u' = g \circ f \circ f' = 0$; let $X \in \mathrm{Ob}\, \mathscr{C}$, and let $h \in \mathrm{Hom}_{\mathscr{C}}(X, E)$ be such that $g \circ u \circ h = 0$; since the middle row is an exact sequence, hence (F', f) is a kernel of g, it follows that there exists $k : X \to F'$ with the property that $f \circ k = u \circ h$; the left upper square being a pullback, it follows that there exists $r : X \to P$ such that

$$u' \circ r = h; \tag{14}$$

the uniqueness of r with the property (14) follows from the fact that u' is a monomorphism: indeed, f is a monomorphism, and it remains to apply Proposition 7, § 12.

Once the exactness of the rows and columns in diagram (13) is proved, assume that $g \circ u$ is an epimorphism; it follows that i is an epimorphism, hence that $S = \mathrm{Coker}\, i = 0$; but then j, having 0 as a cokernel, is an epimorphism, and since q is an epimorphism, it follows that $v \circ f$ is an epimorphism; the converse follows by symmetry.

The other assertions are left to the reader as exercises.∎

PROPOSITION 7. *Let the cartesian square* (14), § 12, *be given in the abelian category* $\mathscr{C}$; *if* u *is an epimorphism of the category* $\mathscr{C}$, *then* v' *is also an epimorphism; likewise for* v *and* u'. *Similarly, given in the abelian category* $\mathscr{C}$ *the cocartesian square* (10), *if* u *is a monomorphism of the category* $\mathscr{C}$, *then* v' *is also a monomorphism; likewise for* v *and* u'.

Proof. We prove only the first assertion, the one concerning cocartesian squares following by duality.

By the proof of Proposition 6, § 12, if (H, p_E, p_F) is a direct product of the family of objects (E, F), then P is obtained from a kernel (P, i) of $u \circ p_E - v \circ p_F$, and u', v' are given by $u' = p_E \circ i$, $v' = p_F \circ i$. Let (H, j_E, j_F) be the direct sum of the family of objects (E, F) associated with the direct product (H, p_E, p_F); by the relations of the type (15), § 7, we have

$$(u \circ p_E - v \circ p_F) \circ j_E = u; \tag{15}$$

consequently, if u is an epimorphism, then $u \circ p_E - v \circ p_F$ is also an epimorphism; making also use of Proposition 7, § 9, it follows that in the diagram

$$\begin{array}{ccccccccc}
 & & & & 0 & & & & \\
 & & & & \downarrow & & & & \\
 & & & & E & & & & \\
 & & & & {\scriptstyle j_E}\downarrow & & & & \\
0 & \longrightarrow & P & \xrightarrow{\ i\ } & H & \xrightarrow{u \circ p_E - v \circ p_F} & G & \longrightarrow & 0 \\
 & & & & {\scriptstyle p_F}\downarrow & & & & \\
 & & & & F & & & & \\
 & & & & \downarrow & & & & \\
 & & & & 0 & & & &
\end{array}$$

the row and the column are exact sequences; since by (15) $(u \circ p_E - v \circ p_F) \circ j_E$ is an epimorphism, it follows, according to Lemma 1, that $p_F \circ i = v'$ is an epimorphism. ▮ *

§ 14. Commutativity Properties

PROPOSITION 1. *Let A be a ring, and let $T = (E_\lambda, u_f)_I$, respectively $U = (F_\mu, v_g)_J$ be an inductive, respectively a projective system of A-modules indexed by the (small) category I, respectively J; then, assigning to each object (λ, μ) of the product category $I \times J$ the abelian group $G_{\lambda,\mu} = \mathrm{Hom}_A(E_\lambda, F_\mu)$, and to each morphism $(f, g) \in \mathrm{Hom}_{I\times J}((\lambda, \mu), (\lambda', \mu'))$ the homomorphism of abelian groups $w_{f,g} = \mathrm{Hom}(u_f, v_g) : G_{\lambda',\mu'} \to G_{\lambda,\mu}$, we obtain a projective system $V = (G_{\lambda,\mu}, w_{f,g})_{\lambda,\mu,f,g}$ of **Z**-modules indexed by the category*[1] *$I \times J$, and we have a canonical isomorphism of abelian groups*

$$\varphi : \mathrm{Hom}_A(\varinjlim_I E_\lambda, \varprojlim_J F_\mu) \xrightarrow{\sim} \varprojlim_{I\times J} V = \varprojlim_{I\times J} \mathrm{Hom}_A(E_\lambda, F_\mu);$$

*if, moreover, $T' = (E'_\lambda, u'_f)_I$, respectively $U' = (F'_\mu, v'_g)_J$, is another inductive, respectively projective system of A-modules indexed by I, respectively J, and if $(h_\lambda)_{\lambda\in I}$, respectively $(l_\mu)_{\mu\in J}$, is an inductive, respectively projective morphism between T' and T, respectively between U and U', then, denoting by $V' = (G'_{\lambda,\mu}, w'_{f,g})_{I\times J}$ the projective system of **Z**-modules obtained from T' and U' in the same way as V is from T and U, the family of **Z**-homomorphisms $(\mathrm{Hom}(h_\lambda, l_\mu))_{(\lambda,\mu)\in I\times J}$ is a projective morphism between V and V', and we have the commutative diagram below, where the horizontal arrows represent the canonical isomorphisms:*

$$\begin{array}{ccc}
\mathrm{Hom}_A(\varinjlim_I E_\lambda, \varprojlim_J F_\mu) & \xrightarrow[\sim]{\varphi} & \varprojlim_{I\times J} \mathrm{Hom}_A(E_\lambda, F_\mu) \\
{\scriptstyle \mathrm{Hom}(\varinjlim_I h_\lambda, \varprojlim_J l_\mu)}\big\downarrow & & \big\downarrow{\scriptstyle \varprojlim_{I\times J} \mathrm{Hom}(h_\lambda, l_\mu)} \\
\mathrm{Hom}_A(\varinjlim_I E'_\lambda, \varprojlim_J F'_\mu) & \xrightarrow[\sim]{\varphi'} & \varprojlim_{I\times J} \mathrm{Hom}_A(E'_\lambda, F'_\mu).
\end{array}$$

Proof. For the first assertion we have to show that, given the diagram in $I \times J$

$$(\lambda, \mu) \xrightarrow{(f,g)} (\lambda', \mu') \xrightarrow{(f',g')} (\lambda'', \mu''),$$

[1] The category $I \times J$ is a small category, its class of objects being the cartesian product of wo sets.

we have in **Ab**

$$\text{(1)} \qquad \operatorname{Hom}(u_{f'\circ f}, v_{g'\circ g}) = \operatorname{Hom}(u_f, v_g)\circ\operatorname{Hom}(u_{f'}, v_{g'}),$$

and that for any $(\lambda, \mu) \in I \times J$ we have

$$\text{(2)} \qquad \operatorname{Hom}(u_{1_\lambda}, v_{1_\mu}) = 1_{\operatorname{Hom}(E_\lambda, F_\mu)};$$

but we have

$$u_{f'\circ f} = u_{f'}\circ u_f \text{ and } v_{g'\circ g} = v_g \circ v_{g'},$$

so we need only apply relation (3), § 2, to get (1); relation (2) is obtained similarly.

For the construction of the canonical isomorphism φ, let j_λ be the canonical mapping $E_\lambda \to \varinjlim_I E_\lambda$, $\lambda \in I$, and let p_μ be the canonical mapping $\varprojlim_J F_\mu \to F_\mu$, $\mu \in J$; if $r \in \operatorname{Hom}_A(\varinjlim_I E_\lambda, \varprojlim_J F_\mu)$, then let us consider the element

$$(p_\mu \circ r \circ j_\lambda)_{(\lambda, \mu)\in I\times J} \in \prod_{(\lambda, \mu)\in I\times J} \operatorname{Hom}_A(E_\lambda, F_\mu);$$

if $(f, g) \in \operatorname{Hom}_{I\times J}((\lambda, \mu), (\lambda', \mu'))$, then

$$w_{f,g}(p_{\mu'}\circ r \circ j_{\lambda'}) = \operatorname{Hom}(u_f, v_g)(p_{\mu'}\circ r\circ j_{\lambda'}) = v_g \circ p_{\mu'}\circ r \circ j_{\lambda'}\circ u_f = p_\mu \circ r \circ j_\lambda,$$

hence $(p_\mu \circ r \circ j_\lambda)_{(\lambda,\mu)\in I\times J} \in \varprojlim_{I\times J} \operatorname{Hom}_A(E_\lambda, F_\mu)$; we define φ by

$$\varphi(r) = (p_\mu \circ r \circ j_\lambda)_{(\lambda, \mu)\in I\times J}.$$

The fact that φ is a homomorphism of abelian groups follows from the distributivity of the composition of homomorphisms over their addition. To check the surjectivity of φ, let $(r_{\lambda,\mu})_{(\lambda, \mu)} \in \varprojlim_{I\times J} \operatorname{Hom}_A(E_\lambda, F_\mu)$; fix the object μ in J, and consider the family of linear mappings $(r_{\lambda,\mu})_{\lambda\in I}$; if $f \in \operatorname{Hom}_I(\lambda, \lambda')$, then, as $w_{f,1_\mu}(r_{\lambda', \mu}) = r_{\lambda, \mu}$, it follows that the following diagram is commutative:

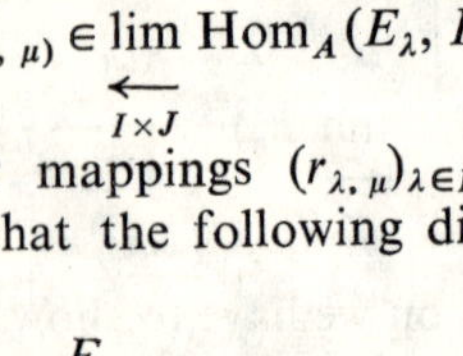

consequently, using the universal property of inductive limits, there exists an A-homomorphism $s_\mu : \varinjlim_{I} E_\lambda \to F_\mu$ uniquely determined by the property that

(3) $$s_\mu \circ j_\lambda = r_{\lambda\mu}$$

for any $\lambda \in I$; applying this to any $\mu \in J$, the family $(s_\mu)_{\mu \in J}$ has the property that for any $g \in \mathrm{Hom}_J(\mu, \mu')$ we have

$$v_g \circ s_{\mu'} = s_\mu:$$

indeed, it suffices to show that for any $\lambda \in I$ we have $v_g \circ s_{\mu'} \circ j_\lambda = s_\mu \circ j_\lambda$; but this follows from (3) taking into account that $w_{1_\lambda, g}(r_{\lambda,\mu'}) = r_{\lambda,\mu}$; consequently, by the universal property of the projective limit, there exists a unique homomorphism $r : \varinjlim_{I} E_\lambda \to \varprojlim_{J} F_\mu$ such that for any $\mu \in J$ we have $p_\mu \circ r = s_\mu$; in view of the preceding considerations, we have $\varphi(r) = (r_{\lambda,\mu})_{(\lambda,\mu)}$. The injectivity of φ follows from the uniqueness of the above construction.

The last assertion is proved as follows: first, we have to show that if $(f, g) \in \mathrm{Hom}_{I\times J}((\lambda, \mu), (\nu, \rho))$, then the following diagram is commutative:

$$\begin{array}{ccc}
\mathrm{Hom}_A(E_\nu, F_\rho) & \xrightarrow{\mathrm{Hom}\,(u_f,\, v_g)} & \mathrm{Hom}_A(E_\lambda, F_\mu) \\
\Big\downarrow{\scriptstyle \mathrm{Hom}\,(h_\nu,\, l_\rho)} & & \Big\downarrow{\scriptstyle \mathrm{Hom}\,(h_\lambda,\, l_\mu)} \\
\mathrm{Hom}_A(E'_\nu, F'_\rho) & \xrightarrow{\mathrm{Hom}\,(u'_f,\, v'_g)} & \mathrm{Hom}_A(E'_\lambda, F'_\mu);
\end{array}$$

but this is a consequence of relations (3), § 2, and of the commutativity of the following diagrams:

$$\begin{array}{ccc}
E'_\lambda & \xrightarrow{u'_f} & E'_\nu \\
\Big\downarrow{\scriptstyle h_\lambda} & & \Big\downarrow{\scriptstyle h_\nu} \\
E_\lambda & \xrightarrow{u_f} & E_\nu\,,
\end{array}
\qquad
\begin{array}{ccc}
F_\rho & \xrightarrow{v_g} & F_\mu \\
\Big\downarrow{\scriptstyle l_\rho} & & \Big\downarrow{\scriptstyle l_\mu} \\
F'_\rho & \xrightarrow{v'_g} & F'_\mu.
\end{array}$$

Next, to prove the commutativity of the last diagram in the statement it suffices, denoting by $q'_{\lambda,\mu}$ the canonical mapping of index (λ, μ) of $\varprojlim_{I\times J} \mathrm{Hom}_A(E'_\lambda, F'_\mu)$ into $\mathrm{Hom}_A(E'_\lambda, F'_\mu)$, to show that for any $(\lambda, \mu) \in I \times J$,

(4) $$q'_{\lambda,\mu} \circ \varprojlim_{I\times J} \mathrm{Hom}\,(h_\lambda, l_\mu) \circ \varphi = q'_{\lambda,\mu} \circ \varphi' \circ \mathrm{Hom}\,(\varinjlim_{I} h_\lambda, \varprojlim_{J} l_\mu);$$

but this is obtained as follows: let $j'_\lambda : E'_\lambda \to \varinjlim_I E'_\lambda$, $\lambda \in I$, $p'_\mu : \varprojlim_J F'_\mu \to F'_\mu$, $\mu \in J$, $q_{\lambda,\mu} : \varprojlim_{I\times J} \operatorname{Hom}_A(E_\lambda, F_\mu) \to \operatorname{Hom}_A(E_\lambda, F_\mu)$, $(\lambda, \mu) \in I \times J$, be the canonical mappings; then, by Corollary 1 to Proposition 1, § 12, and by the definition of φ, we have for any $r \in \operatorname{Hom}_A(\varinjlim E_\lambda, \varprojlim F_\mu)$,

$$(q'_{\lambda,\mu}\circ\varprojlim \operatorname{Hom}(h_\lambda, l_\mu)\circ\varphi)(r) = (\operatorname{Hom}(h_\lambda, l_\mu)\circ q_{\lambda,\mu})(\varphi(r)) =$$

$$= \operatorname{Hom}(h_\lambda, l_\mu)(p_\mu\circ r\circ j_\lambda) = l_\mu\circ p_\mu\circ r\circ j_\lambda\circ h_\lambda;$$

on the other hand,

$$(q'_{\lambda,\mu}\circ\varphi'\circ\operatorname{Hom}(\varinjlim h_\lambda, \varprojlim l_\mu))(r) = (q'_{\lambda,\mu}\circ\varphi')(\varprojlim l_\mu\circ r\circ\varinjlim h_\lambda) =$$

$$= p'_\mu\circ\varprojlim l_\mu\circ r\circ\varinjlim h_\lambda\circ j'_\lambda = l_\mu\circ p_\mu\circ r\circ j_\lambda\circ h_\lambda,$$

and this proves relation (4).▮

* In an abelian category $\mathcal{C}$ with inductive and projective limits, Proposition 1 with the appropriate replacements remains valid; first, we have to choose a "determination" of the inductive limit and of the projective limit.

Given two categories $\mathcal{C}$, $\mathcal{C}'$ and a functor $H : \mathcal{C} \to \mathcal{C}'$, then, obviously, if I is a small category, and if $(E_\lambda, u_f)_I$ is a projective system with values in the category $\mathcal{C}$ indexed by the category I, the assignment

$$\lambda \mapsto H(E_\lambda), \qquad f \mapsto H(u_f)$$

is a projective system with values in the category $\mathcal{C}'$ indexed by the category I; in fact, this is the composition of the functors

$$I^\circ \to \mathcal{C} \xrightarrow{H} \mathcal{C}',$$

where the first arrow represents the functor giving the first projective system; we shall denote this new projective system by $(H(E_\lambda), H(u_f))_{\lambda, f}$ or by $(H(E_\lambda), H(u_f))_I$; this is still valid if we replace "projective system" by "inductive system"; then, of course, in the above composition of functors, I° is replaced by I.

Definition 1. Let $\mathcal{C}$, $\mathcal{C}'$ be two categories, and let $H : \mathcal{C} \to \mathcal{C}'$ be a functor; the functor H is said *to commute with projective limits* or *to preserve projective limits* or to be *continuous* or *left continuous* if from the fact that the system $(E, (p_\lambda)_{\lambda\in I})$ is a projective limit in the category $\mathcal{C}$, in the sense of Definition 9, § 12, of a projective system $(E_\lambda, u_f)_I$ with values in the category $\mathcal{C}$ indexed by a category I it follows that $(H(E), (H(p_\lambda))_{\lambda \in I})$ is, in the same sense, a projective limit in the category $\mathcal{C}'$ of the projective system $(H(E_\lambda), H(u_f))_I$.

Likewise, the functor H is said *to commute with inductive limits* or *to preserve inductive limits* or to be *cocontinuous* or *right continuous* if from the fact that $(F, (j_\lambda)_{\lambda \in I})$ is an inductive limit in the category $\mathcal{C}$, in the sense of Definition 3, § 13, of an inductive system $(F_\lambda, v_f)_I$ with values in the category $\mathcal{C}$ indexed by a category I it follows that $(H(F), (H(j_\lambda))_{\lambda \in I})$ is, in the same sense, an inductive limit in the category $\mathcal{C}'$ of the inductive system $(H(F_\lambda), H(v_f))_I$.

If in the preceding statement we confine ourselves to finite categories I, then the functor H is said *to commute with* (or *to preserve*) *finite projective limits*, respectively *finite inductive limits*.∎

By Proposition 1 it follows that if $\mathcal{C}$ is an abelian category, then *for any* $E \in \operatorname{Ob} \mathcal{C}$, *the functor* $h^E: \mathcal{C} \to \mathbf{Ab}$ *commutes with projective limits*: for let $U = (F_\mu, v_g)_J$ be a projective system with values in the category $\mathcal{C}$ indexed by a category J; assume that $(F, (p_\mu)_{\mu \in J})$ is a projective limit in the category $\mathcal{C}$ of the projective system[1] U; we have to show that the system $(h^E(F), (h^E(p_\mu))_{\mu \in J})$ is in $\mathbf{Ab}$ a projective limit, in the sense of Definition 9, § 12, of the projective system $h^E \circ U = (h^E(F_\mu), h^E(v_g))_J$; it is clear that if $g \in \operatorname{Hom}_J(\mu, \mu')$, then the following diagram commutes:

$$\begin{array}{ccccc} & & h^E(F) & & \\ & \swarrow^{h^E(p_\mu)} & & \searrow^{h^E(p_{\mu'})} & \\ h^E(F_\mu) & & \xleftarrow[h^E(v_g)]{} & & h^E(F_{\mu'}); \end{array}$$

on the other hand, taking for I the category having a single object, say λ, and a single morphism 1_λ, it is clear that $I \times J$ may be identified with J by means of $(\lambda, \mu) \mapsto \mu$, $\mu \in J$; if we consider the inductive system (with values in the category $\mathcal{C}$ indexed by the category I) assigning E to $\lambda \in I$, then, obviously, the inductive limit of this system is $(E, 1_E)$; consequently, the isomorphism φ given by Proposition 1 acts, with these identifications, as follows:

$$h^E(F) = \operatorname{Hom}_{\mathcal{C}}(E, \varprojlim_J F_\mu) \xrightarrow[\sim]{\varphi} \varprojlim_J \operatorname{Hom}_{\mathcal{C}}(E, F_\mu) = \varprojlim_J h^E(F_\mu),$$

and if $r \in \operatorname{Hom}_{\mathcal{C}}(E, \varprojlim_J F_\mu)$, then, in the above notation, $\varphi(r) = (p_\mu \circ r)_{\mu \in J}$;

[1] It is worthwhile remarking that we do not assume $\mathcal{C}$ to be a category with projective limits, but only the fact that the projective system we are concerned with has at least one projective limit in the category $\mathcal{C}$.

consequently, φ renders commutative the diagram

$$\begin{array}{ccc} h^E(F) & \xrightarrow[\sim]{\varphi} & \varprojlim_J h^E(F_\mu) \\ & {\scriptstyle h^E(p_\mu)} \searrow \quad \swarrow {\scriptstyle q_\mu} & \\ & h^E(F_\mu) & \end{array}$$

for any $\mu \in J$; hence, if X is an abelian group and $(t_\mu)_{\mu \in J}$ is a family of homomorphisms of abelian groups, $t_\mu : X \to h^E(F_\mu)$, such that for $g \in \mathrm{Hom}_J(\mu, \mu')$ we have $h^E(v_g) \circ t_{\mu'} = t_\mu$, then there exists the unique homomorphism $t : X \to \varprojlim_J h^E(F_\mu)$ with the property that $q_\mu \circ t = t_\mu$ for any $\mu \in J$; hence the **Z**-homomorphism

$$\varphi^{-1} \circ t : X \to h^E(F)$$

has the property

$$h^E(p_\mu) \circ \varphi^{-1} \circ t = q_\mu \circ t = t_\mu$$

for any $\mu \in J$, and it is uniquely determined by this property.

Likewise, taking into account that the inductive limit is the dual of the projective limit, it follows that *for any object F of the abelian category $\mathcal{C}$, the functor $h_F : \mathcal{C}^\circ \to$* **Ab** *commutes with projective limits*[1]: indeed, given the projective system $T : I^\circ \to \mathcal{C}^\circ$, we get the inductive system $T^\circ : I \to \mathcal{C}$ (see the Remark following Definition 3, § 13); if $T^\circ = (E_\lambda, u_f)_I$, then consider the category J with one object μ and one morphism 1_μ, and the projective system $(F, 1_F)$ with values in the category $\mathcal{C}$ indexed by J; in this case, with the obvious identifications, the isomorphism φ of Proposition 1 acts as follows:

$$h_F((\varinjlim_I E_\lambda)^\circ) = \mathrm{Hom}_{\mathcal{C}}(\varinjlim_I E_\lambda, F) \xrightarrow[\sim]{\varphi} \varprojlim_I \mathrm{Hom}_{\mathcal{C}}(E_\lambda, F) = \varprojlim_I h_F(E_\lambda^\circ),$$

where, if $r \in \mathrm{Hom}_{\mathcal{C}}(\varinjlim_I E_\lambda, F)$, then

$$\varphi(r) = (r \circ j_\lambda)_{\lambda \in I},$$

j_λ, $\lambda \in I$, being the canonical morphisms $E_\lambda \to \varinjlim_I E_\lambda$; further one argues as for h^E.

---

[1] Notice that to prove these results it is not necessary for $\mathcal{C}$ to be an abelian category; in this case, however, the functors h^E and h_F take their values in the category **Ens**.

Concerning general cases of functors commuting with projective or inductive limits, we may prove the following Proposition:

PROPOSITION 2. *Let* $\mathcal{C}, \mathcal{C}'$ *be two categories, and let* $H : \mathcal{C} \to \mathcal{C}'$ *be a functor having a left adjoint* $L : \mathcal{C}' \to \mathcal{C}$; *then* H *commutes with projective limits; likewise, if* $L : \mathcal{C}' \to \mathcal{C}$ *has a right adjoint* $H : \mathcal{C} \to \mathcal{C}'$, *then* L *commutes with inductive limits.*

Proof. We shall prove only the first part of the Proposition, the second one following by dual considerations.

Let $T = (E_\lambda, u_f)_I$ then be a projective system with values in the category $\mathcal{C}$ indexed by the (small) category I; suppose that $(E, (p_\lambda)_{\lambda \in I})$ is a projective limit in $\mathcal{C}$, in the sense of Definition 9, § 12, of the projective system T; consider the projective system $H \circ T = (H(E_\lambda), H(u_f))_I$ with values in $\mathcal{C}'$ indexed by I; then for any $f \in \mathrm{Hom}_I(\lambda, \mu)$, where $\lambda, \mu \in I$, we obviously have

$$H(u_f) \circ H(p_\mu) = H(p_\lambda);$$

let $X' \in \mathrm{Ob}\, \mathcal{C}'$, and let $(r'_\lambda)_{\lambda \in I}$ be a family of morphisms, where $r'_\lambda \in \mathrm{Hom}_{\mathcal{C}'}(X', H(E_\lambda))$, $\lambda \in I$, such that for any f as above we have the commutative diagram

$$\begin{array}{ccc} H(E_\mu) & \xrightarrow{\;H(u_f)\;} & H(E_\lambda). \\ & {\scriptstyle r'_\mu}\nwarrow \quad \nearrow{\scriptstyle r'_\lambda} & \\ & X' & \end{array}$$

By hypothesis, for any $E' \in \mathrm{Ob}\, \mathcal{C}'$ and any $F \in \mathrm{Ob}\, \mathcal{C}$ there exists a bijection of sets

$$\theta(E', F) : \mathrm{Hom}_{\mathcal{C}}(L(E'), F) \xrightarrow{\sim} \mathrm{Hom}_{\mathcal{C}'}(E', H(F))$$

such that any diagram of the type (5), § 10, is commutative; in particular, the following diagram commutes:

$$(5) \qquad \begin{array}{ccc} \mathrm{Hom}_{\mathcal{C}}(L(X'), E_\mu) & \xrightarrow[\sim]{\;\theta(X', E_\mu)\;} & \mathrm{Hom}_{\mathcal{C}'}(X', H(E_\mu)) \\ {\scriptstyle \mathrm{Hom}\,(1_{L(X')}, u_f)} \downarrow & & \downarrow {\scriptstyle \mathrm{Hom}\,(1_{X'}, H(u_f))} \\ \mathrm{Hom}_{\mathcal{C}}(L(X'), E_\lambda) & \xrightarrow[\sim]{\;\theta(X', E_\lambda)\;} & \mathrm{Hom}_{\mathcal{C}'}(X', H(E_\lambda)). \end{array}$$

Denoting for any $\lambda \in I$,

$$r_\lambda = (\theta(X', E_\lambda))^{-1}(r'_\lambda),$$

we get a family of morphisms $(r_\lambda)_{\lambda\in I}$, with $r_\lambda \in \mathrm{Hom}_{\mathscr{C}}(L(X'), E_\lambda)$, $\lambda\in I$, and such that if $f\in \mathrm{Hom}_I(\lambda, \mu)$, then by virtue of the commutativity of diagram (5),

$$u_f \circ r_\mu = \mathrm{Hom}\,(1_{L(X')}, u_f)\,(\theta(X', E_\mu)^{-1}(r'_\mu)) =$$

$$= (\theta(X', E_\lambda))^{-1}(\mathrm{Hom}\,(1_{X'}, H(u_f))\,(r'_\mu)) = (\theta(X', E_\lambda))^{-1}(H(u_f)\circ r'_\mu) = r_\lambda.$$

Consequently, there exists a unique morphism $r\in \mathrm{Hom}_{\mathscr{C}}(L(X'), E)$ with the property that $p_\lambda \circ r = r_\lambda$ for any $\lambda \in I$; let

$$r' = \theta(X', E)(r) \in \mathrm{Hom}_{\mathscr{C}'}(X', H(E));$$

from the commutativity for any $\lambda\in I$ of the diagram obtained from (5) when replacing E_μ by E and u_f by p_λ it follows as above that $H(p_\lambda)\circ r' = \theta(X', E_\lambda)(p_\lambda \circ r) = r'_\lambda$ for any $\lambda\in I$; if $r'^* \in \mathrm{Hom}_{\mathscr{C}'}(X', H(E))$ had the property that

$$H(p_\lambda)\circ r'^* = r'_\lambda$$

for any $\lambda\in I$, then, denoting $r^* = (\theta(X', E))^{-1}(r'^*)$, the commutativity of the diagram obtained from (5) when replacing E_μ by E and u_f by p_λ would imply

$$p_\lambda \circ r^* = (\theta(X', E_\lambda))^{-1}(H(p_\lambda)\circ r'^*) = r_\lambda$$

for any $\lambda\in I$, and this in turn would imply $r^* = r$ hence $r'^* = r'$.

Consequently, the system $(H(E), H(p_\lambda)_{\lambda\in I})$ is a projective limit in $\mathscr{C}$ of the projective system $(H(E), H(u_f))_I$. ∎

REMARK. We could have employed for any $X'\in \mathrm{Ob}\,\mathscr{C}'$ the sequence of bijections

$$\mathrm{Hom}_{\mathscr{C}'}(X', H(E)) \xrightarrow{(\theta(X', E))^{-1}} \mathrm{Hom}_{\mathscr{C}}(L(X'), E) = h^{L(X')}\,(E) =$$

$$= h^{L(X')}\,(\varprojlim_{\lambda\in I} E_\lambda) \underset{\sim}{\rightarrow} \varprojlim_{\lambda\in I} h^{L(X')}(E_\lambda) = \varprojlim_{\lambda\in I} \mathrm{Hom}_{\mathscr{C}}(L(X'), E_\lambda) \underset{\sim}{\xrightarrow{\chi}} \varprojlim_{\lambda\in I} \mathrm{Hom}_{\mathscr{C}'}(X', H(E_\lambda)) =$$

$$= \varprojlim_{\lambda\in I} h^{X'}(H(E_\lambda)) \underset{\sim}{\rightarrow} h^{X'}(\varprojlim_{\lambda\in I} H(E_\lambda)) = \mathrm{Hom}_{\mathscr{C}'}(X', \varprojlim_{\lambda\in I} H(E_\lambda))$$

(where $\chi = \varprojlim_{\lambda\in I} \theta(X', E_\lambda)$), making use of the fact that $h^{L(X')}$ and $h^{X'}$ commute with projective limits; from this and from the fact that these bijections are components of a natural transformation "of argument" X' we could have deduced that $H(E) \approx \varprojlim_{\lambda\in I} H(E_\lambda)$;

however, in order that the proof be complete we should have shown that the constructed isomorphism is compatible with the canonical morphisms, that is, its composition with the canonical morphism $\varprojlim_{\lambda \in I} H(E_\lambda) \to H(E_\lambda)$ is $H(p_\lambda)$ for any $\lambda \in I$. ▮

Proposition 2 has the following

COROLLARY. *Given the category $\mathcal{C}$ with projective limits, and an arbitrary small category I, the projective limit functor* $\varprojlim_I : \mathcal{C}^{I^\circ} \to \mathcal{C}$ *commutes with projective limits; likewise, if $\mathcal{C}$ is a category with inductive limits, then for any small category I, the inductive limit functor* $\varinjlim_I : \mathcal{C}^I \to \mathcal{C}$ *commutes with inductive limits.*

Proof. We have only to apply the generalization for categories with projective, respectively inductive limits of Proposition 3, § 12, respectively of Proposition 4, § 13, stating that the projective, respectively inductive limit functor has a left, respectively right adjoint, namely the immersion functor $\mathcal{C} \to \mathcal{C}^{I^\circ}$, respectively $\mathcal{C} \to \mathcal{C}^I$. ▮

REMARK. The distinction between the property of projective limits of commuting with projective limits given by Corollaries 3, 4 to Proposition 1, § 12, and by the generalization of these Corollaries to categories with projective limits, and the first assertion of the above Corollary to Proposition 2 is only a difference of formulation; similarly, between the property of inductive limits of commuting with inductive limits (see Corollary 3 to Proposition 1, § 13, and its generalization to categories with inductive limits) and the second assertion of the above Corollary to Proposition 2; hence, the assertions of this Corollary could be deduced without necessarily using Proposition 2. ▮*

As an application of the above results, let us consider an integral domain A and the full subcategory, denoted **Tormod** (A), of **Mod**(A) whose objects are the torsion modules (see § 10); since the torsion functor t defined in § 10 takes its values in **Tormod** (A), it defines a uniquely determined functor **Mod**$(A) \to$ $\to$ **Tormod** (A), which, by an abuse of notation, will be denoted also by t.

Let E be a torsion A-module, and let F be an arbitrary A-module; if $u : E \to F$ is an A-homomorphism, then, as shown in § 10, we have $u(E) \subseteq t(F)$; hence u induces by restriction to E and $t(F)$ a (unique) A-homomorphism $v : E \to t(F)$; this means that the homomorphism of abelian groups

$$\mathrm{Hom}_{\mathbf{Tormod}\,(A)}(E, t(F)) \to \mathrm{Hom}_A(E, F)$$

given by

$$v \mapsto i \circ v, \qquad v \in \mathrm{Hom}_{\mathbf{Tormod}\,(A)}(E, t(F)),$$

where i is the canonical injection $t(F) \to F$, is an isomorphism; for a given homomorphism $k : E_1 \to E$ we obtain a commutative diagram of the type (22), § 10 (but with the first argument varying instead of the second); Proposition 13, § 10 (dualized) then shows that the functor $t : \mathbf{Mod}(A) \to \mathbf{Tormod}(A)$ is a right adjoint of the immersion functor $\mathbf{Tormod}(A) \to \mathbf{Mod}(A)$.

The above Proposition 2 then implies that the functor t commutes with projective limits, in particular, with direct products; consequently, if $(E_\lambda)_{\lambda \in I}$ is a family of torsion A-modules, and if $(E, (\pi_\lambda)_{\lambda \in I})$ is the direct product of this family in the category $\mathbf{Mod}(A)$, then $(t(E), (t(\pi_\lambda))_{\lambda \in I})$ is a direct product of the same family in $\mathbf{Tormod}(A)$ (see also Example 7), § 10).

As an example, let $A = \mathbf{Z}$, let $E_p = \mathbf{Z}/p\mathbf{Z}$ for any prime number $p > 0$, and let $E = \prod_p E_p$ in the category $\mathbf{Ab}$, the canonical projections being π_p for any prime p; if $(\hat{x}_p)_p \in t(E)$, where $\hat{x}_p$ is the coset modulo $p\mathbf{Z}$ of $x_p \in \mathbf{Z}$ for any p, then, if $n \in \mathbf{Z}$, $n \neq 0$, has the property that $n((\hat{x}_p)_p) = 0$, it follows that for any prime number p we have $nx_p \in p\mathbf{Z}$; if $p \nmid n$, this means that $x_p \in p\mathbf{Z}$, hence that $\hat{x}_p = 0$; consequently, only a finite number of components $\hat{x}_p$, p prime, are non-zero, and $(\hat{x}_p)_p$ is an fs-family; so $t(E) \subseteq \bigoplus_p E_p$; the converse is true by (10), § 10; it follows that $t(\prod_p E_p) = \bigoplus_p E_p$; hence $(\bigoplus_p E_p, (\pi'_p)_p)$, where $\pi'_p = t(\pi_p)$ for any p is the homomorphism $\bigoplus_p E_p \to E_p$ induced by π_p, is a direct product in the category $\mathbf{Tormod}(\mathbf{Z})$ of the family $(E_p)_p$ of torsion $\mathbf{Z}$-modules.

* Now let $\mathcal{C}, \mathcal{C}'$ be abelian categories, and let $H : \mathcal{C} \to \mathcal{C}'$ be a left exact functor; if $E \in \operatorname{Ob} \mathcal{C}$ and (M, i) is a subobject of E, then, as in $\mathcal{C}$ we have the exact sequence

$$0 \to M \xrightarrow{i} E \xrightarrow{p} P,$$

where (P, p) is a cokernel of i, it turns out that in $\mathcal{C}'$ we have the exact sequence

$$0 \to H(M) \xrightarrow{H(i)} H(E) \xrightarrow{H(p)} H(P),$$

hence that $(H(M), H(i))$ is a subobject of $H(E)$.

Let $(M_k, i_k)_{k \in [1, n]}$ be a family of subobjects of the object E of $\mathcal{C}$; we may consider the intersection $\left(\bigcap_{k=1}^{n} M_k, i\right)$ of this family of subobjects, which is also a subobject of E; applying the functor H we have the two subobjects of $H(E)$ in $\mathcal{C}'$, namely $\left(\bigcap_{k=1}^{n} H(M_k), i'\right)$ and $\left(H\left(\bigcap_{k=1}^{n} M_k\right), H(i)\right)$; let us show that *these two subobjects coincide*:

Indeed, for any $k = 1, \ldots, n$ there exists $v_k : \bigcap_{k=1}^{n} M_k \to M_k$ such that $i_k \circ v_k = i$, $k = 1, \ldots, n$; then $H(v_k) : H(\bigcap_k M_k) \to H(M_k)$ satisfies the condition

$H(i_k) \circ H(v_k) = H(i)$, hence we have

$$(H(\bigcap_k M_k), H(i)) \leqslant (H(M_k), H(i_k))$$

for any $k \in [1, n]$ (see § 8); since the intersection of subobjects is the greatest lower bound of these subobjects, it follows that

$$(6) \qquad (H(\bigcap_k M_k), H(i)) \leqslant (\bigcap_k H(M_k), i').$$

On the other hand, taking into account the definition of the intersection of subobjects (see § 8), if (P_k, p_k) is a cokernel of i_k, $k = 1, \ldots, n$, then $(\bigcap_k M_k, i)$ is a kernel of $p : E \to \prod_{k=1}^{n} P_k$, p being the unique morphism with the property that $q_k \circ p = p_k$, $k = 1, \ldots, n$, where q_k is the canonical projection $\prod_k P_k \to P_k$; hence we have the commutative diagram in which the rows are exact sequences:

$$\begin{array}{ccccccc}
0 & \longrightarrow & M_k & \xrightarrow{i_k} & E & \xrightarrow{p_k} & P_k \\
 & & \uparrow{\scriptstyle v_k} & & \uparrow{\scriptstyle 1_E} & & \uparrow{\scriptstyle q_k} \\
0 & \longrightarrow & \bigcap_k M_k & \xrightarrow{i} & E & \xrightarrow{p} & \prod_k P_k;
\end{array}$$

let now (P'_k, p'_k) be a cokernel of $H(i_k)$, $k = 1, \ldots, n$; let $k \in [1, n]$; since $H(p_k) \circ H(i_k) = 0$, it follows that there exists a morphism $r'_k : P'_k \to H(P_k)$ such that $r'_k \circ p'_k = H(p_k)$; as H commutes with finite direct products (see the Corollary to Proposition 4, § 7), it follows that there exists a unique morphism $\prod_k P'_k \to H(\prod_k P_k)$, which we shall denote by an abuse of notation $\prod_k r'_k$, with the property that[1]

$$H(q_k) \circ \prod_k r'_k = r'_k \circ q'_k$$

for any $k = 1, \ldots, n$, q'_k being the canonical projection $\prod_k P'_k \to P'_k$; let p' be the unique morphism $H(E) \to \prod_k P'_k$ satisfying the condition $q'_k \circ p' = p'_k$ for any

---

[1] In fact, the above mentioned morphism is obtained by composition of the canonical isomorphism $\prod_k H(P_k) \xrightarrow{\sim} H(\prod_k P_k)$ with $\prod_k r'_k$.

$k \in [1, n]$; $(\bigcap_k H(M_k), i')$ is defined as being the kernel of p'; the situation is described by the following diagram:

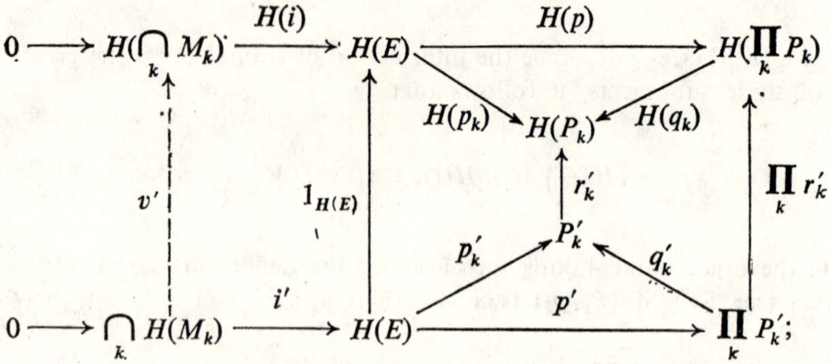

since H is a left exact functor, the first row in the diagram is an exact sequence; on the other hand, we have

$$\prod_k r'_k \circ p' = H(p), \tag{7}$$

which will follow provided the composition of $H(q_k)$ with the morphisms in the two sides of (7) gives the same morphism for any $k = 1, \ldots, n$; but, for any k,

$$H(q_k) \circ \prod_k r'_k \circ p' = r'_k \circ q'_k \circ p' = r'_k \circ p'_k = H(p_k) = H(q_k) \circ H(p);$$

consequently, $H(p) \circ i' = 0$, hence, as $(H(\bigcap_k M_k), H(i))$ is a kernel of $H(p)$, there exists a unique morphism v' as in the diagram having the property that $H(i) \circ v' = i'$; but this means that

$$(\bigcap_k H(M_k), i') \leqslant (H(\bigcap_k M_k), H(i)),$$

which together with (6) proves our assertion.

Assuming, moreover, that $\mathcal{C}, \mathcal{C}'$ are categories with direct products and that the functor H commutes with direct products, those proved for a finite family of subobjects of an object E of the category $\mathcal{C}$ could have been proved for a family indexed by an arbitrary set.

Now we can prove the following Proposition:

PROPOSITION 3. *Let $\mathcal{C}, \mathcal{C}'$ be abelian categories, and let $H : \mathcal{C} \to \mathcal{C}'$ be an additive functor; then H is a left exact functor if and only if it commutes with finite projective limits; likewise, H is a right exact functor if and only if it commutes with finite inductive limits.*

Proof. Assume that H commutes with finite projective limits, and let the exact sequence

$$0 \to E \xrightarrow{u} F \xrightarrow{v} G$$

be given in $\mathcal{C}$; this is equivalent to saying that (E, u) is a kernel of v; but this is equivalent to saying that $(E, 0, u)$ is a projective limit of the projective system T with values in the category $\mathcal{C}$ indexed by the finite category I, where I has two objects λ, μ and two morphisms $f, g \in \mathrm{Hom}_I(\lambda, \mu)$ (besides the identity morphisms), and where

$$T(\lambda) = G, \quad T(\mu) = F, \quad T(f) = v, \quad T(g) = 0;$$

the last assertion follows from the arguments of § 12 when dealing with the third particular case, or also from the more explicit discussion of the dual case in § 13; but then $(H(E), H(0), H(u))$ is a projective limit of the projective system $H \circ T$ with values in $\mathcal{C}'$, hence $(H(E), H(u))$ is a kernel of $H(v)$; hence the diagram in $\mathcal{C}'$

$$0 \to H(E) \xrightarrow{H(u)} H(F) \xrightarrow{H(v)} H(G)$$

is an exact sequence. Thus, H is a left exact functor.

Conversely, let H be a left exact functor, let I be a finite category, let $T = (E_\lambda, u_f)_I$ be a projective system with values in the category $\mathcal{C}$ indexed by the category I, and let $(E, (p_\lambda)_{\lambda \in I})$ be a projective limit of this projective system; to show that H commutes with finite projective limits, we may, with no loss of generality, restrict ourselves to those E and p_λ, $\lambda \in I$, which are constructed in the proof of Proposition 5, § 12 (without assuming, however, now that $\mathcal{C}$ is a category with arbitrary direct products). We use the notations in the proof of Proposition 5, § 12; as H is an additive functor, it follows (see the Corollary to Proposition 4, § 7) that for any $\lambda, \mu \in I$, $(H(E_\mu \times E_\lambda), H(j_\mu^\lambda), H(j_\lambda^\mu))$ is a direct sum in the category $\mathcal{C}'$ of the family of objects $(H(E_\mu), H(E_\lambda))$, and, consequently, that if $f \in \mathrm{Hom}_I(\lambda, \mu)$, $H(s_f)$ is the unique morphism $H(E_\mu \times E_\lambda) \to H(E_\lambda)$ for which

$$H(s_f) \circ H(j_\mu^\lambda) = H(u_f), \qquad H(s_f) \circ H(j_\lambda^\mu) = -1_{H(E_\lambda)};$$

likewise, since the set of objects of I is finite, it follows that $(H(\prod_{\nu \in I} E_\nu), (H(p'_\nu))_{\nu \in I})$ is a direct product of the family of objects $(H(E_\nu))_{\nu \in I}$, and that $H(t_f)$ is the unique morphism $H(\prod_{\nu \in I} E_\nu) \to H(E_\mu \times E_\lambda)$ obtained by the definition of the direct product from the family of morphisms $H(p'_\mu)$ and $H(p'_\lambda)$; further, H being a left exact functor, it follows that $(H(\mathrm{Ker}\, w_f), H(i_f))$ is a kernel of $H(w_f) = H(s_f) \circ H(t_f)$; finally, by the considerations preceding our Proposition, as M is a finite set, it follows that $(H(\bigcap_{f \in M} \mathrm{Ker}\, w_f), H(i))$ is an intersection of the family of subobjects

$(H(\text{Ker } w_f), H(i_f))_{f\in M}$ of $H(\prod_{\lambda\in I} E_\lambda)$; since $H(p_\lambda)=H(p'_\lambda)\circ H(i)$ for any $\lambda\in I$, it follows that $(H(E), H(p_\lambda)_{\lambda\in I})$ is a projective limit in the category $\mathcal{C}'$ of the projective system $H\circ T$.

The second part of the Proposition follows from the first one by duality. ▌

COROLLARY. *If $\mathcal{C}$ is an abelian category with projective limits and I is an arbitrary small category, then the projective limit functor* $\varprojlim_I : \mathcal{C}^{I^\circ} \to \mathcal{C}$ *is left exact; likewise, if $\mathcal{C}$ is an abelian category with inductive limits and I is an arbitrary small category, then the inductive limit functor* $\varinjlim_I : \mathcal{C}^I \to \mathcal{C}$ *is right exact.*

Proof. We have to apply Proposition 3 and the Corollary to Proposition 2. ▌

For an arbitrary small category of indices we have the following result:

PROPOSITION 4. *Let $\mathcal{C}, \mathcal{C}'$ be abelian categories, and let $H : \mathcal{C} \to \mathcal{C}'$ be an additive functor; if $\mathcal{C}, \mathcal{C}'$ are categories with direct products, the functor H commutes with projective limits if and only if it is left exact and commutes with direct products; if $\mathcal{C}, \mathcal{C}'$ are categories with direct sums, the functor H commutes with inductive limits if and only if it is right exact and commutes with direct sums.*

Proof. The second assertion follows from the first one by duality; as for the first assertion, the sufficiency is obtained exactly in the same way as the analogous result of Proposition 3; only now we have to do with an arbitrary (small) category of indices, and in exchange H has the additional property of commuting with arbitrary direct products; concerning the necessity, the fact that H is left exact follows from Proposition 3, while the fact that H commutes with arbitrary direct products follows by hypothesis, since according to § 12, the direct products are particular cases of projective limits. ▌*

PROPOSITION 5. *Let A be a ring, and let I be the category associated with the preordered set I directed to the right; if $T = (E_\lambda, u_{\lambda\mu})_I$, $U = (F_\lambda, v_{\lambda\mu})_I$ are inductive systems of A-modules indexed by the category I, and if $(h_\lambda)_{\lambda\in I}$ is an inductive morphism from T to U such that h_λ is a monomorphism for any $\lambda \in I$, then* $\varinjlim_I h_\lambda : \varinjlim_I E_\lambda \to \varinjlim_I F_\lambda$ *is a monomorphism.*

Proof. Let $x \in \varinjlim_I E_\lambda$ be such that $h(x) = 0$, where $h = \varinjlim_I h_\lambda$; since I is a small pseudofiltered connected category, it follows by Proposition 3, *a*), § 13, that there exists $\lambda \in I$ and $x_\lambda \in E_\lambda$ such that $x = j_\lambda(x_\lambda)$, j_μ being the canonical homomorphism $E_\mu \to \varinjlim_I E_\mu$, $\mu \in I$; it follows that

$$0 = h(x) = h(j_\lambda(x_\lambda)) = k_\lambda(h_\lambda(x_\lambda)),$$

k_μ being the canonical homomorphism $F_\mu \to \varinjlim_I F_\mu$, $\mu \in I$; by Proposition 3, *b*), §13, there exists $\rho \geqslant \lambda$ such that $v_{\lambda\rho}(h_\lambda(x_\lambda)) = 0$; but then $h_\rho(u_{\lambda\rho}(x_\lambda)) = 0$, and since h_ρ is injective, $u_{\lambda\rho}(x_\lambda) = 0$; it follows that $x = j_\lambda(x_\lambda) = j_\rho(u_{\lambda\rho}(x_\lambda)) = 0$. ▌

COROLLARY. *If I is the category associated with a preordered set directed to the right, then the inductive limit functor* $(\mathbf{Mod}(A))^I \to \mathbf{Mod}(A)$ *is exact.*

Proof. This is straightforward consequence of the fact that the inductive limit functor is right exact (see Proposition 5, § 13) and of the above Proposition 5. ▮

* It may be shown that *if $\mathcal{C}$ is an abelian category satisfying Axiom* AB5), *then for any category I associated with a preordered set I directed to the right, the inductive limit functor*[1] $\varinjlim_I : \mathcal{C}^I \to \mathcal{C}$ *is left exact*, hence, by Proposition 3, *it commutes with finite projective limits*; moreover, by the Corollary to Proposition 3, the functor $\varinjlim_I$ for such a category I is exact. It may be shown that if $\mathcal{C}$ is an abelian category satisfying Axiom AB5), then for any index set I, the direct sum functor is exact[2], hence that the following Axiom is satisfied[3]:

AB4) *In the abelian category $\mathcal{C}$ Axiom* AB3) *is satisfied, and for any set I, the direct sum functor $\mathcal{C}^I \to \mathcal{C}$ is exact* (*or*, what is the same thing for this functor, *it is left exact*, *or*, what is the same thing for this functor, *it carries the monomorphisms of the category $\mathcal{C}^I$ into monomorphisms of the category $\mathcal{C}$*).*

The fact that the dual of the above Proposition 5 is not valid, in other words, that the projective limit of a projective system of surjective linear mappings is not necessarily a surjective mapping may be seen from the following

EXAMPLE. 1) Let I be the category associated with the totally ordered set $\mathbf{N}$ (with the natural ordering); define two projective systems of $\mathbf{Z}$-modules indexed by the category I, $(E_n, u_{nm})_{n\in\mathbf{N}}$, $(F_n, v_{nm})_{n\in\mathbf{N}}$, as follows: for any $n \in \mathbf{N}$, $E_n = \mathbf{Z}$, $F_n = \mathbf{Z}/2\mathbf{Z}$; for $n \leqslant m$ in $\mathbf{N}$ define u_{nm}, respectively v_{nm}, as being the multiplication by 3^{m-n} in $\mathbf{Z}$, respectively in $\mathbf{Z}/2\mathbf{Z}$; for any $n \in \mathbf{N}$ define the mapping $k_n : E_n \to F_n$ as being the canonical surjection $\mathbf{Z} \to \mathbf{Z}/2\mathbf{Z}$; let us show that $\varprojlim_{\mathbf{N}} k_n : \varprojlim_{\mathbf{N}} E_n \to \varprojlim_{\mathbf{N}} F_n$ is not surjective: indeed, if $(x_n)_{n\in\mathbf{N}} \in \varprojlim_{\mathbf{N}} E_n$, then by the definition of the projective limit of modules,

$$x_0 = 3x_1 = 3^2x_2 = \ldots = 3^n x_n = \ldots,$$

hence x_0, being an element of $\mathbf{Z}$ divisible by any power of 3, is 0; this is valid for any x_n, $n \in \mathbf{N}$; hence $\varprojlim_{\mathbf{N}} E_n = 0$; on the other hand, the element of $\prod_{n\in\mathbf{N}} F_n$ having all components equal to the coset $\hat{1}$ of 1 modulo 2 is a non-zero element from $\varprojlim_{\mathbf{N}} F_n$, since for $n \leqslant m$ we have $3^{m-n}\hat{1} = \hat{1}$; hence $\varprojlim_{\mathbf{N}} F_n \neq 0$, and $\varprojlim_{\mathbf{N}} k_n$ cannot be surjective. ▮

---

[1] The validity of Axiom AB5), which is stronger then Axiom AB3), implies, by virtue of Proposition 2, § 13, that $\mathcal{C}$ is a category with inductive limits.

[2] We know (see the Corollary to Proposition 3) that, since the direct sum is a particular case of inductive limit (see § 13), in general, the direct sum functor is right exact.

[3] See also Corollary 5 to Proposition 2, § 7, proving directly that in the category $\mathbf{Mod}(A)$, where A is a ring, Axiom AB4) is satisfied.

§ 15. Change of the Ring of Scalars

So far the ring of scalars A has been fixed for a given problem; we proceed to study what happens when the ring of scalars changes (in a certain way).

Let A and B be two rings, and let $\rho : B \to A$ be a homomorphism of rings (sending the unit element of B into the unit element of A); to each A-module E a B-module denoted $\rho_*(E)$ may be assigned in the following way: the set of elements of $\rho_*(E)$ is E, and the addition is the one in E; if $\beta \in B$, $x \in E$, then by definition

$$\beta x = \rho(\beta)x,$$

the operation in the right side being the scalar multiplication in E by elements of A; it is not difficult to see that a B-module is thus obtained: for instance, if $\beta_1, \beta_2 \in B$, $x \in E$, then

$$(\beta_1\beta_2)x = \rho(\beta_1\beta_2)x = (\rho(\beta_1)\rho(\beta_2))x = \rho(\beta_1)(\rho(\beta_2)\cdot x) = \beta_1(\beta_2 x).$$

Of course, if right A-modules were considered instead of left A-modules, we would obtain in this way a right B-module from each right A-module.

Returning to the case of left modules, let E and F be A-modules, and let $u : E \to F$ be an A-homomorphism; the mapping u considered as a mapping from $\rho_*(E)$ to $\rho_*(F)$ is a B-homomorphism: indeed, if $\beta \in B$ and $x \in \rho_*(E)$, then

$$u(\beta \cdot x) = u(\rho(\beta)\cdot x) = \rho(\beta)\cdot u(x) = \beta \cdot u(x);$$

u considered as a B-homomorphism $\rho_*(E) \to \rho_*(F)$ will be denoted $\rho_*(u)$. Now, it is easy to see that the mappings

$$E \mapsto \rho_*(E), \qquad u \mapsto \rho_*(u)$$

define a functor

$$\rho_* : \mathbf{Mod}(A) \to \mathbf{Mod}(B).$$

DEFINITION 1. The above defined functor ρ_* is called the *functor of restriction to B of the ring of scalars A* or also, if confusion is not possible, the *functor of restriction of scalars.*

Sometimes, if there is no danger of confusion, for an A-module E one writes also $E_{[B]}$ or even E instead of $\rho_*(E)$.

The B-module structure obtained on an A-module E by means of ρ_* will be called the B-module structure of E *obtained* from the given A-module structure of E *by restriction to* B (by means of ρ). ▮

REMARK. If B is a subring of A (B containing the unit element of A), and if ρ is the inclusion mapping of B into A, then we have to do with a proper restriction of the scalars of B; the terminology is maintained in the case of an arbitrary ρ, being justified by the particular case described. ▮

EXAMPLES. 1) Let A be a ring, let E be an abelian group; consider the ring $\mathrm{End}_{\mathbf{Z}}(E)$; assume the homomorphism of rings $\rho : A \to \mathrm{End}_{\mathbf{Z}}(E)$ is given; since under the addition in E and the scalar multiplication by elements of $\mathrm{End}_{\mathbf{Z}}(E)$ given by

$$(f, x) \mapsto f(x),$$

where $f \in \mathrm{End}_{\mathbf{Z}}(E)$, $x \in E$, E is an $\mathrm{End}_{\mathbf{Z}}(E)$-module, it follows that by the restriction of scalars we get an A-module, namely $\rho_*(E)$.

Likewise, given a homomorphism σ from the opposite ring A° to $\mathrm{End}_{\mathbf{Z}}(E)$, we get on E the structure of A°-module $\sigma_*(E)$; consequently, taking into account those shown in § 1 and the identification $(A^\circ)^\circ = A$, we get on E a right A-module structure.

In fact, in this way all (left, respectively right) A-module structures of the abelian group E are obtained: for if E is a (left) A-module, then the mapping

$$\alpha \mapsto h_\alpha,$$

where $\alpha \in A$ and h_α is the homothety of ratio α in E, defines a homomorphism of rings (with unit element) from A to $\mathrm{End}_{\mathbf{Z}}(E)$; indeed, for any $\alpha, \beta \in A$ we have

$$h_{\alpha+\beta} = h_\alpha + h_\beta, \quad h_{\alpha\beta} = h_\alpha \circ h_\beta, \quad h_1 = 1_E;$$

on the other hand, if E is a right A-module, then the mapping

$$\alpha^\circ \mapsto h_\alpha$$

is a homomorphism of rings $A^\circ \to \mathrm{End}_{\mathbf{Z}}(E)$, since in this case

$$h_{\beta\alpha} = h_\alpha \circ h_\beta$$

for any $\alpha, \beta \in A$.

These two correspondences, from a (left, respectively right) A-module structure defined on E to the corresponding homomorphism of rings $A \to \mathrm{End}_{\mathbf{Z}}(E)$, respectively $A^\circ \to \mathrm{End}_{\mathbf{Z}}(E)$, and from such a homomorphism to the corresponding structure of (left, respectively right) A-module, are inverse to each other. We leave to the reader as an exercise to verify this.

2) Let A be a ring; we define a mapping $\rho : \mathbf{Z} \to A$ as follows: to each $n \in \mathbf{Z}$ the element $n \cdot 1$ of A is assigned, where 1 is the unit element of A, while $n \cdot 1$ means the multiplication of $1 \in A$ by $n \in \mathbf{Z}$ according to the $\mathbf{Z}$-module structure of the additive group of the ring A; ρ is a homomorphism of rings with unit

element, since, for instance, $(m\cdot 1)(n\cdot 1) = (mn)\cdot 1$ for any $m, n \in \mathbf{Z}$; in this way, to each A-module E the $\mathbf{Z}$-module $\rho_*(E)$ is assigned such that, if $n \in \mathbf{Z}$ and $x \in E$, we have

$$n\cdot x = (n\cdot 1)x;$$

it is easily seen that this is exactly the $\mathbf{Z}$-module structure assigned to the additive group of the A-module E (see Example 4), § 1). ▮

Returning to the general situation of a homomorphism of rings $\rho : B \to A$, the functor ρ_* of restriction of scalars preserves the abelian group structure of any A-module E, as well as any A-homomorphism u as mapping of sets and even as homomorphism of abelian groups; it follows that *ρ_* is an exact additive functor* and that *for any pair of A-modules E, F, the mapping*

$$(1) \qquad \rho_*(E, F) : \mathrm{Hom}_A(E, F) \to \mathrm{Hom}_B(\rho_*(E), \ \rho_*(F))$$

given by the functor ρ_* is injective.

The mapping $\rho_*(E, F)$ is not, in general, surjective, as shown by the following

EXAMPLE. 3) Let $B = \mathbf{Z}$, $A = \mathbf{C}$, let $\rho : \mathbf{Z} \to \mathbf{C}$ be the inclusion homomorphism, and let $E = F$ be the $\mathbf{C}$-vector space defined on $\mathbf{C}$ in the usual manner (see Example 1), § 1); let $v : E \to F$ be the mapping sending any complex number into its conjugate; we have $v \in \mathrm{Hom}_B(\rho_*(E), \rho_*(F))$, but if there existed $u \in \mathrm{Hom}_A(E,F)$ such that $\rho_*(E, F)(u) = v$, then, as mappings of sets we would have $u = v$; but

$$v(i\cdot i) = -1 \neq +1 = i(-i) = i\cdot v(i),$$

proving that v is not $\mathbf{C}$-linear. ▮

* DEFINITION 2. Let $\mathcal{C}$, $\mathcal{C}'$ be two categories, and let $T : \mathcal{C} \to \mathcal{C}'$ be a functor; T is said to be a *faithful*, respectively a *fully faithful* functor if for any pair of objects E, F of $\mathcal{C}$, the mapping of sets

$$T(E, F) : \mathrm{Hom}_{\mathcal{C}}(E, F) \to \mathrm{Hom}_{\mathcal{C}'}(T(E), T(F))$$

is injective, respectively bijective. ▮

Obviously, to say that $T : \mathcal{C} \to \mathcal{C}'$ is a faithful functor is equivalent to saying that T transforms any non-commutative diagram of $\mathcal{C}$ into a non-commutative diagram of $\mathcal{C}'$.

If $\mathcal{C}$, $\mathcal{C}'$ are additive categories and T is an additive functor, to say that T is faithful is equivalent to saying that for any non-zero morphism u of the category $\mathcal{C}$, the morphism $T(u)$ is non-zero. This implies, in particular, that if X is a non-zero object of the category $\mathcal{C}$, then $T(X)$ is a non-zero object of the category $\mathcal{C}'$; when, moreover, $\mathcal{C}$, $\mathcal{C}'$ are abelian categories and T is an exact functor, the fact that T carries the non-zero objects of $\mathcal{C}$ into non-zero objects of $\mathcal{C}'$ is also sufficient for T to be faithful: for let $u \in \mathrm{Hom}_{\mathcal{C}}(E, F)$, $u \neq 0$; if

$$E \xrightarrow{q} C \xrightarrow{\overline{u}} I \xrightarrow{j} F$$

is the canonical factorization of u (hence $u = j \circ \bar{u} \circ q$), it follows that I is a non-zero object; $T(u) = 0$ would imply

$$T(j) \circ 1_{T(I)} \circ T(\bar{u} \circ q) = 0,$$

and as $T(j)$ is a monomorphism (since j is a monomorphism) and $T(\bar{u} \circ q)$ is an epimorphism (since $\bar{u} \circ q$ is an epimorphism), it would follow that $1_{T(I)} = 0$, hence the fact that $T(I)$ is a zero object of $\mathscr{C}'$, a contradiction.

We also notice that a composition of two or several faithful (respectively fully faithful) functors is a faithul (respectively fully faithful) functor.*

From the preceding considerations it follows that for a homomorphism of rings $\rho : B \to A$, *the functor* ρ_* *of restriction of scalars is faithful, but, in general, it is not fully faithful.*

There are, however, in certain particular cases additional properties given by the following Proposition:

PROPOSITION 1. *Let* $\rho : B \to A$ *be a homomorphism of rings; in what follows* E *and* F *are arbitrary* A*-modules.*

a) If ρ *is surjective, then the mapping* $\rho_*(E, F)$ *in* (1) *is bijective; this means that in this case* ρ_* *is a fully faithful functor.*

b) If ρ *is surjective, then, if* M *is a* B*-submodule of* $\rho_*(E)$, *it follows that* M *(as subset of* E*) is an* A*-submodule of* E.

c) If ρ *is surjective, then, if* $(a_\lambda)_{\lambda \in I}$ *is a system of generators of* E *over* A, *it follows that* $(a_\lambda)_{\lambda \in I}$ *is a system of generators of* $\rho_*(E)$ *over* B.

d) If ρ *is injective, then, if* $(a_\lambda)_{\lambda \in I}$ *is a linearly independent family of elements of* E *over* A, *it follows that* $(a_\lambda)_{\lambda \in I}$ *is a linearly independent family of elements of* $\rho_*(E)$ *over* B.

e) If ρ *is surjective, then, if* $(a_\lambda)_{\lambda \in I}$ *is a linearly independent family of elements of* $\rho_*(E)$ *over* B, *it follows that* $(a_\lambda)_{\lambda \in I}$ *is a linearly independent family of elements of* E *over* A.

Proof. a) If u is a B-homomorphism $\rho_*(E) \to \rho_*(F)$, then, since for any $\alpha \in A$ there exists $\beta \in B$ such that $\rho(\beta) = \alpha$, it follows that

$$u(\alpha x) = u(\rho(\beta) \cdot x) = u(\beta x) = \beta \cdot u(x) = \rho(\beta) \cdot u(x) = \alpha \cdot u(x)$$

for any $x \in E$, and hence $u \in \mathrm{Hom}_A(E, F)$.

b) If $\alpha \in A$ and $x \in M$, then, in the notation of the proof of *a*), we have $\alpha x = \rho(\beta) \cdot x = \beta x \in M$.

c) Let $x \in \rho_*(E)$; there exists an fs-family $(\alpha_\lambda)_{\lambda \in I}$ of elements of A such that $x = \sum_{\lambda \in I} \alpha_\lambda a_\lambda$; if for any $\lambda \in I$, $\beta_\lambda \in B$ has the property that $\rho(\beta_\lambda) = \alpha_\lambda$ and, moreover, that for $\alpha_\lambda = 0$ we have $\beta_\lambda = 0$, then it follows that $(\beta_\lambda)_{\lambda \in I}$ is an fs-family and that

$$x = \sum_{\lambda \in I} \rho(\beta_\lambda) \cdot a_\lambda = \sum_{\lambda \in I} \beta_\lambda \cdot a_\lambda.$$

d) Let $(\beta_\lambda)_{\lambda\in I}$ be an fs-family of elements of B such that

$$0 = \sum_{\lambda\in I} \beta_\lambda \cdot a_\lambda = \sum_{\lambda\in I} \rho(\beta_\lambda)\cdot a_\lambda;$$

then, the linear independence over A of the family $(a_\lambda)_{\lambda\in I}$ implies that $\rho(\beta_\lambda) = 0$ for any $\lambda \in I$, hence, by the injectivity of ρ, that $\beta_\lambda = 0$ for any $\lambda \in I$.

e) Let $(\alpha_\lambda)_{\lambda\in I}$ be an fs-family of elements of A such that

$$0 = \sum_{\lambda\in I} \alpha_\lambda a_\lambda;$$

for any $\lambda \in I$ we choose a $\beta_\lambda \in B$ such that $\rho(\beta_\lambda) = \alpha_\lambda$ and, moreover, $\beta_\lambda = 0$ if $\alpha_\lambda = 0$; $(\beta_\lambda)_{\lambda\in I}$ is an fs-family, and we have

$$0 = \sum_{\lambda\in I} \rho(\beta_\lambda)\cdot a_\lambda = \sum_{\lambda\in I} \beta_\lambda \cdot a_\lambda;$$

by hypothesis it follows that $\beta_\lambda = 0$ for any $\lambda \in I$, hence that $\alpha_\lambda = \rho(\beta_\lambda) = 0$ for any λ. ∎

If ρ is not surjective, then the last assertion is no more valid in general even assuming ρ to be injective, as shown by the following

EXAMPLE. 4) Take $B = \mathbf{Q}$, $A = \mathbf{R}$, and ρ to be the canonical injection; let $E = \mathbf{R}$ considered as $\mathbf{R}$-vector space; then the subset $\{1, \sqrt{2}\}$ is linearly independent over $\mathbf{Q}$ (since a relation $\lambda\cdot 1 + \mu\cdot\sqrt{2} = 0$, with $\lambda, \mu \in \mathbf{Q}$ and $\lambda, \mu \neq 0$, would imply $\sqrt{2}$ to be rational), but is linearly dependent over $\mathbf{R}$ (since $\sqrt{2}\cdot 1 - 1\cdot\sqrt{2} = 0$). ∎

Returning to the general case, we may prove the following Proposition:

PROPOSITION 2. *If $\rho : B \to A$ is a homomorphism of rings, then the functor ρ_* of restriction of scalars commutes with direct products and with direct sums.*

Proof. We shall prove only the commutativity with direct products, leaving the other property to the reader as an exercise.

Let $(E_\lambda)_{\lambda\in I}$ then be a family of A-modules; of course, we may consider only the product $\prod_{\lambda\in I} E_\lambda$ and the canonical projections $p_\lambda : \prod_\lambda E_\lambda \to E_\lambda$, $\lambda \in I$, constructed at the beginning of § 6, any other direct product of the family $(E_\lambda)_{\lambda\in I}$ being isomorphic to this one; but the B-module structure of $\rho_*(\prod_{\lambda\in I} E_\lambda)$ is exactly the product structure of the family of B-modules $(\rho_*(E_\lambda))_{\lambda\in I}$: since, for instance, for any $\beta \in B$ and $(x_\lambda)_\lambda \in \rho_*(\prod_\lambda E_\lambda)$ we have

$$\beta\cdot(x_\lambda)_\lambda = \rho(\beta)\cdot(x_\lambda)_\lambda = (\rho(\beta)\cdot x_\lambda)_\lambda = (\beta\cdot x_\lambda)_\lambda;$$

as $\rho_*(p_\lambda)$, $\lambda \in I$, are the canonical projections, the result is proved. ∎

We mention also that *for a homomorphism of rings $\rho : B \to A$, if E is an A-module and $(a_\lambda)_{\lambda\in I}$ is a system of generators of $\rho_*(E)$ (over B), then $(a_\lambda)_{\lambda\in I}$ is* *a system of generators of* E *(over* A*)*: for if $x \in E$, then $x = \sum_{\lambda \in I} \beta_\lambda a_\lambda = \sum_{\lambda \in I} \rho(\beta_\lambda) a_\lambda$, where $(\beta_\lambda)_{\lambda \in I}$ and hence also $(\rho(\beta_\lambda))_{\lambda \in I}$ are fs-families; we have seen that the converse may be proved when ρ is surjective (Proposition 1, *c*)).

The following Proposition involves the B-module structure of $\rho_*(A_s)$ when a homomorphism of rings $\rho : B \to A$ is given.

PROPOSITION 3. *Let* $\rho : B \to A$ *be a homomorphism of rings, and let* E *be an* A*-module; if* $(\alpha_\lambda)_{\lambda \in I}$ *is a system of generators, respectively a linearly independent family, respectively a basis of* $\rho_*(A_s)$ *over* B*, and if* $(a_\mu)_{\mu \in J}$ *is a system of generators, respectively a linearly independent family, respectively a basis of* E *over* A*, then the family* $(\alpha_\lambda a_\mu)_{(\lambda, \mu) \in I \times J}$ *is a system of generators, respectively a linearly independent family, respectively a basis of* $\rho_*(E)$ *over* B*.*

Proof. In fact the Proposition consists of three assertions: the first one concerning systems of generators, the second one concerning linearly independent families, the third one concerning bases; these assertions will be proved successively.

For the first assertion we notice that for any $x \in E$ there exists an fs-family $(\xi_\mu)_{\mu \in J}$ of elements of A such that $x = \sum_{\mu \in J} \xi_\mu a_\mu$; but for any $\mu \in J$ there exists an fs-family $(\beta_{\lambda\mu})_{\lambda \in I}$ of elements of B such that

$$\xi_\mu = \sum_{\lambda \in I} \beta_{\lambda\mu} \alpha_\lambda = \sum_{\lambda \in I} \rho(\beta_{\lambda\mu}) \alpha_\lambda,$$

with the additional condition that $\beta_{\lambda\mu} = 0$ for any $\lambda \in I$ if $\xi_\mu = 0$; but $(\beta_{\lambda\mu})_{(\lambda, \mu) \in I \times J}$ is an fs-family of elements of B, its support being a finite union of finite sets, and we have

$$x = \sum_{\mu \in J} \Big(\sum_{\lambda \in I} \rho(\beta_{\lambda\mu}) \alpha_\lambda\Big) a_\mu = \sum_{(\lambda, \mu) \in I \times J} \rho(\beta_{\lambda\mu})(\alpha_\lambda a_\mu) = \sum_{(\lambda, \mu) \in I \times J} \beta_{\lambda\mu}(\alpha_\lambda a_\mu).$$

For the second assertion we notice that if $(\beta_{\lambda\mu})_{(\lambda, \mu) \in I \times J}$ is an fs-family of elements of B for which

$$\sum_{(\lambda, \mu) \in I \times J} \beta_{\lambda\mu}(\alpha_\lambda a_\mu) = 0, \tag{2}$$

then the family $\big(\sum_{\lambda \in I} \beta_{\lambda\mu} \alpha_\lambda\big)_{\mu \in J}$ is an fs-family of elements of A, and since relation (2) is written as

$$\sum_{\mu \in J} \Big(\sum_{\lambda \in I} \beta_{\lambda\mu} \alpha_\lambda\Big) a_\mu = 0,$$

it follows that $\sum_{\lambda \in I} \beta_{\lambda\mu} \alpha_\lambda = 0$ for any $\mu \in J$; but since for any $\mu \in J$, $(\beta_{\lambda\mu})_{\lambda \in I}$ is an fs-family of elements of B, the last equality implies that $\beta_{\lambda\mu} = 0$ for any $\lambda \in I$ and any $\mu \in J$.

The third assertion is a consequence of the first two ones. ▮

COROLLARY. *Under the conditions of Proposition* 3, *if* E *is a finitely generated* A*-module and* $\rho_*(A_s)$ *is a finitely generated* B*-module, then* $\rho_*(E)$ *is a finitely generated* B*-module.*

Proof. We have only to apply the first assertion of Proposition 3 under the assumption that I and J are finite sets; since $I \times J$ is in this case a finite set, the assertion follows. ∎

Assume now that three rings A, B, C and two homomorphisms of rings

$$\sigma : C \to B, \qquad \rho : B \to A$$

are given; by composition the homomorphism of rings $\rho \circ \sigma : C \to A$ is obtained; let E be an A-module; we have

$$\sigma_*(\rho_*(E)) = (\rho \circ \sigma)_*(E),.$$

i.e., the C-module structure obtained from the initial A-module structure of E by successive restrictions to B and to C coincides with the C-module structure obtained from the same A-module structure by direct restriction to C by means of $\rho \circ \sigma$: indeed, if $\gamma \in C$ and $x \in E$, then according to the first structure we have $\gamma x = \sigma(\gamma) \cdot x = (\rho(\sigma(\gamma))) \cdot x$, while according to the second one we have $\gamma x = ((\rho \circ \sigma)(\gamma)) \cdot x$, and the two elements obtained coincide.

This means that the functor $(\rho \circ \sigma)_* : \mathbf{Mod}(A) \to \mathbf{Mod}(C)$ is the composition of the functors $\sigma_* : \mathbf{Mod}(B) \to \mathbf{Mod}(C)$ and $\rho_* : \mathbf{Mod}(A) \to \mathbf{Mod}(B)$.

Obviously, if 1_A is the identity homomorphism of the ring A, then $(1_A)_*$ is the identity functor of the category $\mathbf{Mod}(A)$.

Now let $\rho : B \to A$ again be a homomorphism of rings; let E be an A-module, let F be a B-module, and let $u : E \to F$ be a homomorphism of abelian groups which, moreover, is a B-homomorphism $\rho_*(E) \to F$; in this case one says that ρ and u are *compatible morphisms* or that *u is compatible with ρ*; this amounts to saying, besides the fact that u is a $\mathbf{Z}$-homomorphism, that for any $x \in E$ and $\beta \in B$ we have

$$u(\rho(\beta) \cdot x) = \beta \cdot u(x); \tag{3}$$

if no confusion may arise, one also says in this situation that "u is a B-homomorphism from E to F".

In the particular case where ρ is an isomorphism of rings, and if $\rho' : A \to B$ is the isomorphism inverse to ρ, then to say that ρ and u are compatible morphisms is equivalent, as relation (3) shows, to saying that for any $\alpha \in A$ and $x \in E$ we have

$$u(\alpha \cdot x) = \rho'(\alpha) \cdot u(x),$$

which means that u is an A-homomorphism $E \to \rho'_*(F)$; in the general case, given a homomorphism of rings $\rho' : A \to B$ (not necessarily an isomorphism), an A-module E, a B-module F, and an A-homomorphism $u : E \to \rho'_*(F)$, one says that u is a *semilinear mapping* (relative to ρ') from the A-module E to the B-module F or that the pair (u, ρ') is a *dimorphism* from E to F.

EXAMPLES. 5) Let $\rho : B \to A$ be a homomorphism of rings, let E be an A-module; consider the B-module $F = \rho_*(E)$; the identity mapping 1_E is compatible with ρ, since if $x \in E$ and $\beta \in B$, then

$$1_E(\rho(\beta)\cdot x) = \rho(\beta)\cdot x = \beta\cdot x = \beta\cdot 1_E(x).$$

6) Let A be a ring, and let E be an A-module; if $\mathfrak{a}$ is the annihilator of E and $B = A/\mathfrak{a}$, then on E a B-module structure is defined, namely that of the faithful module associated with E (see § 10); if $\rho : A \to A/\mathfrak{a} = B$ is the canonical surjection, then 1_E is compatible with ρ, which is easily seen taking into account the way in which B acts in the faithful module associated with E.

7) Let A be a ring, and let α be an invertible element of A; denote by t_α the mapping $A \to A$ given by

$$t_\alpha(\xi) = \alpha^{-1}\xi\alpha,$$

$\xi \in A$; it is easily seen that t_α is a homomorphism of rings $A \to A$; let E be an A-module, and let h_α be the homothety of ratio α in E (see Example 9), § 1); h_α is a homomorphism of abelian groups; moreover, t_α and h_α are compatible, since if $x \in E$ and $\xi \in A$, then

$$h_\alpha(t_\alpha(\xi)\cdot x) = h_\alpha((\alpha^{-1}\xi\alpha)\cdot x) = \alpha\cdot(\alpha^{-1}\xi\alpha)\cdot x = \xi(\alpha x) = \xi\cdot h_\alpha(x). \blacksquare$$

Returning to the general case, let the homomorphisms of rings $\rho : B \to A$ and $\sigma : C \to B$ be given; let E be an A-module, let F be a B-module, let G be a C-module, and let $u : E \to F$, $v : F \to G$ be homomorphisms between the corresponding underlying abelian groups such that u is compatible with ρ and v is compatible with σ; then $v\circ u : E \to G$ is compatible with $\rho\circ\sigma : C \to A$: since if $x \in E$ and $\gamma \in C$, then

$$(v\circ u)((\rho\circ\sigma)(\gamma)\cdot x) = v(u[(\rho(\sigma(\gamma)))\cdot x]) = v[\sigma(\gamma)\cdot u(x)] = \gamma\cdot(v\circ u)x.$$

From the above Example 5) it follows that for any ring A and any A-module E, the mapping 1_E is compatible with 1_A.

These results show that we may define a new category as follows: as objects of this category we take the pairs (A, E), where A is a ring (with unit element) and E is an A-module; if (A, E) and (B, F) are two such pairs, a morphism from (A, E) to (B, F) is a pair (ρ, u), where ρ is a homomorphism of rings (with unit element) from B to A, u is a homomorphism of abelian groups from E to F, and u is compatible with ρ; if (C, G) is a third object and (σ, v) is a morphism $(B, F) \to (C, G)$, then, by definition, the composition $(\sigma, v)\circ(\rho, u)$ is the pair $(\rho\circ\sigma, v\circ u)$; moreover, the identity of the object (A, E) is the morphism $(1_A, 1_E)$. This category will be denoted in the sequel by **Cop.**[1]

PROPOSITION 4. *Let I be the category associated with the preordered set I directed to the right; let $(A_\lambda, \rho_{\lambda\mu})_I$ be a projective system with values in the category* **Ann** *of rings indexed by I, and let $(E_\lambda, u_{\lambda\mu})_I$ be an inductive system of* **Z**-*modules*

---

[1] **Cop** is from "couple".

indexed by the same category I; assume that for any pair λ, μ *of objects of I for which* $\lambda \leqslant \mu$, *the homomorphism* $u_{\lambda\mu}$ *is compatible with the homomorphism* $\rho_{\lambda\mu}$; let $A = \varprojlim_{I} A_\lambda$ *in the category* **Ann**, *and let* $E = \varinjlim_{I} E$ *in the category* **Ab**; *then on E a unique A-module structure* (*the addition being the one in the abelian group E*) *may be defined with the property that, denoting by* $\rho_\lambda : A \to A_\lambda$, *respectively* $j_\lambda : E_\lambda \to E$, *the canonical morphisms,* ρ_λ *and* j_λ *are compatible morphisms for any* $\lambda \in I$.

Proof. Let $\alpha \in A$ and $x \in E$; since I is a pseudofiltered connected category, it follows, by Proposition 3, *a*), § 13, that there exist $\lambda \in I$ and $x_\lambda \in E_\lambda$ such that $j_\lambda(x_\lambda) = x$; we claim that the element $j_\lambda(\rho_\lambda(\alpha)\cdot x_\lambda)$ does not depend on the chosen λ and x_λ: indeed, let $\mu \in I$ and $y_\mu \in E_\mu$ be such that $x = j_\mu(y_\mu)$; by Proposition 3, *c*), § 13, it follows that there exists $\nu \in I$ with the property $\lambda \leqslant \nu$, $\mu \leqslant \nu$ and such that

(4) $$u_{\lambda\nu}(x_\lambda) = u_{\mu\nu}(y_\mu);$$

but taking into account the compatibility of $u_{\lambda\nu}$ with $\rho_{\lambda\nu}$, we have

$$j_\lambda[\rho_\lambda(\alpha)\cdot x_\lambda] = j_\nu(u_{\lambda\nu}[\rho_{\lambda\nu}(\rho_\nu(\alpha))\cdot x_\lambda]) = j_\nu[\rho_\nu(\alpha)\cdot u_{\lambda\nu}(x_\lambda)]$$

and, similarly, $j_\mu[\rho_\mu(\alpha)\cdot y_\mu] = j_\nu[\rho_\nu(\alpha)\cdot u_{\mu\nu}(y_\mu)]$; taking (4) into account, our assertion follows; next, define

(5) $$\alpha x = j_\lambda(\rho_\lambda(\alpha)\cdot x_\lambda).$$

It is easy to verify that in this way an A-module structure was defined on E: for instance, if $\alpha \in A$ and $x, y \in E$, we may assume, since I is directed, that for the same $\lambda \in I$ we have $x = j_\lambda(x_\lambda)$, $y = j_\lambda(y_\lambda)$, with $x_\lambda, y_\lambda \in E_\lambda$; hence $x + y = j_\lambda(x_\lambda + y_\lambda)$, and

$$\alpha(x + y) = j_\lambda[\rho_\lambda(\alpha)\cdot(x_\lambda + y_\lambda)] = j_\lambda(\rho_\lambda(\alpha)\cdot x_\lambda) + j_\lambda(\rho_\lambda(\alpha)\cdot y_\lambda) = \alpha x + \alpha y.$$

The fact that for any $\lambda \in I$, ρ_λ and j_λ are compatible morphisms follows directly from (5) taking into account that $x = j_\lambda(x_\lambda)$. If another A-module structure on E existed with the addition as in the abelian group E and with the external law of composition denoted by $(\alpha, x) \mapsto \alpha \perp x$ such that for any $\lambda \in I$, ρ_λ and j_λ be compatible morphisms also according to this law, then, if $x = j_\lambda(x_\lambda)$, $x_\lambda \in E_\lambda$, we would have

$$\alpha \perp x = \alpha \perp j_\lambda(x_\lambda) = j_\lambda(\rho_\lambda(\alpha)\cdot x_\lambda) = \alpha x;$$

hence the external law of composition defined by (5) is unique under these conditions. ▮

COROLLARY. *If I is the category associated with the preordered set I directed to the right, and* $T = ((A_\lambda, E_\lambda), (\rho_{\lambda\mu}, u_{\lambda\mu}))_I$ *is an inductive system with values in the category* **Cop** *indexed by the category I, then this inductive system has an inductive limit in the category* **Cop**.

Proof. By hypothesis it follows that $(A_\lambda, \rho_{\lambda\mu})_I$, respectively $(E_\lambda, u_{\lambda\mu})_I$ is a projective, respectively an inductive system with values in the category **Ann**, respectively in the category **Ab** indexed by the category I; since for any pair λ, μ for which $\lambda \leqslant \mu$ in I, the pair $(\rho_{\lambda\mu}, u_{\lambda\mu})$ is a morphism of the category **Cop**, we are under the hypotheses of Proposition 4; with the same meaning as there for A, E, ρ_λ, and j_λ, $\lambda \in I$, it follows that there exists a unique A-module structure on E with the property that ρ_λ and j_λ are compatible morphisms for any $\lambda \in I$. Then the system $((A, E), (\rho_\lambda, j_\lambda)_{\lambda \in I})$ is an inductive limit in the category **Cop** of the inductive system T: since, first of all, for any pair of objects λ, μ for which $\lambda \leqslant \mu$ we obviously have

$$(\rho_\mu, j_\mu)\circ(\rho_{\lambda\mu}, u_{\lambda\mu}) = (\rho_\lambda, j_\lambda);$$

next, if (B, X) is an object of **Cop** with the property that for any $\lambda \in I$ there exists a morphism

$$(\sigma_\lambda, k_\lambda) : (A_\lambda, E_\lambda) \to (B, X)$$

such that if $\lambda \leqslant \mu$ in I we have

$$(\sigma_\mu, k_\mu)\circ(\rho_{\lambda\mu}, u_{\lambda\mu}) = (\sigma_\lambda, k_\lambda),$$

then it follows $\rho_{\lambda\mu}\circ\sigma_\mu = \sigma_\lambda$ and $k_\mu \circ u_{\lambda\mu} = k_\lambda$ for $\lambda \leqslant \mu$; consequently, there exists a homomorphism of rings $\sigma : B \to A$ and a homomorphism of abelian groups $k : E \to X$ uniquely determined, respectively, by the relations

$$\rho_\lambda\circ\sigma = \sigma_\lambda, \quad k\circ j_\lambda = k_\lambda \tag{6}$$

for any $\lambda \in I$; σ and k are compatible morphisms: for if $x \in E$ and $\beta \in B$, then $x = j_\lambda(x_\lambda)$ for some $\lambda \in I$ and $x_\lambda \in E_\lambda$; consequently, we have

$$k(\sigma(\beta)\cdot x) = k(j_\lambda[\rho_\lambda(\sigma(\beta))\cdot x_\lambda]) = k_\lambda(\sigma_\lambda(\beta)\cdot x_\lambda) = \beta\cdot k_\lambda(x_\lambda) = \beta\cdot k(x).$$

Hence $(\sigma, k) \in \mathrm{Hom}_{\mathbf{Cop}}((A, E), (B, X))$, and we have in the category **Cop**

$$(\sigma, k)\circ(\rho_\lambda, j_\lambda) = (\sigma_\lambda, k_\lambda)$$

for any $\lambda \in I$; if there existed also $(\sigma', k') : (A, E) \to (B, X)$ with the same properties as (σ, k), then σ', respectively k', would satisfy (6) for any $\lambda \in I$ when replacing σ, respectively k, and hence we would have $(\sigma', k') = (\sigma, k)$. ∎

§ 16. Multimodules

DEFINITION 1. Let A and B be two rings, and let E be a set endowed with a structure of A-module and with one of B-module; the two module structures defined on E are said to be *compatible* provided the following two conditions are satisfied:

a) the addition in the A-module structure coincides with the addition in the B-module structure;

b) the scalar multiplication by elements of A commutes with the scalar multiplication by elements of B, i.e., for any $\alpha \in A$, $\beta \in B$, and $x \in E$ we have

$$\alpha(\beta x) = \beta(\alpha x).$$

In this case one also says that E is a *bimodule over A and B* or an *(A, B)-bimodule* or, simply, a *bimodule*. ▮

If instead of being a left A-module and a left B-module, E is in one of the other three possible situations, the terminology is maintained, and its meaning may be deduced *mutatis mutandis* from the above case; for instance, if E is a left A-module and a right B-module, the commutativity condition on the two scalar multiplications becomes

$$\alpha(x\beta) = (\alpha x)\beta,$$

where $\alpha \in A$, $\beta \in B$, $x \in E$. However, all cases may be reduced to the first one by replacing A or B or both of them by their opposite rings, respectively (see the Remark following Definition 1, § 1).

It is clear that the commutativity condition of the two scalar multiplications in a bimodule E is equivalent to the condition that for any $\alpha \in A$ and any $\beta \in B$ we have in the ring of endomorphisms of the additive group of E, $h_\alpha \circ h'_\beta = h'_\beta \circ h_\alpha$, where h_α, respectively h'_β, represent the homothety of ratio α, respectively β, given by the A-module, respectively B-module structure of E; furthermore, the same condition is equivalent to the fact that for any $\beta \in B$, h'_β with the above meaning is an endomorphism of the A-module E, that is, an element of $\mathrm{Hom}_A(E, E)$, or also to the fact that for any $\alpha \in A$, h_α is in $\mathrm{Hom}_B(E, E)$.

EXAMPLES. 1) If A is a ring, then on A two A-module structures are defined, one on the left, giving the A-module A_s, the other on the right, giving the A-module A_d (see Example 1), § 1); the two structures are compatible, since both in A_s and in A_d the addition is the one in the ring A and, moreover, if $\alpha, \beta, \gamma \in A$, then by the associativity of the multiplication in A we have

$$\alpha(\gamma\beta) = (\alpha\gamma)\beta.$$

2) Let A be a ring, and let E be an A-module; then the underlying abelian group of E is endowed with a **Z**-module structure (see Example 4), § 1); the two structures defined on E, the one of A-module and the one of **Z**-module, are compatible: the addition is the same, and for any $\alpha \in A$, $n \in \mathbf{Z}$, and $x \in E$ we have $\alpha(nx) = n(\alpha x)$ (see § 1).

3) Let A be a commutative ring; if E is an A-module, then, considering the same structure twice, two A-module structures are defined on E; since the addition is the same, and since for any $\alpha, \alpha' \in A$ and $x \in E$ we have

$$\alpha(\alpha' x) = (\alpha\alpha')x = (\alpha'\alpha)x = \alpha'(\alpha x),$$

it follows that the two structures are compatible, hence that E is an (A, A)-bimodule.

4) If A is a ring and E is an A-module, then the addition in E together with the scalar multiplication by elements of the ring $B = \mathrm{End}_A(E)$ given by

$$(f, x) \mapsto f(x),$$

where $f \in \mathrm{End}_A(E)$, $x \in E$, provide us with a B-module structure on E (see also Example 1), § 15). The two structures, the one of A-module and the other of B-module, are compatible, since if f and x have the above meaning and $\alpha \in A$, then, as f is an A-homomorphism, we have

$$f(\alpha x) = \alpha f(x).$$

5) Let A be a ring, and let E, F be A-modules; as shown in § 2, on $\mathrm{Hom}_A(E, F)$ a left $\mathrm{End}_A(F)$-module structure and a right $\mathrm{End}_A(E)$-module structure are defined; these are compatible, since the addition is the same (namely the addition in $\mathrm{Hom}_A(E, F)$) and the two scalar multiplication laws are given by the composition of linear mappings, and this is associative. ▮

The notion of bimodule is generalized as follows:

Definition 2. Let $(A_\lambda)_{\lambda \in I}$, $(B_\mu)_{\mu \in J}$ be two families of rings; a *multimodule over the families of rings* $(A_\lambda)_{\lambda \in I}$, $(B_\mu)_{\mu \in J}$ or an $((A_\lambda)_{\lambda \in I}, (B_\mu)_{\mu \in J})$*-multimodule* is a set E endowed for each $\lambda \in I$ with a left A_λ-module structure and for each $\mu \in J$ with a right B_μ-module structure such that, if $\lambda, \lambda' \in I$, $\lambda \neq \lambda'$, the A_λ-module structure is compatible with the $A_{\lambda'}$-module structure, if $\mu, \mu' \in J$, $\mu \neq \mu'$, the right B_μ-module structure is compatible with the right $B_{\mu'}$-module structure, and if $\lambda \in I$, $\mu \in J$, the A_λ-module structure is compatible with the right B_μ-module structure. ▮

Since the typical situations concerning the notions and properties related to multimodules may be described on bimodules, we shall confine ourselves to the latter ones; we leave to the reader as an exercise to generalize for multimodules the subsequent considerations for bimodules.

Definition 3. Let A, B be two rings, and let E, F be two (A, B)-bimodules (we assume that A acts in E and F on the same (otherwise arbitrary) side; likewise for B); a *homomorphism of* (A, B)-*bimodules* (or an (A, B)-*homomorphism*) *from* E *to* F is a mapping $u : E \to F$ which is both an A-homomorphism and a B-homomorphism. ∎

Let A and B be two fixed rings; suppose we take as class of objects the class of (A, B)-bimodules (where A acts on a determined side, and similarly for B, say both on the left); for any pair of such bimodules E and F we take as morphisms between E and F the homomorphisms of (A, B)-bimodules from E to F, the composition being as for modules (or sets); in this way we get a category, called the *category of* (A, B)-*bimodules* and denoted[1] **Bimod**(A, B); defining for any pair of bimodules E and F the sum of two homomorphisms of (A, B)-bimodules from E to F in the same way as in the case of modules, we get on $\mathrm{Hom}_{\mathbf{Bimod}(A, B)}(E, F)$ an abelian group structure; moreover, the category **Bimod** (A, B) is, with this definition of the addition of morphisms, an abelian category for which the zero object, the kernel, the cokernel, etc. are defined as for modules; in particular, in the same way as for modules we get the notions of *subbimodule* and of *quotient bimodule*; likewise, defining the direct sum and the direct product of bimodules in the same way as for modules, we obtain an abelian category with direct products and direct sums; according to Proposition 5, § 12, and to Proposition 2, § 13, **Bimod**(A, B) is a category with projective and inductive limits; for this category Axiom AB5) is al so valid.

If in an (A, B)-bimodule the structure of B-module is left aside, we get an A-module; this yields a functor

$$\mathbf{Bimod}(A, B) \to \mathbf{Mod}(A),$$

namely the *forgetful functor*; we have also a forgetful functor

$$\mathbf{Bimod}(A, B) \to \mathbf{Mod}(B);$$

naturally, the A-module, respectively the B-module obtained in this way from an (A, B)-bimodule E will be called the *underlying* A-*module*, respectively B-*module of* E.

If $\rho : A' \to A$ and $\sigma : B' \to B$ are homomorphisms of rings (with unit element), then, applying the forgetful functor **Bimod**(A, B)→**Mod**(A), to any (A, B)-bimodule E is assigned an A-module with which, by means of the functor ρ_* of restriction of scalars, an A'-module is associated; likewise, applying the forgetful functor **Bimod**$(A, B) \to$ **Mod**(B) and then the functor of restriction of scalars σ_*, we get a B'-module; both the A'-module and the B'-module obtained are defined on E; the two module structures on E are compatible: first, the addition is the same, namely the one in the (A, B)-module E; next, if $\alpha' \in A'$, $\beta' \in B'$, $x \in E$, we have

$$\alpha'(\beta' x) = \rho(\alpha') \cdot (\sigma(\beta') x) = \sigma(\beta') \cdot (\rho(\alpha') x) = \beta'(\alpha' x);$$

[1] The notation is somewhat ambiguous, since it is not specified on which side A and B act.

assigning to the (A, B)-bimodule E the (A', B')-bimodule thus obtained, and to each homomorphism of (A, B)-bimodules the homomorphism itself considered as an (A', B')-homomorphism, we get a functor

$$(\rho, \sigma)_*: \mathbf{Bimod}(A, B) \to \mathbf{Bimod}(A', B'),$$

which will be also called the *functor of restriction of scalars* (see § 15).

We proceed now to study the way in which an additional structure on a module making from it a bimodule is carried over when applying to this module a functor defined on the respective category of modules with values in another category of modules:

First, let A be a ring; a category, denoted $\mathscr{C}(A)$, may be assigned to A as follows: as class of objects of $\mathscr{C}(A)$ we take a set consisting of one element, also denoted by A; as elements of $\mathrm{Hom}_{\mathscr{C}(A)}(A, A)$ we take the elements of A; as composition of two such elements we take their product in A; all the axioms of categories turn out to be satisfied, the identity morphism of the unique object in $\mathscr{C}(A)$ being the unit element of A; moreover, in $\mathrm{Hom}_{\mathscr{C}(A)}(A, A)$ an addition is introduced, namely the addition in A; with these definitions, $\mathscr{C}(A)$ satisfies conditions *a*) and *b*) of Definition 6, § 2. It is easily seen that denoting by A° the opposite ring of A, then

$$\mathscr{C}(A^\circ) = (\mathscr{C}(A))^\circ.$$

Next, let A and B be two rings, and let E be an (A, B)-bimodule (in which both A and B act, say, on the left); to E is assigned a functor

$$U_E: \mathscr{C}(B) \to \mathbf{Mod}(A)$$

defined as follows: as $U_E(B)$ (B being the unique object of $\mathscr{C}(B)$) the underlying A-module of E is taken; if $\beta \in B$, then $U_E(\beta) = h_\beta$, where h_β is the homothety of ratio β in E, and which is an A-homomorphism $U_E(B) \to U_E(B)$ since the two structures defined on E, the one of A-module and the other of B-module, are compatible; as for any $\beta, \beta' \in B$ we have

$$h_{\beta\beta'} = h_\beta \circ h_{\beta'}, \quad h_1 = 1_E, \quad h_{\beta+\beta'} = h_\beta + h_{\beta'},$$

where the first 1 in the second equality represents the unit element of B (see § 10), it follows that U_E is a functor, namely an additive one (see the Remark following Definition 7, § 2).

Conversely, let

$$U: \mathscr{C}(B) \to \mathbf{Mod}(A)$$

be an additive functor, and let $U(B) = E$; on E we define a (left) B-module structure, taking as addition the one in E and as scalar multiplication by elements of B the external law of composition given by

$$\beta \cdot x = U(\beta)(x),$$

$\beta \in B$, $x \in E$; one easily verifies that the axioms for modules are satisfied: for instance, if $\beta, \beta' \in B$, $x \in E$, then

$$(\beta\beta')\cdot x = U(\beta\beta')(x) = (U(\beta)\circ U(\beta'))(x) = \beta\cdot(\beta'\cdot x);$$

moreover, if $\alpha \in A$, $\beta \in B$, $x \in E$, then, as $U(\beta)$ is an A-homomorphism, we have

$$\alpha(\beta x) = \alpha\cdot(U(\beta)(x)) = U(\beta)(\alpha x) = \beta\cdot(\alpha x),$$

hence E is with this additional structure an (A, B)-bimodule. Of course, the functor U_E assigned as above to the bimodule E coincides with the initial functor U, since $U_E(B)$ and $U(B)$ coincide with the A-module E and for $\beta \in B$ we have, using the preceding notations,

$$U_E(\beta)(x) = h_\beta(x) = \beta x = U(\beta)(x)$$

for any $x \in E$.

Similarly, given the (A, B)-bimodule E, the (A, B)-bimodule assigned to U_E is exactly the (A, B)-bimodule E, as follows from a straightforward verification.

Hence, the two mappings, the first assigning to an (A, B)-bimodule E the additive functor U_E, the other assigning to an additive functor $U : \mathcal{C}(B) \to \mathbf{Mod}(A)$ an (A, B)-bimodule in the manner described above, are inverse to each other.

The preceding considerations show that if we considered (A, B)-bimodules with A acting on the right and B on the left, then in the same way we would assign to them in a one-to-one manner additive functors defined on $\mathcal{C}(B)$ and taking values in $\mathbf{Mod}_d(A)$.

If, however, we consider bimodules with the ring B acting on the right, the things are somewhat different: let then E be an (A, B)-bimodule in which B acts on the right and A, say, on the left; this time to E is assigned an additive functor

$$U_E : (\mathcal{C}(B))^\circ \to \mathbf{Mod}(A)$$

formally defined as above; only now we have to take $(\mathcal{C}(B))^\circ$ as domain of U_E, since denoting by h_β, $\beta \in B$, the homothety of ratio β in E, we have for any $\beta, \beta' \in B$, $h_{\beta\beta'} = h_{\beta'}\circ h_\beta$, hence

$$U_E(\beta'^\circ\beta^\circ) = U_E((\beta\beta')^\circ) = h_{\beta\beta'} = U_E(\beta'^\circ)\circ U_E(\beta^\circ).$$

For an additive functor

$$U : (\mathcal{C}(B))^\circ \to \mathbf{Mod}(A)$$

an (A, B)-bimodule with B acting on the right is set off by defining on the A-module $U(B^\circ) = E$ the right B-module structure given by

$$x\beta = U(\beta^\circ)(x)$$

for any $\beta \in B$ and $x \in E$.

As in the case of bimodules with B acting on the left, the two mappings are inverse to each other; as there, the considerations are mainly the same if A acts on the right in the bimodules and, consequently, the category $\mathbf{Mod}(A)$ is replaced by $\mathbf{Mod}_d(A)$.

As a particular case of the preceding considerations we can take $A = \mathbf{Z}$, and we obtain that, for a given ring B, the (left) B-modules are in a one-to-one correspondence with the additive functors $\mathcal{C}(B) \to \mathbf{Ab}$, while the right B-modules, with the additive functors $(\mathcal{C}(B))^\circ \to \mathbf{Ab}$.

Now, as at the beginning, let A, B be two rings, and let E, F be two (A, B)-bimodules in which A as well as B act on the left; let $u : E \to F$ be a homomorphism of (A, B)-bimodules; then to u is assigned a natural transformation

$$\Phi_u : U_E \to U_F$$

defined by taking as $\Phi_u(B)$ the homomorphism u; indeed, $u \in \mathrm{Hom}_A(E, F)$, and the fact that u is a B-homomorphism implies the commutativity of the following diagram for each $\beta \in B$:

$$\begin{array}{ccc} U_E(B) & \xrightarrow{U_E(\beta)} & U_E(B) \\ {\scriptstyle \Phi_u(B)}\downarrow & & \downarrow{\scriptstyle \Phi_u(B)} \\ U_F(B) & \xrightarrow{U_F(\beta)} & U_F(B), \end{array}$$

ensuring the character of natural transformation of Φ_u; conversely, if Φ is a natural transformation $U \to V$, where U, V are additive functors $\mathcal{C}(B) \to \to \mathbf{Mod}(A)$, then, by the preceding considerations, there exist two (A, B)-bimodules E and F uniquely determined such that $U = U_E$ and $V = U_F$; then Φ defines a homomorphism of (A, B)-bimodules $u : E \to F$, setting $u = \Phi(B)$. It is easily seen that the two mappings, the one assigning to an (A, B)-homomorphism $u : E \to F$ a natural transformation $\Phi_u : U_E \to U_F$, and the one assigning to a natural transformation Φ: $U \to V$ an (A, B)-homomorphism as above, are inverse to each other. The preceding considerations are mainly the same if one or both of A and B act in the two modules on the right; but, in case B acts on the right, we have to do with natural transformations between two functors defined on $(\mathcal{C}(B))^\circ$ and taking values in $\mathbf{Mod}(A)$.

Let A, B, C be three rings, and let T be an additive functor

$$\mathbf{Mod}(A) \to \mathbf{Mod}(C);$$

if E is an (A, B)-bimodule (in which both A and B act, say, on the left), then, taking the composition of T with U_E defined above, we get an additive functor

$$T \circ U_E : \mathcal{C}(B) \to \mathbf{Mod}(C),$$

and we have [1] $(T \circ U_E)(B) = T(E)$; by the preceding considerations, on $T(E)$ a structure of (C, B)-bimodule is obtained in which B as well as C act on the left. Hence T

---

[1] Here and subsequently, an (A, B)-bimodule and the underlying A-module will be denoted by the same letter.

carries an (A, B)-bimodule into a (C, B)-bimodule. We notice that if $y \in T(E)$ and $\beta \in B$, then, with the considered structure, we have

$$\beta y = [(T \circ U_E)(\beta)]\,(y) = T(h_\beta)(y),$$

h_β being the homothety of ratio β in E.

Maintaining the categories **Mod**(A) and **Mod**(C), but letting E be an (A, B)-bimodule in which A acts on the left and B *on the right*, then, by the composition of the additive functor $T : \mathbf{Mod}(A) \to \mathbf{Mod}(C)$ with U_E defined for this case we get an additive functor

$$T \circ U_E : (\mathscr{C}(B))^\circ \to \mathbf{Mod}(C);$$

hence we get a (B, C)-bimodule structure on $T(E)$ in which C acts on the left while B acts on the right; here, if $y \in T(E)$ and $\beta \in B$, then, with this structure, we have

$$y\beta = ((T \circ U_E)(\beta^\circ))(y) = (T(h_\beta))(y),$$

where h_β is the homothety of ratio β in E.

Assume now that T is an additive functor

$$(\mathbf{Mod}(A))^\circ \to \mathbf{Mod}(C)$$

and that E is an (A, B)-bimodule in which A and B act on the left; then, considering the dual U_E° of the functor $U_E : \mathscr{C}(B) \to \mathbf{Mod}(A)$, and taking the composition of T and U_E°, we get an additive functor

$$T \circ U_E^\circ : (\mathscr{C}(B))^\circ \to \mathbf{Mod}(C)$$

inducing on $T(E^\circ)$ a right B-module structure which makes a (C, B)-bimodule of it (with C acting on the left).

If E is an (A, B)-bimodule in which A acts on the left and B on the right, then to it is assigned the functor $U_E : (\mathscr{C}(B))^\circ \to \mathbf{Mod}(A)$; if T is an additive functor

$$(\mathbf{Mod}(A))^\circ \to \mathbf{Mod}(C),$$

then, taking the composition of T with the dual U_E° of U_E, we get an additive functor

$$T \circ U_E^\circ : \mathscr{C}(B) \to \mathbf{Mod}(C)$$

inducing on $T(E^\circ)$ a left B-module structure which makes a (C, B)-bimodule of it, with C acting on the left.

The preceding considerations are essentially the same if in each of the four cases treated above, one or both of the categories **Mod**(A) and **Mod**(C) are replaced by $\mathbf{Mod}_d(A)$, respectively by $\mathbf{Mod}_d(C)$; for instance, if T is an additive functor $\mathbf{Mod}(A) \to \mathbf{Mod}_d(C)$, then T carries an (A, B)-bimodule in which A and B act on the left into a (C, B)-bimodule in which B acts on the left and C on the right; but if T is an additive functor $(\mathbf{Mod}_d(A))^\circ \to \mathbf{Mod}_d(C)$, and if E is an (A, B)-bimodule in which both A and B act on the right, then on $T(E^\circ)$ we have a (C, B)-bimodule structure in which C acts on the right and B on the left, and so on.

In conclusion, *an additive functor* $T: \mathbf{Mod}(A) \to \mathbf{Mod}(C)$ (or $\mathbf{Mod}_d(A) \to \mathbf{Mod}(C)$, and so on) *carries an* (A, B)*-bimodule* E *into a* (C, B)*-bimodule* $T(E)$ *in which* B *acts on the same side as in* E, while C according to the category in which T takes its values; on the contrary, *an additive functor* $T: (\mathbf{Mod}(A))^\circ \to \mathbf{Mod}(C)$ (or $(\mathbf{Mod}_d(A))^\circ \to \mathbf{Mod}(C)$, and so on) *carries the doublet of an* (A, B)*-bimodule* E *into a* (C, B)*-bimodule* $T(E^\circ)$ *in which* B *acts on the opposite side to the one on which it acts in* E.

Assume now that A, B, C are three rings and that an additive functor $T: \mathbf{Mod}(A) \to \mathbf{Mod}(C)$ is given; let E and F be two (A, B)-bimodules in which both A and B act, say, on the left, and let $u: E \to F$ be a homomorphism of (A, B)-bimodules; then, in the above notation, to u is assigned the natural transformation

$$\Phi_u : U_E \to U_F;$$

as $T(\Phi_u)$ is a natural transformation $T \circ U_E \to T \circ U_F$ (see § 10), where $T \circ U_E$ and $T \circ U_F$ are additive functors defined on $\mathcal{C}(B)$ and taking values in $\mathbf{Mod}(C)$, and as to $T \circ U_E$, respectively $T \circ U_F$, the (C, B)-bimodule $T(E)$, respectively $T(F)$, is assigned, to $T(\Phi_u)$ a homomorphism of (C, B)-bimodules $T(E) \to T(F)$ is assigned, which is exactly $T(\Phi_u)(B) = T(\Phi_u(B)) = T(u)$; hence T *carries a homomorphism of* (A, B)*-bimodules into a homomorphism of* (C, B)*-bimodules.* The situation is essentially the same if one of the categories $\mathbf{Mod}(A)$ and $\mathbf{Mod}(C)$ or both are, respectively, replaced by $\mathbf{Mod}_d(A)$ and $\mathbf{Mod}_d(C)$, or if B acts on the right; only now, u, respectively $T(u)$, will be a homomorphism of bimodules with the appropriate meaning.

Modifying the above data only where the additive functor T is concerned, for instance, taking $T: (\mathbf{Mod}(A))^\circ \to \mathbf{Mod}(C)$ (all the other situations being deduced from this one with the corresponding modifications), then $T(\Phi^\circ)$ makes sense, and it is a natural transformation from $T \circ U_F^\circ$ to $T \circ U_E^\circ$; but this means that $(T(\Phi_u^\circ))(B^\circ) = T((\Phi_u(B))^\circ) = T(u^\circ)$ is a (C, B)-homomorphism from the (C, B)-bimodule $T(F^\circ)$ to the (C, B)-bimodule $T(E^\circ)$.

In the particular case where A is a commutative ring, any A-module may be considered an (A, A)-bimodule (see Example 3)), hence, taking in the above cos iderations $B = A$, we see that an additive functor $T: \mathbf{Mod}(A) \to \mathbf{Mod}(C)$ carries an A-module E into a (C, A)-bimodule $T(E)$ and an A-homomorphism $u: E \to F$ into a homomorphism of (C, A)-bimodules; the same does an additive functor $T: (\mathbf{Mod}(A))^\circ \to \mathbf{Mod}(C)$ with the doublet of an A-module, respectively of an A-homomorphism.

Let A, B, C be three rings, let T, T' be two additive functors $\mathbf{Mod}(A) \to \mathbf{Mod}(C)$, and let $\Phi: T \to T'$ be a natural transformation; if E is an (A, B)-bimodule in which both A and B act on the left, then we have a functor $U_E: \mathcal{C}(B) \to \mathbf{Mod}(A)$, and, by virtue of the considerations in § 10, $\Phi(U_E)$ is a natural transformation $T \circ U_E \to T' \circ U_E$; consequently, $\Phi(U_E(B)) = \Phi(E)$ *is a homomorphism of* (C, B)-

bimodules $T(E) \to T'(E)$. Of course, the situation is the same if one of the categories **Mod**(A) and **Mod**(C) or both are, respectively, replaced by **Mod**$_d(A)$ and **Mod**$_d(C)$; likewise, nothing is changed in fact when B acts in E on the right.

In case T, T' are additive functors $(\mathbf{Mod}(A))^\circ \to \mathbf{Mod}(C)$, things are the same as in the above case, with the (in fact formal) modification that $\Phi(E^\circ)$ is a (C, B)-homomorphism from $T(E^\circ)$ to $T'(E^\circ)$.

As an application, let A, B be two rings, and let E be an A-module (A acting, say, on the left); the functor $h^E : \mathbf{Mod}(A) \to \mathbf{Mod}(\mathbf{Z})$ defined in § 2 is an additive functor; consequently, by the precedings arguments, if F is an (A, B)-bimodule in which A acts on the left and B on the left, respectively on the right, then $h^E(F) = \mathrm{Hom}_A(E, F)$ is endowed with a $(\mathbf{Z}, B)$-bimodule structure, or, in fact, with a left, respectively right B-module structure.

If $u \in \mathrm{Hom}_A(E, F)$, $\beta \in B$, $x \in E$, then, with this structure, we have in the first case,

$$(\beta u)(x) = (\mathrm{Hom}(1_E, h_\beta)(u))(x) = (h_\beta \circ u)(x) = \beta \cdot u(x), \tag{1}$$

and in the second one,

$$(u\beta)(x) = u(x) \cdot \beta. \tag{2}$$

If another (A, B)-bimodule F' is also given in which A and B act, respectively, on the same side as in F, and $g : F \to F'$ is a homomorphism of (A, B)-bimodules, then $h^E(g) = \mathrm{Hom}(1_E, g)$ is a homomorphism of $(\mathbf{Z}, B)$-bimodules, in fact a B-homomorphism.

The situation is essentially the same if **Mod**(A) is replaced by the category of right A-modules **Mod**$_d(A)$, of course, E being in this case a right A-module.

Here also formula (1), respectively (2) is still valid for the case in which F is a left, respectively right B-module.

Now, if A and B are again two rings and F is a fixed (left) A-module, the additive functor $h_F : (\mathbf{Mod}(A))^\circ \to \mathbf{Mod}(\mathbf{Z})$ carries the doublet E° of an (A, B)-bimodule E (in which both A and B act on the left) into a right B-module; hence, in this case, $\mathrm{Hom}_A(E, F)$ is endowed with a right B-module structure for which

$$(u\beta)(x) = (\mathrm{Hom}(h_\beta, 1_F)(u))(x) = (u \circ h_\beta)(x) = u(\beta x), \tag{3}$$

$u \in \mathrm{Hom}_A(E, F)$, $\beta \in B$, $x \in E$. Modifying only the fact that B acts in E on the right, $h_F(E^\circ) = \mathrm{Hom}_A(E, F)$ becomes a left B-module for which, using the same notation,

$$(\beta u)(x) = u(x\beta). \tag{4}$$

If E' is a second (A, B)-bimodule in which A and B act respectively on the same side as in E, and $f : E' \to E$ is a homomorphism of (A, B)-bimodules, then $h_F(f^\circ) = \mathrm{Hom}(f, 1_F)$ is a homomorphism of B-modules $\mathrm{Hom}_A(E, F) \to \to \mathrm{Hom}_A(E', F)$ (of course, for each situation, with the operators acting on the corresponding side).

The replacement of **Mod**(A) by $\mathbf{Mod}_d(A)$ and the consideration of a fixed right A-module F does not esentially modify the above results.

Take in particular $A = B$ and F an arbitrary (left) A-module; since A is an (A, A)-bimodule in which A acts both on the left and on the right (see Example 1)), it follows that $h_F(A_s^\circ) = \mathrm{Hom}_A(A_s, F)$ is endowed with a structure of left A-module, since the "additional" ring (B in the general case) acts on the right. Define a mapping

$$\Xi(F) : F \to \mathrm{Hom}_A(A_s, F),$$

taking for any $y \in F$ as $\Xi(F)(y)$ the mapping $A_s \to F$ given by $\alpha \mapsto \alpha y$, $\alpha \in A$; hence we have

$$\Xi(F)(y)(\alpha) = \alpha y;$$

indeed, it is not difficult to see that for any $y \in F$, $\Xi(F)(y)$ is an A-linear mapping $A_s \to F$, hence an element of $\mathrm{Hom}_A(A_s, F)$; moreover, $\Xi(F)$ is a homomorphism of left A-modules for the given structure of F and the one obtained on $\mathrm{Hom}_A(A_s, F)$: for, as it is easily seen, $\Xi(F)$ is a homomorphism of abelian groups; next, if $y \in F$ and $\alpha \in A$, we have by (4) (which corresponds to our case) for any $\alpha' \in A$,

$$[\alpha \cdot (\Xi(F)(y))](\alpha') = (\Xi(F)(y))(\alpha'\alpha) = (\alpha'\alpha)y = \alpha'(\alpha y) = [\Xi(F)(\alpha y)](\alpha'),$$

hence

$$\alpha \cdot (\Xi(F)(y)) = \Xi(F)(\alpha y).$$

Define now a mapping

$$\psi : \mathrm{Hom}_A(A_s, F) \to F$$

by

$$\psi(f) = f(1),$$

where $f \in \mathrm{Hom}_A(A_s, F)$ and 1 is the unit element of A. We have

$$[(\Xi(F) \circ \psi)(f)](\alpha) = [\Xi(F)(f(1))](\alpha) = \alpha f(1) = f(\alpha)$$

for any $f \in \mathrm{Hom}_A(A_s, F)$ and any $\alpha \in A$; hence

$$\Xi(F) \circ \psi = 1_{\mathrm{Hom}(A,\ F)};$$

further, $(\psi \circ \Xi(F))(y) = \Xi(F)(y)(1) = y$ for any $y \in F$; hence

$$\psi \circ \Xi(F) = 1_F,$$

and $\Xi(F)$ is an A-isomorphism having ψ as inverse.

If F is a right A-module, then $h_F(A_d^\circ) = \mathrm{Hom}_A(A_d, F)$ is endowed with a right A-module structure, and we have an isomorphism of right A-modules

$$\Xi(F) : F \xrightarrow{\sim} \mathrm{Hom}_A(A_d, F)$$

given by

$$\Xi(F)(y)(\alpha) = y\alpha,$$

$y \in F$, $\alpha \in A$.

If F, F' are left (respectively right) A-modules, and if $g : F \to F'$ is an A-homomorphism, then a natural transformation

$$h_g : h_F \to h_{F'}$$

may be defined as follows: if $E^\circ \in \mathrm{Ob}\,(\mathbf{Mod}(A))^\circ$ (respectively $E^\circ \in \mathrm{Ob}\,(\mathbf{Mod}_d(A))^\circ$), then

$$h_g(E^\circ) : \mathrm{Hom}_A(E, F) \to \mathrm{Hom}_A(E, F')$$

is, by definition, $\mathrm{Hom}(1_E, g)$; it is easy to verify, using relations (3), § 2, the character of natural transformation of h_g thus defined. But then, according to the preceding arguments, $h_g(A_s^\circ)$ (respectively $h_g(A_d^\circ)$), that is $\mathrm{Hom}(1_A, g)$, is a homomorphism of left (respectively right) A-modules.

It may be easily verified that the following diagram is commutative:

(5)
$$\begin{array}{ccc} F & \xrightarrow{\quad g \quad} & F' \\ {\scriptstyle \Xi(F)}\downarrow & & \downarrow{\scriptstyle \Xi(F')} \\ \mathrm{Hom}_A(A, F) & \xrightarrow{\mathrm{Hom}(1_A, g)} & \mathrm{Hom}_A(A, F'). \end{array}$$

Thus we proved:

PROPOSITION 1. *Let A be a ring; for any left (respectively right) A-module F, $\mathrm{Hom}_A(A_s, F)$ (respectively $\mathrm{Hom}_A(A_d, F)$) is canonically endowed with a left (respectively right) A-module structure, and we have a canonical isomorphism of left (respectively right) A-modules*

$$\Xi(F) : F \xrightarrow{\sim} \mathrm{Hom}_A(A, F);$$

if F' is a second left (respectively right) A-module and $g \in \mathrm{Hom}_A(F, F')$, then diagram (5) is commutative. ▮

REMARK. The fact that for any A-module F, $\mathrm{Hom}_A(A, F)$ is endowed with an A-module structure shows that the functor $h^A : \mathbf{Mod}(A) \to \mathbf{Ab}$ takes its values in the category **Mod**(A), i.e., that it is the composition of the forgetful functor **Mod**(A) → **Ab** with a uniquely determined functor **Mod**(A) → **Mod**(A), denoted also h^A by an abuse of notation; the commutativity of diagram (5) means that the assignment to any A-module F of the homomorphism of A-modules $\Xi(F)$ defines a natural transformation

$$\Xi : 1_{\mathbf{Mod}(A)} \to h^A;$$

the fact that for any F, $\Xi(F)$ is an isomorphism of A-modules means that Ξ is a natural equivalence. ∎

Consider now the particular case when A is a *commutative* ring: in this case, any A-module may be considered as an (A, A)-bimodule (see Example 3)); consequently, if E and F are two A-modules, then by the preceding considerations, $\mathrm{Hom}_A(E, F)$ is endowed in two manners with an A-module structure: once as $h^E(F)$, secondly as $h_F(E^\circ)$; but these two structures coincide: for, denoting by $\alpha \cdot u$, respectively $\alpha \perp u$, the result of the scalar multiplication of $u \in \mathrm{Hom}_A(E, F)$ by $\alpha \in A$ according to the first, respectively to the second structure, we have by relations (1), respectively (3),

$$(\alpha \cdot u)(x) = \alpha \cdot u(x), \text{ respectively } (\alpha \perp u)(x) = u(\alpha x);$$

but u being an A-homomorphism, we have $\alpha \cdot u(x) = u(\alpha x)$; this A-module structure coincides with the one defined by rudimentary means in the Remark following Definition 1, § 2; it will be referred to as the *canonical structure* of A-module of $\mathrm{Hom}_A(E, F)$, with A commutative.

If, under the same conditions, E' is another A-module and $f : E' \to E$ is an A-homomorphism, then, trivially, f is a homomorphism of (A, A)-bimodules, hence $\mathrm{Hom}(f, 1_F)$ is an A-homomorphism; similarly, for an A-homomorphism $g : F \to F'$, the homomorphism $\mathrm{Hom}(1_E, g)$ is an A-homomorphism.

Further, given four A-modules E, E', F, F' and two homomorphisms $f : E' \to E$, $g : F \to F'$, then

$$\mathrm{Hom}(f, g) : \mathrm{Hom}_A(E, F) \to \mathrm{Hom}_A(E', F')$$

is an A-homomorphism: for we have

$$\mathrm{Hom}(f, g) = \mathrm{Hom}(f, 1_{F'}) \circ \mathrm{Hom}(1_E, g),$$

and $\mathrm{Hom}(f, g)$ is the composition of two A-homomorphisms.

With the canonical A-module structure of $\mathrm{Hom}_A(E, F)$, A commutative, it is possible to show that *the canonical isomorphism*

$$\mathrm{Hom}_A\Big(\bigoplus_{\lambda \in I} E_\lambda, \prod_{\mu \in J} F_\mu\Big) \rightleftarrows \prod_{(\lambda, \mu) \in I \times J} \mathrm{Hom}_A(E_\lambda, F_\mu)$$

given by Proposition 3, § 7, *as well as the canonical isomorphism*

$$\varphi : \mathrm{Hom}_A\big(\varinjlim_I E_\lambda, \varprojlim_J F_\mu\big) \rightleftarrows \varprojlim_{I \times J} \mathrm{Hom}_A(E_\lambda, F_\mu)$$

given by Proposition 1, § 14, *are isomorphisms of A-modules*: we shall prove only the second assertion, the first one being a particular case of this (see §§ 12 and 13); but if $r \in \operatorname{Hom}_A(\varinjlim_I E_\lambda, \varprojlim_J F_\mu)$ and $\alpha \in A$, then, in the notation of the proof of Proposition 1, § 14, we have

$$\varphi(\alpha r) = (p_\mu \circ (\alpha r) \circ j_\lambda)_{(\lambda, \mu) \in I \times J};$$

since j_λ and p_μ are A-homomorphisms, it follows that $\operatorname{Hom}(j_\lambda, p_\mu)$ is also an A-homomorphism, hence that

$$p_\mu \circ (\alpha r) \circ j_\lambda = \operatorname{Hom}(j_\lambda, p_\mu)(\alpha r) = \alpha \cdot (\operatorname{Hom}(j_\lambda, p_\mu)(r)) = \alpha \cdot (p_\mu \circ r \circ j_\lambda);$$

consequently,

$$\varphi(\alpha r) = (\alpha \cdot (p_\mu \circ r \circ j_\lambda))_{(\lambda, \mu) \in I \times J} = \alpha \cdot (p_\mu \circ r \circ j_\lambda)_{(\lambda, \mu) \in I \times J} = \alpha \cdot \varphi(r),$$

and the assertion is proved.

* DEFINITION 4. Let $\mathcal{C}$ be a category, respectively an abelian category; we shall say that a functor $T : \mathcal{C} \to \mathbf{Ens}$, respectively an additive functor $T : \mathcal{C} \to \mathbf{Ab}$ is a *representable functor* if there exists a natural equivalence $\Phi : h^E \rightleftarrows T$ for some object E of $\mathcal{C}$; in this case E is said to be a *representative* of the representable functor T; likewise, if U is a functor $\mathcal{C}^\circ \to \mathbf{Ens}$, respectively an additive functor $\mathcal{C}^\circ \to \mathbf{Ab}$, then U is said to be a *representable functor* if there exists a natural equivalence $\Psi : h_F \rightleftarrows U$ for some object F of $\mathcal{C}$; also in this case, F is, by definition, a *representative* of the representable functor U. ∎*

By Proposition 1 it follows (see also the Remark following it) that for any ring A, the forgetful functor $\mathbf{Mod}(A) \to \mathbf{Ab}$ is representable, a representative of it being the object A_s of $\mathbf{Mod}(A)$.

* The problem of the existence of a left (or right) adjoint of a functor is the problem whether certain functors are representable: For let $\mathcal{C}, \mathcal{C}'$ be two categories, and let $S : \mathcal{C} \to \mathcal{C}'$ be a functor; for any object E' of $\mathcal{C}'$ consider the composite functor

$$\mathcal{C} \xrightarrow{S} \mathcal{C}' \xrightarrow{h^{E'}} \mathbf{Ens};$$

to say that the functor $h^{E'} \circ S : \mathcal{C} \to \mathbf{Ens}$ is representable for any $E' \in \operatorname{Ob} \mathcal{C}'$ is to say that for any $E' \in \operatorname{Ob} \mathcal{C}'$ there exists an object $T(E') \in \operatorname{Ob} \mathcal{C}$ and a natural equivalence $h^{T(E')} \rightleftarrows h^{E'} \circ S$; but, according to Proposition 13, § 10, this is equivalent to the existence of a functor $T : \mathcal{C}' \to \mathcal{C}$ left adjoint to S.

Likewise, the fact that the composite functor $h_F \circ T^\circ$ (where T° is the dual of the functor $T : \mathcal{C}' \to \mathcal{C}$) is representable for any $F \in \operatorname{Ob} \mathcal{C}$ is equivalent to the fact that T has a right adjoint.

These considerations are valid also in the case of abelian categories and additive functors, the functors $h^{E'}$, h_F then taking values in the category **Ab**.

Concerning the natural transformations of a functor of the form h^E into an arbitrary functor, we have the following result:

PROPOSITION 2 (**Yoneda Lemma**). *Let $\mathcal{C}$ be a category. For any object E of $\mathcal{C}$ and any functor $T : \mathcal{C} \to$ **Ens** there exists a bijective mapping between the class of natural transformations $h^E \to T$ and the set $T(E)$; in particular, the first class is a set. Likewise, for any object F of the category $\mathcal{C}$ and any functor $U : \mathcal{C}^\circ \to$ **Ens** there exists a bijection between the class of natural transformations of h_F into U and the set $U(F^\circ)$; in particular, the first class is a set.*

Proof. We shall prove only the first part, the second one being proved by dual considerations.

Denote by $\mathrm{Hom}(h^E, T)$ the class of natural transformations from h^E to T, where $E \in \mathrm{Ob}\,\mathcal{C}$ and T is a functor $\mathcal{C} \to$ **Ens**. Define a mapping of classes

$$\rho : \mathrm{Hom}\,(h^E, T) \to T(E)$$

by

$$\rho(\Phi) = \Phi(E)(1_E), \quad \Phi \in \mathrm{Hom}\,(h^E, T).$$

In order to show that ρ thus defined is a bijection, define a mapping of classes

$$\sigma : T(E) \to \mathrm{Hom}\,(h^E, T)$$

as follows: let $\xi \in T(E)$; $\sigma(\xi)$ is the natural transformation $h^E \to T$ having as component at the object X of $\mathcal{C}$ the mapping of sets

$$\sigma(\xi)(X) \colon h^E(X) \to T(X)$$

sending an element $u \in h^E(X)$ into the element $(T(u))(\xi) \in T(X)$. To show that $\sigma(\xi)$ thus defined is indeed a natural transformation, let Y be a second object of $\mathcal{C}$, and let $f \in \mathrm{Hom}_{\mathcal{C}}(X, Y)$; we have to show the commutativity of the following diagram:

$$\text{(6)} \qquad \begin{array}{ccc} h^E(X) & \xrightarrow{\;h^E(f)\;} & h^E(Y) \\ \Big\downarrow{\scriptstyle \sigma(\xi)(X)} & & \Big\downarrow{\scriptstyle \sigma(\xi)(Y)} \\ T(X) & \xrightarrow{\;T(f)\;} & T(Y); \end{array}$$

to do this, let $u \in h^E(X)$; we have

$$(\sigma(\xi)(Y))(h^E(f)(u)) = (\sigma(\xi)(Y))\,(f \circ u) = (T(f \circ u))(\xi) = T(f)(\sigma(\xi)(X)(u)),$$

which proves the commutativity of (6).

To show that ρ and σ are bijections inverse to each other, let first $\xi \in T(E)$; we have $\rho(\sigma(\xi)) = \sigma(\xi)(E)(1_E) = T(1_E)(\xi) = \xi$, hence $\rho \circ \sigma = 1_{TE}$; next, let $\Phi \in \text{Hom}\,(h^E, T)$; to show that $\sigma(\rho(\Phi)) = \Phi$ it suffices to show that for any $X \in \text{Ob}\,\mathscr{C}$ and any $u \in h^E(X)$ we have

$$(\sigma(\rho(\Phi)))(X)(u) = \Phi(X)(u);$$

but we have $(\sigma(\rho(\Phi)))(X)(u) = T(u)(\rho(\Phi)) = T(u)(\Phi(E)(1_E))$; since Φ is a natural transformation, we have $T(u)(\Phi(E)(1_E)) = \Phi(X)(h^E(u)(1_E)) = \Phi(X)(u)$. ∎

In the particular case where $\mathscr{C}$ is a preadditive category, we may consider that h^E is an additive functor $\mathscr{C} \to \mathbf{Ab}$; if T is an additive functor $\mathscr{C} \to \mathbf{Ab}$, then by exactly the same considerations as in the proof of Proposition 2 one proves the existence of a bijection $\rho : \text{Hom}(h^E, T) \to T(E)$; hence, also in this case $\text{Hom}(h^E, T)$ is a set; we may endow this set with an abelian group structure defining in it an addition as follows: if $\Phi_1, \Phi_2 \in \text{Hom}(h^E, T)$, then $\Phi_1 + \Phi_2$ is the natural transformation $h^E \to T$ having $\Phi_1(X) + \Phi_2(X)$ as component at the object X of $\mathscr{C}$, the sum being performed in the abelian group $\text{Hom}_{\mathbf{Z}}(h^E(X), T(X))$; with this structure and the given structure of abelian group of $T(E)$, ρ defined above is an isomorphism of abelian groups; it will be referred to as the *Yoneda isomorphism*; the terminology is maintained for an arbitrary (not necessarily preadditive) category.

Now, if $\mathscr{C}$ is a category and T is a representable functor, then, by Definition 4, there exists an object E of $\mathscr{C}$ and a natural equivalence

$$\Phi : h^E \rightleftarrows T;$$

under the Yoneda isomorphism the natural transformation Φ is carried into an element $x \in T(E)$, namely $x = \Phi(E)(1_E)$; according to Proposition 2, to give $x \in T(E)$ means to have Φ determined; sometimes, one says that *the pair* (E, x) *represents* T or that (E, x) *is a representative* of T, this version of the terminology giving an additional specification compared with Definition 4, since it also gives the natural equivalence Φ. The same terminology is used for representable functors $\mathscr{C}^\circ \to \mathbf{Ens}$, and if $\mathscr{C}$ is preadditive, for representable additive functors $\mathscr{C} \to \mathbf{Ab}$, respectively $\mathscr{C}^\circ \to \mathbf{Ab}$.

We have:

PROPOSITION 3. *Let $\mathscr{C}$ be a category, and let T, T' be representable functors defined on $\mathscr{C}$ and taking values in* **Ens**; *if (E, x), respectively (E', x'), is a representative of T, respectively of T', and $\Psi : T \to T'$ is a natural transformation, then there exists a morphism $u \in \text{Hom}_{\mathscr{C}}(E', E)$ uniquely determined by the property that*

$$\Psi(E)(x) = T'(u)(x'); \tag{7}$$

the assignment $\Psi \mapsto u$ is "functor-like", in the following sense: given the above data, we say that "for the representatives (E, x), respectively (E', x') of T, respectively T', the morphism $u \in \text{Hom}_{\mathscr{C}}(E', E)$ is associated with Ψ"; then, let T'' be a third repre-

sentable functor $\mathcal{C} \to \mathbf{Ens}$, *let* (E'', x'') *be a representative of it, and let* $\Psi' : T' \to T''$ *be a natural transformation with which the morphism* $u' \in \mathrm{Hom}_{\mathcal{C}}(E'', E')$ *is associated for the representatives* (E', x'), *respectively* (E'', x'') *of* T', *respectively* T''; *under these conditions, with the natural transformation* $\Psi' \circ \Psi : T \to T''$ *is associated for the representatives* (E, x), *respectively* (E'', x'') *of* T, *respectively of* T'', *the morphism* $u \circ u' \in \mathrm{Hom}_{\mathcal{C}}(E'', E)$; *moreover, with the identity natural transformation* 1_T *is associated for the representative* (E, x) *of* T *(considered both as domain and as range of* 1_T*) the identity morphism* 1_E.

Likewise, if $\mathcal{C}$ *is a category and* U, U' *are representable functors* $\mathcal{C}^\circ \to \mathbf{Ens}$ *of respective representatives* (F, y), (F', y'), *then with any natural transformation* $\Psi : U \to U'$ *it is associated a morphism* $u \in \mathrm{Hom}_{\mathcal{C}}(F, F')$ *uniquely determined by the property that*

$$\Psi(F^\circ)(y) = U'(u^\circ)(y');$$

if, moreover, U'' *is also a representable functor* $\mathcal{C}^\circ \to \mathbf{Ens}$, *while* (F'', y'') *is a representative of it, and* $\Psi' : U' \to U''$ *is a natural transformation with which, in the above sense, is associated for the representatives* (F', y') *and* (F'', y'') *the morphism* $u' \in \mathrm{Hom}_{\mathcal{C}}(F', F'')$, *then with the natural transformation* $\Psi' \circ \Psi : U \to U''$ *is associated, in the same sense, for the representatives* (F, y) *and* (F'', y'') *the morphism* $u' \circ u$; *moreover, with the identity morphism of* U *is associated for the representatives* (F, y) *and* (F, y) *of* U *the identity morphism of* F.

Proof. As usual, we shall prove only the first part, leaving the second one to the reader.

Denote by Φ the natural equivalence $h^E \rightleftarrows T$ from which we get applying the Yoneda isomorphism the stated element $x \in T(E)$; the same meaning has the notation Φ', Φ'' for T', T'' and x', x'', respectively.

Define a natural transformation $\Theta : h^E \to h^{E'}$ by

$$\Theta = \Phi'^{-1} \circ \Psi \circ \Phi;$$

define u by

$$u = \Theta(E)(1_E) \in h^{E'}(E) = \mathrm{Hom}_{\mathcal{C}}(E', E).$$

Since by definition we have

$$x = \Phi(E)(1_E), \quad x' = \Phi'(E')(1_{E'}),$$

it follows

$$\Psi(E)(x) = (\Psi(E) \circ \Phi(E))(1_E) = (\Phi'(E) \circ \Theta(E))\,(1_E) = \Phi'(E)(u); \tag{8}$$

on the other hand, as Φ' is a natural transformation, it follows

$$T'(u)(x') = (T'(u) \circ \Phi'(E'))(1_{E'}) = (\Phi'(E) \circ h^{E'}(u))(1_{E'}) = \Phi'(E)(u); \tag{9}$$

relations (8) and (9) imply (7).

If $v \in \operatorname{Hom}_{\mathcal{C}}(E', E)$ has when replacing u the property (7), then, from this property for u and v, from (9), and from the relation $T'(v)(x') = \Phi'(E)(v)$ (obtained in the same way as (9)) it follows that $\Phi'(E)(v) = \Phi'(E)(u)$ and, as $\Phi'(E)$ is bijective since Φ' is a natural equivalence, that $u = v$. The existence and the uniqueness of u with the property (7) is proved.

To show, in the notation of the statement, that with $\Psi' \circ \Psi$ is associated $u \circ u'$ it suffices now to show that

$$(\Psi' \circ \Psi)(E)(x) = T''(u \circ u')(x'');$$

from the fact that Ψ' is a natural transformation, from (7), and from the relation $\Psi'(E')(x') = T''(u')(x'')$ (similar to (7) but for Ψ') it follows that

$$(\Psi' \circ \Psi)(E)(x) = \Psi'(E)(T'(u)(x')) = T''(u)(\Psi'(E')(x')) = T''(u \circ u')(x'').$$

Likewise, the fact that with 1_T is associated 1_E follows from the relation

$$1_T(E)(x) = T(1_E)(x),$$

which is obvious. ▮

In Proposition 3 we considered the morphism associated with the identity natural transformation of a representable functor when the representative is preserved; Corollary 1 below shows what happens when the representative is changed.

Corollary 1. *Let $\mathcal{C}$ be a category, and let $T : \mathcal{C} \to$ **Ens** be a representable functor; if (E, x) and (E', x') are two of its representatives, then there exists an isomorphism $u : E' \xrightarrow{\sim} E$ of the category $\mathcal{C}$, and this isomorphism is the unique morphism $E' \to E$ with the property*

$$x = T(u)(x');$$

*if U is a representable functor $\mathcal{C}^\circ \to$ **Ens**, and if (F, y), (F', y') are two of its representatives, then there exists an isomorphism $u : F \xrightarrow{\sim} F'$ of the category $\mathcal{C}$, and this isomorphism is the unique morphism $F \to F'$ with the property*

$$y = U(u^\circ)(y').$$

Proof. We have only to take in Proposition 3 $T' = T$ and $\Psi = 1_T$; u is the one constructed there; to show that u is an isomorphism, we interchange the roles of the representatives: as shown in Proposition 3, there exists a morphism $v : E \to E'$ with the property that with $1_T \circ 1_T = 1_T$ is associated for the representatives (E, x) and (E, x) the composite morphism $u \circ v : E \to E$; but preserving the representative (E, x), with 1_T is associated the identity morphism 1_E; hence $u \circ v = 1_E$; by a similar argument we get $v \circ u = 1_{E'}$. ▮

Remark. Corollary 1 above shows that *the representative of a representable functor is uniquely determined up to an isomorphism.* ▮

We have seen earlier, when discussing modules, that for any ring A and any $g \in \mathrm{Hom}_A(F, F')$, a natural transformation $h_g : h_F \to h_{F'}$ may be defined; the same is true for an arbitrary category $\mathcal{C}$: if $F, F' \in \mathrm{Ob}\,\mathcal{C}$ and $g \in \mathrm{Hom}_{\mathcal{C}}(F, F')$, then a natural transformation

$$h_g : h_F \to h_{F'}$$

is defined by $h_g(X^\circ) = \mathrm{Hom}(1_X, g)$, $X \in \mathrm{Ob}\,\mathcal{C}$; likewise, if $E, E' \in \mathrm{Ob}\,\mathcal{C}$ and $f \in \mathrm{Hom}_{\mathcal{C}}(E', E)$, then, assigning to each object Y of $\mathcal{C}$ the morphism $h^f(Y) = \mathrm{Hom}(f, 1_Y)$, we get a natural transformation

$$h^f : h^E \to h^{E'}.$$

The following Corollary shows that any natural transformation $h^E \to h^{E'}$, respectively $h_F \to h_{F'}$, is of this form.

Corollary 2. *Let $\mathcal{C}$ be a category, let $E, E' \in \mathrm{Ob}\,\mathcal{C}$, and let Ψ be a natural transformation $h^E \to h^{E'}$; then there exists a unique morphism $f \in \mathrm{Hom}_{\mathcal{C}}(E', E)$ with the property that*

$$\Psi = h^f.$$

Similarly, if $F, F' \in \mathrm{Ob}\,\mathcal{C}$, and if Ω is a natural transformation $h_F \to h_{F'}$, then there exists a unique morphism $g \in \mathrm{Hom}_{\mathcal{C}}(F, F')$ with the property that

$$\Omega = h_g.$$

Proof. Using the notations of the statement and of the proof of Proposition 3, let us take $T = h^E$, $T' = h^{E'}$, and the identity natural transformation for Φ and Φ'; then $\Theta = \Psi$, and x, respectively x', are given by $x = 1_E$, $x' = 1_{E'}$; by Proposition 3, there exists $f \in \mathrm{Hom}_{\mathcal{C}}(E', E)$ associated with Ψ for the representatives $(E, 1_E)$ and $(E', 1_{E'})$, hence for which

(10) $$\Psi(E)(1_E) = h^{E'}(f)(1_{E'}) = f = h^f(E)(1_E);$$

in order to show that $\Psi = h^f$ it suffices to show that for any $X \in \mathrm{Ob}\,\mathcal{C}$ and any $s \in h^E(X)$ we have

(11) $$\Psi(X)(s) = h^f(X)(s);$$

since we have the commutative diagrams

$$\begin{array}{ccc} h^E(E) & \xrightarrow{h^E(s)} & h^E(X) \\ \downarrow{\scriptstyle h^f(E)} & & \downarrow{\scriptstyle h^f(X)} \\ h^{E'}(E) & \xrightarrow{h^{E'}(s)} & h^{E'}(X), \end{array} \qquad \begin{array}{ccc} h^E(E) & \xrightarrow{h^E(s)} & h^E(X) \\ \downarrow{\scriptstyle \Psi(E)} & & \downarrow{\scriptstyle \Psi(X)} \\ h^{E'}(E) & \xrightarrow{h^{E'}(s)} & h^{E'}(X), \end{array}$$

it follows, also using (10), that

$$h^f(X)(s) = h^f(X)\,(h^E(s)\,(1_E)) = h^{E'}(s)\,(h^f(E)(1_E)) = h^{E'}(s)\,(\Psi(E)(1_E)) =$$

$$= \Psi(X)\,(h^E(s)(1_E)) = \Psi(X)(s),$$

hence we get relation (11).

If $\bar{f} \in \mathrm{Hom}_{\mathcal{C}}(E', E)$ also has the property $h^{\bar{f}} = \Psi$, then (10) implies

$$\bar{f} = h^{\bar{f}}(E)\,(1_E) = \Psi(E)(1_E) = f. \blacksquare$$

Proposition 3 and the Corollaries to it are still valid if instead of an arbitrary category $\mathcal{C}$ and arbitrary functors $\mathcal{C} \to \mathbf{Ens}$, respectively $\mathcal{C}^\circ \to \mathbf{Ens}$, a preadditive category $\mathcal{C}$ and additive functors $\mathcal{C} \to \mathbf{Ab}$, respectively $\mathcal{C}^\circ \to \mathbf{Ab}$, are considered.

We may now formulate and prove the "naturality" of the Yoneda isomorphism: more precisely, *let* $\mathcal{C}$ *be a category, let* T, T' *be functors* $\mathcal{C} \to \mathbf{Ens}$, *and let* $E, E' \in \mathrm{Ob}\,\mathcal{C}$; *if* $\Psi : T \to T'$ *is a natural transformation, and if* $f \in \mathrm{Hom}_{\mathcal{C}}(E, E')$, *then the following diagrams commute*:

$$(12)\quad \begin{array}{ccc} \mathrm{Hom}(h^E, T) & \xrightarrow[\sim]{} & T(E) \\ {\scriptstyle \mathrm{Hom}(h^f, 1_T)}\downarrow & & \downarrow{\scriptstyle T(f)} \\ \mathrm{Hom}(h^{E'}, T) & \xrightarrow[\sim]{} & T(E'), \end{array} \qquad \begin{array}{ccc} \mathrm{Hom}(h^E, T) & \xrightarrow[\sim]{} & T(E) \\ {\scriptstyle \mathrm{Hom}(1_{h^E}, \Psi)}\downarrow & & \downarrow{\scriptstyle \Psi(E)} \\ \mathrm{Hom}(h^E, T') & \xrightarrow[\sim]{} & T'(E), \end{array}$$

where the horizontal arrows represent the Yoneda isomorphisms, and, as in "legitimate" categories [1], $\mathrm{Hom}(h^f, 1_T)$ assigns to each $\Phi \in \mathrm{Hom}(h^E, T)$ the natural transformation $\Phi \circ h^f : h^{E'} \to T$ while $\mathrm{Hom}\,(1_{h^E}, \Psi)$ assigns to Φ the natural transformation $\Psi \circ \Phi : h^E \to T'$. Similarly, *let* $\mathcal{C}$ *be a category, let* U, U' *be functors* $\mathcal{C}^\circ \to \mathbf{Ens}$, *and let* $F, F' \in \mathrm{Ob}\,\mathcal{C}$; *if* $\Omega : U \to U'$ *is a natural transformation, and if* $g \in \mathrm{Hom}_{\mathcal{C}}(F', F)$, *then the following diagrams commute*:

$$(13)\quad \begin{array}{ccc} \mathrm{Hom}(h_F, U) & \xrightarrow[\sim]{} & U(F^\circ) \\ {\scriptstyle \mathrm{Hom}(h_g, 1_U)}\downarrow & & \downarrow{\scriptstyle U(g^\circ)} \\ \mathrm{Hom}(h_{F'}, U) & \xrightarrow[\sim]{} & U(F'^\circ), \end{array} \qquad \begin{array}{ccc} \mathrm{Hom}(h_F, U) & \xrightarrow[\sim]{} & U(F^\circ) \\ {\scriptstyle \mathrm{Hom}(1_{h_F}, \Omega)}\downarrow & & \downarrow{\scriptstyle \Omega(F^\circ)} \\ \mathrm{Hom}(h_F, U') & \xrightarrow[\sim]{} & U'(F^\circ), \end{array}$$

where the meaning of the notations is similar to that defined above.

---

[1] The functors $\mathcal{C} \to \mathbf{Ens}$ and the natural transformations between them do not form generally a category (see the Remark following Definition 7, § 10).

The commutativity of diagrams (12) and (13) is a straightforward consequence of the definitions, and its verification is an easy exercise for the reader.

If in the above statements the category $\mathcal{C}$ is a preadditive category and T, T', U, U' are additive functors $\mathcal{C} \to \mathbf{Ab}$, respectively $\mathcal{C}^\circ \to \mathbf{Ab}$, the result is still valid.

The commutativity of the first diagram (12), respectively (13), will be said to express the "naturality of the Yoneda isomorphism in the first argument", while the commutativity of the second diagram (12), respectively (13), the "naturality of the Yoneda isomorphism in the second argument".*

REMARK. Using the Yoneda Lemma, it is possible to give another (partial) proof of Proposition 1: for let A be a ring, and let F be an A-module; by the preceding considerations, the modules A_s and F define two functors $U_{A_s}, U_F : \mathcal{C}(A) \to \mathbf{Mod}(\mathbf{Z}) = \mathbf{Ab}$; according to the definitions of h^A and U_{A_s}, we have $h^A = U_{A_s}$. Again by the preceding considerations, between the set of all A-homomorphisms $A_s \to F$ and the set of natural transformations $h^A = U_{A_s} \to U_F$ there exists a bijection assigning to $u \in \mathrm{Hom}_A(A_s, F)$ the natural transformation $\Phi_u : h^A \to U_F$ for which $\Phi_u(A)$ is the mapping u considered as a $\mathbf{Z}$-homomorphism; this bijection is a $\mathbf{Z}$-isomorphism. Finally, applying the Yoneda Lemma, we get a $\mathbf{Z}$-isomorphism between the group of natural transformations $h^A \to U_F$ and the abelian group $U_F(A)$ (which is the underlying $\mathbf{Z}$-module of F) which assigns $u(1)$ to Φ_u. Hence we find a $\mathbf{Z}$-isomorphism $\mathrm{Hom}_A(A_s, F) \xrightarrow{\sim} F$ which assigns $u(1)$ to $u \in \mathrm{Hom}_A(A_s, F)$, i.e., which is exactly ψ in the proof of Proposition 1; but now ψ is proved to be only a $\mathbf{Z}$-isomorphism. ▮

Let now A and C be two rings; if $T : (\mathbf{Mod}(A))^\circ \to \mathbf{Mod}(C)$ is an additive functor, then $T(A_s^\circ)$ has besides the C-module structure an A-module structure (on the left, since, using the above general notations, the ring $B = A$ acts on the right [1]), the two structures being compatible; consequently, if E is an A-module, we may talk of the abelian group

$$\mathrm{Hom}_A(E, T(A_s^\circ)) = h_{T(A_s^\circ)}(E^\circ),$$

which is endowed with a C-module structure, since it may be also viewed as $h^E(T(A_s^\circ))$; moreover, if $u : E \to F$ is an A-homomorphism, then, by virtue of the preceding considerations, the $\mathbf{Z}$-homomorphism $h_{T(A_s^\circ)}(u^\circ) = \mathrm{Hom}(u, 1_{T(A_s^\circ)})$ is in fact a C-homomorphism, since it may be viewed as being $h^u(T(A_s^\circ))$, where h^u is the natural transformation $h^F \to h^E$ defined by $h^u(X) = \mathrm{Hom}(u, 1_X)$ (see the considerations preceding Corollary 2 to Proposition 3). It follows that the functor $h_{T(A_s^\circ)}$ may be, by an abuse, considered as a functor from the category $(\mathbf{Mod}(A))^\circ$ to the category $\mathbf{Mod}(C)$, and *it is in this way that we shall consider it subsequently*; for convenience, we shall denote $T(A_s^\circ) = M$. *The functor* h_M as a functor $(\mathbf{Mod}(A))^\circ \to$

---

[1] We have considered the situation concerning $\mathbf{Mod}(A)$ and $\mathbf{Mod}(C)$; we could have considered, with the corresponding modification of the results, the situation of $\mathbf{Mod}_d(A)$ and $\mathbf{Mod}(C)$, etc.

$\to$ **Mod**(C) *commutes with direct products*; in other words, if $(E_\lambda)_{\lambda\in I}$ is a family of A-modules, then the canonical isomorphism

$$\varphi : \mathrm{Hom}_A(\bigoplus_{\lambda\in I} E_\lambda, M) \rightleftarrows \prod_{\lambda\in I} \mathrm{Hom}_A(E_\lambda, M)$$

constructed in Proposition 3, § 7, is a C-isomorphism: indeed, if $\gamma\in C$ and $r\in \mathrm{Hom}_A(\bigoplus_{\lambda\in I} E_\lambda, M)$, then, denoting by j_λ the canonical injection $E_\lambda \to \bigoplus_{\lambda\in I} E_\lambda$, $\lambda\in I$, we have $\varphi(\gamma r)=((\gamma r)\circ j_\lambda)_{\lambda\in I}$; since $(\gamma r)\circ j_\lambda = \gamma(r\circ j_\lambda)$ for any $\lambda\in I$, we get the desired result.

Obviously, if instead of $M = T(A_s^\circ)$ we choose another (A, C)-bimodule N (in which A and C act on the same side as in M), then the same considerations remain valid for the functor $h_N : (\mathbf{Mod}(A))^\circ \to \mathbf{Mod}(C)$; we shall extend the term of "*representable functor*" for such a functor (inclusively for h_M) as well as for a functor $R : (\mathbf{Mod}(A))^\circ \to \mathbf{Mod}(C)$ isomorphic to such a functor.

Subsequently an essential role will be played by the natural transformation denoted Ψ,

$$\Psi : T \to h_M,$$

whose construction is as follows: if E is an A-module, then, for any $a\in E$, the A-homomorphism $\theta_a : A_s \to E$ defined by

$$\theta_a(\alpha) = \alpha a,\ \alpha\in A$$

(see also Example 6), § 1) is considered; by definition, $\Psi(E^\circ)$ is the homomorphism $T(E^\circ) \to h_M(E^\circ) = \mathrm{Hom}_A(E, M)$ sending $x\in T(E^\circ)$ into the mapping $E\to M = T(A_s^\circ)$ given by $a \mapsto T(\theta_a^\circ)(x)$, $a\in E$; consequently,

$$\Psi(E^\circ)(x)(a) = T(\theta_a^\circ)(x),\quad x\in T(E^\circ),\ a\in E.$$

To show that $\Psi(E^\circ)(x)$ is indeed an A-linear mapping, we take into account the fact that, according to the preceding considerations, the A-module structure of M is obtained as follows: if $\alpha\in A$, then h_α denotes the homothety of ratio α in A_d (i.e., $h_\alpha(\alpha') = \alpha'\alpha$, $\alpha'\in A$), and then

$$\alpha\cdot y = T(h_\alpha^\circ)(y),\qquad y\in M;$$

if $a\in E$ and $\alpha\in A$, we have $\theta_{\alpha a} = \theta_a\circ h_\alpha$, and, consequently,

$$\Psi(E^\circ)(x)(\alpha a) = T(\theta_{\alpha a}^\circ)(x) = (T(h_\alpha^\circ)\circ T(\theta_a^\circ))(x) = \alpha\cdot(\Psi(E^\circ)(x)(a));$$

the fact that $\Psi(E^\circ)(x)$ is a homomorphism of abelian groups follows from the fact that T is an additive functor.

Likewise, to show that $\Psi(E^\circ)$ is a C-homomorphism for any $E\in \mathrm{Ob}(\mathbf{Mod}(A))$, let $\gamma\in C$, and let $x\in T(E^\circ)$; since the C-module structure of $\mathrm{Hom}_A(E, M)$ is given by a relation of the type (1), and since for $a \in E$, $T(\theta_a^\circ)$ is a C-homomorphism, it follows

$$\Psi(E^\circ)(\gamma x)(a) = T(\theta_a^\circ)(\gamma x) = \gamma(T(\theta_a^\circ)(x)) = [\gamma \cdot (\Psi(E^\circ)(x))]\,(a).$$

To complete the proof of the fact that Ψ is indeed a natural transformation, we have to show that, if $u : E \to F$ is an A-homomorphism, the following diagram is commutative :

$$\begin{array}{ccc} T(F^\circ) & \xrightarrow{T(u^\circ)} & T(E^\circ) \\ \Psi(F^\circ) \downarrow & & \downarrow \Psi(E^\circ) \\ h_M(F^\circ) & \xrightarrow{h_M(u^\circ)} & h_M(E^\circ); \end{array}$$

then let $y \in T(F^\circ)$, and let $a \in E$; we have $\theta_{u(a)} = u \circ \theta_a$, hence

$$(\Psi(E^\circ) \circ T(u^\circ))(y)(a) = T(\theta_a^\circ)(T(u^\circ)(y)) = T((u \circ \theta_a)^\circ)(y) =$$
$$= T(\theta_{ua}^\circ)(y) = \Psi(F^\circ)(y)(ua) = (h_M(u^\circ) \circ \Psi(F^\circ))(y)(a).$$

Proposition 4. *If* $T : (\mathbf{Mod}(A))^\circ \to \mathbf{Mod}(C)$ *is an additive functor, then the following assertions concerning* T *are equivalent*:

i) For any two A*-modules* E, F *and any family* $(u_\nu)_{\nu \in J}$ *of* A*-homomorphisms, where* $u_\nu : E \to F$, $\nu \in J$, *from* $\sum_{\nu \in J} \mathrm{Im}(u_\nu) = F$ *it follows* $\bigcap_{\nu \in J} \mathrm{Ker}\,(T(u_\nu^\circ)) = 0$;

ii) All components of the above constructed natural transformation $\Psi : T \to h_M$ *are monomorphisms of the category* $\mathbf{Mod}(C)$;

iii) There exists an (A, C)*-bimodule* M' *(in which* A *and* C *act on the same side as in* M*) and a natural transformation* $\Psi' : T \to h_{M'}$ *all components of which are monomorphisms.*

Proof. i) $\Rightarrow$ *ii)*: Let E be an A-module; let $x \in T(E^\circ)$ be such that

$$\Psi(E^\circ)(x) = 0; \tag{14}$$

then, considering the family of A-homomorphisms $(\theta_a)_{a \in E}$, we obviously have $\sum_{a \in E} \mathrm{Im}(\theta_a) = E$; from (14) it follows

$$0 = \Psi(E^\circ)(x)(a) = T(\theta_a^\circ)\,(x)$$

for any $a \in E$, hence $x \in \bigcap_{a \in E} \mathrm{Ker}\,(T(\theta_a^\circ))$, and this last submodule is, by *i*), the zero submodule; hence $\Psi(E^\circ)$ is a monomorphism.

ii) $\Rightarrow$ *iii)*: Obvious.

iii) $\Rightarrow$ *i)*: Let E, F, and $(u_\nu)_{\nu \in J}$ be as in the statement of *i*), hence such that $\sum_{\nu \in J} \mathrm{Im}(u_\nu) = F$; let $y \in \bigcap_{\nu \in J} \mathrm{Ker}\,(T(u_\nu^\circ))$; let M' and Ψ' be as in the statement of *iii*); for any $\nu \in J$ we have

$$(\Psi'(F^\circ)(y)) \circ u_\nu = \Psi'(E^\circ)\,(T(u_\nu^\circ)(y)) = 0;$$

but for any $b \in F$ there exist some $v_1, \ldots, v_k \in J$ such that $b = \sum_{i=1}^{k} u_{v_i}(a_i)$ for some $a_i \in E$, $i = 1, \ldots, k$; hence $\Psi'(F^\circ)(y) = 0$, and, since $\Psi'(F^\circ)$ is a monomorphism, $y = 0$. ∎

A functor T like the one above satisfying the equivalent conditions of Proposition 4 will be called a *subrepresentable functor*; by *i*) of Proposition 4, a subrepresentable functor carries the monomorphisms of the category $(\mathbf{Mod}(A))^\circ$ into monomorphisms of the category $\mathbf{Mod}(C)$.

PROPOSITION 5. *Let* $T : (\mathbf{Mod}(A))^\circ \to \mathbf{Mod}(C)$ *be an additive functor which commutes with direct products and which carries monomorphisms into monomorphisms; then* T *is a subrepresentable functor.*

Proof. We will show that for any A-module E, $\Psi(E^\circ)$ is a monomorphism. If $E = A_s$, then from the fact that for any $\alpha \in A$, θ_α is the homothety of ratio α in A_d, and from the way in which the A-module structure of M is defined it follows

$$\Psi(A_s^\circ)(x)(\alpha) = T(\theta_\alpha^\circ)(x) = \alpha x$$

for any $x \in T(A_s^\circ) = M$; but then $\Psi(A_s^\circ)$ is exactly the above constructed canonical isomorphism $\Xi(M) : M \rightleftarrows \mathrm{Hom}_A(A_s, M)$. [1]

Now let $(E_\lambda)_{\lambda \in I}$ be a family of A-modules; if E is the direct sum of this family in the category $\mathbf{Mod}(A)$, with the canonical injections $j_\lambda : E_\lambda \to E$, $\lambda \in I$, then $(E^\circ, (j_\lambda^\circ)_{\lambda \in I})$ is a direct product in the category $(\mathbf{Mod}(A))^\circ$ of the family of objects $(E_\lambda^\circ)_{\lambda \in I}$; since T commutes with direct products, there exists a C-isomorphism

$$\chi : T(E^\circ) \rightleftarrows \prod_{\lambda \in I} T(E_\lambda^\circ)$$

such that, if $p_\lambda : \prod_{\lambda \in I} T(E_\lambda^\circ) \to T(E_\lambda^\circ)$ is the canonical projection of index λ, $\lambda \in I$, $p_\lambda \circ \chi = T(j_\lambda^\circ)$; moreover, for the canonical isomorphism $\varphi : h_M(E^\circ) \rightleftarrows \prod_{\lambda \in I} h_M(E_\lambda^\circ)$ we have $q_\lambda \circ \varphi = h_M(j_\lambda^\circ)$ for any $\lambda \in I$, where q_λ is the canonical projection $\prod_{\lambda \in I} h_M(E_\lambda^\circ) \to h_M(E_\lambda^\circ)$. Then

$$q_\lambda \circ (\prod_{\lambda \in I} \Psi(E_\lambda^\circ)) \circ \chi = \Psi(E^\circ) \circ p_\lambda \circ \chi = \Psi(E_\lambda^\circ) \circ T(j_\lambda^\circ) = h_M(j_\lambda^\circ) \circ \Psi(E^\circ) = q_\lambda \circ \varphi \circ \Psi(E^\circ)$$

for any $\lambda \in I$; this proves, using the uniqueness in the universal property of the direct product, that

$$(\prod_{\lambda \in I} \Psi(E_\lambda^\circ)) \circ \chi = \varphi \circ \Psi(E^\circ). \tag{15}$$

If L is a free A-module, than L may be considered as the direct sum of the family $(E_\lambda)_{\lambda \in I}$ for some set of indices I, where $E_\lambda = A_s$ for any $\lambda \in I$; as, by the preceding considerations, $\Psi(E_\lambda^\circ)$ is a C-isomorphism for any $\lambda \in I$, it follows by (15) that $\Psi(L^\circ)$ is also a C-isomorphism (see Corollary 4 to Proposition 2, § 6).

Now let E be an arbitrary A-module; by Proposition 6, § 10, there exists a free A-module L and an epimorphism $p : L \to E$; as $p^\circ : E^\circ \to L^\circ$ is a mono-

[1] $\Xi(F)$ was there an isomorphism of A-modules; here we are interested in the bijection character o$\Xi(M)$, since we know that $\Xi(M) = \Psi(A_s^\circ)$ is a C-homomorphism.

morphism of the category $(\mathbf{Mod}(A))^\circ$, it follows by hypothesis that $T(p^\circ)$ is a monomorphism; from the commutativity of the diagram

$$\begin{array}{ccc} T(E^\circ) & \xrightarrow{T(p^\circ)} & T(L^\circ) \\ \Psi(E^\circ)\downarrow & & \downarrow\Psi(L^\circ) \\ h_M(E^\circ) & \xrightarrow{h_M(p^\circ)} & h_M(L^\circ) \end{array}$$

it follows that $\Psi(E^\circ)$ is a monomorphism (see § 3). ∎

PROPOSITION 6 **(Eilenberg-Watts Representability Theorem).** *Let* $T : (\mathbf{Mod}(A))^\circ \to \mathbf{Mod}(C)$ *be an additive functor; then there exists an* (A, C)*-bimodule* N *such that* T *is naturally equivalent to* h_N *(i.e.,* T *is a representable functor) if and only if* T *is left exact and commutes with direct products, hence if and only if* T *commutes with projective limits* [1].

Proof. By the Corollary to Proposition 4, § 5, and the preceding considerations, the condition is necessary. For the sufficiency, we shall prove that in this case the natural transformation $\Psi : T \to h_M$ constructed above is a natural equivalence; since T carries monomorphisms into monomorphisms, to complete the proof, we have, according to the proof of Proposition 5, to show that $\Psi(E^\circ)$ is an isomorphism for an arbitrary E. Using the notation of that proof, we have the exact sequence in $\mathbf{Mod}(A)$

$$G = \operatorname{Ker} p \xrightarrow{i} L \xrightarrow{p} E \to 0,$$

where i is the canonical injection; since T and h_M are left exact, in the commutative diagram

$$(16)\qquad \begin{array}{ccccccc} 0 & \longrightarrow & T(E^\circ) & \xrightarrow{T(p^\circ)} & T(L^\circ) & \xrightarrow{T(i^\circ)} & T(G^\circ) \\ & & \Psi(E^\circ)\downarrow & & \downarrow\Psi(L^\circ) & & \downarrow\Psi(G^\circ) \\ 0 & \longrightarrow & h_M(E^\circ) & \xrightarrow{h_M(p^\circ)} & h_M(L^\circ) & \xrightarrow{h_M(i^\circ)} & h_M(G^\circ) \end{array}$$

the rows are exact sequences; moreover, according to the proof of Proposition 5, $\Psi(L^\circ)$ is an isomorphism while $\Psi(E^\circ)$ and $\Psi(G^\circ)$ are monomorphisms. Then, applying The Five Lemma (Proposition 5, *b*), § 5) to diagram (16) enlarged with zeros to the left, we obtain that $\Psi(E^\circ)$ is also an epimorphism. ∎

It is also possible to prove that *an additive functor* $T : \mathbf{Mod}(A) \to \mathbf{Ab}$ *which commutes with projective limits is representable.* [2]

---

[1] See Proposition 4, § 14.
[2] See [100].

§ 17. Duality

Let A be a ring, and let E be a left, respectively a right A-module; since A is an (A, A)-bimodule (Example 1), § 16), it follows that $\mathrm{Hom}_A(E, A_s)$, respectively $\mathrm{Hom}_A(E, A_d)$ is canonically endowed with a right, respectively left A-module structure for which, if $\alpha \in A$ and $x^* \in \mathrm{Hom}_A(E, A_s)$, respectively $x^* \in \mathrm{Hom}_A(E, A_d)$, we have

$$(x^*\alpha)(x) = x^*(x)\cdot \alpha, \quad \text{respectively} \quad (\alpha x^*)(x) = \alpha \cdot x^*(x)$$

for any $x \in E$ (see relation (2), respectively (1), § 16).

Definition 1. If E is a left, respectively right A-module, then the right, respectively left A-module

$$E^* = \mathrm{Hom}_A(E, A_s), \quad \text{respectively} \quad E^* = \mathrm{Hom}_A(E, A_d)$$

is called the *dual module* or the *dual*[1] of E; an element of E^*, i.e. an A-linear mapping of E into A_s, respectively A_d, is called a *linear form* on E. ∎

If E is a (left) A-module and E^* is its dual, then for any $x \in E$ and any $x^* \in E^*$ one also writes $\langle x, x^* \rangle$ instead of $x^*(x)$; hence we have

$$\langle x + y, x^* \rangle = x^*(x + y) = x^*(x) + x^*(y) = \langle x, x^* \rangle + \langle y, x^* \rangle,$$

$$\langle x, x^* + y^* \rangle = (x^* + y^*)(x) = x^*(x) + y^*(x) = \langle x, x^* \rangle + \langle x, y^* \rangle,$$

(1) $$\langle \alpha x, x^* \rangle = x^*(\alpha x) = \alpha x^*(x) = \alpha \langle x, x^* \rangle,$$

$$\langle x, x^*\alpha \rangle = (x^*\alpha)(x) = x^*(x)\cdot\alpha = \langle x, x^* \rangle\, \alpha$$

for any $x, y \in E$, $x^*, y^* \in E^*$, and $\alpha \in A$. The mapping

$$E \times E^* \to A$$

given by

$$(x, x^*) \mapsto \langle x, x^* \rangle, \quad x \in E,\ x^* \in E^*,$$

is called the *canonical bilinear form* on $E \times E^*$.

---

[1] Not to be confounded with the doublet E° of E in the dual category $(\mathbf{Mod}(A))^\circ$, respectively $(\mathbf{Mod}_d(A))^\circ$.

For a right A-module E, if E^* is its dual, then for any $x \in E$ and any $x^* \in E^*$ one writes $\langle x^*, x\rangle$ instead of $x^*(x)$; in this case, with the same meaning for the letters, relations (1) are replaced by

$$\langle x^*, x + y\rangle = \langle x^*, x\rangle + \langle x^*, y\rangle,$$

$$\langle x^* + y^*, x\rangle = \langle x^*, x\rangle + \langle y^*, x\rangle,$$

$$\langle x^*, x\alpha\rangle = \langle x^*, x\rangle\alpha, \tag{1'}$$

$$\langle \alpha x^*, x\rangle = \alpha\langle x^*, x\rangle.$$

Of course, here too we have the *canonical bilinear form*, but this time on $E^* \times E$.

Let E and F be two A-modules, let E^* and F^* be their duals, and let $u : E \to F$ be an A-linear mapping. As shown in § 16, there exists a natural transformation

$$h^u : h^F \to h^E,$$

where h^E and h^F are the functors $\mathbf{Mod}(A) \to \mathbf{Ab}$ defined in § 2; since A_s is an (A, A)-bimodule, it follows, as shown in § 16, that $h^u(A_s) = \text{Hom}(u, 1_A)$ is a homomorphism of right A-modules from $\text{Hom}_A(F, A_s) = F^*$ to $\text{Hom}_A(E, A_s) = E^*$.

The same result is obtained if the left A-modules are replaced by right A-modules; of course, then $\text{Hom}(u, 1_A)$ is a homomorphism of left A-modules.

DEFINITION 2. If E and F are two left, respectively right A-modules, and $u: E \to F$ is an A-linear mapping, the homomorphism of right, respectively left A-modules

$$\text{Hom}(u, 1_A) : F^* \to E^*$$

is called the *transpose* (*linear mapping*) of u, and it is denoted by ${}^t u$. ▮

It follows from the definition that if $u : E \to F$ is a homomorphism of left A-modules, then for $x \in E$, $y^* \in F^*$, we have

$$\langle x, {}^t u(y^*)\rangle = ({}^t u(y^*))(x) = (\text{Hom}(u, 1_A)(y^*))(x) = \langle u(x), y^*\rangle; \tag{2}$$

obviously, the validity of relation (2) (considered as between its first and last term) for any $x \in E$ and $y^* \in F^*$ characterizes ${}^t u$.

For a homomorphism $u : E \to F$ of right A-modules, relation (2) is to be replaced by

$$\langle {}^t u(y^*), x\rangle = \langle y^*, u(x)\rangle \tag{2'}$$

for any $x \in E$ and $y^* \in F^*$.

From relations (6), (7), (8), § 2, it follows that, given the diagram of A-modules

$$E \xrightarrow{u} F \xrightarrow{v} G,$$

then

$$^t(v \circ u) = {}^t u \circ {}^t v, \quad {}^t(1_E) = 1_{E^*},$$

and that if u_1 is another homomorphism $E \to F$, then

$$^t(u + u_1) = {}^t u + {}^t u_1.$$

This shows that the mappings

$$E \mapsto E^*, \quad u \mapsto {}^t u$$

define an additive functor

$$(\mathbf{Mod}(A))^\circ \to \mathbf{Mod}_d(A),$$

called the *functor of passing to the dual* (in the category $\mathbf{Mod}(A)$).

Of course, there also exists a functor of passing to the dual in the category $\mathbf{Mod}_d(A)$, which is a functor

$$(\mathbf{Mod}_d(A))^\circ \to \mathbf{Mod}(A).$$

This result may also be obtained by considering the functor h_{A_s}: $(\mathbf{Mod}\,(A))^\circ \to$ $\to \mathbf{Ab}$; namely, by the preceding arguments, this functor takes its values in the category $\mathbf{Mod}_d(A)$, and hence it is the composition of the forgetful functor $\mathbf{Mod}_d(A) \to \mathbf{Ab}$ with a uniquely determined additive functor $(\mathbf{Mod}(A))^\circ \to \mathbf{Mod}_d(A)$; this last one is exactly the functor of passing to the dual.

By the preceding considerations it follows that if $u : E \to F$ is an isomorphism of A-modules, then $^t u : F^* \to E^*$ is an ismorphism of right A-modules, and that

$$(^t u)^{-1} = {}^t(u^{-1});$$

this isomorphism from E^* to F^* is called the *contragredient isomorphism* of u, and is denoted $\check{u}$; for any $x \in E$ and $x^* \in E^*$ we have

$$\langle u(x), \check{u}(x^*)\rangle = \langle u(x), {}^t(u^{-1})(x^*)\rangle = \langle u^{-1}(u(x)), x^*\rangle = \langle x, x^*\rangle,$$

and if $u' : E^* \to F^*$ has the property that for any $x \in E$ and any $x^* \in E^*$ we have

(3) $$\langle u(x), u'(x^*)\rangle = \langle x, x^*\rangle,$$

then $u' = \check{u}$: for, any element $y \in F$ may be written as $y = u(x)$, hence by (3) we have

$$(u'(x^*))\,(u(x)) = \langle x, x^*\rangle = (\check{u}(x^*))\,(u(x));$$

it follows $u'(x^*) = \check{u}(x^*)$ for any $x^* \in E^*$, hence $u' = \check{u}$.

If $v : F \to G$ is another isomorphism of A-modules, then

$$\check{v} \circ \check{u} = {}^t(v^{-1}) \circ {}^t(u^{-1}) = {}^t(u^{-1} \circ v^{-1}) = {}^t((v \circ u)^{-1}) = \widecheck{v \circ u}.$$

PROPOSITION 1. *The functor of passing to the dual commutes with projective limits.*

Proof. Let $T = (E_\lambda^\circ, u_f^\circ)_{\lambda, f}$ be a projective system with values in the category $(\mathbf{Mod}(A))^\circ$ indexed by the small category I, and let $(E^\circ, (j_\lambda^\circ)_{\lambda \in I})$ be a projective limit of it in the category $(\mathbf{Mod}(A))^\circ$; then, by the Remark following Definition 3, § 13, $T^\circ = (E_\lambda, u_f)_{\lambda, f}$ is an inductive system of A-modules indexed by the same category I, and $(E, (j_\lambda)_{\lambda \in I})$ is an inductive limit of it in the category $\mathbf{Mod}(A)$; we have to show that $(E^*, ({}^t(j_\lambda))_{\lambda \in I})$ is a projective limit in the category $\mathbf{Mod}_d(A)$ of the projective system of right A-modules $(E_\lambda^*, {}^t(u_f))_{\lambda, f}$.

By the considerations following Definition 1, § 14, there exists an isomorphism of $\mathbf{Z}$-modules

$$\varphi : \operatorname{Hom}_A(E, A_s) = E^* \xrightarrow{\sim} \varprojlim_I \operatorname{Hom}_A(E_\lambda, A_s) = \varprojlim_I E_\lambda^*$$

for which, if $x^* \in E^*$, we have

$$\varphi(x^*) = (x^* \circ j_\lambda)_{\lambda \in I}.$$

Obviously, the Proposition is proved if we show that φ is an isomorphism of right A-modules and that

$$p_\lambda \circ \varphi = {}^t(j_\lambda)$$

for any $\lambda \in I$, where p_λ is the canonical mapping $\varprojlim E_\lambda^* \to E_\lambda^*$, $\lambda \in I$; the second assertion is obtained as follows: if $x^* \in E^*$, then

$$(p_\lambda \circ \varphi)(x^*) = x^* \circ j_\lambda = \operatorname{Hom}(j_\lambda, 1_A)(x^*) = {}^t(j_\lambda)(x^*);$$

concerning the first assertion, if $\alpha \in A$ and $x^* \in E^*$, then for any $\lambda \in I$ and any $x_\lambda \in E_\lambda$ we have

$$((x^*\alpha) \circ j_\lambda)(x_\lambda) = x^*(j_\lambda(x_\lambda)) \cdot \alpha = ((x^* \circ j_\lambda) \cdot \alpha)(x_\lambda),$$

hence $(x^*\alpha) \circ j_\lambda = (x^* \circ j_\lambda)\alpha$; consequently, with the right A-module structure of $\varprojlim_I E_\lambda^*$ given in § 12,

$$\varphi(x^*\alpha) = ((x^*\alpha) \circ j_\lambda)_\lambda = ((x^* \circ j_\lambda)\alpha)_\lambda = ((x^* \circ j_\lambda)_\lambda)\alpha = \varphi(x^*)\alpha,$$

and the result is proved. ▮

COROLLARY 1. *Let the exact sequence of A-modules*

$$E \xrightarrow{u} F \xrightarrow{v} G \to 0 \tag{4}$$

be given; then the diagram of right A-modules

$$0 \to G^* \xrightarrow{{}^t v} F^* \xrightarrow{{}^t u} E^* \tag{5}$$

is an exact sequence.

Proof. By Proposition 1 above it follows, using Proposition 4, § 14, that the functor of passing to the dual is left exact; consequently, the exactness of the sequence (4) implies the exactness of the sequence (5). ∎

COROLLARY 2. *Let $(E_\lambda)_{\lambda \in I}$ be a family of A-modules; then there exists a canonical isomorphism of right A-modules*

$$(\bigoplus_{\lambda \in I} E_\lambda)^* \simeq \prod_{\lambda \in I} E_\lambda^*;$$

identifying by means of this isomorphism the dual of a direct sum with the product of the duals, the canonical bilinear form on $(\bigoplus_\lambda E_\lambda) \times (\prod_\lambda E_\lambda^*)$ acts as follows:

$$\langle (x_\lambda)_{\lambda \in I}, (x_\lambda^*)_{\lambda \in I} \rangle = \sum_{\lambda \in I} \langle x_\lambda, x_\lambda^* \rangle,$$

where $(x_\lambda)_\lambda \in \bigoplus_\lambda E_\lambda$ and $(x_\lambda^*)_\lambda \in \prod_\lambda E_\lambda^*$.[1]

Proof. We have only to take in Proposition 1 the particular case of a discrete category of indices, denoted also I, in which the objects are the elements of the given set I; then, in the notation of the statement and of the proof of Proposition 1, the inductive limit E is $\bigoplus_{\lambda \in I} E_\lambda$, and the projective limit $\varprojlim_I E_\lambda^*$ is $\prod_{\lambda \in I} E_\lambda^*$ (see the particular cases corresponding to this category of indices in §§ 12 and 13); the stated canonical isomorphism is the isomorphism φ in the proof of Proposition 1; if $(x_\lambda)_\lambda$ and $(x_\lambda^*)_\lambda$ are as in the statement, then, denoting $\varphi^{-1}((x_\lambda^*)_\lambda) = x^*$, we have $(x_\lambda^*)_\lambda = \varphi(x^*) = (x^* \circ j_\lambda)_\lambda$; hence $x_\lambda^* = x^* \circ j_\lambda$ for any $\lambda \in I$; taking Lemma 1, § 7, into account, we get, with the identification by means of φ,

$$\langle (x_\lambda)_\lambda, (x_\lambda^*)_\lambda \rangle = (\varphi^{-1}((x_\lambda^*)_\lambda))((x_\lambda)_\lambda) = x^*((x_\lambda)_\lambda) = x^*(\sum_{\lambda \in I} j_\lambda(x_\lambda)) =$$

$$= \sum_{\lambda \in I} x^*(j_\lambda(x_\lambda)) = \sum_{\lambda \in I} x_\lambda^*(x_\lambda). \ ∎$$

---

[1] Notice that the sum in the right side makes sense, since $(x_\lambda)_\lambda$ being an fs-family, it follows that $(\langle x_\lambda, x_\lambda^* \rangle)_\lambda$ is also an fs-family.

Concerning the dual A_s^* of the A-modules A_s, it is known (see Proposition 1, § 16) that there exists a canonical isomorphism of *left* A-modules

$$\psi = (\Xi(A_s))^{-1} : \mathrm{Hom}_A(A_s, A_s) \rightleftarrows A_s$$

given by

$$\psi(\alpha^*) = \alpha^*(1),$$

where $\alpha^* \in \mathrm{Hom}_A(A_s, A_s)$; here the left A-module structure on $\mathrm{Hom}_A(A_s, A_s)$ is obtained from the (A, A)-bimodule structure of the first A_s; applying the forgetful functor $\mathbf{Mod}(A) \to \mathbf{Ab}$, we get an isomorphism of $\mathbf{Z}$-modules

$$\psi : \mathrm{Hom}_A(A_s, A_s) \rightleftarrows A;$$

now, if we endow the abelian group $\mathrm{Hom}_A(A_s, A_s)$ with the right A-module structure making of it the dual A_s^* of A_s (that is, with the A-module structure obtained from the second A_s) and the abelian group A, the range of ψ, with the right A-module structure making of it the module A_d, then ψ is an isomorphism of *right* A-modules: for let $\alpha \in A$ and $\alpha^* \in \mathrm{Hom}_A(A_s, A_s)$; we have

$$\psi(\alpha^*\alpha) = (\alpha^*\alpha)(1) = \alpha^*(1)\cdot\alpha = \psi(\alpha^*)\cdot\alpha.$$

Hence,

PROPOSITION 2. *If A is a ring, then there exists a canonical isomorphism of right A-modules*

$$\psi : A_s^* \rightleftarrows A_d;$$

if by means of this isomorphism we identify the dual A_s^* with A_d, then the canonical bilinear form on $A_s \times A_d = A_s \times A_s^*$ acts as follows:

$$\langle \alpha, \alpha' \rangle = \alpha\alpha',$$

where $\alpha, \alpha' \in A$. Likewise, there exists a canonical isomorphism of left A-modules

$$\psi : A_d^* \rightleftarrows A_s,$$

and with the identification by means of this isomorphism, the canonical bilinear form on $A_s \times A_d = A_d^* \times A_d$ acts as follows:

$$\langle \alpha', \alpha \rangle = \alpha'\alpha,$$

with $\alpha, \alpha' \in A$. [1]

---

[1] From here it follows that the two bilinear forms on $A_s \times A_d$ coincide no matter whether $A_s \times A_d$ is obtained as $A_s \times A_s^*$ or as $A_d^* \times A_d$.

The *proof* requires only the computation of the values of the canonical bilinear form: if $\alpha, \alpha' \in A$, then, by the definition of $\psi^{-1} = \Xi(A_s)$ in the proof of Proposition 1, § 16,

$$\langle \alpha, \alpha' \rangle = \psi^{-1}(\alpha')(\alpha) = \alpha\alpha'. \blacksquare$$

From the Corollary 2 to Proposition 1 and from Proposition 2 it follows that if A is a ring and if I is a set of indices, then *the dual of the left A-module* $A_s^{(I)}$ *is canonically isomorphic to the module* A_d^I: indeed, $A_s^{(I)} = \bigoplus_{\lambda \in I} E_\lambda$, where $E_\lambda = A_s$ for any $\lambda \in I$; hence $(A_s^{(I)})^* \approx \prod_{\lambda \in I} E_\lambda^*$, and as $A_s^* \approx A_d$, it follows that $\prod_{\lambda \in I} E_\lambda^* \approx A_d^I$; if by means of this isomorphism we identify $(A_s^{(I)})^*$ with A_d^I, then the canonical bilinear form on $A_s^{(I)} \times A_d^I$ acts as follows: if $(\alpha_\lambda)_{\lambda \in I} \in A_s^{(I)}$, $(\alpha'_\lambda)_{\lambda \in I} \in A_d^I$, then

$$\langle (\alpha_\lambda)_\lambda, (\alpha'_\lambda)_\lambda \rangle = \sum_{\lambda \in I} \alpha_\lambda \alpha'_\lambda, \tag{6}$$

which is to be seen from the way in which the canonical bilinear form acts in the cases described in Corollary 2 to Proposition 1 and in Proposition 2.

Now let E be a free A-module having a basis $(a_t)_{t \in I}$; by Proposition 2, *c*), § 10, there exists an A-isomorphism $u : A_s^{(I)} \xrightarrow{\sim} E$, namely the linear mapping determined by the family $(a_t)_{t \in I}$; then, identifying the dual of $A_s^{(I)}$ with A_d^I, there exists the contragredient isomorphism of u, $\check{u} : A_d^I \xrightarrow{\sim} E^*$; let, for any $t \in I$, f_t be the element of A_d^I whose component of index t is 1 and whose component of index t', $t' \neq t$, is 0; let $a_t^* = \check{u}(f_t)$, $t \in I$; if we denote by $(e_t)_{t \in I}$ the canonical basis of $A_s^{(I)}$, we have by (6)

$$\langle a_t, a_{t'}^* \rangle = \langle u(e_t), \check{u}(f_{t'}) \rangle = \langle e_t, f_{t'} \rangle = \begin{cases} 0 \text{ if } t \neq t', \\ 1 \text{ if } t = t' \end{cases}$$

for any $t, t' \in I$.

If I is a set, and M is an abelian group, or a module over some ring A, or a ring, etc., then the mapping $\delta : I \times I \to M$, denoted $(t, t') \mapsto \delta_{tt'}$, $t, t' \in I$, defined by

$$\delta_{tt'} = \begin{cases} 0 \text{ if } t \neq t', \\ 1 \text{ if } t = t' \end{cases}$$

is called the *Kronecker symbol* defined on $I \times I$ with values in M.

Hence, returning to the situation when the free A-module E having the basis $(a_t)_{t \in I}$ is given, we have

$$\langle a_t, a_{t'}^* \rangle = \delta_{tt'}, \tag{7}$$

$t, t' \in I$.

Notice that the linear form $a_{t'}^*$ could have been defined directly for each $t' \in I$ by (7): indeed, $(a_t)_{t \in I}$ being a basis of E, we have only to apply Proposition 4, § 10, for $F = A_s$, $b_t = 0$ if $t \neq t'$, and $b_{t'} = 1$.

Let now $x = \sum_{t \in I} \xi_t a_t$ be an arbitrary element of E, where $(\xi_t)_{t \in I}$ is an fs-family of elements of A; since, in fact, the sum giving x is a finite sum, we have, applying the canonical bilinear form to the pair $(x, a_{t'}^*)$ and using relations (1),

$$\langle x, a_{t'}^* \rangle = \langle \sum_{t \in I} \xi_t a_t, a_{t'}^* \rangle = \sum_{t \in I} \xi_t \langle a_t, a_{t'}^* \rangle = \xi_{t'};$$

consequently, $a_{t'}^*$ maps an element x of E into its coordinate of index t' relative to the basis $(a_t)_{t \in I}$; therefore, the linear form $a_{t'}^*$ is called the *coordinate linear form* of index (or order) t', $t' \in I$, with respect to (or relative to) the basis $(a_t)_{t \in I}$.

If $(\eta_{t'})_{t' \in I}$ is an fs-family of elements of A for which we have in E^*

$$\sum_{t' \in I} a_{t'}^* \eta_{t'} = 0,$$

then, again by relations (1), we have for $t \in I$,

$$0 = \langle a_t, \sum_{t' \in I} a_{t'}^* \eta_{t'} \rangle = \sum_{t' \in I} \langle a_t, a_{t'}^* \rangle \eta_{t'} = \eta_t;$$

it follows that *the family $(a_{t'}^*)_{t' \in I}$ is a linearly independent family* of elements from the right A-module E^*.

Assume, moreover, that I is a finite set, hence that E is a free A-module having a finite basis $(a_t)_{t \in I}$; then $A_d^{(I)} = A_d^I$, and $(f_{t'})_{t' \in I}$ is the canonical basis of $A_d^{(I)}$; consequently, $(a_{t'}^*)_{t' \in I} = (\check{u}(f_{t'}))_{t' \in I}$ is a basis of E^*; this may also be seen directly, i.e., without using the contragredient isomorphism and the family $(f_{t'})_{t' \in I}$, in the following manner: let $x^* \in E^*$; denoting

$$\alpha_{t'} = \langle a_{t'}, x^* \rangle$$

for any $t' \in I$, the linear combination $\sum_{t' \in I} a_{t'}^* \alpha_{t'}$ makes sense, I being a finite set, and by (1) we have

$$\langle a_t, \sum_{t' \in I} a_{t'}^* \alpha_{t'} \rangle = \sum_{t' \in I} \langle a_t, a_{t'}^* \rangle \alpha_{t'} = \alpha_t = \langle a_t, x^* \rangle$$

for any $t \in I$; since x^* and $\sum_{t' \in I} a_{t'}^* \alpha_{t'}$ are linear forms on E, and since they coincide on the basis $(a_t)_{t \in I}$ of E, it follows by Proposition 4, § 10, that

$$x^* = \sum_{t' \in I} a_{t'}^* \alpha_{t'};$$

consequently, $(a_{t'}^*)_{t' \in I}$ is a system of generators of E, and since by the preceding arguments it is also a linearly independent family, it follows that it is a basis of E^*. Denoting $\text{Card}(I) = n$, we have

PROPOSITION 3. *If E is a free A-module having a finite basis consisting of n elements, then its dual E^* is a free right A-module having a basis consisting also of n elements.* ▮

DEFINITION 3. Given a free A-module E having a finite basis $(a_t)_{t \in I}$, the basis $(a^*_{t'})_{t' \in I}$ of E^*, where $a_{t'}$, $t' \in I$, is defined by relations (7), is called the *dual basis of* (or the basis *dual to*) the basis $(a_t)_{t \in I}$. ▮

If, in the same notation, $x = \sum_{t \in I} \xi_t a_t$ is an element of E, and if $x^* = \sum_{t' \in I} a_{t'} \eta_{t'}$ is an element of E^*, then, by relations (1), we have

$$\langle x, x^* \rangle = \left\langle \sum_{t \in I} \xi_t a_t, \sum_{t' \in I} a^*_{t'} \eta_{t'} \right\rangle = \sum_{t \in I} \sum_{t' \in I} \xi_t \langle a_t, a^*_{t'} \rangle \eta_{t'} = \sum_{t \in I} \xi_t \eta_t.$$

It is clear that all the preceding considerations may be carried over to right A-modules; in particular, the definition of the basis *dual* to a finite basis of a right A-module is given in the same way.

DEFINITION 4. Given an A-module E, an element $x \in E$, and an element $x^* \in E^*$, where E^* is the dual of E, the elements x and x^* are said to be *orthogonal* (to each other) or, equivalently, x is said to be *orthogonal* to x^* if $\langle x, x^* \rangle = 0$, i.e., if $x^*(x) = 0$.

Given two subsets $M \subseteq E$ and $M' \subseteq E^*$, M and M' are said to be *orthogonal* (to each other) if for any $x \in M$ and any $x^* \in M'$, the elements x and x^* are orthogonal; in particular, the element $x \in E$ is said to be *orthogonal* to the subset $M' \subseteq E^*$ provided it is orthogonal to any element $x^* \in M'$; the same terminology is used for a given element $x^* \in E^*$ and a subset M of E.

Given a subset M of E, the *total orthogonal subset* (or, simply, the *orthogonal*) of M in E^* is called the subset of all $x^* \in E^*$ which are orthogonal to M; likewise, for any subset $M' \subseteq E^*$, the *total orthogonal subset* (or, simply, the *orthogonal*) of M' in E is the subset consisting of all $x \in E$ orthogonal to M'. ▮

It follows directly from the definition that *for any subset M of E, the orthogonal of M in E^* is a submodule of E*: for if x^* and y^* belong to the orthogonal of M in E^*, and if $\alpha \in A$, then by (1) we have for any $x \in M$,

$$\langle x, x^* + y^* \rangle = \langle x, x^* \rangle + \langle x, y^* \rangle = 0 + 0 = 0$$

and

$$\langle x, x^* \alpha \rangle = \langle x, x^* \rangle \alpha = 0 \cdot \alpha = 0,$$

hence $x^* + y^*$ and $x^* \alpha$ belong to the orthogonal of M in E^*; since this orthogonal contains also the zero linear form, we get the required result; likewise, *the orthogonal in E of a subset M' of E^* is a submodule of E.* Therefore, for a subset M of E (respectively a subset M' of E^*), the total orthogonal subset of M in E^* (respectively of M' in E) is called also the *total orthogonal submodule* or also the *orthogonal submodule* of M in E^* (respectively of M' in E).

It also follows from the definition that *the orthogonal of E in E^* is the zero submodule of E* and that *the orthogonal in E^* of the zero submodule of E is E^*.* Next, *if M, N are subsets of E such that $M \subseteq N$, then the orthogonal of N in E is contained in the orthogonal of M in E.*

Now let $(M_\lambda)_{\lambda \in I}$ be a family of subsets of E; then *the orthogonal* of $\bigcup_{\lambda \in I} M_\lambda$ *in* E^* *is the intersection of the orthogonals in* E^* *of the subsets* M_λ, $\lambda \in I$, *and coincides with the orthogonal in* E^* *of the submodule generated by* $\bigcup_{\lambda \in I} M_\lambda$: for, as $\bigcup_{\lambda \in I} M_\lambda \supseteq M_\lambda$ for any $\lambda \in I$, it follows that the orthogonal of the union is contained in the intersection of the orthogonals of M_λ, $\lambda \in I$; conversely, if x^* belongs to the intersection of the orthogonals in E^* of the subsets M_λ, $\lambda \in I$, and if $x \in \bigcup_{\lambda \in I} M_\lambda$, then x belongs to at least one M_λ, and hence $\langle x, x^* \rangle = 0$; hence x^* is in the orthogonal of $\bigcup_{\lambda \in I} M_\lambda$; further, the orthogonal in E^* of $\bigcup_{\lambda \in I} M_\lambda$ contains the orthogonal of the submodule generated by this union, since any subset is contained in the submodule generated by it (see § 8); conversely, if x^* is an element of the orthogonal of $\bigcup_{\lambda \in I} M_\lambda$, then any element of the submodule generated by $\bigcup_{\lambda \in I} M_\lambda$ is orthogonal to x^* by Proposition 2, § 8; hence x^* belongs to the orthogonal of the submodule generated by $\bigcup_{\lambda \in I} M_\lambda$.

If M *is a subset of* E, *then, denoting by* M' *the total orthogonal subset of* M *in* E^* *and by* M_1 *the total orthogonal subset of* M' *in* E, *we have* $M \subseteq M_1$: since if $x \in M$, then for any $x^* \in M'$ we have $\langle x, x^* \rangle = 0$.

The analogous assertions except the one concerning the orthogonal of E^* in E are valid when considering subsets of E^* and their orthogonals in E.

Also, Definition 4 and the results following it are valid for right A-modules.

Let E and F be A-modules, and let $u : E \to F$ be an A-homomorphism; if M is a submodule of E, then to say that an element $y^* \in F^*$ is in the orthogonal of $u(M)$ in F^* is equivalent to saying that for any $x \in M$ we have $\langle u(x), y^* \rangle = 0$, or, by relation (2), that $\langle x, {}^t u(y^*) \rangle = 0$ for any $x \in M$, hence that ${}^t u(y^*)$ is in the orthogonal of M in E^*; hence, denoting this orthogonal by M', *the orthogonal of* $u(M)$ *in* F^* *is* $({}^t u)^{-1}(M')$.

PROPOSITION 4. *Let* E *and* F *be two* A*-modules, and let* $u : E \to F$ *be an* A*-homomorphism; then the orthogonal of* $\mathrm{Im}(u)$ *in* F *is* $\mathrm{Ker}\,({}^t u)$.

Proof. Taking in the preceding considerations $M = E$, the orthogonal of M in E^* is 0; hence, by the preceding result, the orthogonal of $u(\mathrm{E}) = \mathrm{Im}(u)$ in F^* is $({}^t u)^{-1}(0) = \mathrm{Ker}\,({}^t u)$. ▮

From the preceding considerations it follows that for any homomorphism of A-modules $u : E \to F$, *the image of* u *is contained in the orthogonal of* $\mathrm{Ker}({}^t u)$ *in* F: for, as seen before, $\mathrm{Im}(u)$ is contained in the orthogonal of its orthogonal in F^*, i.e., in the orthogonal of $\mathrm{Ker}({}^t u)$ in F.

Again, given a homomorphism of A-modules $u : E \to F$, if N' is a submodule of F^*, then *the orthogonal of* ${}^t u(N')$ *in* E *is* $u^{-1}(N)$, *where* N *is the total orthogonal subset of* N' *in* F: for to say that an element $x \in E$ is in the orthogonal of ${}^t u(N')$ is equivalent to saying that for any $y^* \in N'$ we have $\langle x, {}^t u(y^*) \rangle = 0$, or, by (2), that $\langle u(x), y^* \rangle = 0$ for any $y^* \in N'$, hence that $u(x)$ is an element of the orthogonal of N' in F.

Now let E be an A-module, and let M be a submodule of it; let $i : M \to E$ be the canonical injection, and let $p : E \to E/M$ be the canonical surjection; from the exactness of the sequence

$$M \xrightarrow{i} E \xrightarrow{p} E/M \to 0$$

it follows, by Corollary 1 to Proposition 1, that we have the exact sequence of right A-modules

$$0 \to (E/M)^* \xrightarrow{{}^t p} E^* \xrightarrow{{}^t i} M^*;$$

by Proposition 4, the orthogonal of $\mathrm{Im}(i) = M$ in E^* is $\mathrm{Ker}\,({}^t i) = {}^t p((E/M)^*)$; since ${}^t p$ is injective, it induces, by Corollary 2 to Proposition 1, § 4, an isomorphism of right A-modules

$$(E/M)^* \rightleftarrows {}^t p((E/M)^*).$$

Hence we have the following

COROLLARY. *Let E be an A-module, and let M be a submodule of it; then the orthogonal of M in E^* is canonically isomorphic to $(E/M)^*$.* ∎

Now let E be an A-module, and let M, N be two submodules of it such that E is the internal direct sum of M and N; then, by Corollary 1 to Proposition 2, § 9, if $i : M \to E$ and $j : N \to E$ are the canonical injections, then there exist two uniquely determined linear mappings $p : E \to M$, $q : E \to N$ such that

$$p \circ i = 1_M,\ p \circ j = 0,\ q \circ j = 1_N,\ q \circ i = 0,\ i \circ p + j \circ q = 1_E;$$

consequently, the diagram

$$0 \to M \xrightarrow{i} E \xrightarrow{q} N \to 0$$

put instead of diagram (13), § 9, satisfies the equivalent conditions of Proposition 7, § 9; making use of Remark 3) following Proposition 7, § 9, and of the fact (also proved in § 9) that an additive functor carries a split exact sequence into a split exact sequence, we obtain that the diagram

$$0 \to N^* \xrightarrow{{}^t q} E^* \xrightarrow{{}^t i} M^* \to 0$$

is a split exact sequence of the category $\mathbf{Mod}_d(A)$; consequently, *E is the internal direct sum of its submodules ${}^t p(M^*)$ and ${}^t q(N^*)$*; moreover, we have seen that *the orthogonal of M in E^** is $\mathrm{Ker}({}^t i)$; by the preceding considerations, this *is* ${}^t q(N^*)$, which is isomorphic to N^*; this result is in accord with the one given by the above Corollary to Proposition 4 if we take into account that $N \approx E/M$; likewise, *the orthogonal of N in E^* is ${}^t p(M^*)$*. Sometimes, by means of ${}^t p$ the module M^* is identified with the orthogonal of N in E^*, that is, with ${}^t p(M^*)$; this identification amounts to identifying a linear form x^* on M with the linear form $x^*\circ p$ on E, that is, with the (unique) linear form on E which coincides with x^* at any element of M and vanishes at any element of N.

Let E be a (left) A-module; considering its dual E^*, which is a right A-module, we may take the dual of this last one, namely $(E^*)^*$, which is a left A-module; this is denoted E^{**}, and is called the *bidual* (*module*) of E; obviously, if F is another A-module, and if $u : E \to F$ is a homomorphism of A-modules, then the transpose of the transpose linear mapping of u, that is, the transpose of ${}^tu : F^* \to E^*$, is an A-linear mapping from E^{**} to F^{**}, and is denoted ${}^{tt}u$; it is obvious that the mappings

$$E \mapsto E^{**}, \quad u \mapsto {}^{tt}u$$

define an additive functor from the category $\mathbf{Mod}(A)$ to the category $\mathbf{Mod}(A)$, called the *functor of passing to the bidual* in the category $\mathbf{Mod}(A)$; if we denote by T the functor of passing to the dual in the category $\mathbf{Mod}(A)$, and by T° its dual, in the sense of § 2,[1] and if we denote by U the functor of passing to the dual in the category $\mathbf{Mod}_d(A)$, then the functor of passing to the bidual in the category $\mathbf{Mod}(A)$ is the composition of the functors

$$\mathbf{Mod}(A) \xrightarrow{T^\circ} (\mathbf{Mod}_d(A))^\circ \xrightarrow{U} \mathbf{Mod}(A).$$

Of course, there exists also a functor of passing to the bidual in the category $\mathbf{Mod}_d(A)$, defined in an obvious manner.

Returning to left A-modules, let E be an A-module; let $x \in E$; assigning to any x^* of the dual E^* of E the element $\langle x, x^*\rangle = x^*(x)$, we get a homomorphism of right A-modules $\tilde{x} : E^* \to A_d$: for, relations (1) imply that

$$\tilde{x}(x^* + y^*) = \langle x, x^* + y^*\rangle = \langle x, x^*\rangle + \langle x, y^*\rangle = \tilde{x}(x^*) + \tilde{x}(y^*)$$

for any $x^*, y^* \in E$, and likewise,

$$\tilde{x}(x^*\alpha) = \tilde{x}(x^*)\cdot\alpha$$

for any $x^* \in E^*$ and $\alpha \in A$; hence $\tilde{x}$ is a linear form on E^*, hence an element of E^{**}. Thus, the assignment

$$x \mapsto \tilde{x}, \quad x \in E,$$

defines a mapping $c_E : E \to E^{**}$; hence we have for any $x \in E$ and $x^* \in E^*$,

$$\langle c_E(x), x^*\rangle = \langle x, x^*\rangle,$$

---

[1] The functor of passing to the dual defined in this section should not be confounded with the dual functor of a functor; here, the term "dual" has two distinct meanings.

where in the left side we have in view the canonical bilinear form on $E^{**} \times E^*$ (E^* being a right A-module), while in the right side, the canonical bilinear form on $E \times E^*$ (E being a left A-module). It is easily verified using formulae (1) and (1′) that c_E is a homomorphism of left A-modules: for instance, if $x \in E$, $x^* \in E^*$, $\alpha \in A$, we have

$$(c_E(\alpha x))(x^*) = \langle c_E(\alpha x), x^* \rangle = \langle \alpha x, x^* \rangle = \alpha \langle x, x^* \rangle =$$
$$= \alpha \langle c_E x, x^* \rangle = \langle \alpha \cdot c_E(x), x^* \rangle = (\alpha \cdot c_E(x))(x^*),$$

and since x^* is arbitrary in E^*, it follows that $c_E(\alpha x) = \alpha \cdot c_E(x)$. The mapping c_E is called the *canonical linear mapping* $E \to E^{**}$.

If, moreover, the A-module F and the A-homomorphism $u : E \to F$ are given, the following diagram is commutative:

$$\text{(8)} \qquad \begin{array}{ccc} E & \xrightarrow{\;u\;} & F \\ {\scriptstyle c_E}\downarrow & & \downarrow{\scriptstyle c_F} \\ E^{**} & \xrightarrow{\;{}^{tt}u\;} & F^{**}; \end{array}$$

since if $x \in E$ and $y^* \in F^*$, then, using relations (2) and (2′), we have

$$(c_F(u(x)))(y^*) = \langle c_F(u(x)), y^* \rangle = \langle u(x), y^* \rangle = \langle x, {}^t u(y^*) \rangle =$$
$$= \langle c_E(x), {}^t u(y^*) \rangle = \langle {}^{tt}u(c_E(x)), y^* \rangle = ({}^{tt}u(c_E(x)))(y^*).$$

The commutativity of diagram (8) for any $u \in \mathrm{Hom}_A(E, F)$ means that the assignment $E \mapsto c_E$, $E \in \mathrm{Ob}\,(\mathbf{Mod}(A))$, defines a natural transformation from the identity functor of the category $\mathbf{Mod}(A)$ to the functor of passing to the bidual in the category $\mathbf{Mod}(A)$.

For a right A-module E there also exists a canonical linear mapping $E \to E^{**}$ similarly defined and having the same properties as in the case of left A-modules.

PROPOSITION 5. *Let E be a free A-module, respectively a free A-module having a finite basis; then the canonical linear mapping $c_E : E \to E^{**}$ is injective, respectively bijective.*

Proof. We begin with the first assertion: if E is a free A-module and $(a_\iota)_{\iota \in I}$ is a basis of it, let $x \in E$ be an element such that $c_E(x) = 0$; in the dual E^* of E consider the family $(a_\iota^*)_{\iota \in I}$ of coordinate linear forms relative to the basis $(a_\iota)_{\iota \in I}$; if $x = \sum_{\iota \in I} \xi_\iota a_\iota$ is the expression of x as a linear combination of the elements of the basis $(a_\iota)_{\iota \in I}$, then we saw previously that $\langle x, a_\iota^* \rangle = \xi_\iota$, $\iota \in I$; on the other hand, by the definition of c_E we have

$$\langle x, a_\iota^* \rangle = \langle c_E(x), a_\iota^* \rangle = \langle 0, a_\iota^* \rangle = 0,$$

$\iota \in I$; hence, $\xi_\iota = 0$ for any $\iota \in I$, and $x = 0$.

For the second assertion, assuming, moreover, in the same notation, that I is a finite set, then $(a_t^*)_{t\in I}$ is the dual basis of the basis $(a_t)_{t\in I}$ (see Definition 3); denoting by $(a_t^{**})_{t\in I}$ the dual basis of $(a_t^*)_{t\in I}$, we have for $t, t' \in I$,

$$\langle a_t^{**}, a_{t'}^*\rangle = \delta_{tt'};$$

but we also have

$$\langle c_E(a_t), a_{t'}^*\rangle = \langle a_t, a_{t'}^*\rangle = \delta_{tt'}$$

for any $t, t' \in I$; when t is fixed while t' varies, the linear mappings a_t^{**} and $c_E(a_t)$ between E^* and A_d coincide on the elements of the basis $(a_{t'}^*)_{t'\in I}$ of E^*; by Proposition 4, § 10, it follows that $c_E(a_t) = a_t^{**}$ for any $t \in I$; since c_E carries a basis $(a_t)_{t\in I}$ of E into a basis $(a_t^{**})_{t\in I}$ of E^{**}, it follows by the same Proposition that c_E is an isomorphism. ▮

REMARK. Suppose we identify the free A-module E having a finite basis with its bidual E^{**} by means of the isomorphism c_E, that is, identifying $x \in E$ with $c_E(x)$; then a finite basis $(a_t)_{t\in I}$ is identified with the basis $(c_E(a_t))_{t\in I}$ of E^{**}, which, as shown in the proof of Proposition 5, is the dual basis of $(a_t^*)_{t\in I}$, the dual basis of $(a_t)_{t\in I}$; in this case $(a_t)_{t\in I}$ and $(a_t^*)_{t\in I}$ are said to be *bases dual to each other*. ▮

If E is a free A-module having a finite basis, and if $(a_t^*)_{t\in I}$ is a basis (necessarily a finite one, by Corollary 2 to Proposition 11, § 10) of the dual E^* of E, then, considering the dual basis $(a_t^{**})_{t\in I}$ of $(a_t^*)_{t\in I}$ and denoting $a_t = c_E^{-1}(a_t^{**})$, $t \in I$, the family $(a_t)_{t\in I}$ is a basis of E; denoting by $(b_t^*)_{t\in I}$ the dual basis of $(a_t)_{t\in I}$, we have for each $t' \in I$,

$$b_{t'}^*(a_t) = \delta_{tt'} = \langle a_t^{**}, a_{t'}^*\rangle = \langle c_E(a_t), a_{t'}^*\rangle = \langle a_t, a_{t'}^*\rangle = a_{t'}^*(a_t)$$

for any $t \in I$; it follows, as above, that $b_{t'}^* = a_{t'}^*$ for any $t' \in I$; hence $(a_t^*)_{t\in I}$ is *the dual basis of a basis of* E, namely of $(a_t)_{t\in I}$.

Obviously, the preceding considerations are valid if instead of a left module a right A-module is considered.

§18. Projective and Injective Modules

Proposition 1. *Let A be a ring, and let P be an A-module; the following assertions concerning the module P are equivalent:*

i) For any epimorphism $v : F \to G$ of the category **Mod**(A) *and any homomorphism $g : P \to G$ there exists a homomorphism $f : P \to F$ such that $v \circ f = g$; in other words, for any diagram of A-modules*

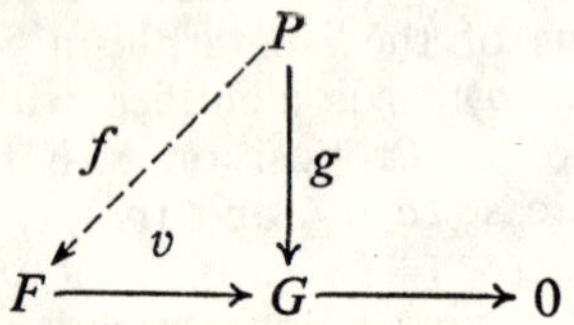

in which the row is an exact sequence there exists a homomorphism $f: P \to F$ rendering it commutative (one also says that the homomorphism $g : P \to G$ may be "lifted" to a homomorphism $f : P \to F$);

ii) For any diagram of A-modules of the form

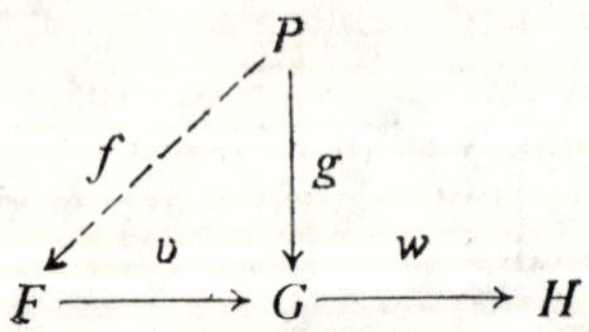

in which the row is an exact sequence and $w \circ g = 0$ there exists a homomorphism $f : P \to F$ rendering it commutative, hence for which $v \circ f = g$;

iii) The functor h^P : **Mod**$(A) \to$ **Ab** *is exact;*

iv) Any short exact sequence of A-modules of the form

$$0 \to E \xrightarrow{u} F \xrightarrow{v} P \to 0$$

is a split exact sequence;

v) P is isomorphic to a direct summand of a suitably chosen free A-module.

Proof. *i*) ⇔ *ii*): Assume that assertion *i*) is valid, and let the diagram of A-modules

$$\begin{array}{ccccc} & & P & & \\ & & \downarrow{\scriptstyle g} & & \\ F & \xrightarrow{v} & G & \xrightarrow{w} & H \end{array}$$

be given such that the row is an exact sequence and $w \circ g = 0$; by the universal property of the kernel (Proposition 1, § 3) there exists a homomorphism $g' : P \to \operatorname{Ker} w = \operatorname{Im} v$ such that, if j is the canonical injection $\operatorname{Ker} w \to G$, then $j \circ g' = g$; but by Corollary 2 to Proposition 1, § 4, there exists an epimorphism $v' : F \to \operatorname{Im}(v)$ such that $j \circ v' = v$; hence by *i*) there exists a homomorphism $f : P \to F$ such that $v' \circ f = g'$; we have

$$v \circ f = j \circ v' \circ f = j \circ g' = g;$$

hence *i*) ⇒ *ii*); the converse is trivial, since we have only to take $H = 0$ in *ii*).

i) ⇔ *iii*): Taking Propositions 1 and 4, § 5, into account, to say that the functor h^P is exact, hence *iii*), is equivalent to saying that for any epimorphism $v : F \to G$, the homomorphism $h^P(v) : \operatorname{Hom}_A(P, F) \to \operatorname{Hom}_A(P, G)$ is an epimorphism of the category **Ab**; this assertion is equivalent to the fact that if v is an epimorphism, then for any $g \in \operatorname{Hom}_A(P, G)$ there exists $f \in \operatorname{Hom}_A(P, F)$ such that

$$g = \operatorname{Hom}(1_P, v)(f) = v \circ f;$$

but this is exactly assertion *i*).

i) ⇒ *iv*): Let the exact sequence in the statement of *iv*) be given; by *i*), there exists a homomorphism $s : P \to F$ such that the following diagram is commutative:

$$\begin{array}{ccccc} & & P & & \\ & \swarrow{\scriptstyle s} & \downarrow{\scriptstyle 1_P} & & \\ F & \xrightarrow{v} & P & \longrightarrow & 0, \end{array}$$

hence for which $v \circ s = 1_P$; by the characterization *iii*) of Proposition 6, § 9, the given sequence is a split exact sequence.

iv) ⇒ *v*): By Proposition 6, § 10, there exists a free module L and an epimorphism $v : L \to P$; hence we have the exact sequence of A-modules

(1) $$0 \to \operatorname{Ker} v \xrightarrow{j} L \xrightarrow{v} P \to 0,$$

where j is the canonical injection; by *iv*), the exact sequence (1) splits; let $s : P \to L$ be a linear section associated with v; since $v \circ s = 1_P$, it follows, by the Corollary to Proposition 6, § 9, that $\mathrm{Im}(s)$ is a direct summand of L; but since s is injective, s induces an isomorphism $P \xrightarrow{\sim} \mathrm{Im}(s)$.

v) ⇒ *i*): Let us show first that if L is a free A-module, then it satisfies condition *i*) when replacing P: then let L be a free A-module having a basis $(a_\lambda)_{\lambda \in I}$; if v and g are as in the statement of *i*), then, as v is surjective, it follows that for any $\lambda \in I$ there exists $y_\lambda \in F$ such that $v(y_\lambda) = g(a_\lambda)$; by Proposition 4, § 10, there exists a linear mapping $f : L \to F$ such that $f(a_\lambda) = y_\lambda$; we have

$$(v \circ f)a_\lambda)(= v(y_\lambda) = g(a_\lambda)$$

for any $\lambda \in I$; again by the Proposition mentioned above it follows that $v \circ f = g$.

Suppose now that P is isomorphic to a direct summand of a free A-module; up to the given isomorphism, we may assume that P itself is a direct summand of a certain free A-module L (see also the Remark below); then, by Corollary 1 to Proposition 2, § 9, there exists a linear mapping $p : L \to P$ such that, denoting by j the canonical injection $P \to L$, we have $p \circ j = 1_P$; then let g and v be as in the statement of *i*) ; hence we have the diagram in which the row is exact

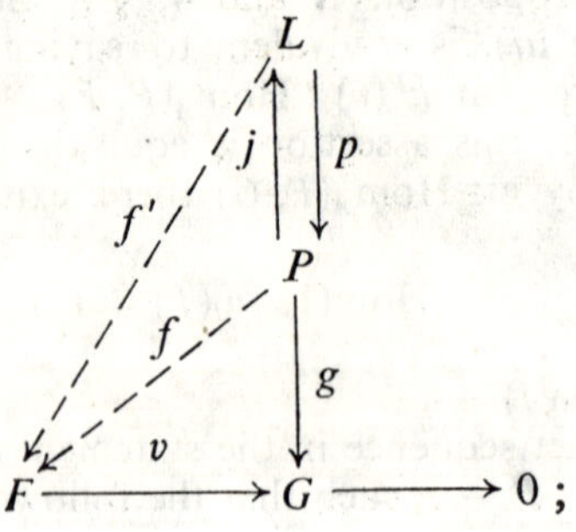

by the preceding arguments, there exists a homomorphism $f' : L \to F$ such that $v \circ f' = g \circ p$; but then, denoting $f = f' \circ j$, we have

$$v \circ f = v \circ f' \circ j = g \circ p \circ j = g. \blacksquare$$

REMARK. We shall show that *if P is isomorphic to a direct summand P' of a certain free A-module L', then P itself is a direct summand of a certain free A-module*; indeed, let $k : P \xrightarrow{\sim} P'$ be an A-isomorphism, and denote by L the disjoint union $(L' \setminus P') \amalg P$; define a mapping $l : L \to L'$ as follows: if $x \in L$, set

$$l(x) = \begin{cases} k(x) & \text{if } x \in P, \\ x & \text{if } x \in L' \setminus P'; \end{cases}$$

the mapping l is obviously bijective; by means of it we may endow L with an A-module structure, setting for $x, y \in L$ and $\alpha \in A$,

$$x + y = l^{-1}(l(x) + l(y)), \quad \alpha x = l^{-1}(\alpha \cdot l(x));$$

it is not difficult to check that these operations verify the axioms of A-module, and that L with this structure contains P as a submodule; moreover, that for the defined A-module structure of L and for the given A-module structure of L', l is an isomorphism of A-modules; it follows thus that L is a free A-module, and that P is a direct summand of it: for instance, a submodule supplementary to P in L is $l^{-1}(Q')$, where Q' is a submodule supplementary to P' in L'. This construction is an example of what is called "transport of structure".

Hence in Proposition 1, condition v) may also be formulated as follows: v') *P is a direct summand of a suitably chosen free A-module.* ▌

* If $\mathcal{C}$ is an abelian category, then assertions i)—iii) of Proposition 1 concerning an object P of the category $\mathcal{C}$ instead of an object P of the category **Mod**(A) are also equivalent.*

DEFINITION 1. Given a ring A, an A-module P is said to be a *projective A-module* provided it satisfies the equivalent conditions i)—v) of Proposition 1.

* Given an abelian category $\mathcal{C}$, an object P of $\mathcal{C}$ is said to be a *projective object* of the category $\mathcal{C}$ if it satisfies the equivalent conditions i)—iii) of the proposition corresponding to Proposition 1 for the object P of the category $\mathcal{C}$.* ▌

Of course, the notion of *projective right A-module* may be defined similarly to that of projective (left) A-module; a projective right A-module is a projective object of the category **Mod**$_d(A)$, but for the projective right A-modules we have also the characterizations iv) and v) of Proposition 1.

From Proposition 1, v) we derive the following result:

COROLLARY 1. *Any free A-module is projective.* ▌

The fact that the converse of Corollary 1 is not generally valid is shown by the following

EXAMPLE. 1) Let $A = \mathbf{Z}/6\mathbf{Z}$ be the ring of integers modulo 6; in A consider the subset E consisting of $\hat{0}$ and $\hat{3}$ (we have denoted by $\hat{n}$ the coset modulo 6 of $n \in \mathbf{Z}$); E is an A-submodule of A; consider the linear mapping $r : A \to E$ defined by virtue of Proposition 4, § 10 (taking into account that A is a free A-module having the basis $\{\hat{1}\}$) by

$$r(\hat{1}) = \hat{3};$$

we have $r(\hat{0}) = \hat{0}$ and $r(\hat{3}) = \hat{3}$, hence $r \circ i = 1_E$, i being the canonical injection $E \to A$; by Proposition 6, § 9, E is a direct summand of A, hence a projective module. But if E were free, it should have a basis; this basis cannot have more then zero elements, since E has two, while A has six elements; thus, since $E \neq 0$, it cannot be free. ▌

The property *iv*) of Proposition 1 for projective modules may be extended to long exact sequences as follows:

COROLLARY 2. *Let*

$$\text{(2)} \qquad \ldots \to E_n \xrightarrow{u_n} E_{n-1} \xrightarrow{u_{n-1}} E_{n-2} \to \ldots \to E_1 \xrightarrow{u_1} E_0 \to 0,$$

$n \geqslant 0$, *be an exact sequence of A-modules and A-homomorphisms having a right extremity* [1], *hence a finite number of terms to the right, and a finite or an infinite number of terms to the left; if all the modules* E_n *are projective, then* (2) *is a split exact sequence* [2].

Proof. We shall construct by induction the homomorphisms $s_n : E_n \to E_{n+1}$, $n \geqslant 0$, such that

$$\text{(3)} \qquad s_{n-1} \circ u_n + u_{n+1} \circ s_n = 1_{E_n} \text{ for } n \geqslant 1 \text{ and } u_1 \circ s_0 = 1_{E_0}.$$

Since u_1 is an epimorphism, and since E_0 is a projective A-module, there exists an A-homomorphism $s_0 : E_0 \to E_1$ such that $u_1 \circ s_0 = 1_{E_0}$; then, since

$$u_1 \circ (1_{E_1} - s_0 \circ u_1) = u_1 - u_1 \circ s_0 \circ u_1 = u_1 - u_1 = 0,$$

and since E_1 is a projective A-module, it follows, by the characterization *ii*) of Proposition 1 for projective modules, that there exists a homomorphism $s_1 : E_1 \to E_2$ such that

$$u_2 \circ s_1 = 1_{E_1} - s_0 \circ u_1;$$

let $n \geqslant 2$, and assume that the homomorphisms s_i, $i = 0, \ldots, n-1$, were constructed such as to verify conditions (3); but then, since by hypothesis

$$u_n \circ (1_{E_n} - s_{n-1} \circ u_n) = u_n - u_n \circ s_{n-1} \circ u_n =$$

$$= u_n - (1_{E_{n-1}} - s_{n-2} \circ u_{n-1}) \circ u_n = u_n - u_n = 0,$$

it follows in the same way as for s_1 that there exists a homomorphism $s_n : E_n \to$ $\to E_{n+1}$ such that

$$u_{n+1} \circ s_n = 1_{E_n} - s_{n-1} \circ u_n;$$

hence we get the induction assertion for n. ▮

From Corollary 1 and Proposition 6, § 10, we derive

---

[1] The fact that we denoted by E_0 the last term to the right is obviously a matter of numbering; on the other hand, we could have used negative indices, directing the arrows in an increasing order.

[2] See Definition 7, § 9.

COROLLARY 3. *Any A-module E is isomorphic to a quotient module of a projective A-module.* ▮

* DEFINITION 2. If $\mathcal{C}$ is an abelian category, $\mathcal{C}$ is said *to have enough projective objects* or, briefly, *enough projectives* if for any object E of $\mathcal{C}$ there exists a projective object P of the category $\mathcal{C}$ and an epimorphism $p : P \to E$. ▮*

From Corollary 3 it follows that for any ring A, the category **Mod**(A) has enough projectives; of course, $\mathbf{Mod}_d(A)$ also has enough projectives.

From Proposition 1 the next Corollary also follows:

COROLLARY 4. *An A-module P is projective and finitely generated if and only if it is a direct summand of a free A-module having a finite basis.*

Proof. By the proof of Proposition 6, § 10, if P is finitely generated, then it is isomorphic to a quotient module of some A_s^n, $n \geqslant 0$; hence there exists an exact sequence of A-modules

$$0 \to \operatorname{Ker} p \to A_s^n \overset{p}{\to} P \to 0,$$

and if P is projective, this exact sequence splits; hence P is isomorphic to a direct summand of A_s^n. The converse follows by Proposition 1, *v*), by Proposition 5, § 9, and by Corollary 2 to Proposition 2, § 8. ▮

COROLLARY 5. *The dual of a projective finitely generated A-module P is a projective finitely generated right A-module.*

Proof. By Corollary 4, there exists a free A-module L having a finite basis such that P is a direct summand of L; as shown in § 17, P^* is isomorphic to a direct summand of L^*, which by Proposition 3, § 17, is a free right A-module having a finite basis; the result is now a consequence of the above Corollary 4 for right A-modules. ▮

COROLLARY 6. *Let A be a commutative ring; if E and F are finitely generated projective A-modules, then* $\operatorname{Hom}_A(E, F)$ *with its canonical structure of A-module*[1] *is a finitely generated projective A-module.*

Proof. By the above Corollary 4, there exist two free finitely generated A-modules L and M such that $L = E \dotplus E'$, $M = F \dotplus F'$; but then there exist some A-isomorphisms

$$u : E \oplus E' \rightleftarrows L, \; v : M \rightleftarrows F \oplus F' = F \times F';$$

applying the functor Hom, we get, as shown in § 16, an A-isomorphism

$$\operatorname{Hom}(u, v) : \operatorname{Hom}_A(L, M) \rightleftarrows \operatorname{Hom}_A(E \oplus E', \; F \times F');$$

[1] See § 16.

again by the considerations in § 16, the canonical isomorphism

$$\mathrm{Hom}_A(E \oplus E', F \times F') \xrightarrow{\sim}$$

(4)

$$\xrightarrow{\sim} G = \mathrm{Hom}_A(E, F) \oplus \mathrm{Hom}_A(E', F) \oplus \mathrm{Hom}_A(E, F') \oplus \mathrm{Hom}_A(E', F')$$

given by Proposition 3, § 7, is an isomorphism of A-modules; on the other hand, we have some A-isomorphisms $f : L \xrightarrow{\sim} A^m$, $g : A^n \xrightarrow{\sim} M$, hence an A-isomorphism

$$\mathrm{Hom}(f, g) : \mathrm{Hom}_A(A^m, A^n) \xrightarrow{\sim} \mathrm{Hom}_A(L, M);$$

applying again Proposition 3, § 7, we have an A-isomorphism

$$\mathrm{Hom}_A(A^m, A^n) \xrightarrow{\sim} (\mathrm{Hom}_A(A, A))^{mn};$$

finally, by Proposition 1, § 16, we have an A-isomorphism $\mathrm{Hom}_A(A, A) \xrightarrow{\sim} A$ from which, by applying the direct product functor $(\mathbf{Mod}(A))^I \to \mathbf{Mod}(A)$, where $I = \{1, \ldots, mn\}$, we get an A-isomorphism

$$(\mathrm{Hom}_A(A, A))^{mn} \xrightarrow{\sim} A^{mn};$$

from all these isomorphisms it follows that G is a free A-module having a finite basis, and by the definition of G (see (4)) it follows that $\mathrm{Hom}_A(E, F)$ is isomorphic to a direct summand of G. ∎

COROLLARY 7. *If P is a projective A-module, respectively a finitely generated projective A-module, then the canonical linear mapping $c_P : P \to P^{**}$ is injective, respectively bijective.*

Proof. If P is projective, then by Proposition 1 there exists a free A-module L such that P is a direct summand of L; let $i : P \to L$ be the canonical injection; denoting by c_L the canonical linear mapping $L \to L^{**}$, we have, according to § 17, the commutative diagram

(5)

$$\begin{array}{ccc} P & \xrightarrow{\;i\;} & L \\ \downarrow{\scriptstyle c_P} & & \downarrow{\scriptstyle c_L} \\ P^{**} & \xrightarrow{\;{}^{tt}i\;} & L^{**}; \end{array}$$

by Proposition 5, § 17, c_L is an injective mapping; since i is also injective, it follows that c_P is also injective.

Assuming, moreover, that P is not only projective but also finitely generated, by Corollary 4 we may take the above L as having finite basis; by Corollary 1 to Proposition 2, § 9, there exists an A-linear mapping $p: L \to P$ such that, in the above notation, $p \circ i = 1_P$; applying the functor of passing to the bidual in the category **Mod**(A), we get ${}^{tt}p \circ {}^{tt}i = 1_{P^{**}}$; next, if $x^{**} \in P^{**}$, then $x^{**} = {}^{tt}p({}^{tt}i(x^{**}))$; since by Proposition 5, § 17, c_L is in this case a bijective linear mapping, it follows that there exists $y \in L$ such that $c_L(y) = {}^{tt}i(x^{**})$; from the commutativity of the diagram obtained from (5) replacing i by p and reversing the horizontal arrows we get

$$x^{**} = {}^{tt}p(c_L(y)) = c_P(p(y));$$

consequently, c_P is surjective; since by the first part of the proof it is also injective, the Corollary is completely proved. ∎

PROPOSITION 2. *Let* $(E_\lambda)_{\lambda \in I}$ *be a family of A-modules, and let* $E = \bigoplus_{\lambda \in I} E_\lambda$; *E is a projective A-module if and only if for any* $\lambda \in I$, E_λ *is a projective A-module.*

Proof. If all E_λ, $\lambda \in I$, are projective modules, let $v : F \to G$ be an epimorphism of the category **Mod**(A); if $f \in \mathrm{Hom}_A(E, G)$, then, denoting by j_λ the canonical injection $E_\lambda \to E$, $\lambda \in I$, we have for any $\lambda \in I$ the following diagram in which the row is exact:

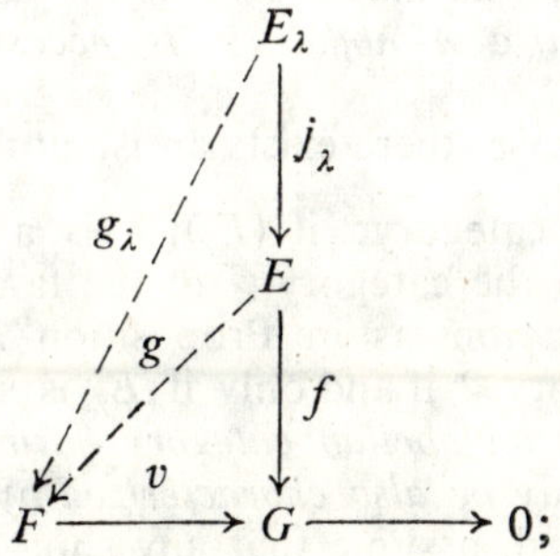

since E_λ is a projective A-module, it follows that there exists a homomorphism $g_\lambda : E_\lambda \to F$ such that $v \circ g_\lambda = f \circ j_\lambda$; the family of homomorphisms $(g_\lambda)_{\lambda \in I}$ determines by the universal property of the direct sum a unique linear mapping $g : E \to F$ such that $g \circ j_\lambda = g_\lambda$ for any $\lambda \in I$; we have for any $\lambda \in I$, $v \circ g \circ j_\lambda = v \circ g_\lambda = f \circ j_\lambda$, whence, by virtue of the uniqueness, $v \circ g = f$; thus E is projective.

Conversely, if E is a projective A-module, then let $p_\lambda : E \to E_\lambda$, $\lambda \in I$, be the unique linear mapping statisfying the conditions

$$p_\lambda \circ j_\mu = \begin{cases} 0 & \text{if } \lambda \neq \mu, \\ 1_{E_\lambda} & \text{if } \lambda = \mu; \end{cases}$$

if, for a given λ, f_λ is a linear mapping $E_\lambda \to G$, while $v : F \to G$ is an epimorphism, then we have the diagram in which the row is exact

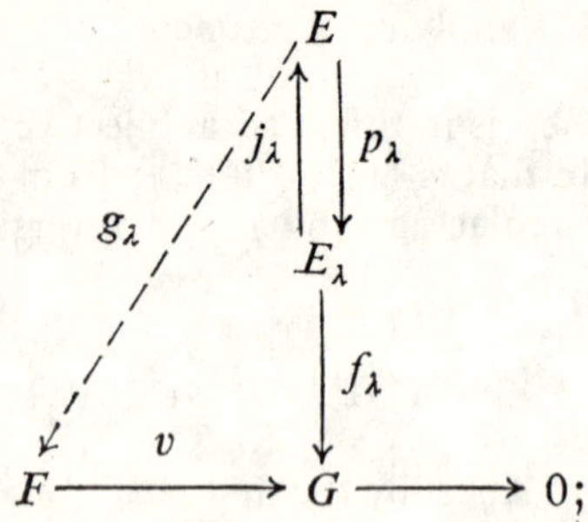

from the fact that E is a projective A-module it follows that there exists a homomorphism $g_\lambda : E \to F$ such that $v \circ g_\lambda = f_\lambda \circ p_\lambda$; then

$$v \circ g_\lambda \circ j_\lambda = f_\lambda \circ p_\lambda \circ j_\lambda = f_\lambda;$$

consequently, E_λ is projective. ▮

COROLLARY. *Let E be an A-module, and let $(M_\lambda)_{\lambda \in I}$ be a family of its submodules such that E is the internal direct sum of the family of submodules $(M_\lambda)_{\lambda \in I}$; in order that E be a projective A-module it is necessary and sufficient that for any $\lambda \in I$, M_λ be projective.*

Proof. Indeed, in this case, there exists an isomorphism $\bigoplus_{\lambda \in I} M_\lambda \xrightarrow{\sim} E$. ▮

*Let $\mathcal{C}$ be an abelian category; if $(E_\lambda)_{\lambda \in I}$ is a family of objects, and if $(E, (j_\lambda)_{\lambda \in I})$ is a direct sum in the category $\mathcal{C}$ of the family of objects $(E_\lambda)_{\lambda \in I}$, then, by exactly the same considerations as in Proposition 2, one proves that E is a projective object of the category $\mathcal{C}$ if and only if E_λ is such an object for any $\lambda \in I$.

It follows that *if $\mathcal{C}$ is an abelian category having enough projective objects, then a projective object P may be also characterized by the assertion corresponding to iv) of Proposition* 1: indeed, assume that any short exact sequence of the category $\mathcal{C}$ of the form

$$0 \to E \xrightarrow{u} F \xrightarrow{v} P \to 0 \tag{6}$$

splits; by the hypothesis that $\mathcal{C}$ has enough projectives there exists such an exact sequence with F projective; this being a split exact sequence, it follows by the characterization given in Proposition 7, § 9, that F is a direct sum of E and P; hence, by Proposition 2, P is also projective.

The converse is valid even in an arbitrary abelian category: if P is a projective object, then given the short exact sequence (6), since v is an epimorphism, it is possible to find a morphism $s : P \to F$ such that $v \circ s = 1_P$; it follows that (6) is a split exact sequence.*

PROPOSITION 3. *Let A be a ring, and let Q be an A-module; the following assertions concerning the module Q are equivalent:*

i) For any monomorphism $u : E \to F$ of the category **Mod**(A) *and any homomorphism $f : E \to Q$ there exists a homomorphism $g : F \to Q$ such that $g \circ u = f$, in other words, whatever the diagram of A-modules in which the row is an exact sequence*

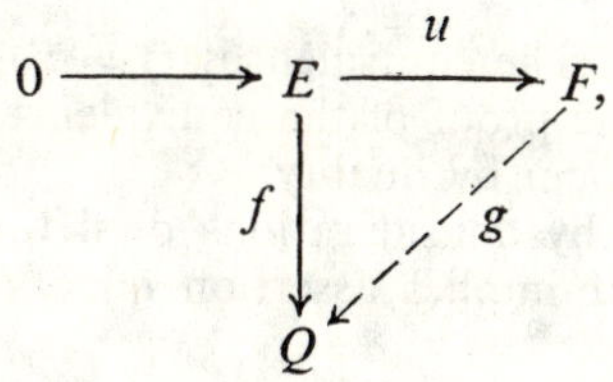

there exists a homomorphism $g : F \to Q$ rendering it commutative;

ii) For any diagram of A-modules of the form

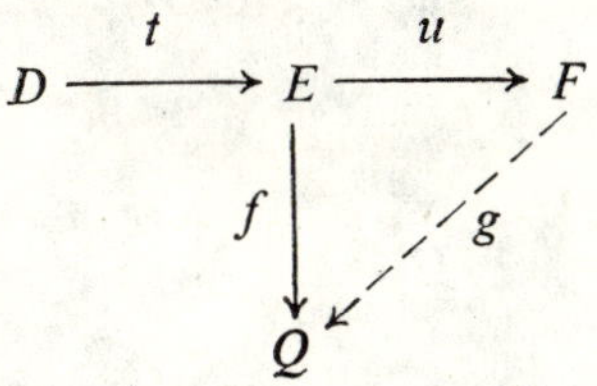

in which the row is an exact sequence and such that $f \circ t = 0$ there exists a homomorphism $g : F \to Q$ rendering the diagram commutative, hence such that $g \circ u = f$;

iii) The functor $h_Q : (\mathbf{Mod}(A))^\circ \to \mathbf{Ab}$ is exact;

iv) Any short exact sequence of A-modules of the form

$$0 \to Q \xrightarrow{u} F \xrightarrow{v} G \to 0$$

is a split exact sequence;

v) If there exists an injective homomorphism of A-modules $u : Q \to E$, and if $E/u(Q)$ is a cyclic A-module, then $u(Q)$ is a direct summand of E;

vi) For any left ideal $\mathfrak{a}$ of A and any linear mapping $f : \mathfrak{a} \to Q$ there exists some $c \in Q$ such that $f(\alpha) = \alpha c$ for any $\alpha \in \mathfrak{a}$;

vii) For any left ideal $\mathfrak{a}$ of A and any linear mapping $f : \mathfrak{a} \to Q$ there exists a linear mapping $g : A_s \to Q$ such that the homomorphism induced by g by restriction to $\mathfrak{a}$ and Q is f.

REMARK. Obviously, assertion *i*) is equivalent to the assertion:

i′) For any A-module F and any submodule E of F, any homomorphism $f : E \to Q$ may be extended to a homomorphism $g : F \to G$, in other words, there *exists $g : F \to Q$ such that the homomorphism induced by g by restriction to E and Q is f.*

Also, assertion *iv*) is equivalent to the assertion:

iv′) *Q is a direct summand of every A-module F having Q as a submodule.*

Finally, *v*) is equivalent to the assertion:

v′) *If E is an A-module having Q as a submodule, and if E is the sum, in the sense of Definition* 3, § 8, *of Q and a cyclic submodule of E, then Q is a direct summand of E.*

Proof of Proposition 3. The equivalences *i*) ⇔ *ii*) ⇔ *iii*) are proved by considerations dual to those in the proof of the equivalence of the assertions of Proposition 1 corresponding to them by duality.

i) ⇒ *iv*): It is proved by considerations dual to the proof of the fact that assertion *i*) of Proposition 1 implies assertion *iv*) of that Proposition.

iv) ⇒ *v*): Obvious.

v) ⇒ *vi*): Consider, under the hypotheses of *vi*), the fibered sum E of A_s and Q with respect to the diagram

$$\begin{array}{ccc} \mathfrak{a} & \xrightarrow{\;i\;} & A_s\,, \\ {\scriptstyle f}\big\downarrow & & \\ Q & & \end{array}$$

where i is the canonical injection; let $v : A_s \to E$ and $u : Q \to E$ be the canonical homomorphisms; since i is injective, by Proposition 7, § 13, u is injective too; since by the definition of E we have $E = v(A_s) + u(Q)$, and since by the Second Isomorphism Theorem for modules (Proposition 5, § 8) we have

$$\frac{E}{u(Q)} = \frac{v(A_s) + u(Q)}{u(Q)} \approx \frac{v(A_s)}{v(A_s) \cap u(Q)},$$

it follows that $E/u(Q)$ is a cyclic A-module; by *v*), $u(Q)$ is a direct summand of E; hence there exists a submodule M of E such that

$$E = u(Q) \dotplus M;$$

by Corollary 1 to Proposition 2, § 9, there exists an A-homomorphism $p : E \to u(Q)$ such that, if $j : u(Q) \to E$ is the canonical injection, then $p \circ j = 1_{u(Q)}$; taking the composition of the isomorphism u'^{-1} (where u' is the isomorphism $Q \xrightarrow{\sim} u(Q)$ induced by u) with p, we get a linear mapping $r : E \to Q$ for which $r \circ u = 1_Q$. We have $f = r \circ u \circ f = r \circ v \circ i$. Define

$$c = r(v(1)),$$

where 1 is the unit element of A; for any $\alpha \in \mathfrak{a}$ we have

$$f(\alpha) = r(v(\alpha)) = \alpha r(v(1)) = \alpha c.$$

vi) $\Rightarrow$ *i')*: Let us show first that *i')* follows from *vi)* in the case where F/E is a cyclic A-module: in this case, if $x + E = \hat{x}$ is a generator of F/E, then for any $y \in F$, $y + E = \hat{y}$ is of the form $\hat{y} = \alpha\hat{x} = \widehat{\alpha x}$, with $\alpha \in A$, hence $y - \alpha x \in E$, and $F = E + Ax$; applying the Second Isomorphism Theorem for modules, we get an isomorphism

$$F/E \rightleftarrows \frac{Ax}{Ax \cap E};$$

but, by the considerations preceding Proposition 7, § 10, there exists an isomorphism

$$\varphi : Ax \rightleftarrows A_s/\mathfrak{b},$$

where $\mathfrak{b} = \mathrm{Ann}(x)$; the image under this isomorphism of the submodule $Ax \cap E$ is a submodule M of $A_s/\mathfrak{b}$; applying the Third Isomorphism Theorem for modules (Proposition 2, § 4) to the canonical surjection $p : A_s \to A_s/\mathfrak{b}$, it follows that to M a left ideal $\mathfrak{a}$ of A corresponds uniquely determined by the properties $\mathfrak{a} \supseteq \mathfrak{b}$ and $\mathfrak{a}/\mathfrak{b} = p(\mathfrak{a}) = M$; denoting by u the composition of the homomorphisms

$$\mathfrak{a} \xrightarrow{q} \mathfrak{a}/\mathfrak{b} \overset{\psi}{\rightleftarrows} Ax \cap E \xrightarrow{j} E \xrightarrow{f} Q,$$

where q is the canonical surjection, ψ is the inverse of the isomorphism $Ax \cap E \rightleftarrows$ $\rightleftarrows \mathfrak{a}/\mathfrak{b}$ induced by φ, and j is the canonical injection, it follows by *vi)* that there exists an element $c \in Q$ such that $u(\gamma) = \gamma c$ for any $\gamma \in \mathfrak{a}$; but for any $\alpha \in A$, $\varphi(\alpha x) = \alpha + \mathfrak{b}$, hence $\psi(\gamma + \mathfrak{b}) = \gamma x$ for any $\gamma \in \mathfrak{a}$; it follows that

$$u(\gamma) = f(j(\psi(q(\gamma)))) = f(j(\gamma x)) = f(\gamma x)$$

for any $\gamma \in \mathfrak{a}$.

As we have already seen, any $y \in F$ may be expressed as $y = z + \alpha x$, with $z \in E$, $\alpha \in A$; if $y = z' + \alpha' x$, with $z' \in E$, $\alpha' \in A$, is another expression of the same type of x, then

$$z - z' = (\alpha' - \alpha)x \in Ax \cap E;$$

denoting $\varphi((\alpha' - \alpha)x) = \gamma + \mathfrak{b}$, where $\gamma \in \mathfrak{a}$, we have $(\alpha' - \alpha) + \mathfrak{b} = \gamma + \mathfrak{b}$, hence $\alpha' - \alpha - \gamma = \varepsilon \in \mathfrak{b} = \mathrm{Ann}(x)$; hence $(\alpha' - \alpha)x = \gamma x$; it follows

$$f(z - z') = f(\gamma x) = u(\gamma) = \gamma c;$$

on the other hand, since $\varepsilon \in \mathfrak{b} \subseteq \mathfrak{a}$, we have $\varepsilon c = u(\varepsilon) = f(\varepsilon x) = 0$, hence $f(z - z') = \gamma c + \varepsilon c = \alpha' c - \alpha c$, that is,

$$f(z) + \alpha c = f(z') + \alpha' c;$$

consequently, the element $f(z) + \alpha c$ does not depend on the particular expression $z + \alpha x$, with $z \in E$, $\alpha \in A$, chosen for y, and we may define a mapping

$$g : F \to Q$$

by

$$g(z + \alpha x) = f(z) + \alpha c,$$

where $z \in E$, $\alpha \in A$. The mapping g thus defined is A-linear: for instance, given also $y_1 \in F$ of the form $y_1 = z_1 + \alpha_1 x$, with $z_1 \in E$, $\alpha_1 \in A$, then $y + y_1 = z + z_1 + (\alpha + \alpha_1)x$, and

$$g(y + y_1) = f(z + z_1) + (\alpha + \alpha_1)c = f(z) + \alpha c + f(z_1) + \alpha_1 c = g(y) + g(y_1).$$

If $z \in E$, then $z = z + 0 \cdot x$, hence

$$g(z) = f(z) + 0 \cdot c = f(z),$$

and g extends f.

We proceed now to verify $i')$ in the general case: consider the set $\mathcal{P}$ of all pairs (M, u), where M is a submodule of F containing E, and u is an A-homomorphism $M \to Q$ extending f; the set $\mathcal{P}$ is not empty, since it contains (E, f); on the set $\mathcal{P}$ we introduce a partial order relation, denoted $<$, as follows: for (M, u), $(M', u') \in \mathcal{P}$ we have

$$(M, u) < (M', u')$$

whenever $M \subseteq M'$ and u' extends u; the set $\mathcal{P}$ partially ordered by $<$ is inductively ordered: for if $((M_\lambda, u_\lambda))_{\lambda \in I}$ is a non-empty family of elements of $\mathcal{P}$ for which, whatever λ, $\mu \in I$, we have either $(M_\lambda, u_\lambda) < (M_\mu, u_\mu)$ or $(M_\mu, u_\mu) < (M_\lambda, u_\lambda)$, then the family of submodules $(M_\lambda)_{\lambda \in I}$ is totally ordered by inclusion, hence, by Corollary 6 to Proposition 2, § 8, the union $M = \bigcup_{\lambda \in I} M_\lambda$ is a submodule of E; if $x \in M$, then there exists $\lambda \in I$ such that $x \in M_\lambda$; the element $u_\lambda(x)$ does not depend on the chosen λ, since if $\mu \in I$ has also the property $x \in M_\mu$, then we have, say, $(M_\lambda, u_\lambda) < (M_\mu, u_\mu)$, hence $u_\mu(x) = u_\lambda(x)$; defining $u(x) = u_\lambda(x)$, and doing this for all $x \in M$, we get a linear mapping $u : M \to Q$; since, for instance, if $x, y \in M$, we may assume, the family $(M_\lambda)_{\lambda \in I}$ being totally ordered by inclusion, that $x \in M_\lambda$, $y \in M_\lambda$ (for the same λ); hence,

$$u(x + y) = u_\lambda(x + y) = u_\lambda(x) + u_\lambda(y) = u(x) + u(y);$$

obviously, $M_\lambda \subseteq M$ and u_λ is an extension of u for any $\lambda \in I$; hence (M, u) is an upper bound in $\mathcal{P}$ of the family $((M_\lambda, u_\lambda))_{\lambda \in I}$.

By Zorn's Lemma there exists in $\mathcal{P}$ a maximal element (M_0, u_0); if $M_0 = F$, everything is clear; assume the contrary, and let $x \in F \setminus M_0$; by the Second Isomorphism Theorem for modules we have

$$\frac{M_0 + Ax}{M_0} \approx \frac{Ax}{M_0 \cap Ax},$$

hence $(M_0 + Ax)/M_0$ is a cyclic module; by the considered particular case there exists an A-homomorphism $v : M_0 + Ax \to Q$ extending u_0; hence $(M_0 + Ax, v) \in \mathcal{P}$ and $(M_0, u_0) < (M_0 + Ax, v)$, and since $M_0 \neq M_0 + Ax$, this contradicts the maximality of (M_0, u_0).

i′) ⇒ *vii*): Obvious.

vii) ⇒ *vi*): Under the hypotheses of *vi*), it follows by *vii*) that there exists an A-homomorphism $g : A_s \to Q$ extending f; let $c = g(1)$; for any $\alpha \in \mathfrak{a}$ we have

$$f(\alpha) = g(\alpha) = \alpha g(1) = \alpha c. \blacksquare$$

* Given an abelian category $\mathcal{C}$ and an object Q of $\mathcal{C}$, one proves by exactly the same considerations the equivalence of the assertions *i*)—*iii*) of Proposition 3 for the object Q and the category $\mathcal{C}$ instead of the A-module Q and the category **Mod**(A).*

Definition 3. Given a ring A and an A-module Q, Q is said to be an *injective A-module* if it satisfies the equivalent conditions *i*)—*vii*) of Proposition 3.

* Given an abelian category $\mathcal{C}$ and an object Q of $\mathcal{C}$, Q is said to be an *injective object* of the category $\mathcal{C}$ if it satisfies the equivalent conditions *i*)—*iii*) of the proposition corresponding to Proposition 3 for the object Q and the category $\mathcal{C}$. ▌

By the preceding considerations (compare, for instance, assertion *i*) of Proposition 1 with assertion *i*) of Proposition 3), the notion of injective object is dual to the one of projective object.*

From Proposition 3 we get the following Corollary:

Corollary 1. *A* **Z**-*module Q is injective if and only if for any* $a \in Q$ *and any* $n \in \mathbf{Z}$, $n \geqslant 1$, *there exists* $b \in Q$ *such that* $a = nb$.[1]

Proof. Assume the **Z**-module Q has the stated property; we want to show that Q satisfies condition *vi*) of Proposition 3: let $\mathfrak{a}$ be an ideal of **Z**, and let $f: \mathfrak{a} \to Q$ be a homomorphism of **Z**-modules; since $\mathfrak{a}$ is also a subgroup of the additive group **Z**, it follows, as shown in Example 1), § 9, that $\mathfrak{a} = n\mathbf{Z}$ for some $n \in \mathbf{Z}$, $n \geqslant 0$; if $n = 0$, then Q trivially satisfies the above condition; if $n \geqslant 1$, then let $f(n) = a \in Q$; by hypothesis there exists $b \in Q$ such that $a = nb$; then, for any element $kn \in \mathfrak{a}$, with $k \in \mathbf{Z}$, we have

$$f(kn) = k \cdot f(n) = ka = knb,$$

and the sufficiency is proved.

---

[1] Such a **Z**-module is said to be a *divisible* **Z**-module.

Conversely, if Q is an injective **Z**-module, and if $a \in Q$ and $n \in \mathbf{Z}$, $n \geqslant 1$, then consider the ideal $n\mathbf{Z}$ of **Z**; $n\mathbf{Z}$ is a free **Z**-module with a basis consisting of n; then, by Proposition 4, § 10, there exists a **Z**-homomorphism $f: n\mathbf{Z} \to Q$ for which $f(n) = a$; by the characterization *vi*), Proposition 3, of injective modules, there exists $b \in Q$ such that $f(nk) = nkb$ for any $k \in \mathbf{Z}$; in particular, $a = f(n) = nb$. ∎

In the same way as for projective modules, the property of injective modules given by *iv*), Proposition 3, has a generalization for long exact sequences as follows:

COROLLARY 2. *Let*

$$0 \to E_0 \xrightarrow{u_0} E_1 \to \ldots \to E_n \xrightarrow{u_n} E_{n+1} \to \ldots,$$

$n \geqslant 0$, *be an exact sequence of A-modules and A-homomorphisms having a finite number of terms to the left, while to the right the situation is arbitrary; if all the modules* E_n *are injective, then the given exact sequence splits.* ∎

From Proposition 2 we get by duality:

PROPOSITION 4. *If* $(E_\lambda)_{\lambda \in I}$ *is a family of A-modules, then* $\prod_{\lambda \in I} E_\lambda$ *is an injective A-module if and only if* E_λ *is an injective A-module for any* $\lambda \in I$. ∎

From Proposition 4 we derive the following

COROLLARY. *A direct summand of an injective A-module is an injective module.*

Proof. If M is a direct summand of the injective module Q, then there exists N such that $Q = M \dot{+} N$; but then, by Remark 2) following Definition 1, § 9, we have an A-isomorphism

$$Q \approx M \oplus N = M \times N;$$

hence $M \times N$ is an injective module, and, by Proposition 4, it follows that M is also injective. ∎

* Proposition 4 is obviously true for a family of objects of an abelian category $\mathscr{C}$ instead of **Mod**(A) provided this family has at least one direct product in $\mathscr{C}$.*

Of course, the above notions and properties may be extended to right A-modules.

In the sequel we need the following Lemma:

LEMMA 1. *Let A be a ring, and let P be a left, respectively a right projective A-module; if Q is an injective* **Z**-*module, then* $\tilde{P} = \mathrm{Hom}_{\mathbf{Z}}(P, Q)$ *with the right, respectively left A-module structure obtained by the considerations of* § 16 *from the left, respectively right A-module structure of P is a right, respectively a left injective A-module.*

Proof. We argue for left A-modules; recall (see formula (3), § 16) that in this case, with the stated structure, if $t \in \mathrm{Hom}_{\mathbf{Z}}(P, Q)$ and $\alpha \in A$, then for $z \in P$ we have

$$(t\alpha)(z) = t(\alpha z).$$

Consider then the diagram in $\mathbf{Mod}_d(A)$ in which the row is exact

$$
\begin{array}{ccccc}
0 & \longrightarrow & E & \xrightarrow{\;u\;} & F; \\
 & & \Big\downarrow{\scriptstyle f} & & \\
 & & \tilde{P} & &
\end{array} \tag{7}
$$

since $0 \to E \xrightarrow{u} F$ is also an exact sequence as a diagram of $\mathbf{Z}$-modules, and since Q is an injective $\mathbf{Z}$-module, it follows that the diagram of $\mathbf{Z}$-modules and $\mathbf{Z}$-homomorphisms

$$h_Q(F^\circ) \xrightarrow{h_Q(u^\circ)} h_Q(E^\circ) \to 0$$

is an exact sequence; this property is not modified by the fact that we endow the modules $h_Q(F^\circ)$ and $h_Q(E^\circ)$ with the left A-module structures obtained by the considerations in § 16; in this case $h_Q(u^\circ)$ is an A-homomorphism. Now let $z \in P$; assign to z the mapping $k(z) : E \to Q$ given by

$$k(z)(x) = f(x)(z)$$

for any $x \in E$; obviously, $k(z)$ is an element of $\mathrm{Hom}_{\mathbf{Z}}(E, Q) = h_Q(E^\circ)$; doing this for all $z \in P$, we get a mapping $k : P \to h_Q(E^\circ)$; k is a homomorphism of (left) A-modules: since, for instance, if $z \in P$, $\alpha \in A$, and $x \in E$, then

$$k(\alpha z)(x) = f(x)(\alpha z) = (f(x)\alpha)(z) = (f(x\alpha))(z) = k(z)(x\alpha) = (\alpha k(z))(x),$$

the last equality being due to relation (4), § 16; since P is a projective left A-module, it follows that there exists an A-homomorphism $l : P \to h_Q(F^\circ)$ rendering commutative the diagram

$$
\begin{array}{ccccc}
h_Q(F^\circ) & \xrightarrow{\;h_Q(u^\circ)\;} & h_Q(E^\circ) & \longrightarrow & 0. \\
 & {\scriptstyle l}\nwarrow & \Big\uparrow{\scriptstyle k} & & \\
 & & P & &
\end{array} \tag{8}
$$

For any $y \in F$ define a mapping $g(y) : P \to Q$ by

$$g(y)(z) = l(z)(y),$$

where $z \in P$; obviously, $g(y)$ is a **Z**-homomorphism, hence an element of $\tilde{P}$; hence we defined a mapping $g: F \to \tilde{P}$ which is a homomorphism of right A-modules: it is obvious that g is a **Z**-homomorphism, and if $y \in F$, $\alpha \in A$, $z \in P$, then

$$g(y\alpha)(z) = l(z)(y\alpha) = (\alpha l(z))(y) = l(\alpha z)(y) = g(y)(\alpha z) = (g(y)\alpha)(z).$$

It remains to show that the diagram obtained from (7) by filling in with $g : F \to \tilde{P}$ is commutative, which will prove that $\tilde{P}$ is injective; but if $x \in E$, then for any $z \in P$ we have $(g(u(x)))(z) = (l(z)\circ u)(x)$; the commutativity of diagram (8) implies

$$l(z)\circ u = h_Q(u^\circ)(lz) = k(z);$$

hence,

$$(g(u(x)))(z) = k(z)(x) = f(x)(z),$$

hence $(g\circ u)(x) = f(x)$. ▮

PROPOSITION 5. *Given a ring A and an A-module E, there exists an injective A-module Q and a monomorphism $u : E \to Q$.*[1]

First proof. Consider the **Z**-module **Q**/**Z**; if $r/s \in \mathbf{Q}$, and if we denote by $\widehat{r/s}$ the coset of r/s modulo **Z**, then, for any $n \in \mathbf{Z}$, $n \geqslant 1$, we have

$$n\widehat{\frac{r}{sn}} = \widehat{\frac{r}{s}},$$

hence **Q**/**Z** is, by Corollary 1 to Proposition 3, an injective **Z**-module.

For any left, respectively right A-module X, denote by $\tilde{X}$ the right, respectively left A-module $\mathrm{Hom}_{\mathbf{Z}}(X,\mathbf{Q}/\mathbf{Z})$ with the structure obtained by the procedure described in § 16.

Now let E be a (left) A-module; by the preceding considerations, the abelian group $\tilde{\tilde{E}} = \mathrm{Hom}_{\mathbf{Z}}(\mathrm{Hom}_{\mathbf{Z}}(E, \mathbf{Q}/\mathbf{Z}), \mathbf{Q}/\mathbf{Z})$ is endowed with a left A-module structure; for any $x \in E$ define a mapping $\varphi(x) : \tilde{E} \to \mathbf{Q}/\mathbf{Z}$ by

$$\varphi(x)(f) = f(x)$$

for any $f \in \tilde{E}$; of course, $\varphi(x)$ is a **Z**-homomorphism, hence an element of $\tilde{\tilde{E}}$ for any $x \in E$. Thus we defined a mapping

$$\varphi : E \to \tilde{\tilde{E}};$$

[1] One also says that "any A-module may be imbedded in an injective module".

let us show that φ *is an injective A-homomorphism*; the fact that φ is a **Z**-homomorphism follows immediately from the definition; if $x \in E$ and $\alpha \in A$, then, for any $f \in \tilde{E}$ we have, because of relations (3) and (4) of § 16,

$$(\varphi(\alpha x))(f) = f(\alpha x) = (f\alpha)(x) = (\varphi(x))(f\alpha) = (\alpha \cdot \varphi(x))(f),$$

which proves that φ is also an A-homomorphism.

To show that φ is an injective mapping, let $x \in E$, $x \neq 0$; since $\mathrm{Ann}_{\mathbf{Z}}(x) = n\mathbf{Z}$, with $n \in \mathbf{Z}$, $n \geqslant 0$, suitably chosen (see Example 1), § 9), we have $\mathbf{Z}x \approx \mathbf{Z}/n\mathbf{Z}$, where $\mathbf{Z}x$ means the **Z**-submodule generated by x in the underlying **Z**-module of E; we cannot, however, have $n = 1$, since this would mean $1 \cdot x = 0$, contradicting our assumption; define a mapping $f' : \mathbf{Z}x \to \mathbf{Q}/\mathbf{Z}$ as follows: if $n = 0$, then $\mathbf{Z}x \approx \mathbf{Z}$, hence considering an arbitrary non-zero element $\widehat{r/s}$ of $\mathbf{Q}/\mathbf{Z}$, we may define

$$f'(kx) = k \cdot \widehat{r/s}, \; k \in \mathbf{Z};$$

if $n > 1$, then define

$$f'(kx) = k \cdot \widehat{1/n}, \; k \in \mathbf{Z};$$

this definition does not depend on the chosen k, i.e., f' is well defined, since if $kx = lx$, then $k - l \in \mathrm{Ann}(x) = n\mathbf{Z}$, hence $k - l = mn$, with $m \in \mathbf{Z}$, and hence $(k - l)\widehat{1/n} = nm.\,\widehat{1/n} = \hat{m} = 0$; in both cases, f' is a **Z**-homomorphism $\mathbf{Z}x \to \mathbf{Q}/\mathbf{Z}$, and we have $f'(x) \neq 0$; since $\mathbf{Q}/\mathbf{Z}$ is an injective **Z**-module, f' may be extended to a **Z**-homomorphism $f_0 : E \to \mathbf{Q}/\mathbf{Z}$ such that, of course, $f_0(x) \neq 0$; if we had $\varphi(x) = 0$, we would get in particular $0 = \varphi(x)(f_0) = f_0(x)$, a contradiction.

On the other hand, by Corollary 3 to Proposition 1 there exists a projective right A-module P and an epimorphism $p : P \to \tilde{E}$; since the sequence

$$P \xrightarrow{p} \tilde{E} \to 0$$

is exact, it follows by Proposition 4, § 5, that the diagram

$$0 \to \tilde{\tilde{E}} \xrightarrow{\tilde{p}} \tilde{P},$$

where $\tilde{p} = \mathrm{Hom}(p, 1_{\mathbf{Q}/\mathbf{Z}})$, is an exact sequence of **Z**-modules, hence also an exact sequence of A-modules and A-homomorphisms; but, by Lemma 1, $\tilde{P}$ is an injective left A-module; hence we have only to take $Q = P$ and $u = \tilde{p} \circ \varphi$, and the result is proved.

Second proof. We prove firstly the Proposition in the particular case $A = \mathbf{Z}$: let E then be an arbitrary **Z**-module; by Proposition 6, § 10, there exists a set I

and a surjective **Z**-homomorphism $f : \mathbf{Z}^{(I)} \to E$; denoting by i the canonical injection $\mathbf{Z} \to \mathbf{Q}$ and by $j : \mathbf{Z}^{(I)} \to \mathbf{Q}^{(I)}$ the direct sum of the family of homomorphisms $(i_\lambda)_{\lambda \in I}$, where $i_\lambda = i$ for any $\lambda \in I$, let $M = j(\operatorname{Ker} f)$; the homomorphism j induces by passing to the quotient a **Z**-homomorphism

$$k : \mathbf{Z}^{(I)}/\operatorname{Ker} f \to \mathbf{Q}^{(I)}/M;$$

if we denote by $\hat{x}$, respectively $\hat{y}$, the coset of $x \in \mathbf{Z}^{(I)}$, respectively of $y \in \mathbf{Q}^{(I)}$, modulo $\operatorname{Ker} f$, respectively M, then we have $k(\hat{x}) = \widehat{j(x)}$; if $\hat{x} \in \operatorname{Ker}(k)$, then $j(x) \in M = j(\operatorname{Ker} f)$, and, since j is injective (see Corollary 1 to Proposition 2, § 7), $x \in \operatorname{Ker} f$, hence $\hat{x} = 0$; hence $\operatorname{Ker}(k) = 0$, and k is injective; since by Corollary 1 to Proposition 1, § 4, we have a **Z**-isomorphism $\hat{f} : \mathbf{Z}^{(I)}/\operatorname{Ker} f \xrightarrow{\sim} E$, this means that taking

$$k \circ \hat{f}^{-1} : E \to \mathbf{Q}^{(I)}/M$$

for u, everything is proved in this particular case provided we show that $\mathbf{Q}^{(I)}/M$ is an injective **Z**-module.

If $(r_\lambda/s_\lambda)_{\lambda \in I}$ is an element of $\mathbf{Q}^{(I)}$ for which $r_\lambda, s_\lambda \in \mathbf{Z}$, $s_\lambda \neq 0$, $\lambda \in I$, while $(r_\lambda)_{\lambda \in I}$ is an fs-family, and if $n \in \mathbf{Z}$, $n \geqslant 1$, then the family $(r_\lambda/ns_\lambda)_{\lambda \in I} \in \mathbf{Q}^{(I)}$ has the property $n((r_\lambda/ns_\lambda)_\lambda) = (r_\lambda/s_\lambda)_\lambda$; hence, for any $y \in \mathbf{Q}^{(I)}$ and for any $n \in \mathbf{Z}$, $n \geqslant 1$, there exists $y_1 \in \mathbf{Q}^{(I)}$ such that $ny_1 = y$; it follows that $n\hat{y}_1 = \hat{y}$, and hence, by Corollary 1 to Proposition 3, $\mathbf{Q}^{(I)}/M$ is an injective **Z**-module.

Now consider an arbitrary ring A, and let E be an A-module; its underlying abelian group will also be denoted E; by the preceding considerations, there exists an injective **Z**-homomorphism $v : E \to Q$, where Q is an injective **Z**-module. By Proposition 4, § 5, it follows that

$$v^* = \operatorname{Hom}_{\mathbf{Z}}(1_A, v) : \operatorname{Hom}_{\mathbf{Z}}(A_d, E) \to \operatorname{Hom}_{\mathbf{Z}}(A_d, Q)$$

is an injective **Z**-homomorphism; since the right A-module structure of A_d is compatible with its **Z**-module structure, it follows, as shown in § 16, that the abelian groups $\operatorname{Hom}_{\mathbf{Z}}(A_d, E)$ and $\operatorname{Hom}_{\mathbf{Z}}(A_d, Q)$ are endowed with left A-module structures in which the scalar multiplication is given by relations of the type (4), § 16; likewise, since h_v defined by $h_v(F^\circ) = \operatorname{Hom}(1_F, v)$, $F \in \operatorname{Ob} \mathbf{Ab}$, is a natural transformation $h_E \to h_Q$, it follows, also by § 16, that $h_v(A_d^\circ) = \operatorname{Hom}_{\mathbf{Z}}(1_A, v) = v^*$ is an (injective) homomorphism of left A-modules.

On the other hand, by Proposition 1, § 16, there exists the isomorphism of A-modules

$$\Xi(E) : E \xrightarrow{\sim} \operatorname{Hom}_A(A_s, E).$$

Next, consider the mapping

$$j : \operatorname{Hom}_A(A_s, E) \to \operatorname{Hom}_{\mathbf{Z}}(A_d, E)$$

assigning to each A-homomorphism u (from A_s to E), u itself considered as **Z**-homomorphism; j is an injective A-homomorphism, the fact that $j(\alpha u) = \alpha j(u)$ for any $\alpha \in A$ and $u \in \mathrm{Hom}_A(A_s, E)$ being a consequence of the fact that the A-module structure of $\mathrm{Hom}_A(A_s, E)$ is given also by a relation of the type (4), § 16.

Consequently, we have the injective A-homomorphism

$$v^* \circ j \circ \Xi(E) : E \to \mathrm{Hom}_{\mathbf{Z}}(A_d, Q);$$

but by Lemma 1 and by Corollary 1 to Proposition 1, $\mathrm{Hom}_{\mathbf{Z}}(A_d, Q)$ is an injective left A-module. ▮

REMARK. Taking in the first proof of Proposition 5 $P = A_d^{(\tilde{E})}$ (see Corollary 1 to Proposition 1 as well as the Remark following Proposition 6, § 10), we get $Q = \tilde{P} = \mathrm{Hom}_{\mathbf{Z}}(A_d^{(\tilde{E})}, \mathbf{Q}/\mathbf{Z})$, hence Q in Proposition 5 may be taken so that the assignment $E \mapsto Q$ has the character of a functor. ▮

* DEFINITION 4. If $\mathcal{C}$ is an abelian category, $\mathcal{C}$ is said *to have enough injective objects* or also *enough injectives* if for any object E of the category $\mathcal{C}$ there exists an injective object Q of the category $\mathcal{C}$ and a monomorphism $u : E \to Q$ (briefly, by an abuse of language, one says "if any object of the category is a subobject of some injective object"). ▮

Of course, if $\mathcal{C}$ is an abelian category having enough projectives, then its dual $\mathcal{C}^\circ$ is an abelian category having enough injectives, and conversely.*

From Proposition 5 it follows that for any ring A, the category **Mod**(A) has enough injectives; in the same way it may be seen that **Mod**$_d(A)$ has enough injectives.

* *In an abelian category $\mathcal{C}$ having enough injectives, an object Q of $\mathcal{C}$ having the property that any short exact sequence of the category $\mathcal{C}$ of the form*

$$0 \to Q \to F \to G \to 0$$

splits is an injective object of the category $\mathcal{C}$; the converse holds in an arbitrary abelian category $\mathcal{C}$; these assertions are proved by considerations dual to those for projective objects.*

§ 19. Generators. Injective Envelopes

* **Proposition** 1. *Let $\mathcal{C}$ be an abelian category, and let U be one of its objects; the following assertions are equivalent:*

i) The functor $h^U : \mathcal{C} \to$ **Ab** *is faithful;*

ii) For any object E of the category $\mathcal{C}$ and any subobject (M, i) of E for which i is not an isomorphism there exists a morphism $u \in \mathrm{Hom}_{\mathcal{C}}(U, E)$ such that u does not factor through i, that is, there is no $v \in \mathrm{Hom}_{\mathcal{C}}(U, M)$ for which $i \circ v = u$.

Proof. *i)* ⇒ *ii)*: Let E, M, i be as in the statement of *ii)*; if for any $u \in \mathrm{Hom}_{\mathcal{C}}(U, E)$ there existed $v \in \mathrm{Hom}_{\mathcal{C}}(U, M)$ rendering commutative the diagram

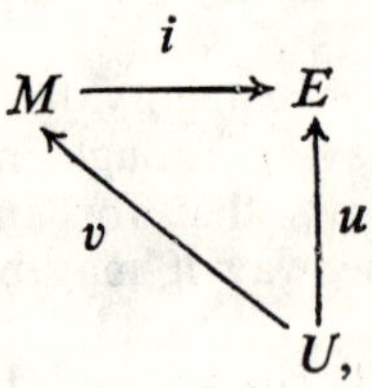

then the homomorphism of abelian groups

$$h^U(i) = \mathrm{Hom}(1_U, i) : \mathrm{Hom}_{\mathcal{C}}(U, M) \to \mathrm{Hom}_{\mathcal{C}}(U, E)$$

would be surjective; but if (C, p) is a cokernel in the category $\mathcal{C}$ of the morphism i, then $p \circ i = 0$, hence $h^U(p) \circ h^U(i) = 0$; since $h^U(i)$ is an epimorphism of the category **Ab**, we would have $h^U(p) = 0$; but h^U is a faithful functor, hence the mapping

$$h^U(E, C) : \mathrm{Hom}_{\mathcal{C}}(E, C) \to \mathrm{Hom}_{\mathbf{Ab}}(h^U(E), h^U(C))$$

is an injective homomorphism of abelian groups; so we would have $p = 0$ and, consequently, also $C = 0$, hence i would be an epimorphism; since i is also a monomorphism, and since $\mathcal{C}$ is an abelian category, i would turn out to be an isomorphism, contradicting the hypothesis.

ii) ⇒ *i)*: Assume under hypothesis *ii)* that there exist two objects E, F of $\mathcal{C}$ and a morphism $f \in \mathrm{Hom}_{\mathcal{C}}(E, F)$ such that $f \neq 0$ and $h^U(f) = 0$; let (M, i) be a kernel of the morphism f in the category $\mathcal{C}$; since $f \neq 0$ and $f \circ i = 0$, i cannot be an isomorphism; hence there exists $u : U \to E$ which does not factor through i; we have however

$$f \circ u = h^U(f)(u) = 0,$$

hence, by the definition of the kernel, u factors through i, which is a contradiction. ▮

DEFINITION 1. Let $\mathcal{C}$ be an abelian category; an object U of the category $\mathcal{C}$ is said to be a *generator* of the category provided it satisfies the equivalent conditions of Proposition 1. ▮

COROLLARY. *Let $\mathcal{C}$ be an abelian category, and let $(U_\lambda)_{\lambda \in I}$, $I \neq \emptyset$, be a family of objects of $\mathcal{C}$ such that U_λ is a generator of $\mathcal{C}$ for any $\lambda \in I$; assume that $(U, (j_\lambda)_{\lambda \in I})$ is a direct sum in $\mathcal{C}$ of the family of objects $(U_\lambda)_{\lambda \in I}$; then U is a generator of the category $\mathcal{C}$.*

Proof. Let $E \in \mathrm{Ob}\,\mathcal{C}$, and let (M, i) be a subobject of E such that i is not an isomorphism; let $\lambda \in I$ be fixed; there exists a morphism $u_\lambda : U_\lambda \to E$ which does not factor through i; let $u : U \to E$ be the unique morphism for which

$$u \circ j_\lambda = u_\lambda, \; u \circ j_\mu = 0 \text{ if } \mu \in I, \; \mu \neq \lambda;$$

if there existed $v \in \mathrm{Hom}_{\mathcal{C}}(U, M)$ such that $i \circ v = u$, then, denoting $v \circ j_\mu = v_\mu$, $\mu \in I$, we would have

$$i \circ v_\lambda = i \circ v \circ j_\lambda = u \circ j_\lambda = u_\lambda,$$

a contradiction. ▮*

If A is a ring, then for any non-empty set I, *the free A-module $A_s^{(I)}$ is a generator of the category* $\mathbf{Mod}(A)$: for let $i : M \to E$ be an injective A-homomorphism which is not an isomorphism; it follows that there exists $x \in E \setminus i(M)$; denote by $(e_\iota)_{\iota \in I}$ the canonical basis of $A_s^{(I)}$, and let ι_0 be a fixed element of I; define a mapping of sets u' from I to E which sends ι_0 into x and otherwise is arbitrary; by Proposition 1, § 10, there exists an A-linear mapping $u : A_s^{(I)} \to E$ obtained from u' by linear extension; hence we have $u(e_{\iota_0}) = x$; if there existed $v \in \mathrm{Hom}_A(A_s^{(I)}, M)$ such that $i \circ v = u$, we would have $x = u(e_{\iota_0}) = i(v(e_{\iota_0})) \in i(M)$, a contradiction. In particular, the A-module A_s is a generator of the category $\mathbf{Mod}(A)$.

* Returning to an abelian category $\mathcal{C}$, assume, moreover, that $\mathcal{C}$ satisfies Axiom AB3); denote, as in the case of modules, for a set I and an object E of $\mathcal{C}$ by $E^{(I)}$ the direct sum (i.e. the object component of it) of the family $(E_\lambda)_{\lambda \in I}$ of objects, where $E_\lambda = E$ for any $\lambda \in I$ (see § 7).

PROPOSITION 2. *Let $\mathcal{C}$ be an abelian category satisfying Axiom* AB3); *if U is a generator of the category $\mathcal{C}$, then any object M of $\mathcal{C}$ is (together with a suitable epimorphism) a quotient object of $U^{(I)}$ for some $I \neq \emptyset$.*

Proof. Let $I = \mathrm{Hom}_{\mathscr{C}}(U, M)$; we have $I \neq \emptyset$; for any $u \in I$ denote by j_u the canonical injection $U \to U^{(I)}$ of index u; the morphisms $u, u \in I$, induce by the definition of the direct sum a unique morphism

$$v : U^{(I)} \to M$$

such that $v \circ j_u = u$ for any $u \in I$; if we show that v is an epimorphism, then (M, v) will be a quotient object of $U^{(I)}$, and the Proposition will be proved; but if v were not an epimorphism, then, denoting by (N, k) an image of v, (N, k) would be a subobject of M, and k would not be an isomorphism (see § 4); since U is a generator, there would exist $u_0 \in I$ which does not factor through k; however, since $\mathscr{C}$ is an abelian category, using the canonical factorization of v, it follows that there exists an epimorphism $q : U^{(I)} \to N$ such that $k \circ q = v$; hence we have

$$k \circ q \circ j_{u_0} = v \circ j_{u_0} = u_0,$$

and u_0 factors through k. ▮

REMARK. Taking into account the preceding result concerning the fact that A is a generator of the category **Mod**(A), Proposition 2 may be viewed as being, in a certain way, the analogous result for abelian categories with direct sums and having a generator of Proposition 6, § 10. ▮*

Now we can give the following characterization of a generator in a category of modules:

PROPOSITION 3. *Let A be a ring; in order that an A-module U be a generator of the category* **Mod**(A) *it is necessary and sufficient that there exists a non-empty set I such that A_s is isomorphic to a direct summand of $U^{(I)}$.*

Proof. Assume that U is a generator of the category **Mod**(A); since **Mod**(A) satisfies Axiom AB3), it follows by Proposition 2 that there exist $I \neq 0$ and an epimorphism $v : U^{(I)} \to A_s$; since A_s is a projective A-module, it follows from the characterization *iv*) of Proposition 1, § 18, of projective modules that A_s is isomorphic to a direct summand of $U^{(I)}$ (see Proposition 6, § 9).

Conversely, if A_s is isomorphic to a direct summand of $U^{(I)}$ for some $I \neq \emptyset$, then there exists an injective homomorphism $j : A_s \to U^{(I)}$ left invertible (see the Corollary to Proposition 6, § 9); hence there exists $r : U^{(I)} \to A_s$ such that $r \circ j = 1_A$; to show that U is a generator of the category **Mod**(A), let E be an A-module, and let $M \neq E$ be a submodule of E, with the canonical injection i; since A_s is a generator of the category **Mod**(A), there exists $u \in \mathrm{Hom}_A(A_s, E)$ such that u does not factor through i; but, denoting by k_λ, $\lambda \in I$, the canonical injection of index λ $U \to U^{(I)}$, if for any $\lambda \in I$ the homomorphism $u \circ r \circ k_\lambda : U \to E$ factored through i, i.e., if there existed $v_\lambda \in \mathrm{Hom}_A(U, M)$ such that $i \circ v_\lambda = u \circ r \circ k_\lambda$, $\lambda \in I$, then, by the universal property of the direct sum there would exist a homomorphism $v : U^{(I)} \to$ $\to M$ such that $v \circ k_\lambda = v_\lambda$ for any $\lambda \in I$, and we would have $i \circ v \circ k_\lambda = i \circ v_\lambda =$ $= u \circ r \circ k_\lambda$, $\lambda \in I$; by virtue of the uniqueness, we would obtain $i \circ v = u \circ r$, hence $i \circ v \circ j = u \circ r \circ j = u$, and u would factor through i. ▮

* The notion dual to that of generator is the one of *cogenerator*: it is an object V of the abelian category $\mathscr{C}$ with the property that for any object E of $\mathscr{C}$ and any quotient object (P, p) of E for which p is not an isomorphism there exists a morphism $v \in \mathrm{Hom}_{\mathscr{C}}(E, V)$ which does not factor through p, or, equivalently, an object V such that the functor $h_V : \mathscr{C}^\circ \to \mathbf{Ab}$ is faithful. We have the result dual to the one given by Proposition 2: *in an abelian category $\mathscr{C}$ with direct products and having a cogenerator V, any object E together with a suitable monomorphism is a subobject of an object of the form V^I, where I is a suitable non-empty set*[1].

If in the abelian category $\mathscr{C}$ with direct sums there exists a generator U which is a projective object of the category (or, briefly, a "*projective generator*"), *then $\mathscr{C}$ has enough projective objects*: for, by Proposition 2 above, there exists a set $I \neq \emptyset$ and an epimorphism $u : U^{(I)} \to E$; but, according to Proposition 2, § 18, adapted for arbitrary abelian categories, $U^{(I)}$ is a projective object. We also formulate the dual result: *if in the abelian category $\mathscr{C}$ with direct products there exists a cogenerator V which is an injective object of the category* (briefly, an "*injective cogenerator*"), *then $\mathscr{C}$ has enough injective objects.*

If $\mathscr{C}$ is an abelian category having a cogenerator V, then, given the morphism $u \in \mathrm{Hom}_{\mathscr{C}}(E, F)$, from the fact that

$$\mathrm{Hom}(u, 1_V) : \mathrm{Hom}_{\mathscr{C}}(F, V) \to \mathrm{Hom}_{\mathscr{C}}(E, V)$$

is a monomorphism of the category $\mathbf{Ab}$ *it follows that u is an epimorphism of the category $\mathscr{C}$*: since let $X \in \mathrm{Ob}\,\mathscr{C}$, and let $r \in \mathrm{Hom}_{\mathscr{C}}(F, X)$ be such that $r \circ u = 0$; V being a cogenerator of $\mathscr{C}$, there exist a set $I \neq \emptyset$ and a monomorphism $j : X \to V^I$; by the result analogous to Proposition 3, § 7, for arbitrary abelian categories, there exist the $\mathbf{Z}$-isomorphisms $\varphi : \mathrm{Hom}_{\mathscr{C}}(E, V^I) \rightleftarrows (\mathrm{Hom}_{\mathscr{C}}(E, V))^I$, $\psi : \mathrm{Hom}_{\mathscr{C}}(F, V^I) \rightleftarrows (\mathrm{Hom}_{\mathscr{C}}(F, V))^I$ such that the following diagram commutes:

$$\begin{array}{ccc}
\mathrm{Hom}_{\mathscr{C}}(F, V^I) & \xrightarrow[\sim]{\ \psi\ } & (\mathrm{Hom}_{\mathscr{C}}(F, V))^I \\
{\scriptstyle \mathrm{Hom}(u,\, 1_{V^I})}\big\downarrow & & \big\downarrow {\scriptstyle \prod_{\lambda \in I} u_\lambda^*} \\
\mathrm{Hom}_{\mathscr{C}}(E, V^I) & \xrightarrow[\sim]{\ \varphi\ } & (\mathrm{Hom}_{\mathscr{C}}(E, V))^I,
\end{array}$$

where $u_\lambda^* = \mathrm{Hom}(u, 1_V)$ for any $\lambda \in I$; as, by hypothesis, $\mathrm{Hom}(u, 1_V)$ is a monomorphism, $\prod_\lambda u_\lambda^*$ is also a monomorphism (see Corollary 4 to Proposition 2, § 6), and, consequently, $\mathrm{Hom}(u, 1_{V^I})$ is also a monomorphism; from $r \circ u = 0$ it follows now that $j \circ r \circ u = 0$, i.e., $\mathrm{Hom}(u, 1_{V^I})(j \circ r) = 0$, hence $j \circ r = 0$; since j is a monomorphism, we obtain $r = 0$; hence u is an epimorphism.

We leave to the reader to formulate and to prove the dual result.

---

[1] As in the dual case, V^I is a direct product (i.e. its object component) of the family $(V_\lambda)_{\lambda \in I}$ of objects, with $V_\lambda = V$ for any $\lambda \in I$.

It was shown in § 8 that if A is a ring and E is an A-module, then the set partially ordered by inclusion of all submodules of E is a complete lattice, the meet, respectively the join, as operations in the lattice being the intersection, respectively the sum of submodules. In attempting to extend this result to the subobjects of an object in an abelian category $\mathcal{C}$, the only difficulty is to see whether the class of these subobjects is a set; for this condition to be satisfied it suffices to require in addition that the category $\mathcal{C}$ has a generator; in fact we have the following result:

PROPOSITION 4. *If the abelian category $\mathcal{C}$ has a generator U, then for any $E \in \mathrm{Ob}\,\mathcal{C}$, the class of subobjects of E is a set*[1].

Proof. Consider two distinct subobjects of E, and let (M_1, i_1), (M_2, i_2) be two pairs representing them, respectively; we want to show that there exists a morphism $U \to E$ which factors through one of the morphisms i_1 or i_2, but does not factor through the other: to do this, consider the intersection of the two subobjects, and let (M, i) be a pair representing it; hence we have the morphisms $j_1 : M \to M_1$, $j_2 : M \to M_2$ for which

$$i_1 \circ j_1 = i_2 \circ j_2 = i;$$

the pair (M, j_1), respectively (M, j_2), defines a subobject of M_1, respectively M_2; if both j_1 and j_2 were isomorphisms, the subobjects of E defined by (M_1, i_1) and (M_2, i_2) would coincide with the subobject defined by (M, i), contradicting the hypothesis; hence, one of j_1 or j_2, say j_1, is not an isomorphism; since U is a generator of the category $\mathcal{C}$, there exists a morphism $u : U \to M_1$ which does not factor through j_1; the morphism $i_1 \circ u$ factors through i_1; let us show that $i_1 \circ u$ does not factor through i_2: If it did, there would exist $v : U \to M_2$ such that

$$i_1 \circ u = i_2 \circ v; \qquad (1)$$

but if (C_1, q_1), respectively (C_2, q_2), is a cokernel of i_1, respectively i_2, then, denoting by (C, π_1, π_2) a direct product in the category $\mathcal{C}$ of the family (C_1, C_2) and by q the morphism $E \to C$ obtained from q_1 and q_2 by the definition of the direct product, (M, i) is exactly the kernel of q (see § 8); the situation is displayed on the commutative diagram below:

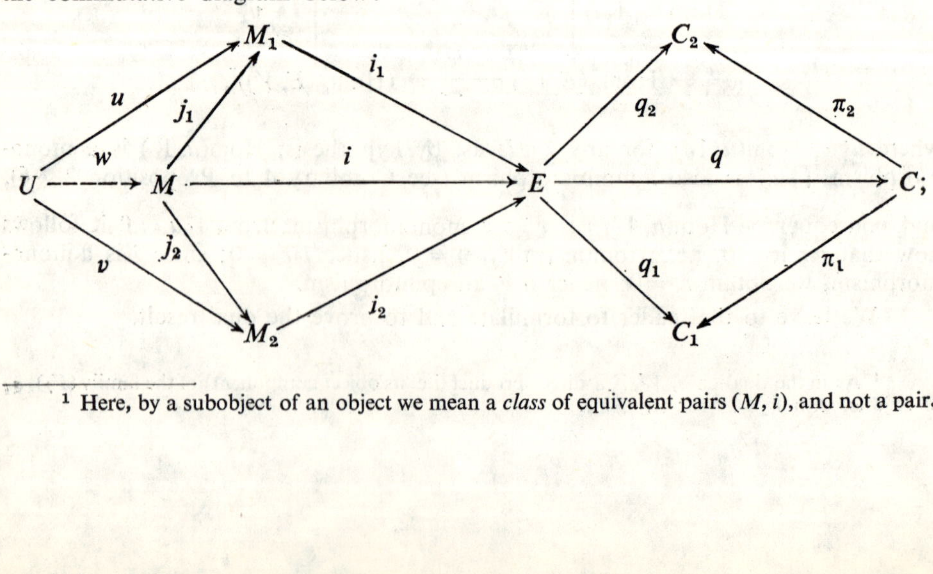

[1] Here, by a subobject of an object we mean a *class* of equivalent pairs (M, i), and not a pair.

we claim that under these conditions we have

$$q \circ i_1 \circ u = 0; \tag{2}$$

for, by (1) we get

$$\pi_1 \circ q \circ i_1 \circ u = q_1 \circ i_1 \circ u = 0, \qquad \pi_2 \circ q \circ i_1 \circ u = q_2 \circ i_2 \circ v = 0,$$

and (2) follows in view of the uniqueness condition in the definition of the direct product; by the definition of the kernel, from (2) the existence of a unique morphism $w : U \to M$ follows with the property that

$$i_1 \circ u = i \circ w = i_1 \circ j_1 \circ w;$$

but then we have $u = j_1 \circ w$, hence u factors through j_1, which was excluded; hence $i_1 \circ u$ does not factor through i_2.

Further, consider for any subobject of E a pair (N, j) representing it and the set $X_{(N,\, j)}$ consisting of those $f \in \mathrm{Hom}_{\mathcal{C}}(U, E)$ which factor through j; it is obvious that this set does not depend on the representative (N, j) chosen, and hence that it defines a mapping from the class of subobjects of E to the class of subsets of $\mathrm{Hom}_{\mathcal{C}}(U, E)$; this last class is a set, since $\mathrm{Hom}_{\mathcal{C}}(U, E)$ is a set by the definition of the category; but by the preceding considerations, the mapping defined is also injective; hence the class of subobjects of E is injectively mapped into a set, and thus it is a set. ▮

COROLLARY. *In an abelian category $\mathcal{C}$ having a generator U, the class of quotient objects of an arbitrary object E of $\mathcal{C}$ is a set.*

Proof. As shown in § 8, between the class of subobjects of E and the class of quotient objects of E there exists a canonical bijection; since the first class is a set, so is the second one. ▮

A category $\mathcal{C}$ is said to be *well-powered* if for any object E of $\mathcal{C}$, the subobjects of E (considered as classes of equivalent pairs) form a set; the dual notion is that of co-well-powered category: a category $\mathcal{C}$ is said to be *co-well-powered* if for any object E of $\mathcal{C}$, the class of quotient objects of E is a set. Proposition 4 and the Corollary to it may be formulated in these terms as follows: *An abelian category having a generator is well-powered and co-well-powered.*

Applying the results obtained in § 8, the subobjects of an object E of an abelian category $\mathcal{C}$ having a generator form a lattice with a first and a last element: the meet in the lattice is the intersection of subobjects, while the join is the sum of subobjects, the first element of the lattice is the zero subobject of E, and the last element of the lattice is the total subobject of E; since the canonical bijection is antimonotone, it also follows that the quotient objects form a lattice, the meet in the lattice of a family of quotient objects being the image under the canonical mapping of the sum of the subobjects corresponding to the quotient objects of the family, and so on. If, moreover, $\mathcal{C}$ is a category with direct sums, then the lattice of subobjects of an arbitrary object is a complete join-semilattice, and if $\mathcal{C}$

is a category with direct products, then the lattice of subobjects of an object is a complete meet-semilattice; we leave it to the reader to formulate the corresponding results for the quotient objects of an object.

We may now prove the following result:

PROPOSITION 5. *Let $\mathcal{C}$ be an abelian category with direct products and having a generator and enough injectives; then $\mathcal{C}$ has an injective cogenerator.*

Proof. Let U be a generator of the category $\mathcal{C}$; by the Corollary to Proposition 4, the class of quotient objects of U is a set; let $\mathcal{M}$ be a chosen set of representatives of these quotient objects, and let $(Q, (q_P)_P)$ be a direct product of the family of all objects P, where (P, p) ranges over $\mathcal{M}$.

Since $\mathcal{C}$ has enough injectives, there exist an injective object V and a monomorphism $j : Q \to V$; we claim that *V is a cogenerator of $\mathcal{C}$*: To prove it, we have to show that the functor $h_V : \mathcal{C}^\circ \to \mathbf{Ab}$ is faithful; let then $u^\circ : F^\circ \to E^\circ$ be a morphism of $\mathcal{C}^\circ$, $u^\circ \neq 0$. Since U is a generator of $\mathcal{C}$, and since $u : E \to F$ is a non-zero morphism of $\mathcal{C}$, we have $h^U(u) \neq 0$; consequently, there exists some $f \in \mathrm{Hom}_{\mathcal{C}}(U, E)$ such that

$$h^U(u)(f) = u \circ f \neq 0;$$

let (I, r) be a coimage of $u \circ f$; the canonical factorization of $u \circ f$ (see § 4) then shows that there exists a monomorphism $k : I \to F$ such that $k \circ r = u \circ f$. Since (I, r) is a representative of a quotient object of U, we may suppose, with no loss of generality, that $(I, r) \in \mathcal{M}$; then, by the definition of $(Q, (q_P)_P)$, there exists a morphism

$$m : I \to Q$$

such that

(3) $$q_I \circ m = 1_I, \; q_P \circ m = 0 \text{ for } (P, p) \neq (I, r) \text{ in } \mathcal{M};$$

from the first relation (3) it follows that m is a monomorphism. On the other hand, V being injective and k being a monomorphism, there exists $l \in \mathrm{Hom}_{\mathcal{C}}(F, V)$ rendering commutative the following diagram:

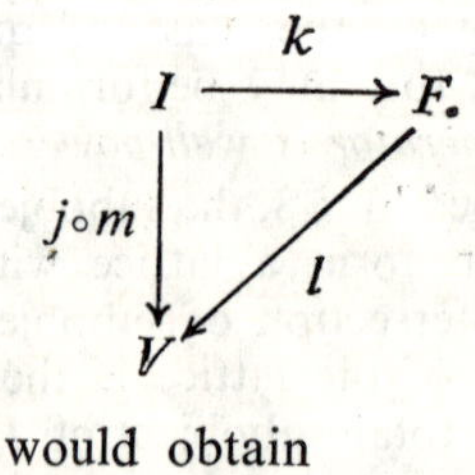

Suppose $h_V(u^\circ) = 0$; then we would obtain

$$0 = h_V(u^\circ)(l) = l \circ u,$$

consequently,

$$0 = l \circ u \circ f = l \circ k \circ r = j \circ m \circ 1_I \circ r,$$

and as $j \circ m$ is a monomorphism and r an epimorphism, it would follow that $1_I = 0$, hence $I = 0$, hence $u \circ f = 0$, which contradicts the choice of f. ∎*

PROPOSITION 6. *Let A be a ring, and let $u : E \to F$ be an injective homomorphism of A-modules; the following assertions concerning u are equivalent:*

i) For any non-zero submodule N of F, $u^{-1}(N)$ is a non-zero submodule of E;

ii) For any non-zero cyclic submodule N of F, $u^{-1}(N)$ is a non-zero submodule of E;

iii) For any A-module X and any $f \in \mathrm{Hom}_A(F, X)$, if $f \circ u$ is a monomorphism, then f is a monomorphism.

Proof. *i)* ⇒ *ii)*: Obvious.

ii) ⇒ *i)*: Let N a be non-zero submodule of E; hence there exists $y \in N$, $y \neq 0$; the submodule Ay generated by y in N is a non-zero cyclic submodule, and we have $Ay \subseteq N$; hence, by *ii)*,

$$u^{-1}(N) \supseteq u^{-1}(Ay) \neq 0.$$

i) ⇒ *iii)*: Let X and f be as in the statement, and let $f \circ u$ be a monomorphism; assume, by way of contradiction, that f is not a monomorphism; then, by Proposition 2, § 3, $\mathrm{Ker} f$ is a non-zero submodule of F; by *i)*, we have $u^{-1}(\mathrm{Ker} f) \neq 0$; but $u^{-1}(\mathrm{Ker} f) = \mathrm{Ker}(f \circ u)$ (see § 4), a contradiction.

iii) ⇒ *i)*: Let N be a submodule of F; if $u^{-1}(N) = 0$, then, considering the canonical surjection $p : F \to F/N$, it follows that

$$0 = u^{-1}(N) = \mathrm{Ker}(p \circ u),$$

and $p \circ u$ is a monomorphism; applying *iii)*, it follows that p is a monomorphism; according to the results of § 3, we have $N = 0$. ∎

* It may be shown that in an abelian category $\mathcal{C}$, the assertions corresponding to *i)* and *iii)* of Proposition 6 are still equivalent (see § 8, the definition of the image, respectively of the inverse image under a morphism of a subobject of the domain, respectively of the range).*

DEFINITION 2. Let A be a ring, and let $u : E \to F$ be a homomorphism of A-modules; u is said to be an *essential homomorphism* if u is injective and satisfies the equivalent conditions of Proposition 6.

If E is a submodule of F, then F is said to be an *essential extension*[1] of E provided the canonical injection $E \to F$ is an essential homomorphism.

* If $\mathcal{C}$ is an abelian category and $u : E \to F$ is a monomorphism of the category, then the pair (F, u) is said to be an *essential extension* of E provided u

---

[1] Note the discrepancy in the terminology: since we have the exact sequence $0 \to E \xrightarrow{u} F \to \mathrm{Coker}\, u \to 0$, u being the canonical injection, we ought to say, according to the Remark following Proposition 3, § 5, that "F is an extension of Coker u by E", and not "F is an (...) extension of E"; we call the reader's attention to this discordance, but we maintain it to be in agreement with the terminology in current use.

satisfies the assertions corresponding for abelian categories to *i*) and *iii*) of Proposition 6; in this case, u is said to be an *essential morphism* or an *essential monomorphism.** ▮

Here are some consequences of the definition:

In the diagram of A-modules

$$E \xrightarrow{u} F \xrightarrow{v} G$$

if u and v are essential homomorphisms, then $v \circ u$ is an essential homomorphism: for if G' is a non-zero submodule of G, then

$$(4) \qquad (v \circ u)^{-1}(G') = u^{-1}(v^{-1}(G')) \neq 0;$$

conversely, *if for the same diagram $v \circ u$ is an essential homomorphism, and if, moreover, v is an injective homomorphism*[1], *then u and v are essential homomorphisms*: for if we had, in the above notation, $v^{-1}(G') = 0$, it would follow, taking the equality in (4) and the injectivity of u into account, that $(v \circ u)^{-1}(G') = u^{-1}(0) = 0$, a contradiction.

In particular, *if the module G is an essential extension of the module E, then for any submodule F of G containing E, F is an essential extension of E, and G is an essential extension of F.*

Obviously, *any isomorphism is an essential homomorphism.*

Let F be an A-module, let E be a submodule of it, and let $(M_\lambda)_{\lambda \in I}$ be a (non-void) family of submodules of F such that $E \subseteq M_\lambda$ and that M_λ be an essential extension of E for any $\lambda \in I$; if $M = \bigcup_{\lambda \in I} M_\lambda$ is a submodule[2] *of F, then M is an essential extension of E*: for if M' is a non-zero submodule of M, then there exists $\lambda_0 \in I$ for which $M' \cap M_{\lambda_0} \neq 0$; otherwise we would have

$$0 = \bigcup_{\lambda \in I} (M' \cap M_\lambda) = M' \cap (\bigcup_{\lambda \in I} M_\lambda) = M';$$

but then, denoting by i the inclusion mapping $E \to M_{\lambda_0}$, we have $0 \neq i^{-1}(M' \cap M_{\lambda_0}) = (M' \cap M_{\lambda_0}) \cap E = M' \cap E$.

From here it follows that *if E is a submodule of the A-module F, then there exists at least a submodule M of F with the property that M is an essential extension of E and that it is maximal with this property*: for, the set of submodules of F which are essential extensions of E contains E, hence it is not empty, and, by the preceding considerations, this set is inductively ordered with respect to inclusion; now, we need only apply Zorn's Lemma.

---

[1] Notice that u follows injective, since $v \circ u$ is so (see § 3).

[2] For instance, if the family $(M_\lambda)_{\lambda \in I}$ of submodules of F is directed to the right with respect to inclusion (see Corollary 6 to Proposition 2, § 8).

Now let $n \geqslant 1$, and let for any $i \in [1, n]$

$$u_i : E_i \to F_i$$

be a given essential homomorphism; then the homomorphism

$$\bigoplus_{i=1}^{n} u_i : \bigoplus_{i=1}^{n} E_i \to \bigoplus_{i=1}^{n} F_i$$

is essential: by the associativity of the direct sum (see § 7) we may reduce the proof to the case $n = 2$, the general case being obtained by induction; denoting by p_i the canonical projection

$$F_1 \oplus F_2 = F_1 \times F_2 \to F_i,$$

where $i = 1, 2$, let G be a non-zero submodule of $F_1 \oplus F_2$; hence we cannot have $p_1(G) = 0$ and $p_2(G) = 0$, since this would imply $G = 0$; assume that, say, $p_1(G) \neq 0$; since u_1 is an essential homomorphism, it follows $u_1^{-1}(p_1(G)) \neq 0$; u_1 being injective, we get

$$0 \neq u_1(u_1^{-1}(p_1(G))) \subseteq p_1(G \cap p_1^{-1}(u_1(E_1)));$$

hence

(5) $$G_1 = G \cap p_1^{-1}(u_1(E_1)) \neq 0.$$

Assume that $p_2(G_1) \neq 0$; then, since u_2 is an essential homomorphism, we have $u_2^{-1}(p_2(G_1)) \neq 0$; arguing as above, we get in this case that

$$G \cap p_1^{-1}(u_1(E_1)) \cap p_2^{-1}(u_2(E_2)) = G_1 \cap p_2^{-1}(u_2(E_2)) \neq 0;$$

hence there exists $(y_1, y_2) \in G$, $(y_1, y_2) \neq 0$, for which

$$y_1 = p_1(y_1, y_2) \in u_1(E_1), \quad y_2 = p_2(y_1, y_2) \in u_2(E_2),$$

hence for which there exist $x_i \in E_i$, $i = 1, 2$, such that

$$(y_1, y_2) = (u_1(x_1), u_2(x_2)) = (u_1 \oplus u_2)(x_1, x_2);$$

it follows that $(x_1, x_2) \neq 0$ and $(x_1, x_2) \in (u_1 \oplus u_2)^{-1}(G)$, hence

(6) $$(u_1 \oplus u_2)^{-1}(G) \neq 0.$$

If $p_2(G_1) = 0$, then by virtue of (5) we may find an element $(y_1, y_2) \in G_1$, $(y_1, y_2) \neq 0$; by the fact that $p_2(G_1) = 0$ we have $y_2 = 0$; but then $y_1 \neq 0$ and

$y_1 = p_1(y_1, y_2) \in u_1(E_1)$, and there exists $x_1 \in E_1$, $x_1 \neq 0$, for which $y_1 = u_1(x_1)$; then $(x_1, 0) \in (u_1 \oplus u_2)^{-1}(G)$, and we recover (6).

Finally, *let F be an A-module, and let E be a submodule of it*; the set $\mathcal{M}$ of those submodules M of F for which $M \cap E = 0$ is not empty, since $0 \in \mathcal{M}$; it is inductively ordered with respect to inclusion, since if $(M_\lambda)_{\lambda \in I}$ is a non-void family of elements of $\mathcal{M}$ totally ordered by inclusion, then $\bigcup_{\lambda \in I} M_\lambda$ is a submodule of F, and we have

$$(\bigcup_{\lambda \in I} M_\lambda) \cap E = \bigcup_{\lambda \in I} (M_\lambda \cap E) = 0,$$

hence $\bigcup_{\lambda \in I} M_\lambda \in \mathcal{M}$; consequently, by Zorn's Lemma, there exists in $\mathcal{M}$ at least a maximal element; *if M is* such an element of $\mathcal{M}$, that is, *a submodule of F maximal with the property that* $M \cap E = 0$, *then the homomorphism obtained by the composition of the homomorphisms*

$$E \xrightarrow{i} F \xrightarrow{p} F/M,$$

where i is the canonical injection and p the canonical surjection, is injective: indeed, let $(p \circ i)(x) = 0$, with $x \in E$; it follows that the coset of x modulo M is zero, hence that $x \in M$; since $M \cap E = 0$, it follows $x = 0$; *moreover, the homomorphism* $p \circ i$ *is essential*: for if G is a non-zero submodule of F/M, then, by the Third Isomorphism Theorem for modules (Proposition 2, § 4), G is of the form $G = F'/M$, where F' is a (of course, non-zero) submodule of F such that $F' \supseteq M$; to say that $(p \circ i)^{-1}(F'/M) = 0$ would mean

$$0 = i^{-1}(p^{-1}(F'/M)) = F' \cap E;$$

hence, it would follow that $F' \in \mathcal{M}$, and by the maximality of M in $\mathcal{M}$ we would have $F' = M$; but then we would obtain $G = F'/M = 0$, a contradiction.

EXAMPLES. 1) Consider the **Z**-module **Q** (the group of rational numbers with respect to addition); **Q** is an essential extension of its submodule **Z**: for let M be a non-zero submodule of **Q**; there exists $r/s \in M$, $r, s \in \mathbf{Z}$, $r/s \neq 0$; we have $r \neq 0$, hence, as $r = s \cdot (r/s) \in M$, it follows that $M \cap \mathbf{Z} \neq 0$. This result and the preceding arguments imply that any submodule of **Q** containing **Z** is an essential extension of **Z**.

2) Consider the **Z**-module **R** (the group of real numbers with respect to the ordinary addition); if $\alpha \in \mathbf{R}$ is an irrational number, then the set $\mathbf{Z}\alpha$ of the integer multiples of α is a non-zero submodule of **R**; but $\mathbf{Z}\alpha \cap \mathbf{Q} = 0$, since if $n\alpha \in \mathbf{Q}$, with $n \in \mathbf{Z}$, we cannot have $n \neq 0$, since this would imply $\alpha \in \mathbf{Q}$, which is not the case; it follows that **R** is not an essential extension of **Q**. ▮

PROPOSITION 7. *Let A be a ring; the following assertions are valid:*

a) An A-module E is injective if and only if any essential homomorphism having E as domain is an isomorphism;

b) If Q is an injective A-module, a submodule E of Q is injective if and only if for any submodule of Q, say M, which is an essential extension of E we have $M = E$;

c) If $u : E \to Q$ is an injective homomorphism of A-modules, where Q is an injective A-module, and if $h: E \to F$ is an essential homomorphism, then there exists an injective homomorphism $l : F \to Q$ rendering commutative the diagram

$$\begin{array}{ccc} E & \xrightarrow{h} & F. \\ {\scriptstyle u}\downarrow & \swarrow{\scriptstyle l} & \\ Q & & \end{array}$$

Proof. a) Assume that E is injective, and let $u: E \to F$ be an essential homomorphism; since u is injective, we may apply the characterization of injective modules given by Proposition 3, *i*), § 18: there exists a homomorphism $f : F \to E$ such that $f \circ u = 1_E$; from this relation it follows that f is an epimorphism; but if we had $\operatorname{Ker} f \neq 0$, since u is an essential homomorphism, it would follow

$$0 \neq u^{-1}(\operatorname{Ker} f) = \operatorname{Ker}(f \circ u)$$

(see § 4), a contradiction; hence $\operatorname{Ker} f = 0$, and f is an isomorphism; but then $u = f^{-1}$, and u is also an isomorphism.

Conversely, if E has the stated property, let $\mathfrak{a}$ be a left ideal of A, and let $f : \mathfrak{a} \to E$ be a linear mapping; let us show that there exists $x \in E$ for which $f(\alpha) = \alpha x$ whatever $\alpha \in \mathfrak{a}$; by the characterization of injective modules given by Proposition 3, *vi*), § 18, it will follow that E is injective; to do this, let S be the fibered sum of E and A_s with respect to the diagram

$$\begin{array}{ccc} \mathfrak{a} & \xrightarrow{i} & A_s, \\ {\scriptstyle f}\downarrow & & \\ E & & \end{array}$$

where i is the canonical injection; there exist homomorphisms $u : E \to S$ and $v : A_s \to S$ which fill in the diagram to a cocartesian square; from the fact that i is an injective mapping it follows by Proposition 7, § 13, that u is an injective mapping too; let M be a submodule of S maximal with the property that $M \cap u(E) = 0$; by the preceding considerations, the composite mapping

$$E \underset{\sim}{\xrightarrow{u'}} u(E) \xrightarrow{j} S \xrightarrow{p} S/M$$

is an essential homomorphism (here u' is induced by u by virtue of Corollary 2 to Proposition 1, § 4, hence it is an isomorphism, j is the canonical injection, and p the canonical surjection); by the hypothesis on E it follows that $p \circ j \circ u'$, hence also $p \circ j$, is an isomorphism; but then any element $z \in S$ has the property that $p(z) = p(j(y)) = p(y)$ for some $y \in u(E)$; hence $z - y \in M$, and $z \in u(E) + M$, that is, $S = u(E) + M$; since $u(E) \cap M = 0$, it follows that

$$S = u(E) \dot{+} M;$$

by Corollary 1 to Proposition 2, § 9, there exists $g \in \mathrm{Hom}_A(S, u(E))$ such that $g \circ j = 1_{u(E)}$; then let

$$x = u'^{-1}(g(v(1))),$$

where 1 is the unit element of A; since we have $j \circ u' = u$, it follows that for any $\alpha \in \mathfrak{a}$,

$$u'(f(\alpha)) = g(j(u'(f(\alpha)))) = g((u \circ f)(\alpha)) = g(v(\alpha)),$$

whence

$$f(\alpha) = u'^{-1}(g(v(\alpha))) = \alpha x.$$

b) If E is injective, and M, where $M \subseteq Q$, is an essential extension of E, then, since the canonical injection $k : E \to M$ is an essential homomorphism, it follows by *a*) that k is an isomorphism, hence that $M = E$.

Conversely, if E satisfies the condition stated, then let N be a submodule of Q maximal with the property that $N \cap E = 0$; if i is the canonical injection $E \to Q$ and p is the canonical surjection $Q \to Q/N$, then, by the preceding considerations, the composition

$$E \xrightarrow{i} Q \xrightarrow{p} Q/N$$

is an essential homomorphism; since $p \circ i$ is injective and Q is an injective module, it follows that there exists an A-homomorphism $u : Q/N \to Q$ rendering commutative the diagram

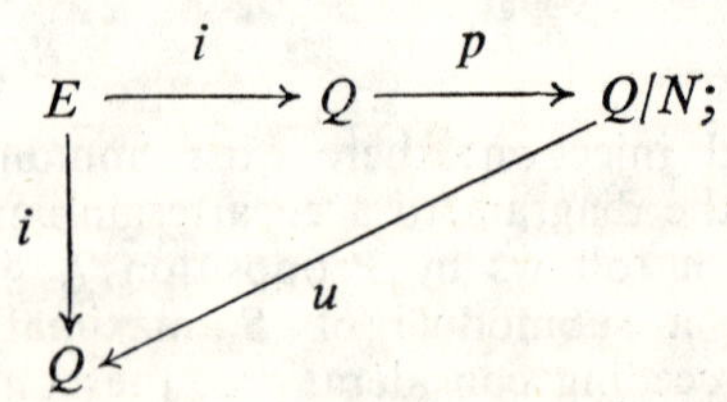

u is injective, since i is injective and $p \circ i$ is an essential homomorphism; let $M = u(Q/N)$, and let $u' : Q/N \xrightarrow{\sim} M$ be the isomorphism induced by u; u' being an

isomorphism, $u' \circ p \circ i : E \to M$ is an essential homomorphism; but for any $x \in E$ we have $(u' \circ p \circ i)(x) = u(p(i(x))) = i(x) = x$, which means, first, that $E \subseteq M$, and next, that $u' \circ p \circ i$ is the inclusion mapping of E into M; hence M is an essential extension of E; by the hypothesis on E, we have $M = E$, and hence $u' \circ p \circ i = 1_E$; hence $u' \circ p$ is a linear retraction associated with i; by the Corollary to Proposition 6, § 9, $E = i(E)$ is a direct summand of Q; since Q is an injective module, it follows that E is also injective (see the Corollary to Proposition 4, § 18).

c) By the characterization of injective modules given by Proposition 3, *i*), § 18, since h is an injective homomorphism, the existence of some $l : F \to Q$ follows rendering commutative the diagram stated; but if we had $\operatorname{Ker}(l) \neq 0$, since h is an essential homomorphism, it would follow that $0 \neq h^{-1}(\operatorname{Ker} l) = \operatorname{Ker}(l \circ h)$ (see § 4); but $l \circ h = u$ and $\operatorname{Ker}(u) = 0$; hence $\operatorname{Ker}(l) = 0$ also, and l is injective. ∎

PROPOSITION 8. *Let A be a ring, and let $u : E \to Q$ be an injective homomorphism of A-modules; the following assertions concerning the triple (E, Q, u) are equivalent:*

i) *Q is an injective A-module, and u is an essential homomorphism;*

ii) *Q is an injective A-module, and it is the only injective submodule of Q containing $u(E)$;*

iii) *Q is an injective A-module, and for any injective homomorphism $v: E \to R$, where R is an injective A-module, there exists an injective homomorphism $w : Q \to R$ rendering commutative the diagram*

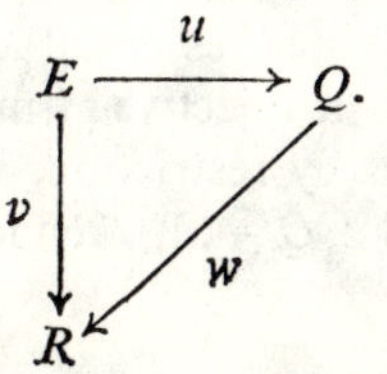

Proof. i) ⇒ *ii*): Let Q' be an injective submodule of Q for which $u(E) \subseteq Q'$; by Proposition 3, *iv*), § 18, the exact sequence of A-modules

$$0 \to Q' \xrightarrow{k} Q \xrightarrow{p} Q/Q' \to 0,$$

where k, respectively p, represents the canonical injection, respectively the canonical surjection, is a split exact sequence; if M is a submodule supplementary to Q' in Q, and if we had $M \neq 0$, as Q is an essential extension of $u(E)$, it would follow that $M \cap u(E) \neq 0$, hence $M \cap Q' \neq 0$, a contradiction; consequently, $M = 0$, and hence $Q = Q'$.

ii) ⇒ *iii*): Let N be a maximal essential extension of $v(E)$ in R; if N' is an essential extension of N contained in R, then N' is an essential extension of $v(E)$ contained in R, and hence, by the maximality of N', we have $N' = N$; by

Proposition 7, *b*), taking into account the fact that R is an injective A-module, it follows that N is an injective A-module; denoting by v' the homomorphism $E \to N$ that v induces by restriction, v' is an essential homomorphism; by Proposition 7, *c*), there exists an injective homomorphism $k : N \to Q$ rendering commutative the diagram

(7)

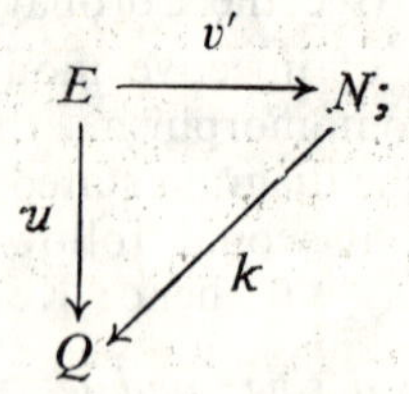

since we have $k(N) \approx N$, it follows that $k(N)$ is an injective submodule of Q; the commutativity of diagram (7) implies $u(E) \subseteq k(N)$; from *ii*) it follows that $k(N) = Q$, hence k is surjective and thus an isomorphism; let j be the canonical injection of N into R, and let

$$w = j \circ k^{-1} : Q \to R;$$

w is injective, and we have $w \circ u = j \circ k^{-1} \circ u = j \circ v' = v$.

iii) $\Rightarrow$ *i*): Let M be a maximal essential extension of $u(E)$ in Q; arguing as above, we deduce that M is an injective A-module; denoting by u' the homomorphism $E \to M$ that u induces by restriction, it follows by *iii*) that there exists an injective homomorphism $w : Q \to M$ rendering commutative the diagram

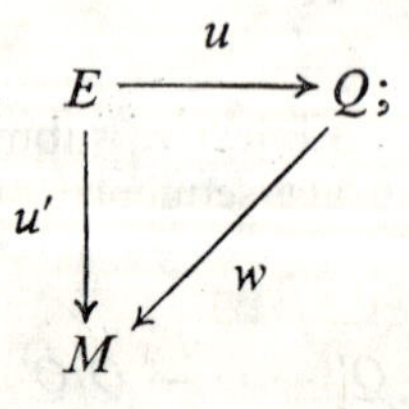

on the other hand, u' is an essential homomorphism, being the composition of the canonical injection $u(E) \to M$ with the isomorphism $E \to u(E)$ that u induces by restriction; since w is an injective homomorphism, it follows that u is an essential homomorphism. ∎

DEFINITION 3. Given an A-module E, a pair (Q, u), where Q is an A-module while u is an injective homomorphism $E \to Q$, is said to be an *injective envelope* of E if the triple (E, Q, u) satisfies the equivalent conditions of Proposition 8. ∎

PROPOSITION 9. *If A is a ring, any A-module E has at least one injective envelope; if (Q, u) and (Q', u') are two injective envelopes of the A-module E, then there exists an A-isomorphism $f : Q \xrightarrow{\sim} Q'$ rendering commutative the diagram*

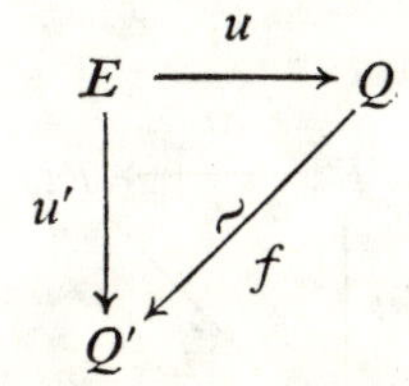

Proof. By Proposition 5, § 18, there exists an injective A-module R and a monomorphism $v : E \to R$; let Q be a maximal essential extension of $v(E)$ in R; Q admits no essential extension contained in R and different from Q, because this would contradict the maximality of Q; hence, by Proposition 7, *b*), it follows that Q is an injective A-module; the composition of the canonical injection $v(E) \to Q$, which is an essential homomorphism, with the isomorphism $E \xrightarrow{\sim} v(E)$ that v induces by restriction gives an essential homomorphism $u : E \to Q$, and by condition *i*) of Proposition 8, the pair (Q, u) is an injective envelope of E.

For the second part, let (Q, u) and (Q', u') be as in the statement; since Q' is an injective A-module and u is an essential homomorphism, it follows by Proposition 7, *c*) that there exists an injective homomorphism $f : Q \to Q'$ rendering commutative the diagram in the statement; if $x \in E$, then $u'(x) = f(u(x)) \in f(Q)$, hence $u'(E) \subseteq f(Q)$; consequently, u' admits the factorization

$$E \xrightarrow[\sim]{\bar{u}'} u'(E) \xrightarrow{i} f(Q) \xrightarrow{j} Q',$$

where $\bar{u}'$ is the isomorphism that u' induces by restriction, while i and j are the canonical injections (see Corollary 2 to Proposition 1, § 4); by the general properties of the essential homomorphisms it follows that Q' is an essential extension of $f(Q)$; but $f(Q) \approx Q$, hence $f(Q)$ is an injective A-module; by Proposition 7, *b*) it follows $f(Q) = Q'$, hence f is surjective, and the result is proved. ▮

COROLLARY. *Under the conditions of Proposition* 8, *the equivalent assertions i*), *ii*), *iii*) *of Proposition* 8 *are equivalent also to the assertion*:

iv) *u is an essential homomorphism, and for any essential homomorphism $h : E \to F$ there exists an injective homomorphism $l : F \to Q$ rendering commutative the diagram*

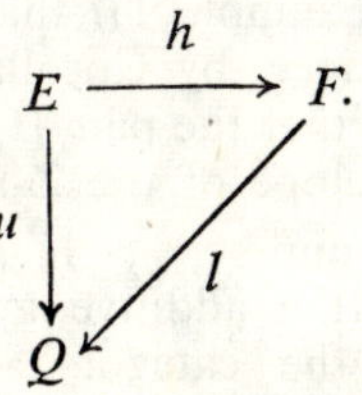

Proof. i) of Proposition 8 *implies iv)*: It follows directly from Proposition 7, *c*).

iv) implies i) of Proposition 8: We have to show that Q is an injective A-module; let (R, v) be an injective envelope of E; this does exist by Proposition 9; by *iv*) there exists an injective homomorphism $l : R \to Q$ rendering commutative the diagram

$$\begin{array}{ccc} E & \xrightarrow{v} & R; \\ {\scriptstyle u}\downarrow & \swarrow {\scriptstyle l} & \\ Q & & \end{array}$$

by the characterization given for injective modules by Proposition 3, *iv'*), § 18, and owing to the fact that $l(R) \approx R$ hence that $l(R)$ is an injective module, it follows that there exists a submodule S of Q such that

$$Q = l(R) \dot{+} S;$$

if we had $S \neq 0$, it would follow, u being an essential homomorphism, that $S \cap u(E) \neq 0$, hence, as $u(E) \subseteq l(R)$, that $S \cap l(R) \neq 0$, a contradiction; it follows $S = 0$ and $Q = l(R) \approx R$, hence Q is injective. ▮

REMARKS. 1) Owing to the fact that, as Proposition 9 shows, the injective envelopes of a module are uniquely determined up to an isomorphism, one sometimes talks, by an abuse of language, of "*the injective envelope*" of a given A-module; sometimes, also by an abuse of language, by this is meant only the component Q of an injective envelope (Q, u).

2) It should not be believed that the injective envelope would define a functor $\mathbf{Mod}(A) \to \mathbf{Mod}(A)$: indeed, there is no question here of a universal property, since in all the above propositions, the homomorphisms, inclusive of the isomorphism f in Proposition 9, may be constructed, but not uniquely (see also the Remark following Proposition 13, § 10).

3) Roughly speaking, assertions *ii*) and *iii*) of Proposition 8 may be interpreted as: "an injective envelope of an A-module E is a minimal injective extension", while assertion *iv*) of the Corollary to Proposition 9 as: "an injective envelope is a maximal essential extension of the module". ▮

EXAMPLES. 3) We saw (Example 1)) that the $\mathbf{Z}$-module $\mathbf{Q}$ is an essential extension of its submodule $\mathbf{Z}$; since, by Corollary 1 to Proposition 3, § 18, $\mathbf{Q}$ is an injective $\mathbf{Z}$-module, it follows that the pair $(\mathbf{Q}, i)$ (where i is the canonical injection $\mathbf{Z} \to \mathbf{Q}$) is an injective envelope of the $\mathbf{Z}$-module $\mathbf{Z}$.

4) Let $p \in \mathbf{Z}$ be a prime number > 1; let $m \geqslant 1$ be a natural number; consider the $\mathbf{Z}$-module $\mathbf{Z}/p^m\mathbf{Z}$ (the additive group of the integers modulo p^m); consider as category of indices the category associated with the totally ordered

set I of all natural numbers $\geqslant m$; for each $k \in I$ define $E_k = \mathbf{Z}/p^k\mathbf{Z}$, and if $k \leqslant l$, $k, l \in I$, then define the mapping

$$u_{kl} : E_k \to E_l$$

by

$$u_{kl}(x + p^k\mathbf{Z}) = p^{l-k}x + p^l\mathbf{Z},$$

where $x \in \mathbf{Z}$; in this way we get an inductive system of $\mathbf{Z}$-modules indexed by the category I; set

$$\mathbf{Z}_{p^\infty} = \varinjlim_{I} E_k;$$

denote by j_k the canonical mapping $E_k \to \mathbf{Z}_{p^\infty}$; since the mappings u_{kl}, $k, l \in I$, are injective, and since the index set I is directed to the right, it follows, by Proposition 3, *e*), § 13, that the j_k, $k \in I$, are also injective linear mappings. But $\mathbf{Z}_{p^\infty}$ is the sum of its family of submodules $(j_k(E_k))_{k \in I}$; if $k, l \in I$, $k \leqslant l$, $y \in E_l$, $y \neq 0$, let r be the smallest natural number for which $p^r y = 0$; we have $r \geqslant 1$, $p^{r-1}y \in u_{kl}(E_k)$, $p^{r-1}y \neq 0$, hence u_{kl} is an essential homomorphism; but then $j_l(E_l)$ is an essential extension of $j_k(E_k)$, hence also of $j_m(E_m)$; consequently, $\mathbf{Z}_{p^\infty}$ is an essential extension of $j_m(E_m)$.

On the other hand, let q be a prime number, $q \neq p$; for a given $z \in \mathbf{Z}_{p^\infty}$ denote by r the smallest natural number r' for which $p^{r'}z = 0$; there exist $\lambda, \mu \in \mathbf{Z}$ such that $\lambda p^r + \mu q = 1$, hence

$$z = \lambda p^r z + \mu q z = q(\mu z);$$

if t is a natural number $\geqslant 1$, then, if $z \in j_k(E_k)$, we have

$$j_k(1 + p^k\mathbf{Z}) = p^t j_{k+t}(1 + p^{k+t}\mathbf{Z}),$$

and since $z = s j_k(1 + p^k\mathbf{Z})$, $s \in \mathbf{Z}$, we have

$$z = p^t j_{k+t}(s + p^{k+t}\mathbf{Z});$$

hence, as any non-zero natural number is a product of powers of prime numbers, it follows that $\mathbf{Z}_{p^\infty}$ is a divisible $\mathbf{Z}$-module, hence by Corollary 1 to Proposition 3, § 18, it is an injective $\mathbf{Z}$-module.

Consequently, $(\mathbf{Z}_{p^\infty}, \varphi)$, where φ is the homomorphism obtained by the composition of the canonical injection of $j_m(E_m)$ into $\mathbf{Z}_{p^\infty}$ with the isomorphism $E_m \xrightarrow{\sim} j_m(E_m)$ that j_m induces by restriction, is an injective envelope of $E_m = \mathbf{Z}/p^m\mathbf{Z}$. ∎

* The notion of injective envelope of an object may be defined in an abelian category $\mathcal{C}$, for instance, by the corresponding assertion of *i*), Proposition 8; thus, if $E \in \operatorname{Ob} \mathcal{C}$, an *injective envelope* of E is a pair (Q, u), where Q is an injective object of $\mathcal{C}$ while $u : E \to Q$ is an essential monomorphism. We have the following result, which we state without proof[1]: *In an abelian category satisfying Axiom* AB5) *and having a generator, each object has an injective envelope.* *

---

[1] For the proof, see for instance, [32], Theorem 6.25, p. 128, or [74], Theorem III, 3.2, p. 89.

§ 20. Representation Theorems

* Enriching a category $\mathcal{C}$ with additional structures inspired by the properties of the categories of modules, we obtain categories which are represented, in a way or another, as coming from module categories either as subcategories of such categories or by means of an equivalence of categories or by the so-called "passing to the quotient". The representation theorems obtained constitute one of the most important directions of research in the Theory of Categories; on the one hand, they give axiomatic characterizations for the categories of modules, while on the other hand, what is more important, they allow us to obtain in this way in the abstract categories endowed with adequate additional structures, results similar to those in the module categories, which would be difficult to prove directly. The aim of this Section is to give some of these representation theorems as well as an example of the manner in which they are used.

PROPOSITION 1. *Let $\mathcal{C}$ and $\mathcal{C}'$ be two categories, and let $T : \mathcal{C} \to \mathcal{C}'$ be a functor; the following assertions concerning these data are equivalent*:

i) There exists a functor $U : \mathcal{C}' \to \mathcal{C}$ and two natural equivalences

$$\Phi : 1_{\mathcal{C}} \rightleftarrows U \circ T, \quad \Psi : 1_{\mathcal{C}'} \rightleftarrows T \circ U$$

such that the natural transformations

$$T(\Phi),\ \Psi(T) : T \to T \circ U \circ T$$

(see § 10) coincide:

$$T(\Phi) = \Psi(T);$$

ii) There exists a functor $U : \mathcal{C}' \to \mathcal{C}$ and two natural equivalences

$$\Phi : 1_{\mathcal{C}} \rightleftarrows U \circ T, \quad \Psi : 1_{\mathcal{C}'} \rightleftarrows T \circ U;$$

iii) The following two assertions are valid:

a) For any pair of objects E, F of $\mathcal{C}$, the mapping

$$T(E, F) : \mathrm{Hom}_{\mathcal{C}}(E, F) \to \mathrm{Hom}_{\mathcal{C}'}(T(E), T(F))$$

is a bijection, that is, T is a fully faithful functor;

b) Any object E' *of* $\mathcal{C}'$ *is isomorphic to an object of the form* $T(E)$ *for some suitably chosen object* E *of* $\mathcal{C}$.

Proof. *i)* $\Rightarrow$ *ii)*: Obvious.

ii) $\Rightarrow$ *iii)*: For assertion *a)*, let $E, F \in \operatorname{Ob} \mathcal{C}$, and let $u, v \in \operatorname{Hom}_{\mathcal{C}}(E, F)$ such that $T(u) = T(v)$; then $U(T(u)) = U(T(v))$, and as we have the commutative diagram[1]

$$\begin{array}{ccc} E & \overset{u}{\underset{v}{\rightrightarrows}} & F \\ \Phi(E)\ \big\downarrow \wr & & \wr \big\downarrow\ \Phi(F) \\ U(T(E)) & \xrightarrow{U(T(u)) = U(T(v))} & U(T(F)), \end{array}$$

it follows that $u = v$; let now $u' \in \operatorname{Hom}_{\mathcal{C}'}(T(E), T(F))$; define

$$u = (\Phi(F))^{-1} \circ U(u') \circ \Phi(E); \tag{1}$$

if we show that $T(u) = u'$, then it will follow that the mapping $T(E, F)$ is bijective; but if we show that

$$T(U(T(u))) = T(U(u')), \tag{2}$$

then the commutativity of the diagram

$$\begin{array}{ccc} T(E) & \overset{T(u)}{\underset{u'}{\rightrightarrows}} & T(F) \\ \Psi(T(E))\ \big\downarrow \wr & & \wr \big\downarrow\ \Psi(T(F)) \\ T(U(T(E))) & \xrightarrow{T(U(T(u))) = T(U(u'))} & T(U(T(F))) \end{array}$$

---

[1] In fact here we should have written two distinct commutative diagrams: one for u, the other for v; we wrote them together in order to have a general view of the situation; one should not think however that *a priori* the commutativity of the diagram $E \overset{u}{\underset{v}{\rightrightarrows}} F$ is also assumed; the same remark is valid for the similar diagrams that follow.

implies $T(u) = u'$; to prove (2), we notice first that from the commutativity of the diagram

$$\begin{array}{ccc} E & \xrightarrow[\sim]{\Phi(E)} & U(T(E)) \\ {\scriptstyle \Phi(E)}\Big\downarrow \wr & & \wr \Big\downarrow {\scriptstyle \Phi(U(T(E)))} \\ U(T(E)) & \xrightarrow{U(T(\Phi(E)))} & U(T(U(T(E)))) \end{array}$$

and from the fact that $\Phi(E)$ is an isomorphism of the category $\mathcal{C}$ it follows that

$$\Phi(U(T(E))) = U(T(\Phi(E))); \tag{3}$$

from (1), from (3) (for E and F), and from the commutativity of the diagram

$$\begin{array}{ccc} U(T(E)) & \xrightarrow{U(u')} & U(T(F)) \\ {\scriptstyle \Phi(U(T(E)))}\Big\downarrow \wr & & \wr \Big\downarrow {\scriptstyle \Phi(U(T(F)))} \\ U(T(U(T(E)))) & \xrightarrow{U(T(U(u')))} & U(T(U(T(F)))) \end{array}$$

it follows that

$$U(T(u)) = (U(T(\Phi(F))))^{-1} \circ U(T(U(u'))) \circ U(T(\Phi(E))) =$$
$$= (\Phi(U(T(F))))^{-1} \circ U(T(U(u'))) \circ \Phi(U(T(E))) = U(u'),$$

whence, applying the functor T, we get (2).

The validity of assertion *b*) is obvious, since for any $E' \in \text{Ob}\, \mathcal{C}'$ we have the isomorphism

$$\Psi(E') : E' \rightleftarrows T(U(E')).$$

iii) $\Rightarrow$ *i*): To define the functor U, we make use of the fact that for any object E' of $\mathcal{C}'$ there exists, according to assertion *b*) of *iii*), an object E of $\mathcal{C}$ and an isomorphism

$$f_{E'} : E' \rightleftarrows T(E);$$

then, by the choice principle for arbitrary classes (see § 0), we choose for any $E' \in \text{Ob}\, \mathcal{C}'$, a well determined $E \in \text{Ob}\, \mathcal{C}$ and a well determined isomorphism $f_{E'}$ with the above properties; define

$$U(E') = E.$$

If F' is a second object of $\mathcal{C}'$ and $u' \in \mathrm{Hom}_{\mathcal{C}'}(E', F')$, consider the composition of morphisms

$$T(E) \xrightarrow[\sim]{f_{E'}^{-1}} E' \xrightarrow{u'} F' \xrightarrow[\sim]{f_{F'}} T(F),$$

where $F = U(F')$; by assertion *a*) of *iii*), there exists a unique morphism $u \in \mathrm{Hom}_{\mathcal{C}}(E, F)$ such that $T(u) = f_{F'} \circ u' \circ f_{E'}^{-1}$; define

$$U(u') = u \in \mathrm{Hom}_{\mathcal{C}}(U(E'), U(F'));$$

if G' is a third object of the category $\mathcal{C}'$ and $v' \in \mathrm{Hom}_{\mathcal{C}'}(F', G')$, then let $G = U(G')$ and $v = U(v')$; we have

$$T(v \circ u) = T(v) \circ T(u) = f_{G'} \circ v' \circ f_{F'}^{-1} \circ f_{F'} \circ u' \circ f_{E'}^{-1} = f_{G'} \circ v' \circ u' \circ f_{E'}^{-1},$$

whence $U(v' \circ u') = v \circ u = U(v') \circ U(u')$; likewise, $T(1_E) = 1_{T(E)} = f_{E'} \circ 1_{E'} \circ f_{E'}^{-1}$, whence $U(1_{E'}) = 1_E = 1_{U(E')}$; hence U is a functor $\mathcal{C}' \to \mathcal{C}$.

From the given definitions it follows that setting

$$\Psi(E') = f_{E'} \tag{4}$$

for any $E' \in \mathrm{Ob}\,\mathcal{C}'$, we get a natural equivalence

$$\Psi : 1_{\mathcal{C}'} \xrightarrow{\sim} T \circ U.$$

On the other hand, considering for any $E \in \mathrm{Ob}\,\mathcal{C}$ the isomorphism $f_{T(E)} : T(E) \to$ $\to T(U(T(E)))$, there exists by assertion *a*) of *iii*) a uniquely determined morphism $g_E \in \mathrm{Hom}_{\mathcal{C}}(E, U(T(E)))$ such that

$$T(g_E) = f_{TE}; \tag{5}$$

if F is a second object of $\mathcal{C}$ and $u \in \mathrm{Hom}_{\mathcal{C}}(E, F)$, then, in view of the fact that Ψ is a natural transformation, the following diagram is commutative:

$$\begin{array}{ccc}
T(E) & \xrightarrow{\quad T(u) \quad} & T(F) \\
\Big\downarrow{\scriptstyle f_{TE}} & & \Big\downarrow{\scriptstyle f_{TF}} \\
T(U(T(E))) & \xrightarrow{\ T(U(T(u)))\ } & T(U(T(F))),
\end{array}$$

i.e., we have

$$T(g_F \circ u) = T(g_F) \circ T(u) = T(U(T(u))) \circ T(g_E) = T[U(T(u)) \circ g_E],$$

and since $T(E, U(T(F)))$ is a bijective mapping, it follows $g_F \circ u = U(T(u)) \circ g_E$; hence, denoting for any $E \in \text{Ob}\, \mathcal{C}$

$$\Phi(E) = g_E, \tag{6}$$

we get a natural transformation

$$\Phi : 1_{\mathcal{C}} \to U \circ T.$$

Also by assertion *a*) of *iii*), for any $E \in \text{Ob}\, \mathcal{C}$ there exists a unique morphism $h_E : U(T(E)) \to E$ with the property $T(h_E) = f_{T(E)}^{-1}$; we have

$$T(g_E) \circ T(h_E) = f_{T(E)} \circ f_{T(E)}^{-1} = 1_{TUTE} = T(1_{UTE}),$$

and by the bijectivity of $T(UTE, UTE)$ it follows $g_E \circ h_E = 1_{UTE}$; likewise, $h_E \circ g_E = 1_E$, and $g_E = \Phi(E)$ is an isomorphism for any $E \in \text{Ob}\, \mathcal{C}$; hence Φ is a natural equivalence.

On the other hand, for any object E of $\mathcal{C}$ we have by (4), (5), (6),

$$\Psi(T(E)) = f_{T(E)} = T(g_E) = T(\Phi(E)),$$

which shows that

$$\Psi(T) = T(\Phi). \blacksquare$$

DEFINITION 1. Given the categories $\mathcal{C}, \mathcal{C}'$ and the functor $T : \mathcal{C} \to \mathcal{C}'$, the functor T is said to be an *equivalence* (*of categories*) *from* $\mathcal{C}$ *to* $\mathcal{C}'$ (or *between* $\mathcal{C}$ *and* $\mathcal{C}'$) provided the equivalent conditions of Proposition 1 are satisfied.

In case there exists an equivalence between the categories $\mathcal{C}$ and $\mathcal{C}'$, one says that the category $\mathcal{C}$ is *equivalent* to the category $\mathcal{C}'$. $\blacksquare$

To justify the term "equivalent", we should show that *any category* $\mathcal{C}$ *is equivalent to itself*, that *if* $\mathcal{C}$ *is equivalent to* $\mathcal{C}'$, *then* $\mathcal{C}'$ *is in turn equivalent to* $\mathcal{C}$, and that *if* $\mathcal{C}$ *is equivalent to* $\mathcal{C}'$, *and* $\mathcal{C}'$ *is equivalent to* $\mathcal{C}''$, *then* $\mathcal{C}$ *is equivalent to* $\mathcal{C}''$: The first assertion is obvious if we take for T in Definition 1 the functor $1_{\mathcal{C}}$; the second assertion is also obvious by virtue of Proposition 1, *ii*); also by that Proposition, the functor $U : \mathcal{C}' \to \mathcal{C}$ having the same properties as in Proposition 1, *ii*) is an equivalence between the category $\mathcal{C}'$ and the category $\mathcal{C}$, of course, with T being the corresponding functor $\mathcal{C} \to \mathcal{C}'$; for the third assertion, let us show that if each of the functors $T : \mathcal{C} \to \mathcal{C}'$ and $T' : \mathcal{C}' \to \mathcal{C}''$ is an equivalence between the corresponding categories, then the functor $T' \circ T : \mathcal{C} \to \mathcal{C}''$ is an equivalence between $\mathcal{C}$ and $\mathcal{C}''$; indeed, if $E, F \in \operatorname{Ob} \mathcal{C}$, then the following diagram is commutative:

$$\begin{array}{ccc}
\operatorname{Hom}_{\mathcal{C}}(E, F) & \xrightarrow{(T' \circ T)(E, F)} & \operatorname{Hom}_{\mathcal{C}''}(T'(T(E)), T'(T(F))), \\
& {\scriptstyle T(E,F)} \searrow \qquad \nearrow {\scriptstyle T'(T(E), T(F))} & \\
& \operatorname{Hom}_{\mathcal{C}'}(T(E), T(F)) &
\end{array}$$

since for $u \in \operatorname{Hom}_{\mathcal{C}}(E, F)$ we have $(T' \circ T)(u) = T'(T(u))$; since $T(E, F)$ and $T'(T(E), T(F))$ are bijections, it follows that their composition $(T' \circ T)(E, F)$ is also a bijection; next, if $E'' \in \operatorname{Ob} \mathcal{C}''$, then there exist $E' \in \operatorname{Ob} \mathcal{C}'$ and an isomorphism $f'' : E'' \rightrightarrows T'(E')$; likewise, there exist $E \in \operatorname{Ob} \mathcal{C}$ and an isomorphism $f' : E' \rightrightarrows T(E)$; then

$$T'(f') \circ f'' : E'' \to (T' \circ T)(E)$$

is an isomorphism, and conditions *iii*) of Proposition 1 are satisfied for the categories $\mathcal{C}, \mathcal{C}''$ and the functor $T' \circ T$.

Having in view the preceding considerations, if a category $\mathcal{C}$ is equivalent to a category $\mathcal{C}'$, the two categories $\mathcal{C}$ and $\mathcal{C}'$ are said to be *equivalent*.

PROPOSITION 2. *Let $T : \mathcal{C} \to \mathcal{C}'$ be an equivalence between the categories $\mathcal{C}$ and $\mathcal{C}'$, and let $U : \mathcal{C}' \to \mathcal{C}$ be a functor satisfying for $\mathcal{C}$, $\mathcal{C}'$, and T the conditions of Proposition* 1, *i*) *or the conditions of Proposition* 1, *ii*); *in this case, U is left and a right adjoint of T.*

Proof. We shall prove the Proposition only for a functor U satisfying the apparently weaker conditions of Proposition 1, *ii*); also, we shall prove only the fact that U is a right adjoint functor of T; by the preceding considerations, to prove that U is also a left adjoint of T we need only interchange the roles of T and U.

Let $E \in \operatorname{Ob} \mathcal{C}$ and $F' \in \operatorname{Ob} \mathcal{C}'$; let $\theta(E, F')$ be the composition of the mappings

$$\operatorname{Hom}_{\mathcal{C}'}(T(E), F') \xrightarrow{U(T(E), F')} \operatorname{Hom}_{\mathcal{C}}(U(T(E)), U(F')) \xrightarrow{\operatorname{Hom}(\Phi(E), 1_{UF'})} \operatorname{Hom}_{\mathcal{C}}(E, U(F'));$$

since U is an equivalence between the categories $\mathcal{C}'$ and $\mathcal{C}$, the first mapping is bijective; further, as $\Phi(E)$ is an isomorphism, it follows that $h_{UF'}((\Phi(E))^{\circ}) = \operatorname{Hom}(\Phi(E), 1_{UF'})$ is an isomorphism of the category **Ens**, hence a bijective mapping; consequently, $\theta(E, F')$ is a bijective mapping.

If, moreover, the object E_1 of $\mathcal{C}$, the object F_1' of $\mathcal{C}'$, and the morphisms $u \in \operatorname{Hom}_{\mathcal{C}}(E_1, E)$, $v' \in \operatorname{Hom}_{\mathcal{C}'}(F', F_1')$ are also given, then consider the diagram[1]

$$
(7)\qquad \begin{array}{ccccc}
\operatorname{Hom}_{\mathcal{C}'}(T(E), F') & \xrightarrow{U(T(E),\, F')} & \operatorname{Hom}_{\mathcal{C}}(U(T(E)), U(F')) & \xrightarrow{\operatorname{Hom}(\Phi(E),\, 1_{UF'})} & \operatorname{Hom}_{\mathcal{C}}(E, U(F')) \\
\Big\downarrow {\scriptstyle \operatorname{Hom}(T(u),\, v')} & & \Big\downarrow {\scriptstyle \operatorname{Hom}(U(T(u)),\, U(v'))} & & \Big\downarrow {\scriptstyle \operatorname{Hom}(u,\, U(v'))} \\
\operatorname{Hom}_{\mathcal{C}'}(T(E_1), F_1') & \xrightarrow{U(T(E_1),\, F_1')} & \operatorname{Hom}_{\mathcal{C}}(U(T(E_1)), U(F_1')) & \xrightarrow{\operatorname{Hom}(\Phi(E_1),\, 1_{UF_1'})} & \operatorname{Hom}_{\mathcal{C}}(E_1, U(F_1'));
\end{array}
$$

the fact that the first square of diagram (7) is commutative is a consequence of the fact that U is a functor; the second square of diagram (7) is also commutative, since Φ is a natural transformation $1_{\mathcal{C}} \to U \circ T$; since the composition of the homomorphisms on the two rows of diagram (7) is $\theta(E, F')$, respectively $\theta(E_1, F_1')$, it follows that U is a right adjoint of T. ∎

By Proposition 2 and the considerations in § 10 concerning the adjoint functors it follows that *the functor U in Proposition* 1, *i*) (*or in Proposition* 1, *ii*)) *is uniquely determined by the functor T up to a natural equivalence.*

From Proposition 2 we get, applying Propositions 2 and 3, § 14, the following

COROLLARY. *If* $\mathcal{C}, \mathcal{C}'$ *are abelian categories and* $T : \mathcal{C} \to \mathcal{C}'$ *is an equivalence between* $\mathcal{C}$ *and* $\mathcal{C}'$, *then T is an exact functor.* ∎

It follows that the notion of "equivalence between two abelian categories" is stronger than the one of "exact and fully faithful functor", since the first satisfies in addition also condition *b*) of Proposition 1, *iii*).

EXAMPLES. 1) Let $\mathcal{C}, \mathcal{C}'$ be two categories, and let $T : \mathcal{C} \to \mathcal{C}'$, $U : \mathcal{C}' \to \mathcal{C}$ be two functors such that

$$U \circ T = 1_{\mathcal{C}}, \quad T \circ U = 1_{\mathcal{C}'};$$

in this case one says that $T : \mathcal{C} \to \mathcal{C}'$ is an *isomorphism of categories* and that U is the *inverse isomorphism* of T; one also says that $\mathcal{C}$ is *isomorphic* to $\mathcal{C}'$ or also that the two categories $\mathcal{C}$ and $\mathcal{C}'$ are *isomorphic.*

Taking as Φ, respectively Ψ, of Proposition 1, *i*) (or of Proposition 1, *ii*)) the identity natural transformation of $1_{\mathcal{C}}$, respectively $1_{\mathcal{C}'}$, we see that in this

---

[1] Here the diagram has been broken only to facilitate printing; the same remark applies to diagram (30) below.

case T, respectively U, is an equivalence between the category $\mathcal{C}$ and the category $\mathcal{C}'$, respectively between the category $\mathcal{C}'$ and the category $\mathcal{C}$.

2) Let $\mathcal{C}'$ be a category; consider the equivalence relation "isomorphic to" on the class of objects of $\mathcal{C}'$; making use of the choice principle for classes, let us choose a representative in each class of isomorphic objects; let $\mathcal{C}$ be the full subcategory of $\mathcal{C}'$ whose class of objects is exactly the class of the chosen representatives; let $T : \mathcal{C} \to \mathcal{C}'$ be the immersion functor (see Definition 6, § 12); the subcategory thus constructed is said to be a *skeleton* of the category $\mathcal{C}'$; the functor T is an equivalence between $\mathcal{C}$ and $\mathcal{C}'$: for if $E, F \in \text{Ob}\,\mathcal{C}$, then

$$T(E, F) = 1_{\text{Hom}_{\mathcal{C}}(E,\, F)},$$

and any $G \in \text{Ob}\,\mathcal{C}'$ is isomorphic to the representative of its class chosen in Ob $\mathcal{C}$.

3) Considering the dual $(\mathbf{Ab})^\circ$ of the category of abelian groups, define for any $G^\circ, H^\circ \in \text{Ob}(\mathbf{Ab})^\circ$ and any $u^\circ \in \text{Hom}_{(\mathbf{Ab})^\circ}(G^\circ, H^\circ)$ (hence G and H are abelian groups, while $u : H \to G$ is a homomorphism of abelian groups)

$$T(G^\circ) = \text{Hom}_{\mathbf{Z}}(G, \mathbf{R}/\mathbf{Z}), \; T(u^\circ) = \text{Hom}(u, 1_{\mathbf{R}/\mathbf{Z}});$$

if $M \subseteq G$ and $N \subseteq \mathbf{R}/\mathbf{Z}$, the set of all $f \in \text{Hom}_{\mathbf{Z}}(G, \mathbf{R}/\mathbf{Z})$ for which $f(M) \subseteq N$ is denoted by $W(M, N)$; taking as a basis of neighbourhoods [1] of $0 \in \text{Hom}_{\mathbf{Z}}(G, \mathbf{R}/\mathbf{Z})$ the family of all $W(F, D)$, where F ranges over all finite subsets of G while D over all open subsets of $\mathbf{R}/\mathbf{Z}$ containing $0 \in \mathbf{R}/\mathbf{Z}$ (the latter with the quotient topology of the natural topology on $\mathbf{R}$), we get on $\text{Hom}_{\mathbf{Z}}(G, \mathbf{R}/\mathbf{Z})$ a structure of compact topological abelian group (the addition being the usual one). Denote by $\mathcal{C}$ the category whose objects are the compact topological abelian groups and, for two objects C, D of $\mathcal{C}$, $\text{Hom}_{\mathcal{C}}(C, D)$ is the set of all continuous homomorphisms from C to D; then the above given mappings define a functor

$$T : (\mathbf{Ab})^\circ \to \mathcal{C};$$

the compact topological abelian group $T(G^\circ) = \text{Hom}_{\mathbf{Z}}(G, \mathbf{R}/\mathbf{Z})$ is called the *group of characters of the group G*. Pontrjagin's theory of character groups shows that the functor T is an equivalence between the category $(\mathbf{Ab})^\circ$ and the category $\mathcal{C}$. ▮

We now proceed to the representation theorems:

PROPOSITION 3 (**Mitchell's weak Full Imbedding Theorem**). *Let $\mathcal{C}$ be an abelian category satisfying Axiom* AB3) *and having a projective generator; suppose $\mathcal{D}$ is a full subcategory of $\mathcal{C}$ which (considered by itself) is a small preadditive category satisfying Axioms* AB1) *and* AB2) *and having the property that the immersion functor $S : \mathcal{D} \to \mathcal{C}$ is exact; then there exist a ring A and an exact fully faithful functor*

$$T : \mathcal{D} \to \mathbf{Mod}_d(A).$$

Proof. Let P_1 be a projective generator of the category $\mathcal{C}$; let

$$P = \bigoplus_{E \in \text{Ob}\,\mathcal{D}} \Big(\bigoplus_{u \in \text{Hom}(P_1,\, E)} P_u \Big),$$

[1] For this Example some knowledge on the Theory of topological groups is assumed; for a detailed account of this theory, the reader may refer to [81].

where for any u, $P_u = P_1$; by the Corollary to Proposition 1, § 19, P is also a generator of the category $\mathcal{C}$; moreover, by the analogue for abelian categories of Proposition 2, § 18, P is a projective object of $\mathcal{C}$; on the other hand, by Proposition 2, § 19, for any $E \in \mathrm{Ob}\,\mathcal{D}$ there exists a suitable epimorphism

$$v_E : \bigoplus_{u \in \mathrm{Hom}(P_1, E)} P_u \to E,$$

where $P_u = P_1$; let $w_E : P \to E$ be the unique morphism for which

$$w_E \circ j_F = \begin{cases} v_E & \text{if } F = E, \\ 0 & \text{if } F \in \mathrm{Ob}\mathcal{D},\ F \neq E, \end{cases}$$

where we denoted by j_F, $F \in \mathrm{Ob}\,\mathcal{D}$, the canonical injection $\bigoplus_{u \in \mathrm{Hom}(P_1, F)} P_u \to P$; by virtue of the relation $w_E \circ j_E = v_E$ it follows that w_E is an epimorphism; hence, any object of $\mathcal{D}$ is (together with a suitable epimorphism) a quotient object of P.

Denote $A = \mathrm{Hom}_{\mathcal{C}}(P, P)$; it is known (see § 2) that A is endowed with a ring structure, the multiplication law being given by the composition, and that for any object E of $\mathcal{C}$ the abelian group $\mathrm{Hom}_{\mathcal{C}}(P, E)$ is endowed, using the composition also as external law, with a right A-module structure; if $E, F \in \mathrm{Ob}\,\mathcal{C}$, and if $u \in \mathrm{Hom}_{\mathcal{C}}(E, F)$, then

$$\mathrm{Hom}_{\mathcal{C}}(1_P, u) : \mathrm{Hom}_{\mathcal{C}}(P, E) \to \mathrm{Hom}_{\mathcal{C}}(P, F)$$

is, with these structures of right A-module of $\mathrm{Hom}_{\mathcal{C}}(P, E)$ and $\mathrm{Hom}_{\mathcal{C}}(P, F)$, a homomorphism of right A-modules: for let $\alpha \in A$ and $f \in \mathrm{Hom}_{\mathcal{C}}(P, E)$; we have

$$\mathrm{Hom}(1_P, u)(f \circ \alpha) = u \circ f \circ \alpha = (\mathrm{Hom}(1_P, u)(f)) \circ \alpha.$$

Consequently, the functor $h^P : \mathcal{C} \to \mathbf{Ab}$ takes its values in $\mathbf{Mod}_d(A)$, and hence induces a uniquely determined additive functor $\mathcal{C} \to \mathbf{Mod}_d(A)$, denoted also h^P; let

$$T : \mathcal{D} \to \mathbf{Mod}_d(A)$$

be the composition $h^P \circ S$; since S is an exact functor and P is a projective object of the category $\mathcal{C}$, it follows that taking the composition of the forgetful functor $\mathbf{Mod}_d(A) \to \mathbf{Ab}$ with T, we get an exact functor (see Proposition 1, § 18); hence T is also an exact functor; since S is a faithful functor and P is a generator of $\mathcal{C}$, it follows that T is a faithful functor (see Proposition 1, § 19). It remains to show that for any $E, F \in \mathrm{Ob}\,\mathcal{D}$, the mapping $T(E, F)$: $\mathrm{Hom}_{\mathcal{D}}(E, F) \to \mathrm{Hom}_A(T(E), T(F))$ is surjective: to do this, let $f \in \mathrm{Hom}_A(T(E), T(F))$; by the preceding arguments, there exist some epimorphisms $p : P \to E$ and $q : P \to F$; let (K, i) be a kernel of p; hence in $\mathcal{C}$ we have the exact sequences

$$0 \to K \xrightarrow{i} P \xrightarrow{p} E \to 0, \quad P \xrightarrow{q} F \to 0;$$

since h^P is an exact functor $\mathcal{C} \to \mathbf{Mod}_d(A)$, we get in $\mathbf{Mod}_d(A)$ the following diagram in which the rows are exact sequences:

$$\begin{array}{ccccccccc} 0 & \longrightarrow & h^P(K) & \xrightarrow{h^P(i)} & h^P(P) = A_d & \xrightarrow{h^P(p)} & h^P(E) & \longrightarrow & 0 \\ & & & & \Big\downarrow{\scriptstyle g} & & \Big\downarrow{\scriptstyle f} & & \\ & & & & A_d & \xrightarrow{h^P(q)} & h^P(F) & \longrightarrow & 0 \end{array} \tag{8}$$

since A_d is a projective object in $\mathbf{Mod}_d(A)$ (see Corollary 1 to Proposition 1, § 18), and since $h^P(q)$ is an epimorphism, it follows that there exists a homomorphism of right A-modules $g\colon A_d \to A_d$ rendering diagram (8) commutative; let $g(1) = \alpha \in A$; since $\alpha \in \mathrm{Hom}_{\mathcal{C}}(P, P)$, we may consider $h^P(\alpha)\colon A_d \to A_d$; if $\beta \in A_d$, then $g(\beta) = g(1)\circ\beta = \alpha\circ\beta = h^P(\alpha)(\beta)$, hence

$$g = h^P(\alpha);$$

thus we have $h^P(q)\circ h^P(\alpha)\circ h^P(i) = h^P(q)\circ g\circ h^P(i) = f\circ h^P(p)\circ h^P(i) = 0$; consequently, in the following diagram (in which the rows are exact sequences)

$$\begin{array}{ccccccccc} 0 & \longrightarrow & K & \xrightarrow{i} & P & \xrightarrow{p} & E & \longrightarrow & 0 \\ & & & & \Big\downarrow{\scriptstyle \alpha} & & \Big\downarrow{\scriptstyle u} & & \\ & & & & P & \xrightarrow{q} & F & \longrightarrow & 0 \end{array} \tag{9}$$

we have $q\circ\alpha\circ i = 0$, since h^P is a faithful functor; since (E, p) is a cokernel of i (see Proposition 2, *b*), § 5), it follows that there exists an $u \in \mathrm{Hom}_{\mathcal{C}}(E, F) = \mathrm{Hom}_{\mathcal{D}}(E, F)$ with the property of making diagram (9) commute; we have

$$h^P(u)\circ h^P(p) = h^P(q)\circ h^P(\alpha) = h^P(q)\circ g = f\circ h^P(p),$$

and since $h^P(p)$ is an epimorphism, it follows that

$$T(u) = h^P(u) = f. \blacksquare$$

We also have

PROPOSITION 4 (**Weak Group Valued Imbedding Theorem**). *Let $\mathcal{C}$ be an abelian category having a projective generator; let $\mathcal{D}$ be a subcategory of $\mathcal{C}$ which (when considered by itself) is a small preadditive category satisfying Axioms* AB1) *and* AB2) *and having the property that the immersion functor* $S : \mathcal{D} \to \mathcal{C}$ *is exact; then there exists an exact faithful functor*

$$T : \mathcal{D} \to \mathbf{Ab}.$$

Proof. If P is a projective generator of the category $\mathcal{C}$, then as T we take the composition of the functor $h^P : \mathcal{C} \to \mathbf{Ab}$ with S; since both h^P and S are exact faithful functors, the same is also true for T. ∎

We shall now prove the following Theorem, which was obtained from the work done by Peter Freyd, Pierre Gabriel, and Barry Mitchell:

PROPOSITION 5 (**Freyd-Gabriel-Mitchell Theorem**). *If* $\mathcal{D}$ *is a small abelian category, then there exist an abelian category* $\mathcal{C}$ *with direct sums and having a projective generator, and an exact fully faithful functor*

$$U : \mathcal{D} \to \mathcal{C}.$$

A stepwise *proof* including several lemmas will be given.

First Step. Since $\mathcal{D}$ is a small category, the functors from $\mathcal{D}$ to **Ab** taken as objects and the natural transformations between them taken as morphisms form a category: indeed, if L, M are two such functors, the class of natural transformations $L \to M$ is a set, being a subclass of $\prod_{E \in \mathrm{Ob}\,\mathcal{D}} \mathrm{Hom}_{\mathcal{D}}(L(E), M(E))$, which is a set (see the considerations following Definition 3, § 12, as well as the Remark following Definition 7, § 10); this category was denoted **Fonct** ($\mathcal{D}$, **Ab**).

Let **Add**($\mathcal{D}$, **Ab**), abbreviated $\mathcal{A}$, be the full subcategory of **Fonct**($\mathcal{D}$, **Ab**) whose objects are the additive functors $\mathcal{D} \to \mathbf{Ab}$; the category **Add**($\mathcal{D}$, **Ab**) $= \mathcal{A}$ may be canonically set up as an abelian category satisfying Axioms AB3), AB3°), and AB5), namely as follows: If $L, M \in \mathrm{Ob}\,\mathcal{A}$ and Φ_1, Φ_2 are natural transformations $L \to M$, then the sum $\Phi_1 + \Phi_2$ is defined "pointwise", i.e.,

$$(\Phi_1 + \Phi_2)(E) = \Phi_1(E) + \Phi_2(E), \qquad E \in \mathrm{Ob}\,\mathcal{D}$$

(see also the considerations following Proposition 2, § 16); a zero object of the category $\mathcal{A}$ is the "zero functor", i.e. the functor sending any object E of $\mathcal{D}$ into the zero abelian group. If $\Phi : L \to M$ is a morphism in $\mathcal{A}$, a kernel, respectively a cokernel of Φ in the category $\mathcal{A}$ may be also taken pointwise, in other words, for each $E \in \mathrm{Ob}\,\mathcal{D}$ the kernel (K_E, i_E), respectively the cokernel (P_E, p_E) of $\Phi(E)$ in **Ab** is considered, and then the pair (K, i), respectively (P, p), where $K(E) = K_E$, $i(E) = i_E$, $P(E) = P_E$, $p(E) = p_E$, is a kernel, respectively a cokernel of Φ in **Add**($\mathcal{D}$, **Ab**) (see the considerations concerning diagram (2), § 4, for the proof of the fact that i and p are natural transformations); since the monomorphisms, respectively the epimorphisms in a preadditive category satisfying Axiom AB1) are characterized by the fact that their kernel, respectively cokernel is zero (see § 4), it follows that in $\mathcal{A}$ a morphism Φ is a monomorphism, respectively an epimorphism if and only if for any $E \in \mathrm{Ob}\,\mathcal{D}$, $\Phi(E)$ is injective, respectively surjective. By the definition of the image and the coimage it follows that the image and the coimage of the morphism Φ may also be taken pointwise, hence that in **Add**($\mathcal{D}$, **Ab**) Axiom AB2) is also satisfied. As for the direct product and the direct sum, these again may be constructed pointwise: if $(L_\lambda)_{\lambda\in I}$ is a family of objects of $\mathcal{A}$, a direct product $(L, (q_\lambda)_{\lambda\in I})$ in this category of the family $(L_\lambda)_{\lambda\in I}$ is given by

$$L(E) = \prod_{\lambda\in I} L_\lambda(E), \qquad E \in \operatorname{Ob} \mathcal{D},$$

$q_\lambda(E)$ being the canonical projection $\prod_{\lambda\in I} L_\lambda(E) \to L_\lambda(E)$, $\lambda \in I$; similarly for direct sums; the fact that in this way additive functors are obtained follows from relations of the type (6), § 6, and (6), § 7. Taking into account the preceding arguments, it follows that the intersection and the sum of subobjects (see § 8) are also done pointwise, that is, if $L \in \operatorname{Ob} \mathcal{A}$ and $((M_\lambda, i_\lambda))_{\lambda\in I}$ is a family of subobjects of L, then, since the intersection and the sum of subgroups respectively represent an intersection and a sum of the respective subobjects in the category **Ab**, the pair (M, i), where $M(E) = \bigcap_{\lambda\in I} M_\lambda(E)$ and $i(E)$ is the canonical injection $M(E) \to L(E)$, $E \in \operatorname{Ob} \mathcal{C}$, is an intersection of the given family of subobjects, and analogously for the sum; since Axiom AB5) is satisfied in **Ab** (see Corollary 7 to Proposition 2, § 8), it follows that Axiom AB5) is satisfied also in **Add**($\mathcal{D}$, **Ab**). By the preceding considerations and by the very construction of projective and inductive limits in an abelian category satisfying AB3°) and AB3), respectively (see Proposition 5, § 12, and Proposition 2, § 13), it follows that these are constructed pointwise too; in particular, this is true for fibered products and fibered sums.

The detailed verification of all the above assertions is quite straightforward, and it is, as well as the verification of other assertions of this type below, a good routine exercise for the reader.

We cannot prove in the manner described above the existence of enough injectives in **Add**($\mathcal{D}$, **Ab**), though this is valid in **Ab** (see Proposition 5, § 18); this is due to the fact that the construction of g from f in the diagram of the statement of i), Proposition 3, § 18, does not define a natural transformation; the result will however follow after the proof of the following Lemma:

LEMMA 1. *If $(H, (j_E)_{E\in \operatorname{Ob}\mathcal{D}})$ is a direct sum in $\mathcal{A}$ of the family of objects $(h^E)_{E\in\operatorname{Ob}\mathcal{D}}$ (where h^E, $E \in \operatorname{Ob}\mathcal{D}$, is the functor defined in § 2), then H is a generator of $\mathcal{A}$.*

Proof. It suffices to show that the functor

$$\chi^H\colon \mathbf{Add}(\mathcal{D}, \mathbf{Ab}) \to \mathbf{Ab}$$

defined by $\chi^H(L) = \operatorname{Hom}_{\mathcal{A}}(H, L)$, $L \in \operatorname{Ob}\mathcal{A}$, is faithful, where, for obvious reasons, in the standard notation used for general categories (see § 2) "h" has been replaced by "χ". Then let $\Phi \in \operatorname{Hom}_{\mathcal{A}}(L, M)$; by Proposition 3, § 7, for arbitrary preadditive categories, we have an isomorphism of abelian groups

$$\chi^H(L) = \operatorname{Hom}_{\mathcal{A}}(H, L) = \operatorname{Hom}_{\mathcal{A}}(\bigoplus_E h^E, L) \rightleftarrows \prod_E \operatorname{Hom}_{\mathcal{A}}(h^E, L),$$

and likewise for M instead of L; by means of these isomorphisms $\chi^H(\Phi)$ becomes $\prod_E \mathrm{Hom}_{\mathcal{A}}(1_{h^E}, \Phi)$ (see the commutative diagram (11), § 7). On the other hand, by the Yoneda Lemma (Proposition 2, § 16) for abelian categories and by the naturality in the second argument of the Yoneda isomorphism (see § 16) we have an isomorphism of abelian groups

$$\prod_E \mathrm{Hom}_{\mathcal{A}}(h^E, L) \rightleftarrows \prod_E L(E),$$

and likewise for M, and by means of these isomorphisms $\prod_E \mathrm{Hom}_{\mathcal{A}}(1_{h^E}, \Phi)$ becomes $\prod_E \Phi(E)$. So that, if $\chi^H(\Phi) = 0$, then $\Phi(E) = 0$ for any $E \in \mathrm{Ob}\,\mathcal{D}$ (see Corollary 4 to Proposition 2, § 6), hence $\Phi = 0$; this proves that χ^H is faithful. ▮

According to the result stated at the end of § 19 it follows that **Add**$(\mathcal{D}, \mathbf{Ab})$ has for each of its objects an injective envelope, in particular, that $\mathcal{A}$ has enough injectives.

The result of the First Step of the proof of Proposition 5 is the fact that **Add**$(\mathcal{D}, \mathbf{Ab})$ *is, with the pointwise constructions, an abelian category satisfying Axioms* AB3°), AB5) (*hence also* AB3)) *and having a generator, hence also enough injectives.*

Second Step. Let us consider the full subcategory, denoted **Lex**$(\mathcal{D}, \mathbf{Ab})$, abbreviated $\mathcal{L}$, of **Add**$(\mathcal{D}, \mathbf{Ab}) = \mathcal{A}$ whose objects are the left exact functors[1] $\mathcal{D} \to \mathbf{Ab}$. Under the addition of morphisms and with the zero object inherited from $\mathcal{A}$, the category **Lex**$(\mathcal{D}, \mathbf{Ab})$ is a preadditive category. If $\Phi: L \to M$ is a natural transformation, where $L, M \in \mathrm{Ob}\,\mathcal{L}$, let (K, i) be a kernel of Φ in $\mathcal{A}$; then K is a left exact functor: for let

$$0 \to E \xrightarrow{u} F \xrightarrow{v} G$$

be an exact sequence in the category $\mathcal{D}$; we get the following diagram in **Ab**:

$$
(10)\qquad
\begin{array}{ccccccc}
 & & 0 & & 0 & & 0\\
 & & \downarrow & & \downarrow & & \downarrow\\
0 & \longrightarrow & K(E) & \xrightarrow{i(E)} & L(E) & \xrightarrow{\Phi(E)} & M(E)\\
 & & {\scriptstyle K(u)}\downarrow & & {\scriptstyle L(u)}\downarrow & & {\scriptstyle M(u)}\downarrow\\
0 & \longrightarrow & K(F) & \xrightarrow{i(F)} & L(F) & \xrightarrow{\Phi(F)} & M(F)\\
 & & {\scriptstyle K(v)}\downarrow & & {\scriptstyle L(v)}\downarrow & & {\scriptstyle M(v)}\downarrow\\
0 & \longrightarrow & K(G) & \xrightarrow{i(G)} & L(G) & \xrightarrow{\Phi(G)} & M(G);
\end{array}
$$

[1] "L" from "left", "ex" from "exact".

diagram (10) is commutative, since i and Φ are natural transformations; the rows in the diagram are exact sequences, since

$$0 \to K \xrightarrow{i} L \xrightarrow{\Phi} M$$

is an exact sequence in $\mathcal{A}$; since L and M are left exact functors, it follows that the middle and the right columns in diagram (10) are exact sequences; now, by an argument analogous to the one used in the proof of 3×3 Lemma (Proposition 6, § 5), whose version is in fact our result, we get that the left column in diagram (10) is also an exact sequence. Consequently, the kernel in $\mathcal{A}$ of a morphism in $\mathcal{L}$ is also a kernel in $\mathcal{L}$ of this morphism; in particular, any monomorphism of $\mathcal{L}$ is also a monomorphism of $\mathcal{A}$.

Next, if $(L_\lambda)_{\lambda \in I}$ is a family of objects of $\mathcal{L}$, then their direct product and their direct sum taken in $\mathcal{A}$ are left exact functors: indeed, due to the fact that the direct product and the direct sum in $\mathcal{A}$ are constructed pointwise, we have only to apply the exactness of the direct product and of the direct sum in **Ab** (Proposition 2, § 6 and Proposition 2, § 7); hence in **Lex**($\mathcal{D}$, **Ab**) Axioms AB3°) and AB3) are satisfied, with the direct products and the direct sums inherited from $\mathcal{A}$.

The result of the Second Step of the proof of Proposition 5 is the fact that **Lex**($\mathcal{D}$, **Ab**) *with the constructions inherited from* $\mathcal{A}$ *is a preadditive category in which any morphism has a kernel and which satisfies Axioms* AB3) and AB3°).

Third Step. Let us consider the full subcategory of **Add**($\mathcal{D}$, **Ab**), denoted **Mon**($\mathcal{D}$, **Ab**), abbreviated $\mathcal{M}$, whose objects are the additive functors $\mathcal{D} \to$ **Ab** which preserve monomorphisms, i.e., send monomorphisms into monomorphisms. We have the following properties:

M1) *If* $\Phi: L \to M$ *is a monomorphism of* $\mathcal{A}$, *with* $M \in \text{Ob}\,\mathcal{M}$, *then* $L \in \text{Ob}\,\mathcal{M}$.

For let $u : E \to F$ be a monomorphism of $\mathcal{D}$; we have the commutative diagram in **Ab**

$$\begin{array}{ccc} L(E) & \xrightarrow{\Phi(E)} & M(E) \\ {\scriptstyle L(u)}\downarrow & & \downarrow{\scriptstyle M(u)} \\ L(F) & \xrightarrow{\Phi(F)} & M(F), \end{array}$$

and since $\Phi(E)$ and $M(u)$ are monomorphisms, it follows that $L(u)$ is also a monomorphism.

M2) *If* $(M_\lambda)_{\lambda \in I}$ *is a family of objects of* $\mathcal{M}$, *and* $(M, (p_\lambda)_{\lambda \in I})$ *is a direct product of this family in* $\mathcal{A}$, *then* $M \in \text{Ob}\,\mathcal{M}$.

This follows by Corollary 4 to Proposition 2, § 6.

M3) *For any* $M \in \text{Ob}\,\mathcal{M}$ *there exist an injective object* Q *of* $\mathcal{A}$, $Q \in \text{Ob}\,\mathcal{M}$, *and a monomorphism* $\Phi : M \to Q$.

To prove M3), we need the following Lemma:

LEMMA 2. *If in the category* $\mathcal{A}$, (Q, Φ) *is an essential extension of* M, *with* $M \in \text{Ob}\,\mathcal{M}$, *then* $Q \in \text{Ob}\,\mathcal{M}$.

Proof. If Q does not carry monomorphisms into monomorphisms, then there exists a monomorphism $u \in \text{Hom}_{\mathcal{D}}(E, F)$, with $E, F \in \text{Ob}\,\mathcal{D}$, such that $Q(u)$ is not injective; hence there exists $x \in Q(E)$, $x \neq 0$, such that $Q(u)(x) = 0$. Now define a functor $N: \mathcal{D} \to \mathbf{Ab}$ as follows: if $X \in \text{Ob}\,\mathcal{D}$, then, by definition, $N(X)$ is the set of all elements of $Q(X)$ of the form $Q(f)(x)$, where f ranges over $\text{Hom}_{\mathcal{D}}(E, X)$; if $t: X \to Y$ is a morphism of $\mathcal{D}$, then

$$Q(t)\,(N(X)) \subseteq N(Y),$$

since $Q(t)(Q(f)(x)) = Q(t \circ f)(x)$; then $Q(t)$ induces by restriction to $N(X)$ and $N(Y)$ a mapping, which we take as $N(t)$; the fact that $N(X)$ thus defined is a subgroup follows from the fact that Q is an additive functor; the fact that $N(t)$ is a homomorphism of groups follows from the fact that N acts on morphisms in the same way as Q; for the same reason, N is a functor, namely an additive one.

If $X \in \text{Ob}\,\mathcal{D}$, denoting by $\Psi(X)$ the inclusion homomorphism $N(X) \to Q(X)$, the mapping $X \mapsto \Psi(X)$, $X \in \text{Ob}\,\mathcal{D}$, defines a natural transformation, again because N acts on morphisms in the same way as Q; moreover, this morphism, denoted Ψ, is a monomorphism of $\mathcal{A}$.

Further, N is a non-zero functor, since $0 \neq x = Q(1_E)(x) \in N(E)$; since (Q, Φ) is an essential extension, it follows that the intersection of (M, Φ) and (N, Ψ) as subobjects of Q is a non-zero functor; in other words, the intersection in $\mathcal{A}$ being done pointwise, there exists $X \in \text{Ob}\,\mathcal{D}$ such that the subgroup $\Phi(X)(M(X)) \cap N(X)$ of $Q(X)$ is different from zero; hence there exists $f \in \text{Hom}_{\mathcal{D}}(E, X)$ such that $Q(f)(x) \neq 0$ and, moreover, $Q(f)(x) \in \Phi(X)(M(X))$, let say $Q(f)(x) = \Phi(X)(y)$, with $y \in M(X)$, $y \neq 0$. Since $\mathcal{D}$ is an abelian category, it has fibered sums (see § 13); consider then the cocartesian square in $\mathcal{D}$

$$\begin{array}{ccc} E & \xrightarrow{\;u\;} & F \\ {\scriptstyle f}\downarrow & & \downarrow{\scriptstyle u'} \\ X & \xrightarrow{\;f'\;} & S; \end{array} \tag{11}$$

since u is a monomorphism of $\mathcal{D}$, it follows by Proposition 7, § 13, that f' is also a monomorphism; from the commutativity of diagram (11) and from $Q(u)(x) = 0$ it follows that

$$0 = Q(u')\,(Q(u)\,(x)) = Q(f')\,(Q(f)(x)). \tag{12}$$

Since Φ is a natural transformation, it follows

$$Q(f')\,(Q(f)(x)) = Q(f')\,(\Phi(X)(y)) = \Phi(S)\,(M(f')\,(y)); \tag{13}$$

but since M carries monomorphisms into monomorphisms, since f' is a monomorphism, and since $\Phi(S)$ is also a monomorphism, it follows, by (13), $Q(f')(Q(f)(x)) \neq \neq 0$, contradicting (12). ▮

Now let (Q, Φ) be an injective envelope of $M \in \mathrm{Ob}\,\mathcal{M}$ in $\mathcal{A}$; since (Q, Φ) is an essential extension of M, Lemma 2 implies that $Q \in \mathrm{Ob}\,\mathcal{M}$; Q being an injective object of $\mathcal{A}$, this proves M3).

By M1), any object isomorphic in $\mathcal{A}$ to an object of $\mathcal{M}$ is also an object of $\mathcal{M}$.

The result of the Third Step of the proof of Proposition 5 is the fact that *the subcategory* $\mathbf{Mon}(\mathcal{D}, \mathbf{Ab}) = \mathcal{M}$ *of* $\mathcal{A}$ *satisfies the conditions* M1), M2), M3).

Fourth Step. We proceed now to prove the following Lemma:

LEMMA 3. *For any object L of $\mathcal{A}$ there exist an object $W'(L)$ of $\mathcal{M}$ and a morphism $\varphi'_L \in \mathrm{Hom}_{\mathcal{A}}(L, W'(L))$ which is an epimorphism of $\mathcal{A}$, such that the pair $(W'(L), \varphi'_L)$ has the following universal property: whatever $M \in \mathrm{Ob}\,\mathcal{M}$ and the morphism $\Phi \in \mathrm{Hom}_{\mathcal{A}}(L, M)$, there exists a unique morphism $\Psi \in \mathrm{Hom}_{\mathcal{M}}(W'(L), M)$ rendering commutative the diagram*

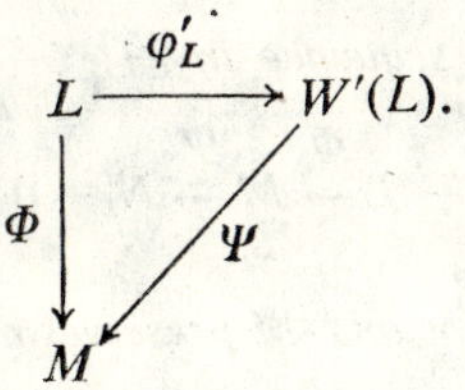

Proof. Since $\mathcal{A}$ has a generator, it follows by the Corollary to Proposition 4, § 19, that the class of quotient objects of an object of $\mathcal{A}$ is a set; let $((M_\lambda, q_\lambda))_{\lambda \in I}$ be a complete system of representatives of those quotient objects of L for which the object component is in $\mathcal{M}$; the set of these quotient objects is non-void, since the quotient object $(0, 0)$ of L has the object component in $\mathcal{M}$. Let $(N, (\pi_\lambda)_{\lambda \in I})$ be a direct product in $\mathcal{A}$ of the family of objects $(M_\lambda)_{\lambda \in I}$; by M2) we have $N \in \mathrm{Ob}\,\mathcal{M}$. Let $\mu : L \to N$ be the morphism satisfying the condition

$$\pi_\lambda \circ \mu = q_\lambda$$

for any $\lambda \in I$; let $(W'(L), \varphi'_L)$ be a coimage in $\mathcal{A}$ of μ; clearly, $W'(L)$ is uniquely determined up to an isomorphism compatible with φ'_L; since by the canonical factorisation of μ there exists a monomorphism $\sigma: W'(L) \to N$ (such that $\sigma \circ \varphi'_L = \mu$), it follows by M1) that $W'(L) \in \mathrm{Ob}\,\mathcal{M}$.

Let us verify the universal property: let M and Φ be as in the statement; if (M', Φ') is a coimage of Φ in $\mathcal{A}$, since $M \in \mathrm{Ob}\,\mathcal{M}$, it follows that $M' \in \mathrm{Ob}\,\mathcal{M}$, and since Φ' is an epimorphism, it follows that there exists $\lambda \in I$ such that, by a change of the representative of the coimage, we have $M' = M_\lambda$, $\Phi' = q_\lambda$; next, denoting by ρ the morphism $M_\lambda \to M$ for which $\rho \circ q_\lambda = \Phi$, we have

$$\Phi = \rho \circ q_\lambda = \rho \circ \pi_\lambda \circ \mu = \rho \circ \pi_\lambda \circ \sigma \circ \varphi'_L\,,$$

and we have only to take $\Psi = \rho\circ\pi_\lambda\circ\sigma$; the uniqueness of Ψ follows from the fact that, by its construction, φ'_L is an epimorphism. ∎

Now let $L \in \operatorname{Ob} \mathcal{A}$ and $M \in \operatorname{Ob} \mathcal{M}$; denoting by V' the immersion functor $\mathcal{M} \to \mathcal{A}$, the mapping of sets

$$\theta'(L, M) : \operatorname{Hom}_{\mathcal{M}}(W'(L), M) \to \operatorname{Hom}_{\mathcal{A}}(L, V'(M))$$

given by

$$\Psi \mapsto \Psi\circ\varphi'_L, \quad \Psi \in \operatorname{Hom}_{\mathcal{M}}(W'(L), M),$$

is, by Lemma 3, a bijection; of course, it is even an isomorphism of abelian groups; moreover, if $M \to M_1$ is a morphism of $\mathcal{M}$, then the corresponding diagram of the type (2?), § 10, is commutative; by Proposition 13, § 10, this means that there exists a functor $W' : \mathcal{A} \to \mathcal{M}$ left adjoint to V' and coinciding on the objects with the mapping $L \mapsto W'(L)$, $L \in \operatorname{Ob} \mathcal{A}$, given by Lemma 3.

The result of the Fourth Step of the proof of Proposition 5 is that *the immersion functor* $V' : \mathbf{Mon}(\mathcal{D}, \mathbf{Ab}) \to \mathbf{Add}(\mathcal{D}, \mathbf{Ab})$ *has a left adjoint* W'.

Fifth Step.

LEMMA 4. *Let the exact sequence in* $\mathcal{A}$

$$0 \to L \xrightarrow{\Phi} M \xrightarrow{\Psi} N \to 0$$

be given.

a) If L is a left exact functor and M preserves monomorphisms, then N preserves monomorphisms.

b) If M is left exact and N preserves monomorphisms, then L is a left exact functor.

Proof. Let

$$0 \to E \xrightarrow{u} F \xrightarrow{v} G \to 0$$

be an exact sequence of the category $\mathcal{D}$; consider the diagram in **Ab**

(14)
$$\begin{array}{ccccccccc}
 & & 0 & & 0 & & 0 & & \\
 & & \downarrow & \Phi(E) & \downarrow & \Psi(E) & \downarrow & & \\
0 & \longrightarrow & L(E) & \longrightarrow & M(E) & \longrightarrow & N(E) & \longrightarrow & 0 \\
 & & \downarrow{\scriptstyle L(u)} & & \downarrow{\scriptstyle M(u)} & & \downarrow{\scriptstyle N(u)} & & \\
 & & & \Phi(F) & & \Psi(F) & & & \\
0 & \longrightarrow & L(F) & \longrightarrow & M(F) & \longrightarrow & N(F) & \longrightarrow & 0 \\
 & & \downarrow{\scriptstyle L(v)} & & \downarrow{\scriptstyle M(v)} & & \downarrow{\scriptstyle N(v)} & & \\
 & & & \Phi(G) & & \Psi(G) & & & \\
0 & \longrightarrow & L(G) & \longrightarrow & M(G) & \longrightarrow & N(G) & \longrightarrow & 0;
\end{array}$$

diagram (14) is commutative, since Φ and Ψ are natural transformations; moreover, the rows in the diagram are exact sequences; for *a*) we have to show that if the left column in (14) is an exact sequence and $M(u)$ is a monomorphism, then $N(u)$ is a monomorphism; for *b*) we have to show that if the middle column in (14) is an exact sequence and $N(u)$ is a monomorphism, then the left column is an exact sequence; these assertions may be proved by considerations analogous to those in the proof of 3 × 3 Lemma (Proposition 6, § 5). ∎

LEMMA 5. *Let* $Q \in \text{Ob}\,\mathcal{A}$ *be an injective object of* $\mathcal{A}$; *then* Q *is a right exact functor.*

Proof. By Proposition 3, § 5, for abelian categories, it suffices to show that for any exact sequence in $\mathcal{D}$ of the form

$$E \xrightarrow{u} F \xrightarrow{v} G \to 0, \tag{15}$$

the diagram

$$Q(E) \xrightarrow{Q(u)} Q(F) \xrightarrow{Q(v)} Q(G) \to 0 \tag{16}$$

is an exact sequence in **Ab**. Consider the diagram in $\mathcal{A}$

$$0 \to h^G \xrightarrow{h^v} h^F \xrightarrow{h^u} h^E, \tag{17}$$

where h^E, h^F, h^G are the additive functors $\mathcal{D} \to$ **Ab** defined in § 2, and h^u, h^v are the natural transformations defined in § 16; diagram (17) is an exact sequence of the category $\mathcal{A}$: a pointwise verification will be sufficient; then let $X \in \text{Ob}\,\mathcal{A}$; from (17) we get the diagram in **Ab**

$$0 \to \text{Hom}_{\mathcal{D}}(G, X) \xrightarrow{\text{Hom}(v,\, 1_X)} \text{Hom}_{\mathcal{D}}(F, X) \xrightarrow{\text{Hom}(u,\, 1_X)} \text{Hom}_{\mathcal{D}}(E, X); \tag{18}$$

but by Proposition 4, § 5, for abelian categories, the exactness of the sequence (15) implies the exactness of the sequence (18).

Since Q is an injective object of $\mathcal{A}$, it follows by Proposition 3, *iii*), § 18, for abelian categories that the diagram in **Ab**

$$\text{Hom}_{\mathcal{A}}(h^E, Q) \xrightarrow{\text{Hom}(h^u,\, 1_Q)} \text{Hom}_{\mathcal{A}}(h^F, Q) \xrightarrow{\text{Hom}(h^v,\, 1_Q)} \text{Hom}_{\mathcal{A}}(h^G, Q) \to 0$$

is an exact sequence; by the Yoneda Lemma (Proposition 2, § 16) and by the naturality of the Yoneda isomorphism in the first argument we obtain that (16) is an exact sequence. ▮

Of course, any left exact functor $\mathcal{D} \to \mathbf{Ab}$ preserves monomorphisms, in other words, $\text{Ob}\,\mathcal{L} \subseteq \text{Ob}\,\mathcal{M}$; the following Lemma provides a characterization in terms of exact sequences of $\mathcal{A}$ for the left exact functors among the functors preserving monomorphisms.

LEMMA 6. *An additive functor* $M : \mathcal{D} \to \mathbf{Ab}$ *preserving monomorphisms is left exact if and only if for any monomorphism* $\Phi: M \to N$ *of* $\mathcal{A}$, *with* $N \in \text{Ob}\,\mathcal{M}$, *it follows that* $\text{Coker}\,\Phi \in \text{Ob}\,\mathcal{M}$.

Proof. Assume that $M \in \text{Ob}\,\mathcal{M}$ is left exact; let $\Phi : M \to N$ be a monomorphism in $\mathcal{A}$, with $N \in \text{Ob}\,\mathcal{M}$, and let $(P, \Psi) = \text{Coker}\,\Phi$ in $\mathcal{A}$; hence we have the exact sequence

$$0 \to M \xrightarrow{\Phi} N \xrightarrow{\Psi} P \to 0;$$

by Lemma 4, *a*) it follows that $P \in \text{Ob}\,\mathcal{M}$.

Conversely, assume that $M \in \text{Ob}\,\mathcal{M}$ has the stated property; let (Q, Φ) be an injective envelope of M in $\mathcal{A}$; let $(P, \Psi) = \text{Coker}\,\Phi$; we have the exact sequence

$$0 \to M \xrightarrow{\Phi} Q \xrightarrow{\Psi} P \to 0; \tag{19}$$

since Φ is an essential monomorphism, it follows by Lemma 2 that $Q \in \text{Ob}\,\mathcal{M}$; taking the hypothesis into account, we get $P \in \text{Ob}\,\mathcal{M}$; since Q is an injective object of $\mathcal{A}$, by Lemma 5 Q is a right exact functor; since Q is right exact and preserves monomorphisms, it follows that Q is an exact functor, hence left exact; in order to prove that M is left exact it remains only to apply Lemma 4, *b*) to the exact sequence (19). ▮

The result of the Fifth Step of the proof of Proposition 5 is exactly the characterization given by Lemma 6 for left exact functors among the functors preserving monomorphisms.

Sixth Step. At the Fourth Step we constructed a left adjoint W' of the immersion functor $V' : \mathcal{M} \to \mathcal{A}$; notice that *for* $L \in \text{Ob}\,\mathcal{A}$, *the fact that* $W'(L) = 0$ *is equivalent to the fact that for any* $M \in \text{Ob}\,\mathcal{M}$ *we have* $\text{Hom}_{\mathcal{A}}(L, M) = 0$; for if $W'(L) = 0$, then, since any morphism $\Phi \in \text{Hom}_{\mathcal{A}}(L, M)$, with $M \in \text{Ob}\,\mathcal{M}$, factors through φ'_L of Lemma 3, it follows that $\Phi = 0$; conversely, if $\text{Hom}_{\mathcal{A}}(L, M) = 0$ for any $M \in \text{Ob}\,\mathcal{M}$, since $W'(L) \in \text{Ob}\,\mathcal{M}$, it follows in particular that $\varphi'_L = 0$; but φ'_L is an epimorphism, hence $W'(L) = 0$.

LEMMA 7. *Let*

$$0 \to L \xrightarrow{\Phi} M \xrightarrow{\Psi} N \to 0$$

be an exact sequence in $\mathcal{A}$; *if* $L, N \in \text{Ob}\,\mathcal{M}$, *then* $M \in \text{Ob}\,\mathcal{M}$.

Proof. By M3) proved at the Third Step, there exists an injective object Q of $\mathcal{A}$, with $Q \in \text{Ob}\,\mathcal{M}$, and a monomorphism $\mu : L \to Q$; consider the cocartesian square in $\mathcal{A}$

$$\begin{array}{ccc} L & \xrightarrow{\Phi} & M \\ {\scriptstyle\mu}\downarrow & & \downarrow{\scriptstyle\Phi'} \\ Q & \xrightarrow{\mu'} & S; \end{array}$$

if $\nu : S \to N$ is obtained according to the universal property of the fibered sum from the morphisms $\Psi : M \to N$ and $0 : Q \to N$, then we have in $\mathcal{A}$ the commutative diagram

$$(20) \qquad \begin{array}{ccccccccc} 0 & \longrightarrow & L & \xrightarrow{\Phi} & M & \xrightarrow{\Psi} & N & \longrightarrow & 0 \\ & & {\scriptstyle\mu}\downarrow & & \downarrow{\scriptstyle\Phi'} & & \downarrow{\scriptstyle 1_N} & & \\ 0 & \longrightarrow & Q & \xrightarrow{\mu'} & S & \xrightarrow{\nu} & N & \longrightarrow & 0; \end{array}$$

we claim that the bottom row in diagram (20) is also an exact sequence in $\mathcal{A}$: indeed, it suffices to show that for any $E \in \text{Ob}\,\mathcal{D}$, in the commutative diagram

$$(21) \qquad \begin{array}{ccccccccc} 0 & \longrightarrow & L(E) & \xrightarrow{\Phi(E)} & M(E) & \xrightarrow{\Psi(E)} & N(E) & \longrightarrow & 0 \\ & & {\scriptstyle\mu(E)}\downarrow & & \downarrow{\scriptstyle\Phi'(E)} & & \downarrow{\scriptstyle 1_N(E)} & & \\ 0 & \longrightarrow & Q(E) & \xrightarrow{\mu'(E)} & S(E) & \xrightarrow{\nu(E)} & N(E) & \longrightarrow & 0, \end{array}$$

the bottom row is an exact sequence of **Ab**; but since fibered sums are done in $\mathcal{A}$ pointwise, it follows that the left square of diagram (21) is a pushout in **Ab**; now the result follows taking into account Proposition 7, § 13, and the fact that any element of $S(E)$ is of the form $\Phi'(E)x+\mu'(E)y$, with $x \in M(E)$, $y \in Q(E)$.

Since Q is an injective object of $\mathcal{A}$, it follows that the bottom row in diagram (20) is a split exact sequence (see § 18); but then, by Proposition 7, § 9, it follows that S is a direct product in $\mathcal{A}$ of the family of two objects (Q, N), with $Q, N \in$

$\in \operatorname{Ob} \mathcal{M}$; hence, by virtue of M2), $S \in \operatorname{Ob} \mathcal{M}$; since μ is a monomorphism, it follows by Proposition 7, § 13, that Φ' is also a monomorphism, and to complete the proof we need only apply M1). ▮

LEMMA 8. *Let $L \in \operatorname{Ob} \mathcal{A}$; in the notation of Lemma 3, let (K, i) be a kernel in $\mathcal{A}$ of φ'_L; then $W'(K) = 0$, and the pair (K, i) has the following universal property: for any pair (M, Φ), with $M \in \operatorname{Ob} \mathcal{A}$ and $\Phi \in \operatorname{Hom}_{\mathcal{A}}(M, L)$, such that $W'(M) = 0$ there exists a unique morphism $\Psi : M \to K$ with the property of rendering commutative the diagram*

$$\begin{array}{ccc} K & \xrightarrow{\;i\;} & L \\ & \nwarrow^{\Psi} & \uparrow{\scriptstyle \Phi} \\ & & M. \end{array}$$

Proof. We start with the second assertion: by the remark at the beginning of this Step, from $W'(M) = 0$ we get $\operatorname{Hom}_{\mathcal{A}}(M, W'(L)) = 0$; in particular, $\varphi'_L \circ \Phi = 0$; then, by the universal property of the kernel, the existence and the uniqueness of Ψ with the stated property follows.

Now, let us show that $W'(K) = 0$: by the preceding arguments, it suffices to show that for any $N \in \operatorname{Ob} \mathcal{M}$ and $\Psi \in \operatorname{Hom}_{\mathcal{A}}(K, N)$ we have $\Psi = 0$; consider the following diagram in $\mathcal{A}$:

$$\begin{array}{ccccccccc} 0 & \longrightarrow & K & \xrightarrow{\;i\;} & L & \xrightarrow{\;\varphi'_L\;} & W'(L) & \longrightarrow & 0 \\ & & \downarrow{\scriptstyle \Psi} & & \downarrow{\scriptstyle i'} & & \downarrow{\scriptstyle 1_{W'(L)}} & & \\ 0 & \longrightarrow & N & \xrightarrow{\;\Psi'\;} & S & \xrightarrow{\;\nu\;} & W'(L) & \longrightarrow & 0 \end{array} \tag{22}$$

in which the left square is a cocartesian square, while ν is obtained by the universal property of the fibered sum from the morphisms $\varphi'_L : L \to W'(L)$ and $0 : N \to W'(L)$; consequently, the right square of diagram (22) is also commutative; by the same argument as for diagram (20) one proves that the bottom row of diagram (22) is an exact sequence. By Lemma 7 it follows $S \in \operatorname{Ob} \mathcal{M}$; then, by Lemma 3, there exists a morphism $\rho \in \operatorname{Hom}_{\mathcal{A}}(W'(L), S)$ such that $\rho \circ \varphi'_L = i'$; we have

$$\Psi' \circ \Psi = i' \circ i = \rho \circ \varphi'_L \circ i = 0,$$

and since by Proposition 7, § 13, Ψ' is a monomorphism, it follows $\Psi = 0$. ▮

LEMMA 9. *Let the exact sequence of the category* $\mathcal{A}$

$$0 \to M \xrightarrow{\Phi} N \xrightarrow{\Psi} P \to 0$$

be given, where $M \in \operatorname{Ob} \mathcal{M}$, $N \in \operatorname{Ob} \mathcal{L}$, *and* $W'(P) = 0$; *then, whatever* $L \in \operatorname{Ob} \mathcal{L}$ *and* $\Omega \in \operatorname{Hom}_{\mathcal{M}}(M, L)$, *there exists a unique morphism* $\Omega' \in \operatorname{Hom}_{\mathcal{L}}(N, L)$ *rendering commutative the diagram*

$$\begin{array}{ccc} M & \xrightarrow{\Phi} & N. \\ {\scriptstyle\Omega}\downarrow & \swarrow{\scriptstyle\Omega'} & \\ L & & \end{array}$$

Proof. By M3), there exist an injective object Q of $\mathcal{A}$, $Q \in \operatorname{Ob} \mathcal{M}$, and a monomorphism $\mu : L \to Q$; if (R, ρ) is a cokernel of μ, then, since $L \in \operatorname{Ob} \mathcal{L}$, it follows by Lemma 6 that $R \in \operatorname{Ob} \mathcal{M}$. Since Q is an injective object of $\mathcal{A}$ and Φ is a monomorphism, it follows that there exists a $\sigma : N \to Q$ such that $\sigma \circ \Phi = \mu \circ \Omega$, and by the universal property of the cokernel it follows that there exists a $\tau : P \to R$ such that the following diagram is commutative:

$$\begin{array}{ccccccccc} 0 & \longrightarrow & M & \xrightarrow{\Phi} & N & \xrightarrow{\Psi} & P & \longrightarrow & 0 \\ & & \downarrow{\scriptstyle\Omega} & & \downarrow{\scriptstyle\sigma} & & \downarrow{\scriptstyle\tau} & & \\ 0 & \longrightarrow & L & \xrightarrow{\mu} & Q & \xrightarrow{\rho} & R & \longrightarrow & 0; \end{array} \tag{23}$$

moreover, the rows in diagram (23) are exact sequences. Since $R \in \operatorname{Ob} \mathcal{M}$, and since $W'(P) = 0$, it follows that $\tau = 0$; hence $\rho \circ \sigma = 0$, and by the definition of the kernel (see also Proposition 2, § 5) the existence of an $\Omega' : N \to L$ with the property $\mu \circ \Omega' = \sigma$ follows; we have

$$\mu \circ \Omega = \sigma \circ \Phi = \mu \circ \Omega' \circ \Phi,$$

and, since μ is a monomorphism, $\Omega = \Omega' \circ \Phi$. Assume that there also exists $\Omega_1' : N \to L$ such that $\Omega_1' \circ \Phi = \Omega$; then $(\Omega_1' - \Omega') \circ \Phi = 0$, and by the definition of the cokernel the existence of a morphism $v : P \to L$ follows such that $v \circ \Psi = \Omega_1' - \Omega'$; but since $W'(P) = 0$, and since $L \in \operatorname{Ob} \mathcal{M}$, we have $v = 0$, hence $\Omega_1' = \Omega'$. ∎

LEMMA 10. *For any object M of* $\mathcal{M}$ *there exist an object* $W''(M)$ *of* $\mathcal{L}$ *and a morphism* $\varphi_M'' : M \to W''(M)$ *which is a monomorphism of* $\mathcal{A}$, *such that the pair* $(W''(M), \varphi_M'')$ *has the following universal property: whatever* $N \in \operatorname{Ob} \mathcal{L}$ *and the morphism* $\Phi \in \mathrm{Hom}_{\mathcal{M}}(M, N)$, *there exists a unique morphism* $\Psi \in \mathrm{Hom}_{\mathcal{L}}(W''(M), N)$ *such that the following diagram is commutative*:

$$\begin{array}{ccc} M & \xrightarrow{\varphi''_M} & W''(M). \\ {\scriptstyle \Phi}\downarrow & \swarrow{\scriptstyle \Psi} & \\ N & & \end{array}$$

Proof. By M3), there exist an injective object Q of $\mathcal{A}$, $Q \in \mathrm{Ob}\ \mathcal{M}$, and a monomorphism $\mu : M \to Q$; let (R, ρ) be a cokernel of μ; consider the morphism $\varphi'_R : R \to W'(R)$ given by Lemma 3, and let (K, i) be a kernel of φ'_R; consider the diagram in $\mathcal{A}$

$$(24)\qquad \begin{array}{ccccccccc} & & & & 0 & & 0 & & \\ & & & & \downarrow & & \downarrow & & \\ 0 & \longrightarrow & M & \xrightarrow{\tau} & P & \xrightarrow{i'} & K & \longrightarrow & 0 \\ & & {\scriptstyle 1_M}\downarrow & & \downarrow{\scriptstyle \rho'} & & \downarrow{\scriptstyle i} & & \\ 0 & \longrightarrow & M & \xrightarrow{\mu} & Q & \xrightarrow{\rho} & R & \longrightarrow & 0, \\ & & & & {\scriptstyle \sigma}\downarrow & & \downarrow{\scriptstyle \varphi'_R} & & \\ & & & & W'(R) & \xrightarrow{1_{W'(R)}} & W'(R) & & \\ & & & & \downarrow & & \downarrow & & \\ & & & & 0 & & 0 & & \end{array}$$

where the right upper square is a pullback, $\sigma = \varphi'_R \circ \rho$, and τ is obtained by the universal property of the fibered product from $\mu : M \to Q$ and $0 : M \to K$; diagram (24) is commutative, the middle row and the right column are exact sequences by definition, the top row follows exact by pointwise considerations and by Proposition 7, § 13, while the middle column turns out to be exact by pointwise considerations and by Proposition 7, § 17.

Since Q is an injective object of $\mathcal{A}$ and $Q \in \mathrm{Ob}\ \mathcal{M}$, it follows by Lemma 5 that Q is an exact functor; hence $Q \in \mathrm{Ob}\ \mathcal{L}$; since $W'(R) \in \mathrm{Ob}\ \mathcal{M}$, it follows by Lemma 4, *b*) that $P \in \mathrm{Ob}\ \mathcal{L}$; by Lemma 8, $W'(K) = 0$; taking $W''(M) = P$, $\varphi''_M = \tau$, Lemma 9 shows that the pair $(W''(M), \varphi''_M)$ satisfies the condition stated. ∎

Now, if $M \in \mathrm{Ob}\ \mathcal{M}$ and $N \in \mathrm{Ob}\ \mathcal{L}$, denoting by V'' the immersion functor $\mathcal{L} \to \mathcal{M}$, the mapping of sets

$$\theta''(M, N) : \mathrm{Hom}_{\mathcal{L}}(W''(M), N) \to \mathrm{Hom}_{\mathcal{M}}(M, V''(N))$$

given by

$$\Psi \mapsto \Psi \circ \varphi''_M, \qquad \Psi \in \mathrm{Hom}_{\mathcal{L}}(W''(M), N),$$

is bijective by Lemma 10; of course, it is even an isomorphism of abelian groups; moreover, if $N \to N_1$ is a morphism in $\mathcal{L}$, then the corresponding diagram of the type (22), § 10, is commutative; this means by Proposition 13, § 10, that there exists a functor $W'' : \mathcal{M} \to \mathcal{L}$ left adjoint to V'' and coinciding on the objects of $\mathcal{M}$ with the mapping $M \mapsto W''(M)$, $M \in \mathrm{Ob}\, \mathcal{M}$, defined in Lemma 10.

The result of the Sixth Step of the proof of Proposition 5 is the fact that *the immersion functor* $V'' : \mathbf{Lex}(\mathcal{D}, \mathbf{Ab}) \to \mathbf{Mon}(\mathcal{D}, \mathbf{Ab})$ *has a left adjoint* W''.

Seventh Step. From the results of the Fourth and Sixth Steps it follows by Proposition 14, § 10, that the immersion functor

$$V : \mathbf{Lex}(\mathcal{D}, \mathbf{Ab}) \to \mathbf{Add}(\mathcal{D}, \mathbf{Ab}),$$

which (in the previous notation) is $V' \circ V''$, has a left adjoint, namely $W'' \circ W'$, henceforth denoted by W; from the proof of Proposition 14, § 10, it follows that the mappings $\theta(L, N)$, $L \in \mathrm{Ob}\, \mathcal{A}$, $N \in \mathrm{Ob}\, \mathcal{L}$, obtained for given L and N as the composition

$$\mathrm{Hom}_{\mathcal{L}}(W''(W'(L)), N) \xrightarrow{\theta''(W'(L), N)} \mathrm{Hom}_{\mathcal{M}}(W'(L), V''(N)) \xrightarrow{\theta'(L, V''(N))}$$
$$\to \mathrm{Hom}_{\mathcal{A}}(L, V'(V''(N)))$$

define the adjointness situation; by the definition of θ' and θ'', if $\Psi \in \mathrm{Hom}_{\mathcal{L}}(W''(W'(L)), N)$, then

$$\theta(L, N)(\Psi) = \Psi \circ \varphi''_{W'(L)} \circ \varphi'_L,$$

or, denoting $\varphi''_{W'(L)} \circ \varphi'_L = \varphi_L$, $\theta(L, N)(\Psi) = \Psi \circ \varphi_L$. Of course, by the preceding considerations it follows that if $L \in \mathrm{Ob}\, \mathcal{A}$, $N \in \mathrm{Ob}\, \mathcal{L}$, and $\Phi \in \mathrm{Hom}_{\mathcal{A}}(L, N)$, then Φ factors uniquely through φ_L.

Let $\Omega \in \mathrm{Hom}_{\mathcal{A}}(K, L)$, with $K, L \in \mathrm{Ob}\, \mathcal{A}$; from the commutativity of the diagram of the type (5), § 10, it follows that W acts on the morphisms of $\mathcal{A}$ according to a relation of the type (23), § 10, in other words, that $W(\Omega)$ is the unique morphism $W(K) \to W(L)$ rendering commutative the diagram

$$\begin{array}{ccc} K & \xrightarrow{\varphi_K} & W(K) \\ {\scriptstyle \Omega}\downarrow & & \downarrow{\scriptstyle W(\Omega)} \\ L & \xrightarrow{\varphi_L} & W(L). \end{array}$$

For this reason, W is an additive functor.

We note that *if* $N \in \mathrm{Ob}\, \mathcal{L}$, *then* φ_N *is an isomorphism* (*in* $\mathcal{A}$ *and in* $\mathcal{L}$): indeed, $1_N \in \mathrm{Hom}_{\mathcal{L}}(N, N)$ factors through φ_N, hence there exists $\Psi : W(N) \to N$ such that

$$\Psi \circ \varphi_N = 1_N; \tag{25}$$

on the other hand, $\varphi_N \in \mathrm{Hom}_{\mathcal{L}}(N, W(N))$ factors uniquely through φ_N, and, since by (25) we have

$$\varphi_N = 1_{W(N)} \circ \varphi_N = \varphi_N \circ \Psi \circ \varphi_N,$$

we get

$$\varphi_N \circ \Psi = 1_{W(N)},$$

which together with (25) proves our assertion.

At the Second Step we proved that $\mathbf{Lex}(\mathcal{D}, \mathbf{Ab}) = \mathcal{L}$ with the constructions inherited from $\mathbf{Add}(\mathcal{D}, \mathbf{Ab}) = \mathcal{A}$, i.e. pointwise, is a preadditive category in which any morphism has a kernel and which is, moreover, with direct products and direct sums.

Now let $\Phi: L \to M$ be a morphism, with $L, M \in \mathrm{Ob}\,\mathcal{L}$; let (P, p) be a cokernel of Φ in $\mathcal{A}$; we claim that *$(W(P), \varphi_P \circ p)$ is a cokernel of Φ in $\mathcal{L}$*: indeed, we have $\varphi_P \circ p \circ \Phi = 0$, and if $\Psi \in \mathrm{Hom}_{\mathcal{L}}(M, N)$ has the property that $\Psi \circ \Phi = 0$, where $N \in \mathrm{Ob}\,\mathcal{L}$, then, since the same is true in $\mathcal{A}$, the existence of $\Psi' \in \mathrm{Hom}_{\mathcal{A}}(P, N)$ follows such that the following diagram is commutative:

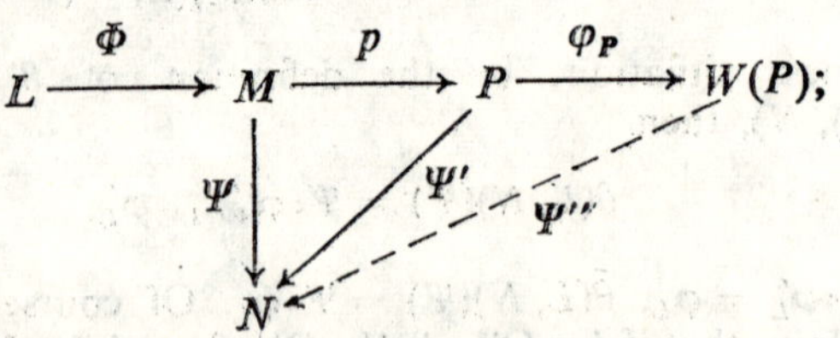

since, however, $N \in \mathrm{Ob}\,\mathcal{L}$, by the universal property of $(W(P), \varphi_P)$ it follows that there exists a unique $\Psi'' \in \mathrm{Hom}_{\mathcal{L}}(W(P), N)$ such that $\Psi'' \circ \varphi_P = \Psi'$; hence,

$$\Psi'' \circ \varphi_P \circ p = \Psi' \circ p = \Psi;$$

if there existed also $\Omega : W(P) \to N$ for which $\Omega \circ \varphi_P \circ p = \Psi$, since p is an epimorphism in $\mathcal{A}$, it would follow $\Omega \circ \varphi_P = \Psi'' \circ \varphi_P$, and, due to the unique factorization through φ_P, $\Omega = \Psi''$.

Consequently, *in $\mathcal{L}$ any morphism also has a cokernel, hence $\mathcal{L}$ satisfies Axiom* AB1).

LEMMA 11. *The additive functor $W : \mathcal{A} \to \mathcal{L}$ preserves kernels, that is, if $\Psi \in \mathrm{Hom}_{\mathcal{A}}(M, N)$, with $M, N \in \mathrm{Ob}\,\mathcal{A}$, and (L, Φ) is a kernel of Ψ in $\mathcal{A}$, then $(W(L), W(\Phi))$ is a kernel of $W(\Psi)$ in $\mathcal{L}$.*

Proof. First, we show that *if $\Phi : L \to M$ is a monomorphism of the category $\mathcal{A}$, then $W(\Phi)$ is a monomorphism of the category $\mathcal{L}$*: since $W(L) \in \mathrm{Ob}\,\mathcal{M}$, it follows by M3) proved at the Third Step that there exists an injective object Q of $\mathcal{A}$, with $Q \in \mathrm{Ob}\,\mathcal{M}$, and a monomorphism $\mu : W(L) \to Q$; since Φ is a monomorphism of $\mathcal{A}$, it follows by the injectivity of Q that there exists $\nu : M \to Q$ such that $\nu \circ \Phi = \mu \circ \varphi_L$; by an argument already employed (see the proof of Lemma 10) we have $Q \in \mathrm{Ob}\, \mathcal{L}$. But then the morphism $\nu: M \to Q$ factors uniquely through φ_M, i.e., there exists a unique $\rho \in \mathrm{Hom}_{\mathcal{L}}(W(M), Q)$ such that $\rho \circ \varphi_M = \nu$; this is displayed on the diagram

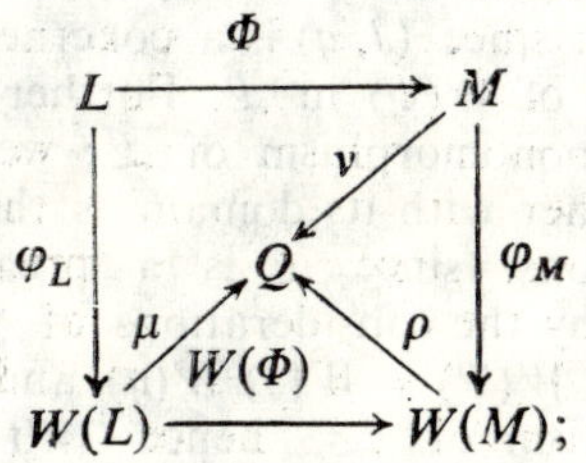

consequently, we have

$$\mu \circ \varphi_L = \nu \circ \Phi = \rho \circ \varphi_M \circ \Phi = \rho \circ W(\Phi) \circ \varphi_L;$$

since $\nu \circ \Phi$ factors uniquely through φ_L, it follows that $\mu = \rho \circ W(\Phi)$; since μ is a monomorphism of $\mathcal{A}$, $W(\Phi)$ is a monomorphism of $\mathcal{A}$, hence also of $\mathcal{L}$.

Next, we show that *if $L, M \in \mathrm{Ob}\, \mathcal{L}$ and $\Phi : L \to M$ is a monomorphism, then there exists a morphism $\Psi: M \to N$, with $N \in \mathrm{Ob}\, \mathcal{L}$, such that (L, Φ) is a kernel of Ψ in $\mathcal{L}$*: indeed, let (P, p) be a cokernel of Φ in $\mathcal{A}$; as shown above, $(W(P), \varphi_P \circ p)$ is a cokernel of Φ in $\mathcal{L}$; but, by Lemma 4, *a*), $P \in \mathrm{Ob}\, \mathcal{M}$; on the other hand, φ_P is the composition

$$P \xrightarrow{\varphi'_P} W'(P) \xrightarrow{\varphi''_{W'(P)}} W''(W'(P)) = W(P);$$

by an argument analogous to the one given in case $N \in \mathrm{Ob}\, \mathcal{L}$ for φ_N, we obtain that φ'_P is an isomorphism of $\mathcal{A}$; since by Lemma 10 $\varphi''_{W'(P)}$ is a monomorphism of $\mathcal{A}$, it follows that φ_P is a monomorphism of $\mathcal{A}$. We obviously have

$$\varphi_P \circ p \circ \Phi = 0;$$

next, if $R \in \mathrm{Ob}\, \mathcal{L}$ and $\Omega : R \to M$ is such that $\varphi_P \circ p \circ \Omega = 0$, then, since φ_P is a monomorphism, it follows $p \circ \Omega = 0$, hence there exists $\Omega' : R \to L$ such that the following diagram is commutative:

the uniqueness of Ω' with this property follows from the fact that Φ is a monomorphism. Thus (L, Φ) is a kernel of $\varphi_P \circ p$ in $\mathcal{L}$.

Finally, *we prove the assertion of the statement*: let L, M, N, Φ, Ψ be as in the statement, let (J, q) be a coimage of Ψ in $\mathcal{A}$, and let $j : J \to N$ be the monomorphism of $\mathcal{A}$ for which $j \circ q = \Psi$. Since W has a right adjoint, it follows by Proposition 2, § 14, that W commutes with inductive limits; in particular, W preserves cokernels (see § 13)[1]: since (J, q) is a cokernel of Φ in $\mathcal{A}$, it follows that $(W(J), W(q))$ is a cokernel of $W(\Phi)$ in $\mathcal{L}$. Further, as shown at the beginning of the proof, $W(\Phi)$ is a monomorphism of $\mathcal{L}$; we have also shown that any monomorphism of $\mathcal{L}$ together with its domain is the kernel of some morphism of $\mathcal{L}$; under these conditions, since $\mathcal{L}$ is a preadditive category satisfying Axiom AB1), it follows by the considerations of § 4 that $(W(L), W(\Phi))$ is a kernel of $W(q)$ in $\mathcal{L}$; but $W(\Psi) = W(j) \circ W(q)$, and since j is a monomorphism of $\mathcal{A}$, $W(j)$ is monomorphism of $\mathcal{L}$; hence $(W(L), W(\Phi))$ is a kernel of $W(\Psi)$ in $\mathcal{L}$. ▮

Now we shall prove that *the category* **Lex** $(\mathcal{D}$, **Ab**$)$ *also satisfies* AB2): Let $\Phi \in \mathrm{Hom}_{\mathcal{L}}(L, M)$, with $L, M \in \mathrm{Ob}\,\mathcal{L}$; let (K, i), (D, q), (J, j), (P, p) be, respectively, a kernel, a coimage, an image, a cokernel of Φ in $\mathcal{A}$; let $\Psi : D \to J$ be the canonical morphism Coim $\Phi \to$ Im Φ corresponding to this choice; we have the commutative diagram in $\mathcal{A}$

$$
\begin{array}{ccccccccccc}
K & \xrightarrow{\;i\;} & L & \xrightarrow{\;q\;} & D & \xrightarrow{\;\Psi\;} & J & \xrightarrow{\;j\;} & M & \xrightarrow{\;p\;} & P \\
{\scriptstyle \varphi_K}\downarrow\wr & & {\scriptstyle \varphi_L}\downarrow\wr & & \downarrow{\scriptstyle \varphi_D} & & \downarrow{\scriptstyle \varphi_J} & & \wr\downarrow{\scriptstyle \varphi_M} & & \downarrow{\scriptstyle \varphi_P} \\
W(K) & \xrightarrow{W(i)} & W(L) & \xrightarrow{W(q)} & W(D) & \xrightarrow{W(\Psi)} & W(J) & \xrightarrow{W(j)} & W(M) & \xrightarrow{W(p)} & W(P);
\end{array}
\tag{26}
$$

since $K, L, M \in \mathrm{Ob}\,\mathcal{L}$, it follows that φ_K, φ_L, and φ_M are isomorphisms.

At the Second Step we have shown that (K, i) is a kernel of Φ also in the category $\mathcal{L}$; we have also seen that $(W(P), \varphi_P \circ p)$ is a cokernel of Φ in $\mathcal{L}$; likewise, $(W(D), \varphi_D \circ q)$ is a cokernel of i in $\mathcal{L}$, hence a coimage of Φ in $\mathcal{L}$; since by Lemma 11 $(W(J), W(j))$ is a kernel of $W(p)$ in $\mathcal{L}$, it follows from the commutativity of diagram (26) that $(W(J), \varphi_M^{-1} \circ W(j))$ is a kernel of $\varphi_P \circ p$ in $\mathcal{L}$, hence an image of Φ in $\mathcal{L}$. Again by the commutativity of diagram (26) it follows that

$$\varphi_M^{-1} \circ W(j) \circ W(\Psi) \circ \varphi_D \circ q = \Phi,$$

which means that in $\mathcal{L}$ the canonical morphism Coim $\Phi \to$ Im Φ for this choice is $W(\Psi)$; but $W(\Psi)$ is an isomorphism of $\mathcal{L}$, since Ψ is an isomorphism of $\mathcal{A}$.,

Since, as proved at the Second Step, $\mathcal{L}$ satisfies also Axioms AB3) and AB3°), it follows that **Lex**$(\mathcal{D}$, **Ab**$)$ *is an abelian category.*

---

[1] The corresponding result of § 13 may be applied to this case also, although $\mathcal{L}$ is not yet proved to be an abelian category, but merely a preadditive category satisfying Axiom AB1).

By M3), if $L \in \text{Ob}\,\mathcal{L}$, there exists an injective object Q of $\mathcal{A}$, $Q \in \text{Ob}\,\mathcal{M}$, and a monomorphism $\Phi : L \to Q$; by the argument employed in the proof of Lemma 10, $Q \in \text{Ob}\,\mathcal{L}$; since the monomorphisms of $\mathcal{L}$ are also monomorphisms of $\mathcal{A}$, we get that Q is also an injective object of the category $\mathcal{L}$; hence **Lex**$(\mathcal{D}$, **Ab**) *has enough injective objects.*

The result of the Seventh Step of the proof of Proposition 5 is the fact that **Lex** $(\mathcal{D}$, **Ab**) *with the constructions described is an abelian category having enough injectives.*

Eighth Step. By the Corollary to Proposition 4, § 5, for abelian categories, if $E \in \text{Ob}\,\mathcal{D}$, then the functor $h^E : \mathcal{D} \to \textbf{Ab}$ is an object of the category $\mathcal{L}$; from the results obtained at the Second Step it follows that the functor H defined at the First Step as the direct sum of the family $(h^E)_{E \in \text{Ob}\mathcal{D}}$ is an object of $\mathcal{L}$; but then, since by Lemma 1 H is a generator of $\mathcal{A}$, it follows that H *is a generator of* $\mathcal{L}$. Hence $\mathcal{L}$ is an abelian category with direct products and having a generator and enough injectives; by Proposition 5, § 19, it follows that $\mathcal{L}$ *has an injective cogenerator.*

Now, from Corollary 2 to Proposition 3, § 16, for abelian categories (hence with **Ab** instead of **Ens**) it follows, in the same notation as there, that the mappings

$$E \mapsto h^E, \qquad f \mapsto h^f,$$

where $E, E' \in \text{Ob}\,\mathcal{D}$, $f \in \text{Hom}_{\mathcal{D}}(E', E)$, define an additive fully faithful functor

$$h : \mathcal{D}^\circ \to \textbf{Add}(\mathcal{D}, \textbf{Ab});$$

taking into account that the kernel in **Add**$(\mathcal{D}$, **Ab**) is constructed pointwise, and making use of Propositions 2, *a*), § 5, and 4, § 5, it follows that *h is a left exact functor*.

On the other hand, for any $E \in \text{Ob}\,\mathcal{D}$, h^E is a left exact functor $\mathcal{D} \to \textbf{Ab}$, that is, an object of **Lex**$(\mathcal{D}$, **Ab**); hence the above h is the composition of the immersion functor V : **Lex**$(\mathcal{D}$, **Ab**) → **Add**$(\mathcal{D}$, **Ab**) with a uniquely determined additive functor $\mathcal{D}^\circ \to$ **Lex**$(\mathcal{D}$, **Ab**), which will be, by an abuse of notation, denoted also by h.

LEMMA 12. *The above defined additive functor*

$$h : \mathcal{D}^\circ \to \textbf{Lex}(\mathcal{D}, \textbf{Ab})$$

is exact and fully faithful.

Proof. The fact that h is fully faithful follows from the corresponding result for **Add**$(\mathcal{D}$, **Ab**). To show that h is exact, let

$$0 \to E \xrightarrow{u} F \xrightarrow{v} G \to 0$$

be a short exact sequence of the category $\mathcal{D}$;[1] we get the diagram in $\mathcal{L}$

$$0 \to h^G \xrightarrow{h^v} h^F \xrightarrow{h^u} h^E \to 0; \tag{27}$$

[1] We should have started with an exact sequence of the category $\mathcal{D}^\circ$; but this one provides an exact sequence of the category $\mathcal{D}$.

considered as a diagram in $\mathcal{A}$, (27) is an exact sequence at h^G and at h^F; since the kernel in $\mathcal{A}$ of a morphism of $\mathcal{L}$ is also a kernel in $\mathcal{L}$ of this morphism, it follows that (27) is an exact sequence at h^G and at h^F when considered also as a diagram in $\mathcal{L}$. We have only to show that h^u is an epimorphism of $\mathcal{L}$: since $\mathcal{L}$ has a cogenerator C, it suffices, as shown in § 19, to prove that

$$\mathrm{Hom}_{\mathcal{L}}(h^u, 1_C) : \mathrm{Hom}_{\mathcal{L}}(h^E, C) \to \mathrm{Hom}_{\mathcal{L}}(h^F, C)$$

is a monomorphism of the category **Ab**; by the Yoneda Lemma for abelian categories (Proposition 2, § 16) and by the naturality of the Yoneda isomorphism in the first argument (see § 16) it follows that there exists an isomorphism of abelian groups

$$\mathrm{Hom}_{\mathcal{L}}(h^E, C) \rightleftarrows C(E),$$

and likewise for F instead of E, and that by means of this isomorphisms $\mathrm{Hom}_{\mathcal{L}}(h^u, 1_C)$ becomes $C(u)$; but C is a left exact functor and u is a monomorphism of $\mathcal{D}$; hence $C(u)$ is a monomorphism of **Ab**, and hence so is $\mathrm{Hom}_{\mathcal{L}}(h^u, 1_C)$. ∎

The result of the Eighth Step of the proof of Proposition 5 is the fact that *the abelian category* **Lex**($\mathcal{D}$, **Ab**) *has an injective cogenerator, and* that *there exists a fully faithful exact functor* $h : \mathcal{D}^\circ \to$ **Lex**($\mathcal{D}$, **Ab**).

We are now in a position to collect the results and to draw the conclusion: Considering the dual h° of the functor h defined at the Eighth Step, and identifying $(\mathcal{D}^\circ)^\circ$ with $\mathcal{D}$ (see § 2), we get the additive functor

$$h^\circ : \mathcal{D} \to (\mathbf{Lex}(\mathcal{D}, \mathbf{Ab}))^\circ;$$

(**Lex**($\mathcal{D}$, **Ab**))$^\circ$ *is an abelian category*, being the dual of an abelian category (see § 7); moreover, since by virtue of the results of the Second and Eighth Steps, respectively, **Lex**($\mathcal{D}$, **Ab**) is a category with direct products and having an injective cogenerator, it follows that (**Lex**($\mathcal{D}$, **Ab**))$^\circ$ *is a category with direct sums and having a projective generator*; since by the results of the Eighth Step the functor h is exact and fully faithful, it follows that h° *is also an exact and fully faithful functor*. We need only take $\mathcal{C} =$ (**Lex**($\mathcal{D}$, **Ab**))$^\circ$ and $U = h^\circ$, and Proposition 5 is completely proved. ∎

REMARKS. 1) The complication determined by the passing from the category **Add**($\mathcal{D}$, **Ab**), where the structures are relatively simple and the results are obtained almost immediately from the similar results in **Ab**, to the category **Lex**($\mathcal{D}$, **Ab**) is necessary due to the fact that the above functor $h : \mathcal{D}^\circ \to$ **Add**($\mathcal{D}$, **Ab**) cannot be proved to be exact but only left exact.

2) If $\mathcal{K}$ is a category and E, F are distinct objects of it, then the functors h^E and h^F defined on $\mathcal{K}$ and taking values in **Ens** are distinct: since, for instance, $h^E(E) = \mathrm{Hom}_{\mathcal{K}}(E, E)$ contains 1_E, while $h^F(E) = \mathrm{Hom}_{\mathcal{K}}(F, E)$ does not; the same remark is valid if $\mathcal{K}$ is an abelian category and the functors h^E, $E \in \mathrm{Ob}\,\mathcal{K}$, are considered as taking values in **Ab**. Consequently, by its construction, the functor U in Proposition 5 carries distinct objects into distinct objects. ∎

If $\mathcal{D}$ is a small abelian category, then, in the notation of the statement of Proposition 5, by the above Remark 2), $\mathcal{D}$ is isomorphic *via* U to a full subcategory $\mathcal{D}'$ of $\mathcal{C}$ which (when considered by itself) is a small abelian category and, moreover, has the property that the immersion functor $\mathcal{D}' \to \mathcal{C}$ is exact.

Using Mitchell's weak Full Imbedding Theorem, we get:

PROPOSITION 6 (**Mitchell's Full Imbedding Theorem**). *For any small abelian category* $\mathcal{D}$ *there exists a ring* A *and an exact fully faithful functor*

$$T : \mathcal{D} \to \mathbf{Mod}_d(A). \blacksquare$$

Using the Weak Group Valued Imbedding Theorem, we get:

PROPOSITION 7 (**Group Valued Imbedding Theorem**). *For any small abelian category* $\mathcal{D}$ *there exists an exact faithful functor*

$$T : \mathcal{D} \to \mathbf{Ab}. \blacksquare$$

REMARK. The Group Valued Imbedding Theorem was obtained simultaneously by different methods by P. Freyd, Heron, and S. Lubkin.

We proceed now to another type of representation theorem; first we prove the following Lemma:

LEMMA 13. *Let* $\mathcal{C}$ *be an abelian category satisfying Axiom* AB3) *and having a generator; let* $\mathcal{D}$ *be a subclass of* Ob $\mathcal{C}$ *with the following properties*:

a) $\mathcal{D}$ *is closed under isomorphisms, i.e., if* $E, F \in \text{Ob}\,\mathcal{C}$, $E \in \mathcal{D}$, *and* $E \approx F$ *in* $\mathcal{C}$, *then* $F \in \mathcal{D}$;

b) $\mathcal{D}$ *contains a generator* U *of the category* $\mathcal{C}$;

c) For any family $(E_\lambda)_{\lambda \in I}$ *of objects of* $\mathcal{D}$, *if* $(E, (j_\lambda)_{\lambda \in I})$ *is a direct sum in* $\mathcal{C}$ *of the family* $(E_\lambda)_{\lambda \in I}$, *then* $E \in \mathcal{D}$;

d) If $E, F \in \mathcal{D}$, *if* $u \in \text{Hom}_{\mathcal{C}}(E, F)$, *and if* (G, p) *is a cokernel in the category* $\mathcal{C}$ *of* u, *then* $G \in \mathcal{D}$;

under these conditions, $\mathcal{D} = \text{Ob}\,\mathcal{C}$.

Proof. By condition *a*), it does not matter which "determination" of the direct sum and of the cokernel we choose. Let $M \in \text{Ob}\,\mathcal{C}$; then, by Proposition 2, § 19, there exist a set I and an epimorphism v of the category $\mathcal{C}$

$$v : U^{(I)} \to M;$$

let (K, i) be a kernel of v; in the same way, there exist a set J and an epimorphism

$$w : U^{(J)} \to K;$$

then, as shown in § 4, (M, v) is a cokernel of i, hence also a cokernel of $i \circ w : U^{(J)} \to U^{(I)}$, since w is an epimorphism; since $U \in \mathcal{D}$, it follows by *c*) that $U^{(I)}$, $U^{(J)} \in \mathcal{D}$; hence, by *d*), $M = \text{Coker}(i \circ w)$ is also an object of $\mathcal{D}$. $\blacksquare$

PROPOSITION 8 (**Gabriel-Mitchell Representation Theorem**). *Let* $\mathcal{C}$ *be an abelian category with direct sums and having a projective generator* P; *denoting by* A *the ring* $\text{Hom}_{\mathcal{C}}(P, P)$ *and by* h^P *the additive functor* $\mathcal{C} \to \mathbf{Mod}_d(A)$ *given by*

$$h^P(E) = \text{Hom}_{\mathcal{C}}(P, E),$$

$E \in \operatorname{Ob} \mathcal{C}$, where $\operatorname{Hom}_{\mathcal{C}}(P, E)$ is endowed with the right A-module structure obtained by composition[1]*, we assume that h^P commutes with direct sums; then h^P is an equivalence between the category $\mathcal{C}$ and the category $\mathbf{Mod}_d(A)$.*

Proof. We shall check, for the given situation, conditions *a*) and *b*) of Proposition 1, *iii*).

To verify condition *a*), fix an object F in the category $\mathcal{C}$; let $T = h_F : \mathcal{C}^\circ \to \mathbf{Ab}$; let U be the composition of the functors

$$\mathcal{C}^\circ \xrightarrow{(h^P)^\circ} (\mathbf{Mod}_d(A))^\circ \xrightarrow{h_N} \mathbf{Ab},$$

where $N = h^P(F)$ and $(h^P)^\circ$ is the dual of the functor h^P (see § 2); hence, if $E, E_1 \in \operatorname{Ob} \mathcal{C}$ and $u \in \operatorname{Hom}_{\mathcal{C}}(E_1, E)$, then

$$T(E^\circ) = \operatorname{Hom}_{\mathcal{C}}(E, F), \qquad T(u^\circ) = \operatorname{Hom}(u, 1_F),$$

$$U(E^\circ) = \operatorname{Hom}_A(h^P(E), h^P(F)), \quad U(u^\circ) = \operatorname{Hom}_A(h^P(u), h^P(1_F)).$$

Defining

$$\Phi_F(E^\circ) = h^P(E, F),$$

$E \in \operatorname{Ob} \mathcal{C}$, we get a natural transformation $\Phi_F : T \to U$; to prove this it suffices to show that, in the above notation, the following diagram in $\mathbf{Ab}$ is commutative:

$$\begin{array}{ccc} \operatorname{Hom}_{\mathcal{C}}(E, F) & \xrightarrow{\operatorname{Hom}(u, 1_F)} & \operatorname{Hom}_{\mathcal{C}}(E_1, F) \\ \Big\downarrow{\scriptstyle h^P(E, F)} & & \Big\downarrow{\scriptstyle h^P(E_1, F)} \\ \operatorname{Hom}_A(h^P(E), h^P(F)) & \xrightarrow{\operatorname{Hom}(h^P(u), h^P(1_F))} & \operatorname{Hom}_A(h^P(E_1), h^P(F)); \end{array} \tag{28}$$

but if $f \in \operatorname{Hom}_{\mathcal{C}}(E, F)$ and $g \in h^P(E_1)$, then

$$[(h^P(E_1, F) \circ \operatorname{Hom}(u, 1_F))(f)](g) = (h^P(f \circ u))(g),$$

$$[(\operatorname{Hom}(h^P(u), h^P(1_F)) \circ h^P(E, F))(f)](g) = (h^P(f) \circ h^P(u))(g),$$

which verifies our assertion.

If we show that $h^P(E, F)$ is an isomorphism of the category $\mathbf{Ab}$ for any $E \in \operatorname{Ob} \mathcal{C}$, then, as F is an arbitrary object of $\mathcal{C}$, it will follow that *a*) of Proposition 1, *iii*) is verified; hence, denote by $\mathcal{D}_F$ the class of those objects E of $\mathcal{C}$ for which $h^P(E, F)$ is a $\mathbf{Z}$-isomorphism; if we show that $\mathcal{D}_F$ satisfies conditions *a*)–*d*) of Lemma 13, then, by this Lemma, we shall have $\mathcal{D}_F = \operatorname{Ob} \mathcal{C}$, and the result will be proved; therefore we proceed to verify these conditions in our notation:

---

[1] See also the proof of Proposition 3.

If $u : E_1 \xrightarrow{\sim} E$ is an isomorphism of the category $\mathcal{C}$ and $E \in \mathcal{D}_F$, then the horizontal arrows in diagram (28) are isomorphisms and the left vertical arrow is also an isomorphism; the commutativity of diagram (28) implies that the right vertical arrow is an isomorphism; hence $E_1 \in \mathcal{D}_F$, and condition *a*) of Lemma 13 is verified.

To verify condition *b*) of Lemma 13, we show that $P \in \mathcal{D}_F$; indeed, $h^P(P, F) : \mathrm{Hom}_{\mathcal{C}}(P, F) \to \mathrm{Hom}_A(h^P(P), h^P(F)) = \mathrm{Hom}_A(A, \mathrm{Hom}_{\mathcal{C}}(P, F))$ acts as follows: if $f \in \mathrm{Hom}_{\mathcal{C}}(P, F)$ and $\xi \in A$, then

$$(h^P(P, F)(f))(\xi) = (h^P(f))(\xi) = f \circ \xi;$$

if we denote by Ξ the canonical isomorphism $\mathrm{Hom}_{\mathcal{C}}(P, F) \xrightarrow{\sim} \mathrm{Hom}_A(A, \mathrm{Hom}_{\mathcal{C}}(P, F))$ (see Proposition 1, § 16), then, in the same notation, $\Xi(f)(\xi) = f \circ \xi$ by the definition of the right A-module structure on $\mathrm{Hom}_{\mathcal{C}}(P, F)$; hence $h^P(P, F) = \Xi$, and $h^P(P, F)$ is an isomorphism; it follows that $P \in \mathcal{D}_F$.

Let us verify condition *c*) of Lemma 13: let $(E_\lambda)_{\lambda \in I}$ be a family of objects of $\mathcal{D}_F$; let $(E, (j_\lambda)_{\lambda \in I})$ be a direct sum in the category $\mathcal{C}$ of this family of objects; by the hypothesis on h^P, the system $(h^P(E), (h^P(j_\lambda))_{\lambda \in I})$ is a direct sum in the category $\mathbf{Mod}_d(A)$ of the family $(h^P(E_\lambda))_{\lambda \in I}$; in **Ab** we have the following diagram:

$$(29) \qquad \begin{array}{ccc} \mathrm{Hom}_{\mathcal{C}}(E, F) & \xrightarrow[\sim]{\rho} & \prod\limits_{\lambda \in I} \mathrm{Hom}_{\mathcal{C}}(E_\lambda, F) \\ {\scriptstyle h^P(E, F)} \downarrow & & \downarrow {\scriptstyle \prod\limits_{\lambda \in I} h^P(E_\lambda, F)} \\ \mathrm{Hom}_A(h^P(E), h^P(F)) & \xrightarrow[\sim]{\rho'} & \prod\limits_{\lambda \in I} \mathrm{Hom}_A(h^P(E_\lambda), h^P(F)), \end{array}$$

where the horizontal arrows represent the canonical isomorphisms obtained from the property of h_F, respectively h_N, where $N = h^P(F)$, of commuting with direct products (see Proposition 3, § 7); hence, for the first isomorphism, for instance,

$$f \mapsto (f \circ i_\lambda)_{\lambda \in I},$$

where $f \in \mathrm{Hom}_{\mathcal{C}}(E, F)$; diagram (29) is commutative, since

$$((\prod_{\lambda \in I} h^P(E_\lambda, F)) \circ \rho)(f) = (\prod_{\lambda \in I} h^P(E_\lambda, F))((f \circ j_\lambda)_{\lambda \in I}) = (h^P(f \circ j_\lambda))_{\lambda \in I} =$$

$$= (h^P(f) \circ h^P(j_\lambda))_{\lambda \in I} = (\rho' \circ h^P(E, F))(f);$$

but since $h^P(E_\lambda, F)$ is a **Z**-isomorphism for any $\lambda \in I$, and since the passing to the direct product is a functor, the right vertical arrow is an isomorphism, which proves that the same is true also for the left vertical arrow, hence that $E \in \mathcal{D}_F$.

Finally, for condition *d*) of Lemma 13, let us show that, given the exact sequence in the category $\mathcal{C}$

$$E_1 \xrightarrow{u} E_2 \xrightarrow{p} G \to 0,$$

with $E_1, E_2 \in \mathcal{D}_F$, then $G \in \mathcal{D}_F$ (see Proposition 2, § 5); but since P is a projective object, the diagram

$$h^P(E_1) \xrightarrow{h^P(u)} h^P(E_2) \xrightarrow{h^P(p)} h^P(G) \rightarrow 0$$

is an exact sequence in the category **Ab**, hence also in $\mathbf{Mod}_d(A)$; by the Corollary to Proposition 4, § 5, for arbitrary abelian categories, the rows in the diagram

$$(30)\quad \begin{array}{ccccc}
0 \longrightarrow & \mathrm{Hom}_{\mathscr{C}}(G, F) & \xrightarrow{\mathrm{Hom}(p, 1_F)} & \mathrm{Hom}_{\mathscr{C}}(E_2, F) & \longrightarrow \\
& \downarrow h^P(G, F) & & \downarrow h^P(E_2, F) & \\
0 \longrightarrow & \mathrm{Hom}_A(h^P(G), h^P(F)) & \xrightarrow{\mathrm{Hom}(h^P(p), h^P(1_F))} & \mathrm{Hom}_A(h^P(E_2), h^P(F)) & \longrightarrow
\end{array}$$

$$\begin{array}{cc}
\xrightarrow{\mathrm{Hom}(u, 1_F)} & \mathrm{Hom}_{\mathscr{C}}(E_1, F) \\
& \downarrow h^P(E_1, F) \\
\xrightarrow{\mathrm{Hom}(h^P(u), h^P(1_F))} & \mathrm{Hom}_A(h^P(E_1), h^P(F))
\end{array}$$

are exact sequences; next, diagram (30) is commutative, since its squares are in fact diagrams of the type (28); diagram (30) may be enlarged with zero **Z**-modules and zero morphisms to the left so that, trivially, in the diagram obtained the rows are exact sequences and the squares are commutative; since $E_1, E_2 \in \mathcal{D}_F$, it follows that $h^P(E_2, F)$ and $h^P(E_1, F)$ are isomorphisms; applying The Five Lemma (Proposition 5, § 5) to the diagram enlarged in this way, we obtain that $h^P(G, F)$ is also a **Z**-isomorphism, hence that $G \in \mathcal{D}_F$.

In order to prove that condition *b*) of Proposition 1, *iii*) is fulfilled, let M be a right A-module; by Proposition 6, § 10, there exists a set J and a surjective linear mapping $p : A_d^{(J)} \rightarrow M$; if $K = \operatorname{Ker} p$ and $j : K \rightarrow A_d^{(J)}$ is the canonical injection, then, as above, there exists a set I and a surjective linear mapping $q : A_d^{(I)} \rightarrow K$; we have the exact sequence of right A-modules

$$A_d^{(I)} \xrightarrow{j \circ q = f} A_d^{(J)} \xrightarrow{p} M \rightarrow 0,$$

because $\operatorname{Im} f = \operatorname{Im} j = \operatorname{Ker} p$; since $A_d = h^P(P)$, since $P^{(I)} = \bigoplus_{\lambda \in I} P_\lambda$, with $P_\lambda = P$ whatever $\lambda \in I$, and likewise for $P^{(J)}$, and since h^P commutes with direct sums, it follows that there exist the isomorphisms of right A-modules

$$\varphi : h^P(P^{(I)}) \xrightarrow{\sim} A_d^{(I)}, \qquad \psi : h^P(P^{(J)}) \xrightarrow{\sim} A_d^{(J)};$$

since the mapping

$$h^P(P^{(I)}, P^{(J)}) : \mathrm{Hom}_{\mathcal{C}}(P^{(I)}, P^{(J)}) \to \mathrm{Hom}_A(h^P(P^{(I)}), h^P(P^{(J)}))$$

is, by the above arguments, a bijection, it follows that there exists a morphism $u \in \mathrm{Hom}_{\mathcal{C}}(P^{(I)}, P^{(J)})$ such that

$$h^P(u) = \psi^{-1} \circ f \circ \varphi;$$

if (R, r) is a cokernel of u in the category $\mathcal{C}$, we have the exact sequence

$$P^{(I)} \xrightarrow{u} P^{(J)} \to R \to 0;$$

hence, since P is a projective object of $\mathcal{C}$, we get in the category $\mathbf{Mod}_d(A)$ the commutative diagram with exact rows

$$\begin{array}{ccccccc} h^P(P^{(I)}) & \xrightarrow{h^P(u)} & h^P(P^{(J)}) & \xrightarrow{h^P(r)} & h^P(R) & \longrightarrow & 0 \\ \varphi \downarrow \wr & & \psi \downarrow \wr & & \vdots\, \chi & & \\ A_d^{(I)} & \xrightarrow{\quad f \quad} & A_d^{(J)} & \xrightarrow{\quad p \quad} & M & \longrightarrow & 0; \end{array} \tag{31}$$

since $p \circ \psi \circ h^P(u) = 0$, and since $(h^P(R), h^P(r))$ is a cokernel of $h^P(u)$, in the sense of Definition 1, § 4 (see Proposition 2, § 5), it follows that there exists a unique homomorphism $\chi : h^P(R) \to M$ rendering commutative diagram (31); enlarging this diagram with zero modules and zero homomorphisms to the right, and applying The Five Lemma, we get that χ is an isomorphism. ▮*

REMARK Let A be a ring; as shown in § 16, for any right A-module E, on the abelian group $h^A(E) = \mathrm{Hom}_A(A_d, E)$ a structure of right A-module is canonically defined, which allows us to consider the functor h^A defined on the category $\mathbf{Mod}_d(A)$ as taking values in the category $\mathbf{Mod}_d(A)$ as well (see the Remark following Proposition 1, § 16); on the other hand, as shown in this Section, on the abelian group $h^A(E)$ a right $\bar{A}$-module structure may also be introduced, where $\bar{A}$ is the ring $\mathrm{Hom}_A(A_d, A_d)$ (in which the multiplication law is given by composition); this allows us to think of the functor h^A as being defined on $\mathbf{Mod}_d(A)$ and taking values in $\mathbf{Mod}_d(\bar{A})$; in fact, we have to do with two different functors denoted, in an abuse of notation, by the same symbol; this abuse is justified by the following facts: By Proposition 1, § 16, there exists a canonical isomorphism of A-modules, in particular of abelian groups,

$$\psi_A : \mathrm{Hom}_A(A_d, A_d) \rightleftarrows A;$$

but ψ_A *is moreover an isomorphism of rings* (with unit element): for if $f, g \in \in \mathrm{Hom}_A(A_d, A_d)$, then, since f is a homomorphism of right A-modules, we get

$$\psi_A(f\circ g) = (f\circ g)(1) = f(g(1)) = f(1)\cdot g(1) = \psi_A(f)\cdot\psi_A(g),$$

and $\psi_A(1_A) = 1$; likewise, *the isomorphism of abelian groups*

$$\psi_E : \mathrm{Hom}_A(A_d, E) \xrightarrow{\sim} E$$

obtained by Proposition 1, § 16, where E is a right A-module, *is compatible with the isomorphism of rings* $\psi_{\bar{A}}^{-1}$ (see § 15): for if $f \in \mathrm{Hom}_A(A_d, E)$ and $\alpha \in A$, then, denoting $g = \psi_{\bar{A}}^{-1}(\alpha)$, we have

$$\psi_E(f\circ\psi_{\bar{A}}^{-1}(\alpha)) = \psi_E(f\circ g) = f(g(1)) = f(1)\cdot g(1) = \psi_E(f)\cdot\psi_A(g) = \psi_E(f)\cdot\alpha.$$

This means that if we identify the rings $\bar{A}$ and A by means of ψ_A, and the abelian groups $\mathrm{Hom}_A(A_d, E)$ and E by means of ψ_E, then the two structures of right A-module on E coincide; this identification is "natural", since if E and F are two right A-modules and $u \in \mathrm{Hom}_A(E, F)$, then the following diagram is commutative:

$$\begin{array}{ccc} \mathrm{Hom}_A(A_d, E) & \xrightarrow[\sim]{\psi_E} & E \\ {\scriptstyle \mathrm{Hom}(1_A, u)}\downarrow & & \downarrow{\scriptstyle u} \\ \mathrm{Hom}_A(A_d, F) & \xrightarrow[\sim]{\psi_F} & F, \end{array} \tag{32}$$

as follows from the commutativity of diagram (5), § 16, or directly; from the commutativity of diagram (32) it follows that if $(E, (j_\lambda)_{\lambda\in I})$ is a direct sum in the category $\mathbf{Mod}_d(A)$ of the family of right A-modules $(E_\lambda)_{\lambda\in I}$, then the system $(h^A(E), (h^A(j_\lambda))_{\lambda\in I})$ is a direct sum in the category $\mathbf{Mod}_d(\bar{A})$ of the family of right $\bar{A}$-modules $(h^A(E_\lambda))_{\lambda\in I}$, that is, the functor $h^A : \mathbf{Mod}_d(A) \to \mathbf{Mod}_d(\bar{A})$ commutes with direct sums; taking into account the fact that A_d is a projective generator of the category $\mathbf{Mod}_d(A)$, this provides an example of an object P satisfying the conditions of Proposition 8. ∎

* We shall present now without proof another type of representation theorem. For this we need the notion of quotient category, which will be outlined below.

Let $\mathcal{C}$ be an additive category, and let Σ be a subclass of the class of all morphisms of $\mathcal{C}$; one proves that, *under certain conditions*, there exist an additive category canonically constructed starting with $\mathcal{C}$ and Σ, denoted $\mathcal{C}_\Sigma$, and an additive functor $H : \mathcal{C} \to \mathcal{C}_\Sigma$ such that the pair $(\mathcal{C}_\Sigma, H)$ satisfies the following conditions:

a) If u is a morphism of the category $\mathcal{C}$ and $u \in \Sigma$, then $H(u)$ is an isomorphism of the category $\mathcal{C}_\Sigma$;

b) For any additive category $\mathcal{C}'$ and any additive functor $T : \mathcal{C} \to \mathcal{C}'$ having the property that the morphism $T(u)$ is an isomorphism of the category $\mathcal{C}'$ whatever the morphism $u \in \Sigma$, there exists a unique additive functor $\overline{T} : \mathcal{C}_\Sigma \to \mathcal{C}'$ with the property that

$$\overline{T} \circ H = T.$$

If $\mathcal{C}_\Sigma$ exists, $\mathcal{C}_\Sigma$ is called the *category of fractions* of $\mathcal{C}$ with respect to Σ, and H is called the *canonical functor* $\mathcal{C} \to \mathcal{C}_\Sigma$. The role of $\mathcal{C}_\Sigma$ and H whenever they exist is exactly that of transforming into isomorphisms the morphisms of $\mathcal{C}$ belonging to Σ, and of being universal with this property.

Now let $\mathcal{C}$ be an abelian category; a full subcategory $\mathcal{D}$ of $\mathcal{C}$ is said to be a *thick subcategory* of $\mathcal{C}$ if Ob $\mathcal{D}$ is a Serre class of the category $\mathcal{C}$ (see Definition 4, § 11). If $\mathcal{D}$ is a thick subcategory of the abelian category $\mathcal{C}$, let $\Sigma(\mathcal{D})$ be the class of all morphisms u of $\mathcal{C}$ for which, if (M, i), respectively (P, p) is a kernel, respectively a cokernel of u in $\mathcal{C}$, then M and P are objects of $\mathcal{D}$;[1] if the category of fractions $\mathcal{C}_{\Sigma(\mathcal{D})}$ exists, then it is called the *quotient category* of $\mathcal{C}$ with respect to $\mathcal{D}$, and is denoted $\mathcal{C}/\mathcal{D}$; if $\mathcal{C}$ is a well-powered abelian category, then $\mathcal{C}/\mathcal{D}$ exists whatever the thick subcategory $\mathcal{D}$. In this last case, $\mathcal{C}/\mathcal{D}$ is abelian, and, denoting by H the canonical functor $\mathcal{C} \to \mathcal{C}/\mathcal{D}$, the object $H(E)$, where $E \in \mathrm{Ob}\,\mathcal{C}$, is a zero object of the category $\mathcal{C}/\mathcal{D}$ if and only if $E \in \mathrm{Ob}\,\mathcal{D}$; next, if u is a morphism of the category $\mathcal{C}$, the morphism $H(u)$ is an isomorphism in $\mathcal{C}/\mathcal{D}$ if and only if, denoting, respectively, by (M, i) and (P, p) a kernel and a cokernel of u in $\mathcal{C}$, M and P are in Ob $\mathcal{D}$.

We are now in a position to state

Gabriel-Popescu Representation Theorem. *Let $\mathcal{C}$ be an abelian category satisfying Axiom* AB5); *let U be a given object of $\mathcal{C}$; if A is the ring* $\mathrm{Hom}_\mathcal{C}(U, U)$, *and if for each object E of $\mathcal{C}$ on the abelian group* $\mathrm{Hom}_\mathcal{C}(U, E)$ *the right A-module structure given by composition is introduced*[2], *let*

$$h^U : \mathcal{C} \to \mathbf{Mod}_d(A)$$

be the functor defined by $h^U(E) = \mathrm{Hom}_\mathcal{C}(U, E)$; *in these conditions, h^U admits a left adjoint* $T : \mathbf{Mod}_d(A) \to \mathcal{C}$; *let* $\Psi : T \circ h^U \to 1_\mathcal{C}$ *be the natural transformation given by the adjointness situation*[3]; *the following assertions are equivalent*:

i) *U is a generator of the category $\mathcal{C}$*;

ii) *h^U is a fully faithful functor*;

iii) *Ψ is a natural equivalence, and T is an exact functor*;

iv) *T is an exact functor, and if* Ker T *is the full subcategory of* $\mathbf{Mod}_d(A)$ *whose objects are those objects E of* $\mathbf{Mod}_d(A)$ *for which $T(E)$ is a zero object of $\mathcal{C}$*,

---

[1] As shown in § 11, it does not matter which kernel or cokernel is considered.

[2] See § 2 and the proof of Proposition 3.

[3] See (20), § 10.

then Ker T *is a thick subcategory of* $\mathbf{Mod}_d(A)$; *moreover, if* $H : \mathbf{Mod}_d(A) \to \to \mathbf{Mod}_d(A)/\text{Ker } T$ *is the canonical functor* [1], *and if*

$$\overline{T} : \mathbf{Mod}_d(A)/\text{Ker } T \to \mathcal{C}$$

is the functor obtained from T *by the above universal property b) of categories of fractions (that is, for which* $\overline{T} \circ H = T$), *then* $\overline{T}$ *is an equivalence between the two categories.* ∎

Consequently, *an abelian category satisfying Axiom* AB5) *and having a generator is equivalent to a quotient category of the category of right modules over a suitable ring.*

In order to give an example of the manner in which the representation theorems are used, we shall prove The Five Lemma (Proposition 5, § 5) for an abelian category satisfying Axiom AB3) and having a projective generator, making use of Mitchell's weak Full Imbedding Theorem (Proposition 3).

First we prove the following result:

PROPOSITION 9. *Let* $\mathcal{C}$ *be an abelian category, and let* $\mathcal{M}$ *be a set of objects of* $\mathcal{C}$; *then there exists a full subcategory* $\mathcal{D}$ *of* $\mathcal{C}$ *which (considered by itself) is a small preadditive category satisfying Axioms* AB1) *and* AB2) *and such that the immersion functor* $\mathcal{D} \to \mathcal{C}$ *is exact and for which, moreover,* Ob $\mathcal{D}$ *contains* $\mathcal{M}$.

Proof. For any $E, F \in \mathcal{M}$ and any $u \in \text{Hom}_{\mathcal{C}}(E, F)$ let (M_u, i_u), (C_u, q_u), (I_u, j_u), (P_u, p_u) be, respectively, a determined kernel, coimage, image, cokernel of u; denote by $\mathcal{M}_1$ the class of objects of $\mathcal{C}$ obtained by joining to $\mathcal{M}$ the objects M_u, C_u, I_u, P_u for all $u \in \coprod_{E,F \in \mathcal{M}} \text{Hom}_{\mathcal{C}}(E, F)$, as well as the zero object of $\mathcal{C}$; obviously, $\mathcal{M}_1$ is a set; hence we may consider the assignment $\mathcal{M} \mapsto \mathcal{M}_1$ as a mapping defined on the class of subclasses of Ob $\mathcal{C}$ which are sets and taking values in the same class. If $n > 1$, define recursively

$$\mathcal{M}_n = (\mathcal{M}_{n-1})_1;$$

let

$$\mathcal{M}^* = \bigcup_{n \in \mathbf{N}} M_n,$$

where, by convention, $\mathcal{M}_0 = \mathcal{M}$; $\mathcal{M}^*$ is a set, being a union of sets indexed by a set. Denote by $\mathcal{D}$ the full subcategory of $\mathcal{C}$ which has $\mathcal{M}^*$ as class of objects; if for any $E, F \in \mathcal{M}^*$ we consider on $\text{Hom}_{\mathcal{D}}(E, F)$ the given abelian group structure of $\text{Hom}_{\mathcal{C}}(E, F)$, then, obviously, $\mathcal{D}$ is a full subcategory which is also pre-

---

[1] $\mathbf{Mod}_d(A)/\text{Ker } T$ exists since $\mathbf{Mod}_d(A)$ is a well-powered abelian category (see § 8).

additive when considered by itself, and for which the forgetful functor is an additive functor. The category $\mathcal{D}$ satisfies also Axioms AB1) and AB2): for if $u \in \mathrm{Hom}_{\mathcal{D}}(E, F)$ is a morphism of the category $\mathcal{D}$, with $E, F \in \mathrm{Ob}\,\mathcal{D}$, then $E, F \in \mathcal{M}^*$, hence there exists some $n \in \mathbf{N}$ such that $E, F \in \mathcal{M}_n$; consequently, by the construction of $\mathcal{M}_{n+1}$, there exists a kernel (M, i), a coimage (C, q), an image (I, j), and a cokernel (P, p) of u in $\mathcal{C}$ such that $M, C, I, P \in \mathcal{M}_{n+1} \subseteq \mathcal{M}^* = \mathrm{Ob}\,\mathcal{D}$; since $\mathcal{D}$ is a full subcategory of $\mathcal{C}$, it follows that i, q, j, p are morphisms of the category $\mathcal{D}$, and obviously (M, i), (C, q), (I, j), (P, p) represent, respectively, a kernel, a coimage, an image, a cokernel of u in $\mathcal{D}$; this shows that $\mathcal{D}$ satisfies Axiom AB1); further, if in the above notation, $\bar{u} : C \to I$ is the canonical morphism from the coimage of u to the image of u in the category $\mathcal{C}$, then, as $\bar{u} \in \mathrm{Hom}_{\mathcal{D}}(C, I)$, it follows that $\bar{u}$ plays the same role in the category $\mathcal{D}$; since $\bar{u}$ is an isomorphism of the category $\mathcal{C}$, and since $\mathcal{D}$ is a full subcategory of $\mathcal{C}$, it follows that $\bar{u}$ is also an isomorphism of $\mathcal{D}$, hence that Axiom AB2) is also satisfied in this category. Finally, for any morphism u of the category $\mathcal{D}$ there exist a kernel and an image of u in the category $\mathcal{C}$ playing the same role in $\mathcal{D}$; it follows that the immersion functor $S: \mathcal{D} \to \mathcal{C}$ is exact, and our Proposition is entirely proved. ▮

REMARK. If in the construction of $\mathcal{M}_1$ described in the proof of Proposition 9 we considered for any pair (E, F) of elements of $\mathcal{M}$ a determined direct sum (S_{EF}, j_E, j_F) of the family (E, F) in the category $\mathcal{C}$, and $\mathcal{M}_1$ were obtained from $\mathcal{M}$ by joining besides the objects M_u, C_u, I_u, P_u for all $u \in \coprod_{E,F \in \mathcal{M}} \mathrm{Hom}_{\mathcal{C}}(E, F)$ also the objects $S_{E.F}$ for all pairs (E, F), we would obtain by the same procedure a subcategory $\mathcal{D}$ which considered by itself would be a small abelian category such that the immersion functor $\mathcal{D} \to \mathcal{C}$ would be exact and for which $\mathrm{Ob}\,\mathcal{D}$ would contain $\mathcal{M}$. ▮

Now, we notice that *if $\mathcal{C}, \mathcal{C}'$ are arbitrary categories and $T : \mathcal{C} \to \mathcal{C}'$ is a fully faithful functor, then a morphism $u \in \mathrm{Hom}_{\mathcal{C}}(E, F)$, with $E, F \in \mathrm{Ob}\,\mathcal{C}$, is an isomorphism of the category $\mathcal{C}$ if and only if $T(u)$ is an isomorphism of the category $\mathcal{C}'$*: for if $T(u)$ is an isomorphism, then there exists $v' \in \mathrm{Hom}_{\mathcal{C}'}(T(F), T(E))$ such that $v' \circ T(u)$ and $T(u) \circ v'$ are the respective identity morphisms; since T is fully faithful, it follows that there exists $v \in \mathrm{Hom}_{\mathcal{C}}(F, E)$ such that $T(v) = v'$; we have

$$T(v \circ u) = T(v) \circ T(u) = v' \circ T(u) = 1_{T(E)} = T(1_E);$$

again since T is a fully faithful functor, it follows $v \circ u = 1_E$; likewise, $u \circ v = 1_F$.

Hence, under these conditions, if $T(E) \approx T(F)$ in the category $\mathcal{C}'$, then $E \approx F$ in the category $\mathcal{C}$.

Next, we will show that *if $\mathcal{C}$ and $\mathcal{C}'$ are preadditive categories satisfying Axioms* AB1) *and* AB2) *and $T : \mathcal{C} \to \mathcal{C}'$ is a fully faithful exact functor, then, given the diagram*

(33) $$E \xrightarrow{u} F \xrightarrow{v} G$$

in the category $\mathcal{C}$, this is an exact sequence if and only if the diagram

$$(34) \qquad T(E) \xrightarrow{T(u)} T(F) \xrightarrow{T(v)} T(G)$$

is an exact sequence in the category $\mathcal{C}'$: for let (34) be an exact sequence, let (I, j) be an image of u, and let (K, i) be a kernel of v in the category $\mathcal{C}$; then, since T is an exact functor, $(T(I), T(j))$ is an image of $T(u)$, and $(T(K), T(i))$ is a kernel of $T(v)$ in the category $\mathcal{C}'$; since (34) is an exact sequence, $(T(I), T(j))$ and $(T(K), T(i))$ define the same subobject of $T(F)$, hence there exists an isomorphism $w' \in \in \mathrm{Hom}_{\mathcal{C}'}(T(I), T(K))$ such that $T(i) \circ w' = T(j)$; since T is fully faithful, it follows that there exists a unique morphism $w \in \mathrm{Hom}_{\mathcal{C}}(I, K)$ with the property that $T(w) = w'$; by the preceding considerations, w is an isomorphism of the category $\mathcal{C}$, and, furthermore, since T is fully faithful, we have $i \circ w = j$; hence (I, j) and (K, i) define the same subobject of F, and diagram (33) is an exact sequence.

Let now $\mathcal{C}$ be an abelian category satisfying Axiom AB3) and having a projective generator; in $\mathcal{C}$ consider the diagram given in the statement of Proposition 5, § 5, diagram which is assumed to be commutative and the rows in it to be exact sequences; then, by Proposition 9, there exists a full subcategory $\mathcal{D}$ of $\mathcal{C}$ which (when considered by itself) is a small preadditive category satisfying Axioms AB1) and AB2) and having the property that the immersion functor $S: \mathcal{D} \to \mathcal{C}$ is an exact functor; moreover, for which $E_i, F_i \in \mathrm{Ob}\, \mathcal{D}, i = 1, \ldots, 5$; by Mitchell's weak Full Imbedding Theorem there exists an exact fully faithful functor

$$T : \mathcal{D} \to \mathbf{Mod}_d\,(A),$$

where A is a certain ring constructed according to that Theorem; assuming f_1 to be an epimorphism and f_2, f_4 to be monomorphisms of the category $\mathcal{C}$, by Proposition 1, *a*) and *b*), § 5 (for preadditive categories satisfying AB1) and AB2)) we have, applying the exact functor T, the following commutative diagram of right A-modules in which the rows and columns are exact sequences:

$$\begin{array}{ccccccccc}
 & & 0 & & & & 0 & & \\
 & & \downarrow & & & & \downarrow & & \\
T(E_1) & \xrightarrow{T(u_1)} & T(E_2) & \xrightarrow{T(u_2)} & T(E_3) & \xrightarrow{T(u_3)} & T(E_4) & \xrightarrow{T(u_4)} & T(E_5) \\
{\scriptstyle T(f_1)}\downarrow & & {\scriptstyle T(f_2)}\downarrow & & {\scriptstyle T(f_3)}\downarrow & & {\scriptstyle T(f_4)}\downarrow & & {\scriptstyle T(f_5)}\downarrow \\
T(F_1) & \xrightarrow{T(v_1)} & T(F_2) & \xrightarrow{T(v_2)} & T(F_3) & \xrightarrow{T(v_3)} & T(F_4) & \xrightarrow{T(v_4)} & T(F_5); \\
\downarrow & & & & & & & & \\
0 & & & & & & & &
\end{array}$$

by The Five Lemma for right A-modules we get that $T(f_3)$ is a monomorphism, hence that the diagram

$$0 \to T(E_3) \xrightarrow{T(f_3)} T(F_3)$$

is an exact sequence in the category $\mathbf{Mod}_d(A)$; since T is exact and fully faithful, it follows by the preceding considerations that the diagram

$$0 \to E_3 \xrightarrow{f_3} F_3$$

is an exact sequence in the category $\mathcal{C}$, hence that f_3 is a monomorphism.

We propose to the reader that he also prove as exercises other assertions of this type, in particular, the other assertions of Proposition 5, § 5, as well as Propositions 6 and 7, § 5.

If we applied Mitchell's Full Imbedding Theorem (Proposition 6), then, also making use of the Remark following Proposition 9, we could dispense with the assumptions that the abelian category $\mathcal{C}$ satisfies Axiom AB3) and has a projective generator; and this gives an idea of the way the power of the tool increases when passing from the weak form to the plenary form of Mitchell's Full Imbedding Theorem. *

Bibliography

1. T. ALBU, *On a Class of Modules* (in Romanian), I., Studii cercet. mat., **24** (1972), 1329–1392; II., Ibid., **24** (1972), 1455–1501.
2. T. ALBU and C. NĂSTĂSESCU, *Modules arithmétiques*, Acta Math. Acad. Sci. Hungaricae, **25**, *3–4* (1974), 299–311.
3. C. ANDREIAN-CAZACU, A. DELEANU, and M. JURCHESCU, *Topology, Categories, Riemann Surfaces* (in Romanian), Edit. Academiei, Bucharest, 1966.
4. R. BAER, *Abelian Groups that are Direct Summands of every containing Abelian Group*, Bull. Amer. Math. Soc., **46** (1940), 800–806.
5. C. BĂNICĂ and N. POPESCU, *Quotient Categories* (in Romanian), Studii cercet. mat., **17** (1965), 951–985.
6. C. BĂNICĂ and N. POPESCU, *Sur les catégories préabeliennes*, Rev. roum. math pures et appl., **10** (1965). 621–633.
7. D. BARBILIAN, *Groups with Operators (Decomposition Theorems of Algebra)* (in Romanian), Edit. Academiei, Bucharest, 1960.
8. H. BASS, *Algebraic K-Theory*, Math. Lecture Notes Series, W. A. Benjamin, Inc., New York, Amsterdam, 1968.
9. G. BIRKHOFF, *Lattice Theory*, Revised Edition, Amer. Math. Soc. Colloquium Publications Vol. 25, Amer. Math. Soc., New York, 1948.
10. N. BOURBAKI, *Algèbre*, Chapitres 1 à 3; Ch. I, Structures algébriques, Ch. II, Algèbre linéaire, Ch. III, Algèbres tensorielles, algèbres extérieures, algèbres symétriques, Hermann, Paris, 1970.
11. N. BOURBAKI, *Algèbre*, Ch. VIII, Modules et anneaux semi-simples, Act. sci. et ind. 1261, Hermann, Paris, 1958.
12. N. BOURBAKI, *Algèbre commutative*, Ch. I, Modules plats, Ch. II, Localisation, Fasc. XXVII, Act. sci. et ind. 1290, Hermann, Paris, 1961.
13. D. A. BUCHSBAUM, *Exact Categories and Duality*, Trans. Amer. Math. Soc., **80** (1955), 1–34.
14. I. BUCUR, *Fonctions définies sur le spectre d'une catégorie et théories de décomposition*, Rev. roum. math. pures et appl., **9** (1964), 583–588.
15. I. BUCUR, *Homological Algebra* (in Romanian), Edit. Didactică şi Pedagogică, Bucharest, 1965.
16. I. BUCUR and A. DELEANU, *Introduction to the Theory of Categories and Functors*, Pure and Appl. Math. Vol. XIX, A Wiley-Interscience Publication, John Wiley & Sons, Ltd., London, New York, Sydney, 1968.
17. H. CARTAN and S. EILENBERG, *Homological Algebra*, Princeton Math. Series, Princeton University Press, Princeton, N. J., 1956.
18. CL. CHEVALLEY, *Fundamental Concepts of Algebra*, Pure and Appl. Math. VII, Academic Press Inc. Publishers, New York, 1956.

19. P. M. Cohn, *Universal Algebra*, Harper's Series in Modern Mathematics, Harper & Row, New York, Evanston, and London, 1965.

20. I. Creangă and C. Reischer, *Linear Algebra* (in Romanian), IInd Edition, Edit. Didactică și Pedagogică, Bucharest, 1970.

21. Ch. W. Curtis and I. Reiner, *Representation Theory of Finite Groups and Associative Algebras*, Pure and Appl. Math. Vol. XI, Interscience Publishers, A Division of John Wiley & Sons, New York, London, Sydney, 1966.

22. S. E. Dickson, *A Torsion Theory for Abelian Categories*, Trans. Amer. Math. Soc., **121** (1966), 223–235.

23. B. Eckmann and P. J. Hilton, *Group-like Structures in general Categories*, I., Math. Ann., **145** (1961–1962), 227–255; II., Ibid., **151** (1963), 150–186; III., Ibid., **150** (1963) 165–187.

24. B. Eckmann and P. J. Hilton, *Commuting Limits with Colimits*, J. of Algebra, **11** (1969), 116–144.

25. B. Eckmann and A. Schopf, *Über injektive Moduln*, Archiv der Math., **4** (1953), 75–78.

26. Ch. Ehresmann, *Catégories et structures*, Travaux et recherches math. X, Dunod, Paris, 1965.

27. S. Eilenberg, *Abstract Description of some Basic Functors*, J. Indian Math. Soc., **24** (1960), 231–234.

28. S. Eilenberg and G. M. Kelly, *Closed Categories*, in Proc. of the Conference on Categorical Algebra, La Jolla, 1965, pp. 421–562, Springer-Verlag, Berlin, Heidelberg, New York, 1966.

29. S. Eilenberg and S. MacLane, *Natural Isomorphisms in Group Theory*, Proc. Nat. Acad. Sci. U.S.A., **28** (1942), 537–543.

30. S. Eilenberg and S. MacLane, *General Theory of Natural Equivalences*, Trans. Amer. Math. Soc., **58** (1945), 231–294.

31. S. Eilenberg and N. Steenrod, *Foundations of Algebraic Topology*, Princeton University Press, Princeton, N.J., 1952.

32. P. Freyd, *Abelian Categories: An Introduction to the Theory of Functors*, Harper's Series in Modern Mathematics, Harper & Row, New York, Evanston, and London, 1964.

33. P. Freyd, *Representations in Abelian Categories*, in Proc. of the Conference on Categorical Algebra, La Jolla, 1965, pp. 95–120, Springer-Verlag, Berlin, Heidelberg, New York, 1966.

34. A. Froda, *Higher Algebra* (in Romanian), Biblioteca Mat. I, Edit. Academiei, Bucharest, 1958.

35. L. Fuchs, *Infinite Abelian Groups*, vols. I, II, Academic Press, New York and London, 1970, 1973.

36. P. Gabriel, *Des catégories abéliennes*, Bull. Soc. Math. France, **90** (1962), 323–448.

37. P. Gabriel and N. Popescu, *Caractérisation des catégories abéliennes avec générateurs et limites inductives exactes*, C.R. Acad. Sci. Paris, **258** (1964), 4188–4190.

38. P. Gabriel and F. Ulmer, *Lokal präsentierbare Kategorien*, Math. Institut der Univ. Zürich, 1971.

39. P. Gabriel and M. Zisman, *Calculus of Fractions and Homotopy Theory*, Ergebnisse der Math. und ihrer Grenzgebiete Bd. 35, Springer-Verlag, Berlin, Heidelberg, New York, 1967.

40. G. Galbură, *Fields of Algebraic Functions and Algebraic Varieties* (in Romanian), Edit. Academiei, Bucharest, 1961.

41. G. GALBURĂ, *Algebra* (in Romanian), Edit. Didactică şi Pedagogică, Bucharest, 1972.
42. R. GODEMENT, *Topologie algébrique et Théorie des faisceaux*, Act. sci. et ind. 1252, Publ. de l'Institut de Math. de l'Univ. de Strasbourg XIII, Hermann, Paris, 1958.
43. M. GREENBERG, *Lectures on Algebraic Topology*, Math. Lecture Notes, W. A. Benjamin, Inc., New York, Amsterdam, 1967.
44. A. GROTHENDIECK, *Sur quelques points d'Algèbre homologique*, Tôhoku Math. Journal, 2nd series, 9 (1957), 119–221.
45. A. GROTHENDIECK and J. DIEUDONNÉ, *Éléments de Géométrie Algébrique*, I–IV, Publ. Math., I.H.E.S., Paris, 1960–1964.
46. M. HALL, JR., *The Theory of Groups*, The Macmillan Co., New York, 1959.
47. P. R. HALMOS, *Finite-dimensional Vector Spaces*, D. van Nostrand Publ. Co., New York, 1958.
48. M. HASSE and L. MICHLER, *Theorie der Kategorien*, Math. Monographien 7, VEB Deutscher Verlag der Wiss., Berlin, 1966.
49. A. HELLER, *Homological Algebra in Abelian Categories*, Ann. of Math., **68** (1958), 484–525.
50. I. D. ION and N. RADU, *Algebra* (in Romanian), IInd edition, Edit. Didactică şi Pedagogică, Bucharest, 1975.
51. J. R. ISBELL, *Structures of Categories*, Bull. Amer. Math. Soc., **72** (1966), 619–655.
52. N. JACOBSON, *Lectures in Abstract Algebra*, The University Series in Higher Math.; vol. I. *Basic Concepts*, vol. II. *Linear Algebra*, D. van Nostrand Co., Princeton (N.J.), Toronto, New York, London, 1951, 1953.
53. M. JURCHESCU and A. LASCU, *Strict Morphisms, Cantorian Categories, and Completion Functors* (in Romanian), Studii cercet. mat., **18** (1966), 219–234.
54. M. JURCHESCU and A. LASCU, *Cantorian Categories and Completion Functors*, Monatsber. Deutsch. Akad. Wiss. Berlin, **8**, 1966, 593–599.
55. M. JURCHESCU and A. LASCU, *On Morphisms in Arbitrary Categories*, Monatsber. Deutsch. Akad. Wiss. Berlin, **8**, 1966, 599–605.
56. D. N. KAN, *Adjoint Functors*, Trans. Amer. Math. Soc., **87** (1958), 294–329.
57. I. KAPLANSKY, *Infinite Abelian Groups*, Univ. of Michigan Press, Ann Arbor, Michigan, 1954.
58. I. KAPLANSKY, *Projective Modules*, Ann. of Math., **68** (1958), 372–377.
59. A. G. KUROSH, *The Theory of Groups*, Chelsea Publ. Co., New York, 1956.
60. A. G. KUROSH, *Lectures in General Algebra*, International Series in Pure and Applied Math. Vol. 70, Pergamon Press, Oxford, London, Edinburgh, New York, Paris, Frankfurt, 1965.
61. J. LAMBEK, *Lectures on Rings and Modules*, Blaisdell Publ. Co., A Division of Ginn and Co., Waltham (Mass.), Toronto, London, 1966.
62. S. LANG, *Algebra*, Addison-Wesley Series in Math., Addison-Wesley Publ. Co., Reading, Mass., 1965.
63. F. E. J. LINTON, Review of [92], #2665, Math. Reviews, **34**, *3*, (Sept. 1967), 474.
64. S. LUBKIN, *Imbedding of Abelian Categories*, Trans. Amer. Math. Soc., **97** (1960), 410–417.
65. S. MACLANE, *Groups, Categories, and Duality*, Proc. Nat. Acad. Sci. U.S.A., **34** (1948), 263–267.
66. S. MACLANE, *Duality for Groups*, Bull. Amer. Math. Soc., **56** (1950), 485–516.
67. S. MACLANE, *Locally small Categories and the Foundations of Set Theory*, in Infinitistic Methods, Proc. Symp. on Foundations of Math., Warsaw, 1959, pp. 25–43, Pergamon Press, New York, Oxford, London, Paris, and P.W.N., Warsaw, 1961.

68. S. MacLane, *Homology*, Die Grundlehren der Math. Wiss. Bd. 114, Springer-Verlag, Berlin, Göttingen, Heidelberg, 1963.

69. S. MacLane, *Categorical Algebra*, Bull. Amer. Math. Soc., **71** (1965), 40–106.

70. S. MacLane, *Categories for the Working Mathematician*, Graduate Texts in Math. Vol. 5, Springer-Verlag, New York, Heidelberg, Berlin, 1971.

71. S. MacLane and G. Birkhoff, *Algebra*, The Macmillan Co., New York, Collier-Macmillan Ltd., London, 1967.

72. A. P. Mishina and L. A. Skornjakov, *Abelian Groups and Modules* (in Russian), Izdatel'stvo "Nauka", Moscow, 1969.

73. B. Mitchell, *The Full Imbedding Theorem*, Amer. J. Math., **86** (1964), 619–637.

74. B. Mitchell, *Theory of Categories*, Pure and Appl. Math. XVII, Academic Press, New York and London, 1965.

75. Gr. C. Moisil, *Introduction to Algebra*. I. *Rings and Ideals*, vol. I (in Romanian), Edit. Academiei, Bucharest, 1954.

76. C. Năstăsescu and N. Popescu, *Quelques observations sur les topos abéliens*, Rev. roum. math. pures et appl., **12** (1967), 553–563.

77. C. Niţă, *L'anneau des endomorphismes d'un module et S-anneau*, C.R. Acad. Sci. Paris, **272** (1971), 940–941.

78. D. G. Northcott, *An Introduction to Homological Algebra*, Cambridge University Press, Cambridge, 1960.

79. B. Pareigis, *Categories and Functors*, Academic Press, London and New York, 1970.

80. G. Pic, *Higher Algebra* (in Romanian), Edit. Didactică şi Pedagogică, Bucharest, 1966.

81. L. Pontrjagin, *Topological Groups*, Princeton Univ. Press, Princeton, N.J., 1958.

82. D. Popescu, *Catégories de Faisceaux*, J. of Algebra, **18** (1971), 343–365.

83. N. Popescu, *Elements of the Theory of Sheaves* (in Romanian), I., Studii cercet. mat., **18** (1966), 267–296; II., Ibid., **18** (1966), 407–456; III., Ibid., **18** (1966), 547–583; IV., Ibid., **18** (1966), 645–669; V., Ibid., **18** (1966), 945–991; VI., Ibid., **19** (1967), 205–240.

84. N. Popescu, *Abelian Categories with Applications to Rings and Modules*, L.M.S. Monographs No. 3, Academic Press, London & New York, 1973.

85. N. Popescu and A. Radu, *Theory of Categories and Sheaves* (in Romanian), Edit. Ştiinţifică, Bucharest, 1971.

86. N. Radu, *Local Rings* (in Romanian), vols. I, II, Edit. Academiei, Bucharest, 1968, 1970.

87. P. Ribenboim, *Rings and Modules*, Interscience Tracts No. 24, Interscience Publishers, A Division of John Wiley & Sons, New York, London, Sydney, Toronto, 1969.

88. J. E. Roos, *Sur la distributivité des foncteurs* $\varprojlim$ *par rapport aux* $\varinjlim$ *dans les catégories des faisceaux (topos)*, C.R. Acad. Sci. Paris, **259** (1964), 969–972; Ibid., **259** (1964), 1605–1608; Ibid., **259** (1964), 1801–1804.

89. J. J. Rotman, *The Theory of Groups; An Introduction*, Second Edition, Allyn and Bacon Series in Advanced Math., Allyn and Bacon, Inc., Boston, 1973.

90. H. Schubert, *Categories*, Springer-Verlag, Berlin, Heidelberg, New York, 1972.

91. A. Solian, *Foncteurs qui conservent les épimorphismes locaux*, C.R. Acad. Sci. Paris, **259** (1964), 1376–1379.

92. A. Solian, *Foncteurs qui transforment les épimorphismes locaux en monomorphismes locaux*, Rev. roum. math. pures et appl., **11** (1966), 401–410.

93. A. Solian, *Characterization of some Functors of Modules*, Rev. roum. math. pures et appl., **11** (1966), 283–285.

94. F. Ulmer, *Properties of Dense and Relative Adjoint Functors*, J. of Algebra, **8** (1968), 77–95.

95. F. Ulmer, *Representable Functors with Values in Arbitrary Categories*, J. of Algebra, **8** (1968), 96–129.

96. P. Vámos, *The Dual of the Notion of "Finitely Generated"*, J. London Math. Soc., **43** (1968), 643–646.

97. J.-L. Verdier, *Topologies et faisceaux*, Exposés I–III, Séminaire de Géométrie algébrique, I.H.E.S., Paris, 1963/64.

98. B. L. van der Waerden, *Modern Algebra*, vols. I, II, F. Ungar, New York, 1949.

99. C. L. Walker and E. A. Walker, *Quotient Categories of Modules*, in Proc. of the Conference on Categorical Algebra, La Jolla, 1965, pp. 404–420, Springer-Verlag, Berlin, Heidelberg, New York, 1966.

100. Ch. E. Watts, *Intrinsic Characterizations of some additive Functors*, Proc. Amer. Math. Soc., **11** (1960), 5–8.

101. N. Yoneda, *On the Homology Theory of Modules*, J. Fac. Sci. Tokyo, **7** (1954), 193–227.

102. M. Sh. Zalenko and E. G. Shul'geĭfer, *Foundations of the Theory of Categories* (in Russian), Sovremennaia Algebra, Izdatel'stvo "Nauka", Moscow, 1974.

103. H. Zassenhaus, *The Theory of Groups*, Chelsea. Publ. Co., New York, 1958.

Subject Index

(The numbers following the Section number represent the pages)

G

H

N

O

T